Surface Engineered Surgical Tools and Medical Devices

Surface Engineered Surgical Tools and Medical Devices

Edited by

Mark J. Jackson
Purdue University
West Lafayette, IN, USA

Waqar Ahmed
University of Ulster
Newtownabbey, UK

 Springer

Mark J. Jackson
Purdue University
Birck Nanotechnology Center
College of Technology
West Lafayette, IN, USA

Waqar Ahmed
University of Ulster
NIBEC
Newtownabbey, UK

Library of Congress Control Number: 2007920041

ISBN 978-0-387-27026-5   e-ISBN 978-0-387-27028-9

Printed on acid-free paper.

9 8 7 6 5 4 3 2 1

springer.com

Foreword by Sir Harold W. Kroto

The new millennium has seen the birth of a new perspective that conflates research in solid-state physics, biological science as well as materials engineering. The perspective is one that recognizes that future new advances in all these areas will be based on a fundamental understanding of the atomic and molecular infrastructure of materials that has resulted from two centuries of chemistry. Major advances will be achieved when the novel behavior, in particular the quantum mechanical behavior, that nanoscale structures possess, can be controlled and harnessed.

To go with this new perspective the conflated fields have acquired a new name - Nanoscience and Nanotechnology (N&N). The promise of developing functional devices at the molecular and atomic scale is now becoming a reality. However, a massive effort is still needed in order to control the fabrication of such novel nano-devices and nanomachines and exploit processes based on quantum mechanical laws. The next decade should see the emergence of new technologies based on nano-systems with not only improved but hopefully also fundamentally new physico-chemical properties produced at reasonable costs. Experimental and theoretical research should lead to industrial applications yielding important breakthroughs. If universities, independent research centers, government agencies and innovative industrial organizations invest time and resources imaginatively in this multidisciplinary adventure, a highly synergistic process will ensue in the development of these new technologies.

Nanotechnology is the ability to manipulate atoms and molecules to produce nanostructured materials and functional nanocoatings on biomedical devices and surgical tools. Nanotechnology is likely to have a significant effect on the global economy and on society in this century, and it promises to make breakthroughs in the biological and medical sciences.

This book contains chapters that focus not only on fundamental advances that are taking place but also on important applications, in particular in the biomedical field that promise to revolutionize a wide range of technologies in the 21st century. The unique properties that emanate from nanoscale structures are immensely varied and in the next decade nanoscience and nanotechnology will give birth to a vast new range of exciting technological applications that promise to help the creation of a sustainable socio-economic environment.

Professor Sir Harold W. Kroto, FRS
Nobel Laureate, Chemistry 1996

Preface

Medical devices and surgical tools that contain micro and nano-scale features allow surgeons to perform clinical procedures with greater precision and safety in addition to monitoring physiological and biomechanical parameters more accurately. While surgeons have started to master the use of nanostructured surgical tools in the operating room, the impact and interaction of nanomaterials and nanostructured coatings has yet to be addressed in a comprehensive manner.

Nobel Laureate Richard Feynman's revolutionary vision on nanotechnology was captured in a paper published in the February 1960 issue of Caltech's journal, '*Engineering and Science*'. In this paper, Feynman speaks about manipulating atoms and constructing products atom-by-atom, and molecule-by-molecule. Feynman describes the scaling down of lathes and drilling machines, and talks about drilling holes, turning, molding, and stamping parts. Even in 1959, Feynman describes the need for micro and nanofabrication as the basis for creating a microscopic world that would benefit mankind. Nanotechnology encompasses technology performed at the nanoscale that has real world applications. Nanotechnology will have a profound effect on our society that will lead to breakthrough discoveries in materials and manufacturing, medicine, healthcare, the environment, sustainability, energy, biotechnology, and information technology.

President Bill Clinton talked about the exciting promise of nanotechnology in January 2000, and later announced an ambitious national nanotechnology initiative (NNI) that was enacted in 2001 with a budget of $497 million to promote nanoscale research that would benefit society. The purpose of this book is to present information and knowledge on the emerging field of surface engineered biomedical devices and surgical tools. The book is written in the spirit of scientific endeavor outlined by Richard Feynman, who stated that one of the greatest challenges to scientists in the field of miniaturization is the manufacture of objects for medical applications using techniques such as turning, molding, stamping, and drilling. The book presents information on surface engineered surgical tools and medical devices that looks at the interaction between nanotechnology, nanomaterials, and tools for surgical applications. Chapters of the book describe developments in coatings for heart valves, stents, hip and knee joints, cardiovascular devices, orthodontic applications, and regenerative materials such as bone substitutes. Chapters are also dedicated to the performance of surgical tools and dental tools and also describe how nanostructured surfaces can be created for the purposes of improving cell adhesion between medical device and the human body.

The structure of the book is based on matter provided by many colleagues and the author wishes to thank the contributors of this book for helping construct a source of knowledge and information on surface engineered medical devices and surgical tools and for granting the editors permission to use such matter.

Mark J. Jackson
Purdue University

Waqar Ahmed
University of Ulster

October 2006

Contents

Foreword   v
Preface   vii
List of Authors   xv
Nomenclature   xxi

1. **Atomic Scale Machining of Surfaces**   1
    1.1   Introduction   1
    1.2   Theoretical Basis of Nanomachining   3
    1.3   Further Developments   18
    References   18

2. **Anodization: A Promising Nano-Modification**
    Technique of Titanium-based Implants
    for Orthopedic Applications   21
    2.1   Introduction   21
    2.2   Anodization of Titanium   23
    2.3   Structure and Properties of Anodized Oxide Film 34
    2.4   Future Directions   44
    References   45

3. **Titanium Dioxide Coatings in Medical Device**
    Applications   49
    3.1   Introduction   49
    References   62

4. **The Effect of Shape and Surface Modification**
    on the Corrosion of Biomedical Nitinol Alloy
    Wires Exposed to Saline Solution   65
    4.1   Introduction   65
    4.2   Experimental Methods   66
    4.3   Summary   78
    References   80

**5.   Cardiovascular Interventional
      and Implantable Devices**   **83**
      5.1   Introduction   83
      5.2   Cardiovascular Interventional Tools   83
      5.3   Key Surface Properties for Cardiovascular
              Interventional Devices   86
      5.4   Cardiovascular Implantable Devices   87
      5.5   Electrical Implantable Devices   88
      5.6   Mechanical Implantables   91
      5.7   Important Surface Properties
              for Implantable Cardiovascular Devices   94
      References   96

**6.   Surface Engineering Artificial Heart Valves
      to Improve Quality of Life and Lifetime
      using Modified Diamond-like Coatings**   **99**
      6.1   Introduction   99
      6.2   History of Mechanical Heart Valves   100
      6.3   Thrombosis   107
      6.4   Hemocompatibility   109
      6.5   Endothelium and Endothelial Cell Seeding   112
      6.6   Surface Engineering Artificial Heart Valves   114
      6.7   Summary   133
      References   135

7.   Diamond Surgical Tools   **141**
      7.1   Introduction   141
      7.2   Properties of Diamond   143
      7.3   History of Diamond   143
      7.4   CVD Diamond Technology   149
      7.5   CVD Diamond Processes   150
      7.6   Treatment of Substrate   154
      7.7   Modification of HFCVD Process   159
      7.8   Nucleation and Growth   162
      7.9   Deposition on 3-D Substrates   171
      7.10   Wear of Diamond   180
      7.11   Time-Modulated CVD Diamond   188
      7.12   Conclusions   196
      References   196

8. Dental Tool Technology   **201**
    8.1   Introduction   201
    8.2   Burs and Abrasive Points   203
    8.3   Classification of Dental Burs   207
    8.4   Coding of Dental Tools   207
    8.5   Dental Devices   212
    8.6   Dental Laboratory Materials   213
    8.7   Dental Cutting Tools   224
    8.8   Health and Safety   229
    References   231

**9. Nanocrystalline Diamond: Deposition Routes
    and Clinical Applications**   **241**
    9.1   Introduction   241
    9.2   Nanocrystalline Diamond   243
    9.3   Clinical Applications   256
    9.4   Summary   264
    References   265

**10. Environmental Engineering Controls
    and Monitoring in Medical Device Manufacturing**   **273**
    10.1   Introduction   273
    10.2   Stressor Source, Properties,
            and Characteristics   275
    10.3   Sterilization   275
    10.4   Cleaning, Etching, and Surface Preparation   284
    10.5   Adhesive Applications   294
    10.6   Coating Applications   295
    10.7   Drilling, Grinding, Cutting, and Machining   296
    10.8   Welding and Soldering   298
    10.9   General Maintenance Activities   299
    10.10   Laboratory Research and Testing   300
    10.11   Environmental and Engineering Controls   301
    10.12   Substitution   302
    10.13   Process Controls   302
    10.14   Enclosure/Isolation   303
    10.15   Process Change or Elimination   304
    10.16   Ventilation Controls   304
    10.17   Personal Protective Equipment and Clothing   312
    10.18   Control Strategies in Device Manufacturing   312
    10.19   Monitoring   314
    10.20   Particle, Fumes, and Aerosol Monitoring   315

10.21   Vapors and Gases   321
10.22   Ionizing Radiation   327
10.23   Non-Ionizing Radiation   329
10.24   Noise and Heat Stress   330
10.25   Microbial Environmental Monitoring   331
10.26   Clean Room Monitoring Requirements   334
10.27   Monitor Selection in Device Manufacturing   335
10.28   Summary   337
        References   337

11. **Biomaterial-Cell-Tissue Interactions In Surface
    Engineered Carbon-Based Biomedical Implants
    and Devices**   **341**
    11.1   Introduction   341
    11.2   Potential Biomedical Applications of DLC   347
    11.3   Definitions and General Aspects
            of Biocompatibility   348
    11.4   Blood   350
    11.5   Cell Culture/Seeding Peculiar to Each Cell   356
    11.6   Statistics and Counting of Cells   359
    11.7   Stereological Investigations   360
    11.8   Photo-Fluorescent Imaging of Cells/Tissues   361
    11.9   Biocompatibility and Hemo-compatibility
            Models   363
    11.10   Carbon-based Materials Interaction
            with Selected Proteins and Cells   367
    11.11   DLC Interactions with Fibroblasts In-Vitro   368
    11.12   Endothelial Pre-seeding on Biomaterials
            for Tissue Engineering   400
    11.13   Bio-Assays and Assessment of Intracellular
            Activities   406
    11.14   In-vivo Studies of Carbon-based Materials:
            Cell-Tissue Interactions In-situ   417
    11.15   On-going and Future Investigations   426
            References   429

12. **Applications of Carbon Nanotubes
    in Bio-Nanotechnology**   **439**
    12.1   Introduction   439
    12.2   Bio-Nanomaterials   440
    12.3   Carbon Nanotubes   441
    12.4   Analysis   464

12.5   Toxicity of Carbon Nanotubes   468
12.6   Conclusions   469
References   469

13. **Bonelike Graft for Regenerative Bone Applications**   **477**
13.1   Introduction   477
13.2   Synthetic Bone Graft Material – Bonelike   486
13.3   Summary   509
References   509

14. **Machining Cancellous Bone Prior to Prosthetic**
    Implantation   **513**
14.1   Introduction   513
14.2   Structure of Cancellous Bone   514
14.3   Theory of Micromachining   515
14.4   Initial Chip Curl Modeling   518
14.5   Experimental   524
14.6   Discussion   529
14.7   Conclusions   530
References   531

15. **Titanium and Titanium Alloy Applications**
    in Medicine   **533**
15.1   Metallurgical Aspects   533
15.2   Principal Requirements of Medical Implants   545
15.3   Shape Memory Alloys   554
15.4   Conclusions   568
References   568

Subject Index   **577**

List of Authors

Wunmi Ademosu
Thin Film Research Center,
School of Engineering and Science,
University of Paisley,
Paisley, PA1 2BE, Scotland, U.K.
E-mail: *wunmi.ademosu@paisley.ac.uk*

Waqar Ahmed
University of Ulster,
NIBEC,
Newtownabbey, BT 37 0QB,
United Kingdom.
E-mail: *w.ahmed@ulster.ac.uk*

Nasar Ali
University of Aveiro,
Department of Mechanical Engineering,
3810-193 Aveiro, Portugal.
E-mail: *n.ali@mec.ua.pt*

Matej Balazic
Machining Laboratory,
Department of Machining Technology Management,
Faculty of Mechanical Engineering,
University of Ljubljana,
Askerceva 6, SI-1000 Ljubljana,
Slovenia.

D. Bombac
Ul. Metoda Mikuza 12,
SI-1000 Ljubljana, Slovenia.

M. Brojan
Laboratory for Nonlinear Mechanics,
Department of Mechanics,
Faculty of Mechanical Engineering,

University of Ljubljana,
Askerceva 6, SI-1000 Ljubljana,
Slovenia.

J. Anthony Byrne
University of Ulster, NIBEC,
Newtownabbey, BT 37 0QB,
United Kingdom.
E-mail: *j.a.byrne@ulster.ac.uk*

Gil Cabral
University of Aveiro,
Department of Mechanical Engineering,
3810-193 Aveiro, Portugal.

R. Caram Jr.
Department of Materials Engineering,
State University of Campinas,
UNICAMP C.P. 6122, 13083-970 Campinas SP,
Brazil.

M. H. Fernandes
FMDUP – Faculdade de Medicina Dentária da Universidade do Porto, Rua,
Dr., Manuel Pereira da Silva,
4200-393 Porto, Portugal.

Juan Gracio
University of Aveiro,
Department of Mechanical Engineering,
3810-193 Aveiro, Portugal.
E-mail: *jgracio@mec.ua.pt*

Rodney G. Handy
Purdue University,
College of Technology,
401 North Grant Street,
West Lafayette, IN 47907-2021, USA.
E-mail: *rhandy@purdue.edu*

N. Sooraj Hussain
INEB - Instituto de Engenharia Biomédica,
Laboratório de Biomateriais, Rua Campo Alegre,
823, 4150 -180, Porto, Portugal.

Mark J. Jackson
Purdue University,
Birck Nanotechnology Center,
College of Technology,
401 North Grant Street,
West Lafayette, IN 47907-2021, USA.
E-mail: *jacksomj@purdue.edu*

Januz Kopac
Machining Laboratory,
Department of Machining Technology Management,
Faculty of Mechanical Engineering,
University of Ljubljana,
Askerceva 6, SI-1000 Ljubljana,
Slovenia.

F. Kosel
Laboratory for Nonlinear Mechanics,
Department of Mechanics,
Faculty of Mechanical Engineering,
University of Ljubljana,
Askerceva 6, SI-1000 Ljubljana,
Slovenia.

Yasmeen Kousar
University of Aveiro,
Department of Mechanical Engineering,
3810-193 Aveiro, Portugal.
E-mail: *y.kousar@mec.ua.pt*

Michael D. Lafreniere
Ohio University,
Chillicothe,
Ohio, USA.
E-mail:

J. C. Madaleno
University of Aveiro,
Department of Mechanical Engineering,
3810-193 Aveiro, Portugal.

Chris Maryan
Manchester Metropolitan University,
Chester Street,

Manchester, M1 5GD, U.K.
E-mail: *c.maryan@mmu.ac.uk*

A. C. Maurício
FEUP – Faculdade de Engenharia da Universidade do Porto,
DEMM, Rua Dr., Roberto Frias, 4200 - 465 Porto, Portugal.

Andrew McLean
School of Engineering and Science,
University of Paisley,
Paisley, PA1 2BE, Scotland, U.K.
E-mail: *andrew.mclean@paisley.ac.uk*

Abraham A. Ogwu
School of Engineering and Science,
University of Paisley,
Paisley, PA1 2BE, Scotland, U.K.
E-mail: *ogwu-ph0@paisley.ac.uk*

Thomas Okpalugo
University of Ulster,
NIBEC,
Newtownabbey, BT 37 0QB,
United Kingdom.
E-mail: *tit.okpalugo@ulster.ac.uk*

Frank Placido
Thin Film Research Center,
School of Engineering and Science,
University of Paisley,
Paisley, PA1 2BE, Scotland, U.K.
E-mail: *frank.placido@paisley.ac.uk*

J. D. Santos
FEUP,
Faculdade de Engenharia da Universidade do Porto,
DEMM, Rua Dr.,
Roberto Frias,
4200 - 465 Porto, Portugal.

Patrick Senarith
Medtronic, Inc,
7000 Central Ave.
Minneapolis, MN 55432
E-mail: *Patrick.Senarith@medtronic.com*

T. Shokuhfar
University of Aveiro,
Department of Mechanical Engineering,
3810-193 Aveiro, Portugal.

A. C. M. Sousa
University of Aveiro,
Department of Mechanical Engineering,
3810-193 Aveiro, Portugal.

Elby Titus
University of Aveiro,
Department of Mechanical Engineering,
3810-193 Aveiro, Portugal.

Thomas J. Webster
Division of Engineering,
Brown University,
Providence, RI USA 02912.
E-mail: *Thomas_Webster@Brown.edu*

Michael D. Whitt
Purdue University,
Mechanical Engineering Technology,
Knoy Hall of Technology, Room 184,
West Lafayette, IN 47907-2021.
E-mail: *mwhitt@purdue.edu*

Chang Yao
Division of Engineering,
Brown University,
Providence, RI USA 02912.
E-mail: *Chang_Yao@Brown.edu*

Nomenclature

AIDS = Acquired Immune Deficiency Syndrome
AFI = Average Fluorescence Intensity
ALP = Alkaline Phosphatase Activity
a-C:H = Hydrogenated amorphous carbon
a-C:H:N = Nitrogen doped hydrogenated amorphous carbon
a-C:H:Si = Silicon doped hydrogenated amorphous carbon
AFM = Atomic Force Microscopy
BMA = Bone Marrow Aspirate
BMP = Bone Morphogenetic Protein
BMU = Basic Multicellular Unit
BSA = Bovine Serum Albumine
BSP = Bone Sialoprotein
CAM = Cell Adhesion Molecule
CCD = Charged Coupled Device
COL I = Collagen type 1
CPD = Contact Potential Difference
PECVD = Plasma Enhanced Chemical Vapour Deposition
DAC = Digital to Analogue Converter
DAS = Data Acquisition System
DC = Direct Current
DCA = Dynamic Contact Angle
DBM = Demineralized Bone Matrix
DLC = a-C:H = Diamond Like Carbon
DMSO = Dimethylsulphoxide
DSP = Data Signal Processor
ECM = Extracellular Matrix
EDS = Energy Dispersive Spectroscopy
EDTA = Ethylene-di-tetra-acetic acid
EGF = Epithelial Growth Factor
ELISA = Enzyme Linked Immuno-Sorbent Assay
EM = Electroforetic Mobility
FAT = Fixed Analyser Transmission
FCS = Foetal Calf Serum

FTIR = Fourier Transform Infra-Red
GPIB = General Purpose Interface Bus
Hela = Henrietta Lacks-cervical carcinoma cell line (the first in-vitro human cell line)
HMDS = Hexamethyldisilaxane
HMEC = Human Microvascular Endothelial Cell
L132 = Human embryonic lungs cell line (Hela-characteristics/contamination)
LASER = Light Amplification by Stimulated Emission of Radiation
LED = Light Emitting Diode
MCDB = Microvascular endothelial cell growth media
MEM = Minimal Essential Media
DMEM = Dulbecco Minimal Essential Media
PPP = Platelet Poor Plasma
MSM = Metal Semiconductor Metal sandwich
MTT = 3-(4, 5-dimethylthiazol-2-yl)-2,5-diphenyl tetrazolium bromide
N1 = N-DLC film obtained with nitrogen ions bombardment for 1 hour
N1h = N-DLC film obtained with nitrogen ions bombardment for 1.5 hours
N2= N-DLC film obtained with nitrogen ions bombardment for 2 hours
N2h = N-DLC film obtained with nitrogen ions bombardment for 2.5 hours
N-DLC = a-C:H:N = Nitrogen doped Diamond Like Carbon
NEAA = Non-Essential Amino Acid
PBS = Phosphate Buffered Saline
PRP = Platelet Rich Plasma
Ra = Arithmetical Mean Roughness
RF = Radio Frequency
RMS = Root Mean Square
Sccm= Standard cubic centimetre
SD10 = Si-DLC film obtained using TMS flow rate of 10sccm
SD15 = Si-DLC film obtained using TMS flow rate of 15sccm
SD20 = Si-DLC film obtained using TMS flow rate of 20sccm
SD5 = Si-DLC film obtained using TMS flow rate of 5sccm
SEM = Scanning Electron Microscope

Si-DLC = a-C:H:Si = Silicon doped Diamond Like Carbon
SN1 = N-DLC film obtained with nitrogen neutrals bombardment
for 1 hour
SN1h = N-DLC film obtained with nitrogen neutrals bombardment
for 1.5 hours
SN2 = N-DLC film obtained with nitrogen neutrals bombardment
for 2 hours
SN2h = N-DLC film obtained with nitrogen neutrals bombardment
for 2.5 hours
SPR = Surface Plasmon Resonance
SW = Silicon wafer (uncoated)
TCPS = Tissue culture polysterene
TMS = Tetramethylsilane
T_R = Room temperature
V79 = Chinese Hamster cell line
WF = Work Function
WR = Web Reference
XPS = X-ray Photoelectron Spectroscopy
ZDOI = Zero Depth of Immersion
σ = Bending strength
θ = Contact Angle
λ = Wavelength

1. Atomic Scale Machining of Surfaces

1.1 Introduction

Molecular dynamic simulations of machining at the atomic scale can reveal a significant amount of information regarding the behavior of machining and grinding processes that cannot be explained easily using classical theory or experimental procedures. This chapter explains how the use of molecular dynamic simulations can be applied to the many problems associated with machining and grinding at the meso, micro, and nanoscales. These include: (a) mechanics of nanoscale machining of ferrous and non-ferrous materials; (b) physics of nanoscale grinding of semiconductor materials; (c) effects of simulating a variety of machining parameters in order to minimize sub-surface damage; (d) modeling of exit failures experienced during machining such as burr formation and other dynamic instabilities during chip formation; (e) simulation of known defects in microstructures using molecular dynamic simulations, statistical mechanical, and Monte Carlo methods; (f) simulation of machining single crystals of known orientation; (g) extremely high speed nanometric cutting; (h) tool wear during machining; and (i) the effects of hardness on the wear of tool and workpiece materials. The nature of wear of the material ahead of the machining and grinding process, the variation of machining forces, and the amount of specific energy induced into the workpiece material using molecular dynamic simulations is discussed in this chapter.

Nanotechnology is the creation and utilization of materials, structures, devices and systems through the control of matter at the nanometer length scale. The essence of nanotechnology is the ability to work at these levels to generate large structures with fundamentally new properties. Although certain applications of nanotechnology, such as giant magnetoresistance (GMR) structures for computer hard disk read head and polymer displays have entered the marketplace, in general nanotechnology is still at a very early stage of development.

The barriers between nanotechnology and the marketplace lie in how to reduce the fabrication cost and how to integrate nanoscale assemblies with functional microscale and macro devices. Therefore, reliable mass production of nanostructures is currently one of the most crucial issues in nanotechnology. The commercialization of nanotechnology has to address the underlying necessities of predictability, repeatability, producibility and productivity in manufacturing at nanometer scale.

Nanometric machining refers to a "top down" nanofabrication approach. To the authors' knowledge the concept of nanometric machining is more concerned with precision rather than the characteristic size of the product. Therefore, nanometric machining is defined as the material removal process in which the dimensional accuracy of a product can be achieved is 100 nm or less. Nanometric machining can be classified into four categories:

- Deterministic mechanical nanometric machining. This method utilizes fixed and controlled cutting tools, which can specify the profiles of three-dimensional components by a well-defined tool surface. The method can remove materials in amounts as small as tens of nanometers. It includes processes such as diamond turning, micro milling, and nano/micro grinding.
- Loose abrasive nanometric machining. This method uses loose abrasive grits to removal a small amount of material. It consists of polishing, lapping, and honing, etc.
- Non-mechanical nanometric machining comprises processes such as focused ion beam machining, micro-EDM, and excimer laser machining.
- Lithographic method. The method employs masks to specify the shape of the product. Two-dimensional shapes are the main outcome; severe limitations occur when three-dimensional products are attempted [1]. Processes include X-ray lithography, LIGA, and electron beam lithography.

The author believes that mechanical nanometric machining has more advantages than other methods since it is capable of machining complex 3D components in a controllable and deterministic way.

The machining of complex surface geometries is just one of the future trends in nanometric machining, which is driven by the integration of multiple functions in one product. For instance, the method can be used to machine micro moulds/dies with complex geometric features and high dimensional and form accuracy, and even nanometric surface features. The method is indispensable to manufacturing complex micro and miniature structures, components, and products in a variety of engineering materials. This chapter focuses on nanometric cutting theory, methods, and its implementation and application perspectives.

1.2 Theoretical Basis of Nanomachining

The scientific study of nanometric machining has been undertaken since the late 1990s. Much attention to the study has been paid especially with the advancement of nanotechnology [2]. The scientific study will result in the formation of the theoretical basis of nanometric machining, which enables the better understanding of nanometric machining physics and the development of controllable techniques to meet the requirements for nanotechnology and nanoscience.

1.2.1 Cutting Force and Energy

In nanomanufacturing, the cutting force and cutting energy are important issues. They are important physical parameters for understanding cutting phenomena as they clearly reflect the chip removal process. From the point of view of atomic structures, cutting forces are the superposition of the interactions of forces between workpiece atoms and cutting tool atoms. Specific energy is an intensive quantity that characterizes the cutting resistance offered by a material [3]. Ikawa et alia [2], and Luo et alia [4] have acquired the cutting forces and cutting energy by molecular dynamics (MD) simulations. Ikawa et alia [2] have carried out experiments to measure the cutting forces in nanometric machining. Table 1.1 shows the simulation and experimental results in nanometric cutting. Table 1.1 (a) illustrates the linear relationship that exists between the cutting force per width and depth of cut in both simulations and experiments. The cutting forces per width increase with the increment of the depth of cut.

The difference in the cutting force between simulations and experiment is caused by the different cutting edge radii applied in the simulations. In nanometric machining the cutting edge radius plays an important role since the depth of cut is similar in scale. Under the same depth of cut higher cutting forces are required for a tool with a large cutting edge radius compared with a tool with a small cutting edge radius. The low cutting force per width is obviously the result of fine cutting conditions, which will decrease the vibration of the cutting system and thus improve machining stability and will also result in better surface roughness.

A linear relationship between the specific energy and the depth of cut can also be observed in Fig. 1.1. The figure shows that the specific energy increases with a decreasing of depth of cut, because the effective rake angle is different under different depths of cut. In small depths of cut the effective rake angle will increase with the decreasing of depth of cut. Large rake angles result in an increase in specific cutting energy. This phenomenon is often called the 'size effect', which can be clearly explained by material data listed in Table 1.1. According to Table 1.1, in nanometric machining only point defects exist in the machining zone in a crystal. Therefore, the material will need more energy to initiate the formation of an atomic-crack or the movement of an atomic-dislocation. The decreasing of depth of cut will decrease the chance for the cutting tool to meet point defects in the material and will result in increasing the specific cutting energy.

If the machining unit is reduced to 1 nm, the workpiece material structure at the machining zone may approach atomic perfection, hence more energy will be required to break the atomic bonds. Alternatively, when the machining unit is higher than 0.1 μm, the machining points will fall into the distribution distances of some defects such as dislocations, cracks, and grain boundaries. The pre-existing defects will ease the deformation of workpiece material and result in a comparatively low specific cutting energy.

Nanometric cutting is also characterized by the high ratio of the normal to the tangential component in the cutting force [3,4], as the depth of cut is very small in nanometric cutting and the workpiece is

Table 1.1. Material properties under different machining units [5].

	1 nm – 0.1 μm	0.1 μm – 10 μm	10 μm – 1 mm
Defects/Impurities	Point defect	Dislocation/crack	Crack/grain boundary
Chip removal unit	Atomic cluster	Sub-crystal	Multi-crystals
Brittle fracture limit	10^4 J/m^3 - 10^3 J/m^3 Atomic-crack	10^3 J/m^3 - 10^2 J/m^3 Micro-crack	10^2 J/m^3 - 10^1 J/m^3 Brittle crack
Shear failure limit	10^4 J/m^3 - 10^3 J/m^3 Atomic-dislocation	10^3 J/m^3 - 10^2 J/m^3 Dislocation slip	10^2 J/m^3 - 10^1 J/m^3 Shear deformation

mainly processed by the cutting edge. The compressive interactions will thus become dominant in the deformation of the workpiece material, which will therefore result in the increase of friction force at the tool-chip interface with a relatively high cutting ratio.

Usually, the cutting force in nanometric machining is very difficult to measure due to its small amplitude compared with the noise generated (mechanical or electronic) [2]. A piezoelectric dynamometer, or load cell, is used to measure the cutting forces because of their characteristic high sensitivity and natural frequency. Figure 1.2 shows an experimental force measuring system in micromilling process carried out by Dow et alia [6]. The three-axis load cell, Kistler 9251, is mounted in a specially designed mount on the Y-axis of a Nanoform 600 diamond turning machine. A piece of S-7 steel that has been ground flat on both sides is used as the workpiece and secured through the top of the load cell with a bolt preloaded to 30 N. The tool was moved in the +Z direction to set the depth of cut and the workpiece was fed in the +y direction to cut the groove.

The milling tool is mounted in a Westwind D1090-01 air bearing turbine spindle capable of speeds up to 60,000 rpm. The spindle is attached to the Z-axis of the Nanoform 600. To determine the rotational speed of the tool and the orientation of each flute, an optical detector (Angstrom resolver) was used to indicate a single rotation of the spindle by reading a tool revolution marker aligned with one flute. The measured 3D cutting forces under depth of cut of 25 μm, feed rate of 18.75 μm/flute, are of the order of several Newtons.

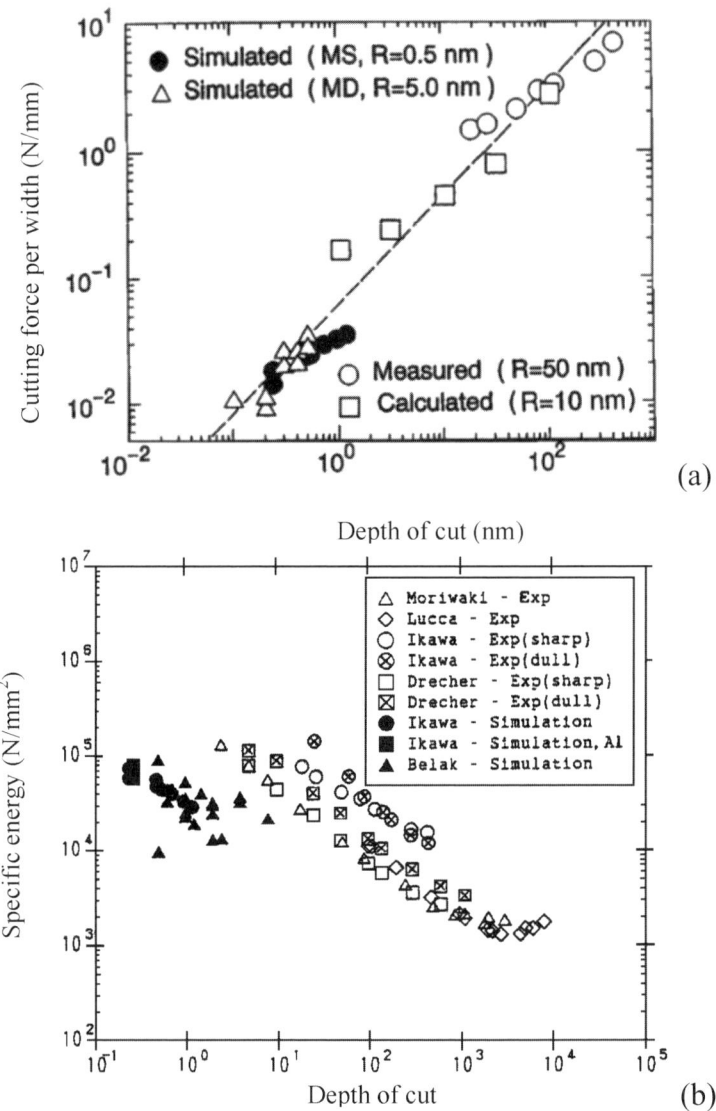

Fig. 1.1. The comparison of results between simulations and experiments: (a) cutting force per width against depth of cut; (b) specific energy against depth of cut [2].

1.2.2 Cutting Temperature

In MD simulations, the cutting temperature can be calculated under the assumption that the cutting energy totally transforms into cutting

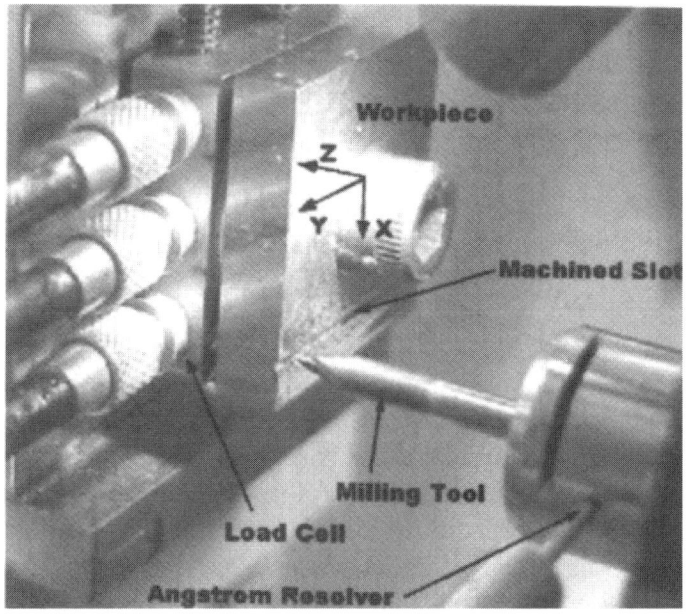

Fig. 1.2. Experimental force measurement system in micro-milling process [6].

heat and results in increasing the cutting temperature and kinetic energy of system. The lattice vibration is the major form of thermal motion of atoms. Each atom has three degrees of freedom. According to the theorem of equi-partition of energy, the average kinetic energy of the system can be expressed as:

$$\bar{E}_k = \frac{3}{2} N k_B T = \sum_i \frac{1}{2} m(V_i^2) \qquad (1.1)$$

Where $\bar{E}_k$ is average kinetic energy in equilibrium state, K_B is Boltzmann's constant, T is absolute temperature, and m_i and V_i are the mass and velocity of an atom, respectively. N is the number of atoms.

The cutting temperature can be calculated using the following equation:

$$T = \frac{2 \bar{E}_k}{3 N k_B} \qquad (1.2)$$

Figure 1.3 shows the variation of cutting temperature on the cutting tool in MD simulation of nanometric cutting of single crystal aluminium [7].

The highest temperature is observed at cutting edge although the temperature at the flank face is also higher than that at the rake face. The temperature distribution suggests that a major source of heat exists at the interface between the cutting edge and the workpiece, and that the heat be conducted from there to the rest of the cutting zone. This is because that most of cutting action takes place at the cutting edge of the tool and the resulting dislocation deformation in the workpiece material will transfer their potential energy into kinetic energy and result in the observed temperature rise. The comparatively high temperature exhibited at the flank face is caused by the friction between the flank face and the workpiece. The released energy due to the elastic recovery of the machined surface also contributes to the incremental increase in temperature at flank face. Although there is friction between the rake face and the chip, the heat will be taken away from the rake face by the removal of the chip.

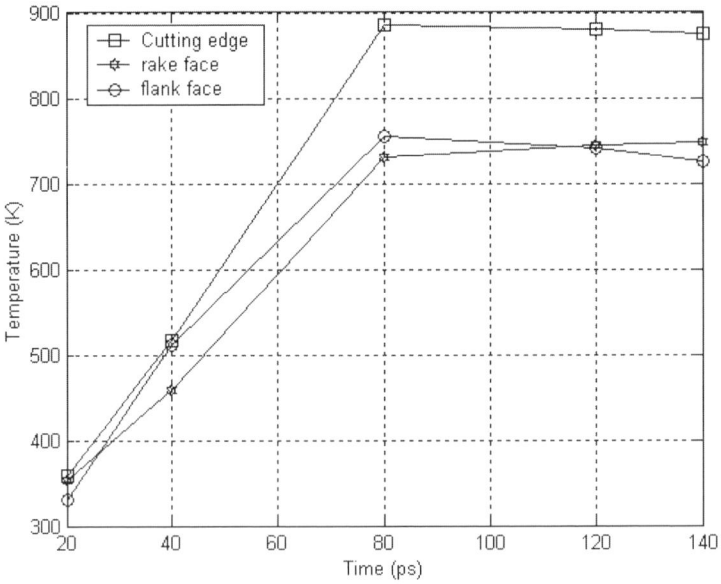

Fig. 1.3. Cutting temperature distribution of cutting tool in nanometric cutting (cutting speed = 20 m/s, depth of cut = 1.5 nm, cutting edge radius = 1.57 nm) [7].

Therefore, the temperature at tool rake face is lower than that at the tool cutting edge and tool flank face. The temperature value shows that the cutting temperature in diamond machining is quite low in comparison with that in conventional cutting, due to low cutting energy in addition to high thermal conductivity of diamond and the workpiece material. The cutting temperature is considered to govern the wear of a diamond tool as shown in the MD simulation study by Cheng et alia [8]. In-depth experimental and theoretical studies are needed to find out the quantitative relationship between cutting temperature and tool wear although there is considerable evidence of chemical damage on the surface of diamond in which increases in temperature tends to plays a significant role [8].

1.2.3 Chip Formation and Surface Generation

Chip formation and surface generation can be simulated by MD simulations. Figure 1.4 shows an MD simulation of a nanometric cutting process on single crystal aluminium [7]. From Fig. 1.4(a) it is shown that after the initial plough of the cutting edge the workpiece, atoms are compressed in the cutting zone near to the rake face and the cutting edge. The disturbed crystal lattices of the workpiece and even the initiation of dislocations can be observed in Fig. 1.4(b). Figure 1.4(c) shows the dislocations have piled up to form a chip. The chip is removed with the unit of an atomic cluster as shown in Fig. 1.4(d). Lattice disturbed workpiece material is observed on the machined surface.

Based on the visualisation of the nanometric machining process, the mechanism of chip formation and surface generation in nanometric cutting can be explained. Owing to the ploughing of the cutting edge, the attractive force between the workpiece atoms and the diamond tool atoms becomes repulsive. Because the cohesion energy of diamond atoms is much larger than that of Al atoms, the lattice of the workpiece is compressed. When the strain energy stored in the compressed lattice exceeds a specific level, the atoms begin to rearrange so as to release the strain energy. When the energy is not sufficient to perform the rearrangement, some dislocation activity is generated. Repulsive forces between compressed atoms in the upper

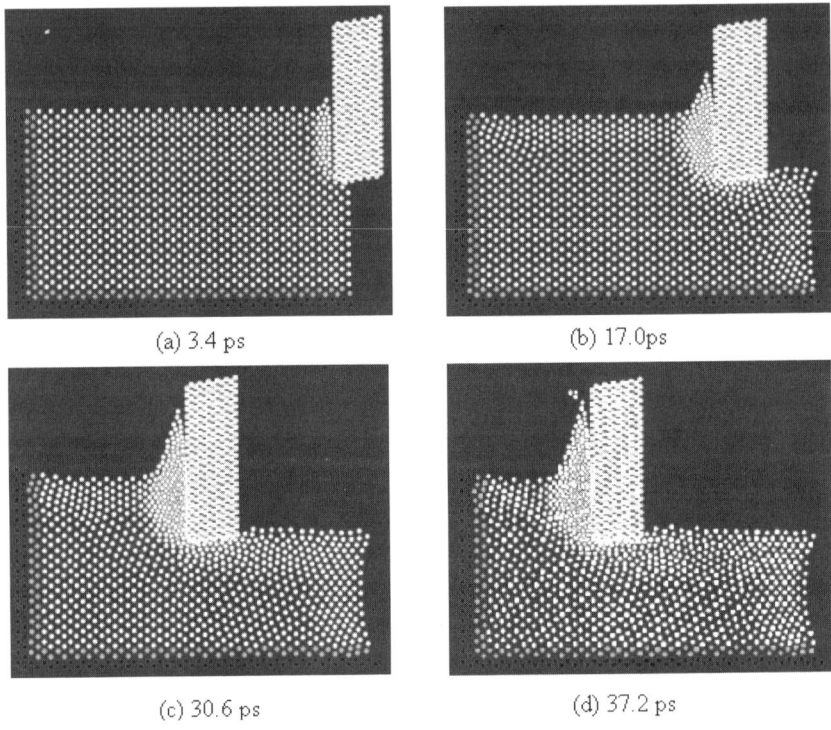

(a) 3.4 ps

(b) 17.0ps

(c) 30.6 ps

(d) 37.2 ps

Fig. 1.4. MD simulations of the nanometric machining process (Cutting speed = 20m/s, depth of cut = 1.4 nm, cutting edge radius = 0.35 nm) [7].

layer and the atoms in the lower layer are increasing, so the upper atoms move along the cutting edge, and at the same time the repulsive forces from the tool atoms cause the resistance for the upward chip flow to press the atoms under the cutting line. With the movement of the cutting edge, some dislocations move upward and disappear from the free surface as they approach the surface.

This phenomenon corresponds to the process of chip formation. As a result of the successive generation and disappearance of dislocations, the chip seems to be removed steadily. After the passing of the tool, the pressure at the flank face is released. The layers of atoms move upwards and result in elastic recovery, so that the machined surface is generated.

The conclusion can therefore be drawn that chip removal and machined surface generation are, in nature, dislocation slip motion inside the workpiece material grains. In conventional cutting, dislocations are initiated from the existing defects between the crystal grains, which will ease the movement of dislocations and result in smaller specific cutting forces compared with that in nanometric cutting.

The height of the atoms on the surface create the observed surface roughness. For this, 2-D MD simulation roughness average (R_a) can be used to assess the machined surface roughness. The surface integrity parameters can also be calculated based on the simulation results. For example, the residual stress of the machined surface can be estimated by averaging the forces acting on the atoms in a unit area on the upper layer of the machined surface.

MD simulation has been proved to be a useful tool for the theoretical study of nanometric machining [9]. At present, MD simulation studies on nanometric machining are limited by the computing memory size and speed of the computer. It is therefore difficult to enlarge the dimension of the current MD model on a personal computer. In fact, the machined surface topography is produced as a result of the copy of the tool profile on a workpiece surface that has a specific motion relative to the tool. The degree of the surface roughness is governed by both the controllability of machine tool motions (or relative motion between tool and workpiece) and the transfer characteristics (or the fidelity) of tool profile to the workpiece [2]. A multi-scale analysis model, which can fully model the machine tool and cutting tool motion, environmental effects and the tool-workpiece interactions, is needed to predict and control the nanometric machining process in a deterministic manner.

1.2.4 Minimum Undeformed Chip Thickness

Minimum undeformed chip thickness is an important issue in nanometric machining because it relates to the ultimate machining accuracy. In principle, the minimum undeformed chip thickness will be determined by the minimum atomic distance within the workpiece. However, in ultra-precision machining practices it depends on the sharpness of the diamond cutting tool, the capability of the ultra-precision machine tool and the machining environment. Diamond turning experiments of non-ferrous work materials carried out at Lawrence Livermore National Laboratory in the United States of America show the minimum undeformed chip thickness, down to 1 nm, is attainable with a specially prepared fine diamond cutting tool on a very stiff ultra-precision machine tool.

Figure 1.5 illustrates chip formation of single crystal aluminium with the tool cutting edge radius of 1.57 nm [7]. No chip formation is observed when the undeformed chip thickness is 0.25 nm. But the initial stage of chip formation is apparent when the undeformed chip thickness reaches 0.26 nm. In nanometric cutting, as the depth of cut is very small, the chip formation is related to the force conditions on the cutting edge of the tool. Generally, chip formation is mainly a function of tangential cutting force.

The normal cutting force makes little contribution to the chip formation since it has the tendency to penetrate the atoms of the surface into the bulk of the workpiece. The chip is formed on condition that the tangential cutting force is larger than the normal cutting force. The relationships between minimum undeformed chip thickness, cutting edge radius, and cutting forces are studied by MD simulations. The results are highlighted in Table 1.2 [7]. The data shows that the minimum undeformed chip thickness is about 1/3 to 1/6 magnitude of the tool cutting edge radius. Chip formation will be initiated when the ratio of tangential cutting force to normal cutting force is larger than 0.92.

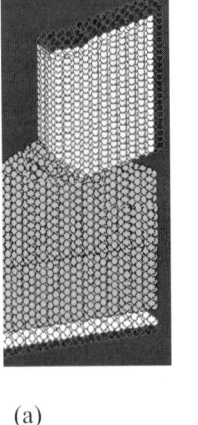
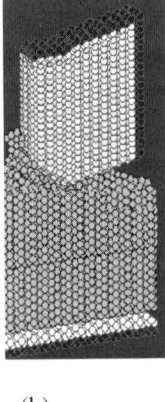

(a)   (b)

Fig. 1.5. Study of minimum undeformed chip thickness by MD simulation. (a) Undeformed chip thickness = 0.25nm; (b) Undeformed chip thickness = 0.26 nm [7].

1.2.5 Critical Cutting Edge Radius

It is widely accepted that the sharpness of the cutting edge of a diamond cutting tool directly affects the machined surface quality. Previous MD simulations show that the sharper the cutting edge, the smoother the machined surface becomes. But this conclusion is based on zero tool wear. To study the real effects of cutting edge radius, the MD simulations on nanometric cutting of single crystal aluminium are carried out using a tool wear model [9].

In the simulations the cutting edge radius of the diamond cutting tool varies from 1.57 nm to 3.14 nm with depth of cut of 1.5 nm, 2.2 nm and 3.1 nm, respectively. The cutting distance is fixed at 6 nm. The root-mean–square deviation of the machined surface and mean stress on the cutting edge are listed in Table 1.3.

Table 1.2. Minimum undeformed chip thickness against the tool cutting edge radius and cutting forces [7].

Cutting edge radius (nm)	1.57	1.89	2.31	2.51	2.83	3.14
Minimum undeformed chip thickness (nm)	0.26	0.33	0.42	0.52	0.73	0.97
Ratio of minimum undeformed chip thickness to tool cutting edge radius	0.17	0.175	0.191	0.207	0.258	0.309
Ratio of tangential cutting force to normal cutting force	0.92	0.93	0.92	0.92	0.94	0.93

Table 1.3. The relationship between cutting edge radius and machined surface quality [7].

Cutting edge radius (nm)	1.57	1.89	2.31	2.51	2.83	3.14
Depth of S_q (nm) cut: 1.5 nm		0.89	0.92	0.78	0.86	0.98	1.06
Depth of S_q (nm) cut: 2.2 nm		0.95	0.91	0.77	0.88	0.96	1.07
Depth of S_q (nm) cut: 3.1 nm		0.97	0.93	0.79	0.87	0.99	1.08
Mean stress at cutting edge (GPa)		0.91	0.92	–0.24	–0.31	–0.38	–0.44

Figure 1.6 shows the visualization of the simulated data, which clearly indicates that surface roughness increases with the decreasing cutting edge radius when the cutting edge radius is smaller than 2.31 nm. The tendency is caused by the rapid tool wear when a cutting tool with small cutting edge radius is used. But when the cutting edge is larger than 2.31 nm, the cutting edge is under compressive stress and no tool wear happens. Therefore, it shows the same tendency that the surface roughness increases with decreasing the tool cutting edge radius as in the previous MD simulations.

The MD simulation results also illustrate that it is not true that the sharper the cutting edge, the better the machined surface quality. The cutting edge will wear and results in the degradation of the machined surface quality if its radius is smaller than the critical value for cutting. But when the cutting edge radius is higher than the critical value, the compressive stress will take place at the tool edge and the tool condition is more stable. As a result a high quality machined surface can be achieved. Therefore, there is a critical cutting edge radius for achieving a high quality machined surface.

For cutting single crystal aluminium, the critical cutting edge radius is 2.31 nm. The MD simulation approach is applicable for acquiring the critical cutting edge radius for nanometric cutting of other materials.

1.2.6 Properties of Workpiece Materials

In nanometric machining the microstructure of the workpiece material will play an important role in affecting the machining accuracy and machined surface quality. For example, when machining polycrystalline materials, the difference in the elastic coefficients at the grain boundary and interior of the grain causes small steps to be formed on the cut surface since the respective elastic 'rebound' will vary [10]. The study by Lee and Chueng [10] shows the shear angle varies with the crystallographic orientation of the materials being cut. This will produce a self-excited vibration between cutting tool and workpiece and result in a local variation of surface roughness of a diamond turned surface [11].

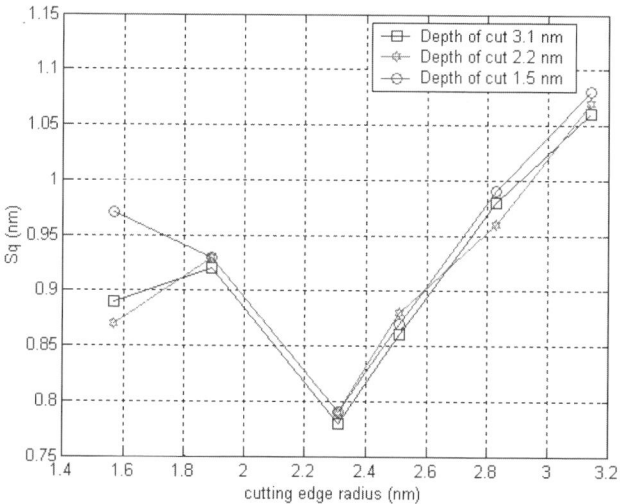

Fig. 1.6. Cutting edge radius against machined surface quality.

A material's destructive behaviour can also be affected by nano-metric machining. In nanometric machining of brittle materials it is possible to produce plastically deformed chips if the depth of cut is sufficiently small [11]. It has been shown that a "brittle-to-ductile" transition exists when cutting brittle materials at low load and pene-tration levels. The transition from ductile to brittle fracture has been widely reported and is usually described as the "critical depth of cut", i.e., generally small up to 0.1 to 0.3 μm. These small depths of cut will result in relatively low material removal rates.

However, it is a cost-effective technique for producing high quality spherical and non-spherical optical surfaces, with or without the need for lapping and polishing.

The workpiece materials should also have a low affinity with the cutting tool material. If pieces of the workpiece material are deposited onto the tool, this will cause tool wear and will adversely affect the surface in terms of surface roughness and integrity. Therefore, work-piece materials chosen must possess an acceptable machinability on which nanometric surface finishes can thus be achieved.

Diamond tools are widely used in nanometric machining because of their excellent characteristics. The materials currently turned with diamond tools are listed in Table 1.4. Materials that can be processed using ductile mode grinding with diamond wheels are listed in Table 1.5. Table 1.6 summarises the comparison of nanometric machining and conventional machining in all major aspects of cutting mechanics and physics. The comparison highlighted in the table is by no means comprehensive, but rather provides a starting point for further study on the physics of nanometric machining.

A material's destructive behaviour can also be affected by nanometric machining. In nanometric machining of brittle materials it is possible to produce plastically deformed chips if the depth of cut is sufficiently small [11]. It has been shown that a "brittle-to-ductile" transition exists when cutting brittle materials at low load and penetration levels. The transition from ductile to brittle fracture has been widely reported and is usually described as the "critical depth of cut", i.e., generally small up to 0.1 to 0.3 μm. These small depths of cut will result in relatively low material removal rates.

However, it is a cost-effective technique for producing high quality spherical and non-spherical optical surfaces, with or without the need for lapping and polishing.

The workpiece materials should also have a low affinity with the cutting tool material. If pieces of the workpiece material are deposited

Table 1.4. Diamond turned materials [11].

Semiconductors	Metals	Plastics
Cadmium telluride	Aluminium and alloys	Acrylic
Gallium arsenide	Copper and alloys	Fluoroplastics
Germanium	Electroless nickel	Nylon
Lithium niobate	Gold	Polycarbonate
Silicon	Magnesium	Polymethylmethacrylate
Zinc selenide	Silver	Propylene
Zinc sulphide	Zinc	Styrene

Table 1.5. Materials that can be processed using ductile mode diamond grinding [11].

Ceramics/Intermetallics		Glasses
Aluminium oxide	Tatanium aluminide	BK7 or equivalent
Nickel aluminide	Tatanium carbide	SF10 or equivalent
Silicon carbide	Tungsten carbide	ULE or equivalent
Silicon nitride	Zirconia	Zerodur or equivalent

Table 1.6. The comparison of nanometric machining with conventional machining [4].

		Nanometric machining	Conventional machining
Fundamental cutting principles		Discrete molecular mechanics/micromechanics	Continuum elastic/plastic/ fracture mechanics
Workpiece material		Heterogeneous (presence of microstructure)	Homogeneous (ideal element)
Cutting physics		Atomic cluster or microelement model $$\dot{q}_i = \frac{\partial H}{\partial p_i} \quad i=1, 2, \ldots N$$ $$p_i = -\frac{\partial H}{\partial q_i}$$	Shear plane model (continuous points in material)
		First principal stress $$\sigma = \frac{1}{S}\sum_{i=1}^{N_A}\sum_{j=1}^{N_B} f_{ij} - \frac{1}{S}\sum_{i=1}^{N_A}\sum_{j=1}^{N_B} f_{0ij}$$ (crystal deformation included)	Cauchy stress principle $$\tau_S = \frac{F_s}{A}$$ (constant)
Cutting force and energy	Energy consideration	Interatomic potential functional $$U(r^N) = \sum_i \sum_{<i} u(r_{ij})$$	Shear/friction power $$P_s = F_s \cdot V_s$$ $$P_u = F_u \cdot V_c$$
	Specific energy	High	Low
	Cutting force	Interatomic forces $$F_I = \sum_{j\neq i}^{N} F_{ij} = \sum_{j\neq i}^{N} -\frac{du(r_{ij})}{dr_{ij}}$$	Plastic deformation/friction $$F_c = F(b, d_c, \tau_s, \beta_a, \phi_c, \alpha_r)$$
Chip formation	Chip initiation	Inner crystal deformation (point defects or dislocation)	Inter crystal deformation (grain boundary void)
	Deformation and stress	Discontinuous	Continuous
Cutting tool	Cutting edge radius	Significant	Ignored
	Tool wear	Clearance face and Cutting edge	Rake face
Surface generation		Elastic recovery	Transfer of tool profile

onto the tool, this will cause tool wear and will adversely affect the surface in terms of surface roughness and integrity. Therefore, workpiece materials chosen must possess an acceptable machinability on which nanometric surface finishes can thus be achieved. Diamond tools are widely used in nanometric machining because of their excellent characteristics. The materials currently turned with diamond tools are listed in Table 1.4. Materials that can be processed using ductile mode grinding with diamond wheels are listed in Table 1.5. Table 1.6 summarises the comparison of nanometric machining and conventional machining in all major aspects of cutting mechanics and physics. The comparison highlighted in the table is by no means comprehensive, but rather provides a starting point for further study on the physics of nanometric machining.

1.3 Further Developments

The further developments associated with nanometric machining is concerned with two aspects. The first aspect is concerned with MD simulations of machining different materials with coated and uncoated cutting tool materials, and the simulation of problems that are commonly encountered in machining such as tool wear during machining, thermal deformation of certain engineering alloys during machining, burr formation during machining, and the modelling of exit failures when the cutting tool has completed its task. The second aspect is focused on developing meso-microscale mechanical machine tools that are in the process of being developed. Here, MD simulations of the machining process will contribute to the design of machine tools dedicated to machining at the nanoscale, and will hopefully illuminate the need for better machine tool design at the meso and macroscales.

References

1. El-Hofy H., Khairy A., Masuzawa T., McGeough J. Introduction. In: McGeough J., Editor, Micromachining of Engineering Materials, *Marcel Dekker*, New York, 2002.

2. Ikawa N., Donaldson R., Komanduri R., König W., Mckeown P.A., Moriwaki T., Stowers I., Ultraprecision metal cutting – the past, the present and the future, *Annals of the CIRP*, 1991, 40 (2), 587-594.
3. Shaw M.C., Metal Cutting Principles, New York, *Oxford University Press*, 2005.
4. Luo X., Cheng K., Guo X., Holt R., An investigation on the mechanics of nanometric cutting and the development of its test-bed, *International Journal of Production Research*, 2003, 41 (7), 1449-1465.
5. Taniguchi N., Nanotechnology. New York, *Oxford University Press*, 1996.
6. Dow T., Miller E., Garrard K., Tool force and deflection compensation for small milling tools, *Precision Engineering*, 2004, 28 (1), 31-45.
7. Cheng K., Luo X., Jackson M.J., In, Jackson, M.J., Editor, Microfabrication and Nanomanufacturing, *Taylor and Francis, CRC Press*, Florida, 2006. pp. 311-338.
8. Cheng K., Luo X., Ward R., Holt R., Modelling and simulation of the tool wear in nanometric cutting, *Wear*, 2003, 255, 1427-1432.
9. Shimada S., Molecular dynamics simulation of the atomic processes in micro-cutting. In, McGeough J., Editor, Micromachining of Engineering Materials. New York: *Marcel Dekker*, 2002, pp. 63-84.
10. Lee W., and Cheung C., A dynamic surface topography model for the prediction of nano-surface generation in ultra-precision machining. *International Journal of Mechanical Sciences*, 2001, 43, 961-991.
11. Corbett J., Diamond Micromachining, In, McGeough J., Editor, Micro-machining of Engineering Materials, New York, *Marcel Dekker*, 2002, pp. 125-146.

2. Anodization: A Promising Nano-Modification Technique of Titanium-based Implants for Orthopedic Applications

2.1 Introduction

As one of the valve metals (including Ti, Al, Ta, Nb, V, Hf, W), titanium is protected by a thin titanium oxide layer which spontaneously forms on its surface when exposed to air or other oxygen containing environments. This oxide passive layer is typically 2 to 5 nm thick and is responsible for the well-documented corrosion resistance property of titanium and its alloys.[1] Because of this and their excellent mechanical properties, titanium and its alloys are widely used in orthopedic and dental applications. However, the native TiO_2 layer is not bioactive enough to form a direct bonding with bone, which means the lack of osseointegration to juxtaposed bone might lead to long term failure after implantation.[2] Specifically, the 10 to 15 year lifetime of current titanium-based orthopedic implants is not as long as expected by many patients. [3]

As a result, many attempts have been made to improve the surface properties of titanium-based implants (e.g., topography, chemistry and surface energy), which directly determine the implant-environment interactions after implantation. These surface modification techniques include mechanical methods (e.g., sand-blasting), chemical methods (e.g., acid etching), coatings (e.g., plasma spraying), etc.[4-9] Through these conventional approaches, a better bonding ability with bone has been achieved due to the creation of a optimum micro-scale surface roughness, a more favorable surface chemistry, and/or a new morphology preferred by bone-forming cells (or osteoblasts). However, neither these mechanical nor chemical methods have the ability to produce controlled surface topographies. Moreover, these methods have the potential to form surface residuals. Thus, alternative methods to modify titanium surfaces are highly desirable for promoting new bone growth.

Other attempts to improve bone-bonding involves coating titanium-based implants with hydroxyapatite (HA) or other calcium phosphates, which is commonly accomplished by plasma spraying.[2] This is based on the fact that HA and other calcium phosphates are the main inorganic components of bone and they have been shown by many to directly bond to juxtaposed bone.[10-13] Unfortunately, such coatings have long-term failures due to weak adhesion to the metal substrate and dissolution once implanted. Therefore, an alternative method to deposit HA firmly onto titanium surfaces with optimal bioactivity is highly desirable for orthopedic applications.

In this light, a current strategy is to modify titanium-based implants to possess nanometer surface features considering that natural bone is a nanostructured material. It is important to note that type I collagen (organic matrix of bone) is a triple helix 300 nm in length, 0.5 nm in width, and periodicity of 67 nm while HA (inorganic mineral phase of bone) are approximately 20 to 40 nm long. Besides, HA crystals are uniquely patterned within the collagen network.[14] These indicate that bone cells may be used to an environment in nano-scale rather than micro-scale. Recently, human osteoblasts were observed to initially adhere to grain boundaries on both nanophase and conventional titanium; greater osteoblast adhesion was found on nanophase titanium that possessed more grain boundaries on the surface.[15] However, the mechanical strength of this nanophase titanium (compacts of nanoparticles) was not high enough for use as a bulk material like titanium alloys through metallurgy techniques. Proper nanometer surface modification methods for current titanium-based implants are, thus, being actively pursued.

An electrochemical method known as anodization or anodic oxidation is a well-established surface modification technique for valve metals to produce protective layers.[4] It has been successfully used as a surface treatment for orthopedic implants in the past few decades and it has some new advances recently. This chapter will present an overview of anodization and discuss processing parameters, microstructure and composition, biological responses of anodized titanium, which are pertinent for orthopedic applications. Finally, this chapter will also discuss mechanisms of enhanced osteoblast functions on anodized titanium that possesses nanometer structures.

2.2 Anodization of Titanium

2.2.1 Basics of Anodization Process

Typical anodization procedures include alkaline cleaning, acid activation, and electrolyte anodizing. Acid activation is performed in a mixture of nitric acid and hydrofluoric acid (HF) to remove the natural titanium oxide layer and surface contaminants. The electrolyte anodization is carried out in an electrochemical cell, which usually has a three-electrode configuration (titanium anode, platinum cathode and Ag/AgCl reference electrode). When a constant voltage or current is applied between the anode and cathode, electrode reactions (oxidation and reduction) in combination of field-driven ion diffusion lead to the formation of an oxide layer on the anode surface.

The main chemical reactions specifically for anodizing titanium are listed below (Equation (1) to (5) adapted from [4]).

At the Ti/Ti oxide interface: $Ti \Leftrightarrow Ti^{2+} + 2e^-$ (2.1)

At the Ti oxide/electrolyte interface: $2H_2O \Leftrightarrow 2O^{2-} + 4H^+$ (2.2)

$$2H_2O \Leftrightarrow O_2 + 4H^+ + 4e^-$$ (2.3)

At both interfaces: $Ti^{2+} + 2O^{2-} \Leftrightarrow TiO_2 + 2e^-$ (2.4)

Because titanium oxides have higher resistivity than the electrolyte and the metallic substrate, the applied voltage will mainly drop over the oxide film on the anode. As long as the electrical field is strong enough to drive the ion conduction through the oxide, the oxide film will keep growing. This explains why the final oxide thickness, d, is almost linearly dependent on the applied voltage, U:

$$d \approx aU$$ (2.5)

Where a is usually a constant within the range 1.5-3 nm/V.[4]

2.2.2 Influences of Processing Parameters

The resulting oxide film properties (such as degree of nanometer roughness, morphology, chemistry, etc.) after anodization varies over a wide range according to different process parameters such as applied potential (voltage), current density, electrolyte composition, pH, and temperature. Different acids (phosphoric acid-H_3PO_4, sulfuric acid-H_2SO_4, acetic acid-CH_3COOH, and others), neutral salts, and alkaline solutions are widely-used electrolytes for the anodization of titanium. Their detailed electrochemical oxide growth behavior on titanium was studied by Sul et al.[16] Generally, it was found that among all the electrolytes (including H_3PO_4, H_2SO_4, CH_3COOH, and NaOH, $Ca(OH)_2$) the anodic oxide thickness in H_2SO_4 was the highest. Importantly, the oxide formation ability in acidic electrolytes exceeded that in hydroxide solutions. Usually, H_3PO_4 and H_2SO_4 were used to produce thick (tens of microns) and micro-porous oxide layers at high voltages. In contrast, fluoride solutions were found to have the ability of producing biologically-inspired nano-tubular structures in the past few years[30-38]. Due to the importance of nanostructures for biological applications as discussed above, this will be discussed in this section.

The anodization process can be done either at constant voltage (poteniostatic) or constant current (galvanostatic). If the applied voltage exceeds the dielectric breakdown limit of the oxide, the oxide will no longer be resistive to prevent further current flow and oxide growth, which will lead to more gas revolution and sparking. This technique is, thus, known as Anodic Spark Deposition (ASD) or Micro-Arc Oxidation (MAO). For example, it has been reported that the breakdown potentials for H_3PO_4 and H_2SO_4 were around 80 and 100 V, respectively.[17] Below the breakdown limit, the anodic oxide film was relatively thin and usually non-porous using non-fluorine electrolytes.

A constant temperature during the anodization process is usually required to maintain a homogeneous field-enhanced dissolution over the entire area. Since increased temperature will accelerate the chemical dissolution rate, the working temperature is often kept relatively low to prevent the oxide from totally dissolving.[17]

2.2.3 Creation of Rough Surfaces

The anodization technique was discovered in the early 1930's and was widely studied in the 1960's to enhance titanium implant osseointegration.[18] These studies usually adopted a high voltage anodization (i.e., ASD) of titanium in electrolyte solutions whose ions would be embedded into the oxide coating, resulting in a micro-porous structure.[18-22, 24-27] Table 2.1 shows the anodizing parameters of some ASD studies.

The mechanism of the ASD is usually described by the avalanche theory. During anodization, the newly-formed oxide layer on the anode is a dielectric barrier to the current flow and it keeps growing until reaching the dielectric breakdown limit. Generally, the anodized layer is not uniform due to the existence of flaws, defects, local stress, and non-uniform oxide thickness. When the applied voltage increases, the potential drop at the weak points exceeds the dielectric limit so that sparking happens. The local temperature at these points can be up to several thousand Kelvin and lead to a local melting process. Thermal stressing of these anodized titanium leads to the multiplication of weak points, i.e., a cascading process, and consequently breakdown of the dielectric. Figure 2.1 shows a schematic diagram of porous titanium oxide formation proposed by Choi et al.[17]

Basically, the anodic film growth is determined by a balance between the oxide film formation rate and the oxide dissolution rate given by the nature of the electrolyte. Meanwhile, the nature of the electrolyte is closely connected with other processing parameters

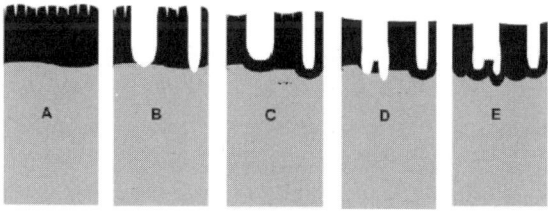

Fig. 2.1. Schematic diagram of porous titanium oxide formation above the breakdown potential: (A) oxide growth to maximal thickness, (B) burst of oxide by the formation of crystallites (pore formation), (C) immediate repassivation of pore tips, (D) burst of repassivated oxide, and (E) dissolution of the formed oxide and second repassivation. (Adapted from Ref. 17)

such as electrolyte concentration, applied voltage, current density, pH, etc. The explanations in detail could be found elsewhere.[16]

2.2.4　Creation of Nano-Roughness

While the generation of micro-structures through titanium anodization is well-established, current research efforts focus on creating biologically-inspired nanometer surface structures. Studies has shown that nanoporous structures can be created by titanium anodization in chromic acid at 10–40 V.[28] Another unique surface morphologie obtained through titanium anodization is self-ordered nano-tubular structures.[30-38] For these studies, fluorine electrolyte solutions are used and the applied voltage must be much lower than the dielectric

Table 2.1. Experimental parameters of some ASD studies.

Electrolyte Composition	Molarity	Voltage (Current)	Time (s)	Temperature (°C)	Reference
Sulfuric acid	1	125	-	-	19
	0.5, 1, 3	90, 155, 180	-	-	20
	1	80	-	RT	21
Acetic acid	0.1	40-80	8-67	17	16
	2	175	-	25	19
Phosphoric acid	0.2	200, 300, 350 (70 A/cm^2)	-	20	22
	1	40-80	10-47	17	16
Sodium tripoly-phosphate	0.15	(210 A/cm^2)	-	-	18
Sodium hydroxide	5	10-20	-	25	23
	0.1	40-80	22-110	17	16
Calcium hydroxide	0.1	40-80	13-53	17	16
	0.02/0.1	(70 A/cm^2)	1530	4.1-4.5	18
Calcium glycero-phosphate and calcium acetate	0.03/0.15	200, 260, 300 (70 A/cm^2)		20	22
	0.15/0.02	190-600	180	-	24
	0.02/0.15	(70 A/cm^2)	1800	-	25
	0.02/0.15	350	1200	20	26
β-glycerophosphate and sodium acetate	0.06/0.3	250-350 (50 A/cm^2)	-	-	27

RT = room temperature

breakdown. Some of the specific anodizing parameters to create titanium nanometer structures are listed in Table 2.2.

The need of fluoride ions to form nano-porous titania structures on a titanium surface under relatively low voltages was first reported by Zwilling et al.[29] However, the nano-tubular structures were not reported here. In 1999, Grimes and co-workers successfully fabricated self-ordered nano-tube arrays after anodizing titanium between 10 and 40 V in dilute (0.5–1.5 wt%) aqueous HF solutions.[30] It was found that the diameters of nano-tubes were determined by applied voltage while the final length of tubes were independent of the anodization time.

The tube diameter was approximately 60 nm and tube length was 200 nm at 20 V in 0.5% HF solution for 20 min (Fig. 2.2). Later, they developed a method to produce tapered, conical-shape titania nano-tubes in 0.5% HF by linearly changing the voltage from 10 to 23 V (Fig. 4).[31] Schmuki and co-workers also observed self-ordered

Table 2.2. Survey of different fluorine solutions to produce titania nano-tubular structures with different size and thickness.

Electrolyte Composition (pH)	Voltage (V)	Time (h)	Temperature (°C)	Thickness (nm)	Pore diameter (nm)	Ref.
CH_3COOH and 0.5 M HF	10	4	-	60	500 (inter)	17
0.5 or 1.5 % HF	10-40	<1	18	250	25-65	30
0.5 % HF	10-23	<1	-	300	22-76	31
KF and NaF (4.5)	25	20	-	4400	115	34
DMSO and CH_3COOH and 4 % HF	20	70	RT	2300	60	35
1 M H_2SO_4 and 0.15 % HF	30	24	-	540	140	32
CH_3COOH and 0.5 % NH_4F	20	1	-	200	30	33
1 M $(NH_4)_2SO4$ and 0.5 % NH_4F	20	-	-	2500	100	36
1 M $(NH_4)H_2PO_4$ and 1M H_3PO_4 and 0.5 % HF	20	40	-	4070	50	37
0.138 M HF or NaF + 0.5 M H_3PO_4	20	-	24	500	100-120 (outer)	38

RT = room temperature

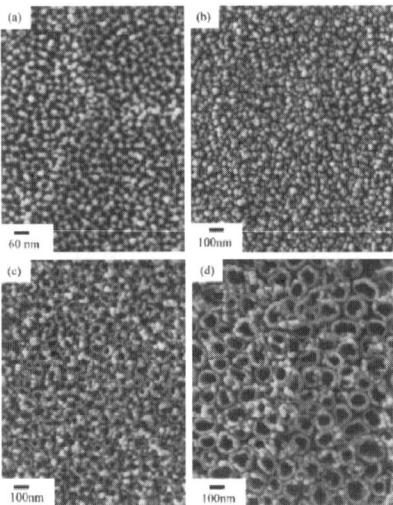

Fig. 2.2. FE-SEM top-view images of porous titanium oxide films anodized in 0.5 wt% HF solution for 20 min under different voltages: (a) 3 V, (b) 5 V, (c) 10 V, and (d) 20 V. (Adapted from Ref. 30)

nano-tubular titanium oxide films in HF/H_2SO_4 or CH_3COOH/NH_4F electrolyte solutions.[32]

Moreover, the nano-tubular (nano-pore) structure was also achieved in organic electrolytes. Choi et al. used nano-indented titanium for anodization in ethanolic HF and produced a pore lattice with a 500 nm inter-pore distance (Figs. 2.3-2.4).[17] Schmuki's group reported nano-tubular structures using non-aqueous mixtures of ethanol and ammonium fluoride without an imprinting treatment.[33] However, in these studies, the depth of titania nano-tubes was limited to a few hundred of nanometers. Recently, high-aspect-ratio titania nano-tubes up to several micrometers were reported by both Grime's and Schmuki's groups.[34-37] Grime's group reported the formation of 4.4 μm long titania tube arrays by anodizing titanium in NaF or KF of pH 4.5 (Fig. 2.5).[34] They also reported formation of 2.3 μm thick nano-tubular structures using DMSO/ethanol/HF electrolyte (Fig. 2.6).[35] Meanwhile, Schmuki's group succeeded in using neutral fluoride solutions to produce nano-tubular structures up to 2.5 μm.[36] They achieved this by controlling the electrochemical parameters to enchance acidification at the bottom of tubes.

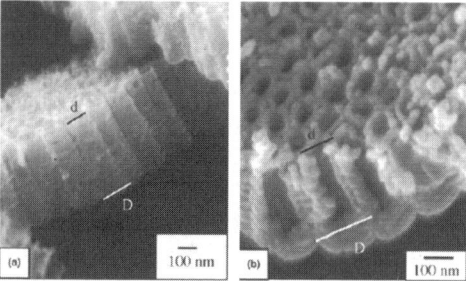

Fig. 2.3. FE-SEM cross-sectional views of the titania nano-tubes obtained by using a time-varying anodization voltages; d denotes diameter of apex and D diameter of cone base. (a) Tapered nano-tubes obtained using a ramp rate of 0.43 V/min to raise the voltage from 10–23 V within 30 min and then holding the voltage at 23 V for 10 min. (b) Tapered nano-tubes obtained by initially anodizing the sample at 10 V for 20 min and then increasing the voltage linearly at a rate of 1.0 V/min to 23V, and finally keeping the voltage at 23 V for 2 min (Ref. 31).

Fig. 2.4. SEM of (a) nano-indented surface of titanium substrate and (b) anodized titanium at 10V in ethanolic 0.5M HF for 240 min.(Adapted from Ref. 17)

Fig. 2.5. Lateral view of the nano-tubes formed in a KF and NaF solution at pH 4.5 under 25 V for 20 h. (Adapted from Ref. 34)

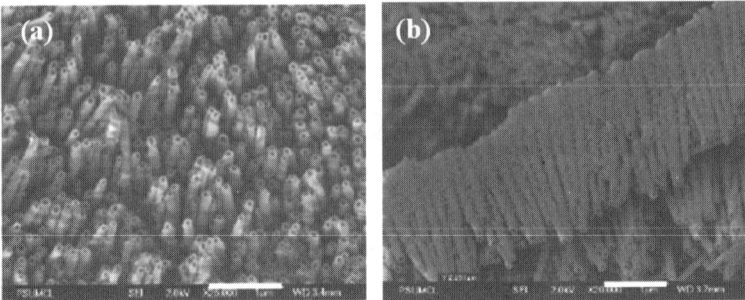

Fig. 2.6. FE-SEM images of titanium foil sample anodized in DMSO and ethanol mixture solution (1:1) containing 4% HF at 20 V for 70 h. (a) is top-view and (b) is cross-section. Scale bars = 1 μm. (Adapted and redrawn from Ref. 35)

Chemical dissolution, field-assisted dissolution and oxidation are the three main reactions in fluorine electrolyte anodization. Among these, field-enhanced dissolution has been considered as the predominant mechanism of titania tubular structure formation by many researchers.[30-37] The evolution of nano-tube structures is shown in Fig. 2.7. Grime and co-workers proposed a mechanism based on a point defect model.[31] Grime proposed that the initial pore formation was due to

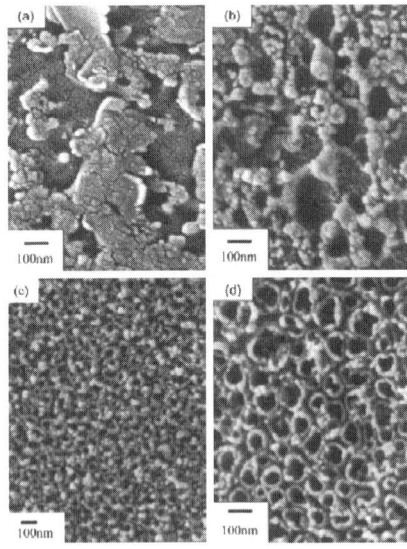

Fig. 2.7. Evolution of nano-tube structures. Porous titanium oxide films anodized in 1.5% HF at 20 V for (a) 10s, (b) 30s, (c) 120s, and (d) 8 min. (Adapted from Ref. 30)

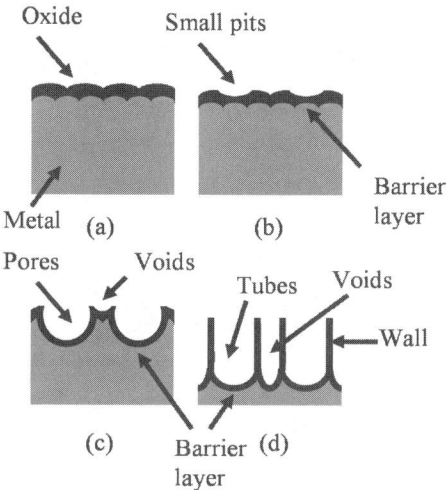

Fig. 2.8. Schematic diagram of the evolution of nano-tube-like structures on the titanium surface during anodization in aqueous HF under constant voltage: (a) oxide formation; (b) pit formation in some concave sites; (c) pore formation and growth under field-enhanced dissolution leading to voids formation; and (d) fully developed tubes. (Based on the model from Ref. 31 and modified according to experimental observations)

localized dissolution at weak points and the unanodized metallic portions would exist between the pores. Later, voids were formed in these inter-pore regions by field-enhanced oxidation/dissolution (Fig. 2.8). The growth of voids in equilibrium with the pores would form the final nano-tubular structures. However, it didn't explain how voids are created and lead pores to be well-separated, individual tubes. Recently, Raja et al. suggested that the instability of the oxide layer and the self-ordered structures could be explained by the perturbation theory; separation of individual nano-tubes of titanium oxide layer from the inter-connected nano-pores could be attributed to the repulsion forces of the cation vacancies (Fig. 2.9).[38]

2.2.5 Control of Chemical Composition

The ions contained in the electrolyte are usually present in the thick, porous ASD film and the concentration of these elements decreases from the outer layer towards the substrate.[22] For example, phosphorous

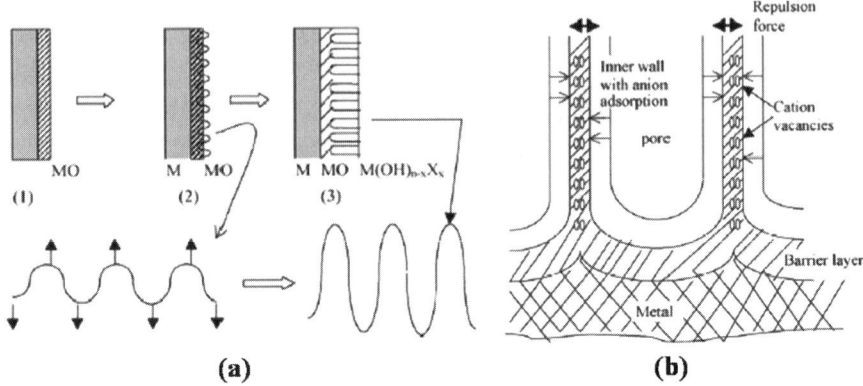

Fig. 2.9. Schematic diagram of (a) pore formation (fluoride addition) during anodization of Ti. The barrier film is intact during porous anodic film formation and
substrate metal is not attacked locally. Perturbation of the surface shown in (2) can
lead to adsorption of fluorides at the valleys and develop into nano-tubular structure.
Higher strain energy density at the valleys drives the mass flow to the lower energy
crests; (b) pore separation mechanism. Cation vacancies generated by dissolution
of Ti cations are transported radially from the two sides of common wall of the
neighbor pores. Charges of similar polarity repel and when electrical neutrality is
not maintained this repulsion causes separation of pores into individual nano-tubes.
(Adapted and redrawn from Ref. 38)

was found to be embedded in titanium oxide layer after anodization
with a H_3PO_4 electrolyte.[39] For electrolytes containing Ca and P,
such as calcium glycerophosphate (Ca-GP) and calcium acetate
(CA), both Ca and P were contained in the oxide layer with a Ca/P
ratio close to HA (1.67).[40] After an additional hydrothermal treatment (e.g., high pressure streaming), HA crystals were randomly
precipitated on the anodic oxide film. These HA crystals were usually columnar or need-like (Fig. 2.10). This could be another way to
create HA coatings as opposed to plasma spraying. The advantages
of such HA coatings compared to plasma sprayed HA will be discussed in following sections.

Another approach reported to introduce apatite layers onto the anodized titanium is simply by soaking crystalline titania in simulated
body fluid (SBF) because anodized titanium with anatase and rutile
titania surfaces were shown to induce apatite formation in vitro.
Yang et al. soaked titanium metal in SBF for 6 days after H_2SO_4
anodization and observed uniform apatite formation (Fig. 2.11).[20] One

Fig. 2.10. Needle-like or columnar hydroxyapatite crystals deposited on anodized titanium after a hydrothermal treatment. (a) ASD followed by Ishizawa's procedure[44] (Adapted and redrawn from reference 18) and (b) ASD followed by Suh's procedure[25]. (Adapted and redrawn from Ref. 25). Scale bars = 10 μm.

advantage of this method is that the composition and surface morphology of the resulting apatite layer is very close to those in natural bone, but the adhesive strength of such coatings are not clear yet.

Similarly, a two-step procedure was used to produce nano-scale HA for anodized titanium with nano-tubular structures.[41] Specifically, the anodized titanium was treated with NaOH to form nano-fibers of bioactive sodium titanate structures on the top edge of the nano-tube wall, which was then immersed in a SBF to induce the formation of nano-scale HA (Fig. 2.12). This technique could be useful as well-adherent bioactive nano-HA layers on titanium-based implants are created which simulate the size and shape of natural HA in bone. The advantage of introducing nano-HA onto titanium anodized structures was supported by previous work revealing greater osteoblast functions on nano-HA compared to conventional, or micron grain size, HA.[14]

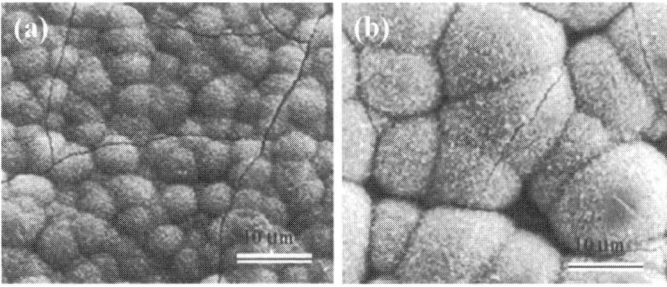

Fig. 2.11. SEM images of titanium metal soaked in SBF for 6 days after they were anodized in 1 M H2SO4 at (a) 155 V and (b) 180 V. (Adapted and redrawn from Ref. 20)

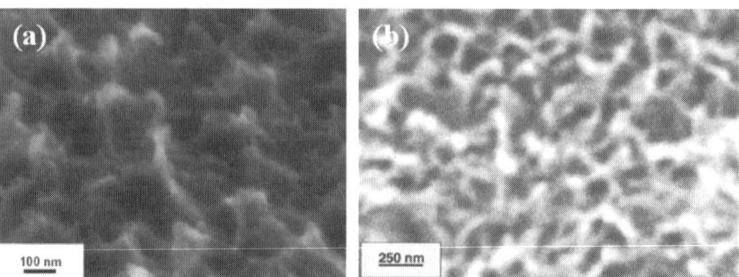

Fig. 2.12. SEM images of (a) nano-inspired sodium titanate nano-fibers and (b) nanoscale HA phase. (Adapted and redrawn from Ref. 41)

2.3 Structure and Properties of Anodized Oxide Film

2.3.1 Structure

The structures and properties of ASD films were widely investigated by Kurze et al.[39] The typical morphology of the titania layer resulting from ASD is a rough, porous texture with cracks on it (Figs. 2.13 and 2.14). The dimensions of the pores varied from a few hundred nanometers to a few micrometers depending on the processing parameters and are not uniform within the same anodized surface. Moreover, these pores were interconnected and had a layered structure, i.e., they overlapped with each other. The shapes of the pores were mostly round or irregular. The diameter of the pores and the film roughness were reported to increase with greater current densities (Fig. 2.13),[18, 42] applied potential (Fig. 2.14),[20] and electrolyte concentrations.[20] The thickness of oxide film increases with time up to tens of micrometers.

In contrast, the biological-inspired nano-tubular structures were highly ordered. The pore size is determined by the voltage and can be varied from a few tens of nanometers to around 100 nanometers. The thickness of the tubular-structured oxide was formed to be a few hundred nanometers but has been elongated to a few microns by controlling pH and electrolytes. Generally, the dimensions of nano-tubular structures within one sample are uniform but might be variable due to differences (e.g., surface defects) on a substrate.

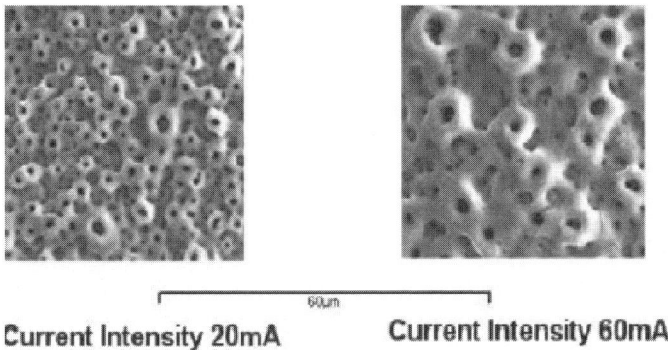

Current Intensity 20mA   **Current Intensity 60mA**

Fig. 2.13. SEM micrographs of an ASD formed film on c.p. grade 2 titanium from an electrolyte containing 0.015 M calcium glycerophosphate and 0.1 M calcium acetate. Increased current density from 20 to 60 mA/cm^2 led to a larger pore size in the ASD porous structure. (Adapted from Ref. 18)

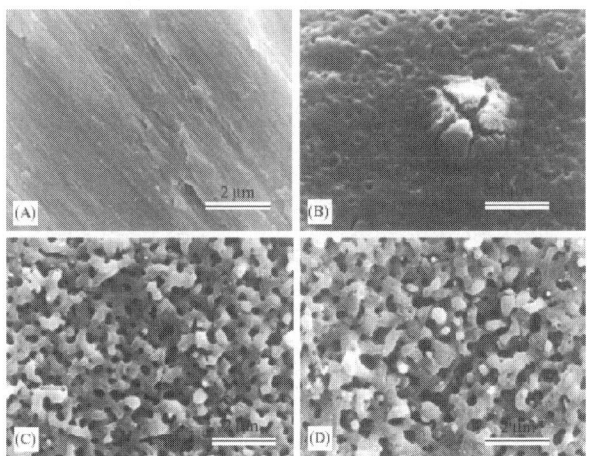

Fig. 2.14. SEM micrographs of ASD formed film on c.p. titanium from an electrolyte containing 1 M H_2SO_4 at voltages of: (a) 0 V, (b) 90 V, (c) 155 V, and (d) 180 V for 1 min. Different voltages from 90 to 155 V led to a larger pore size in the ASD porous structure. (Adapted from Ref. 20)

2.3.2 Corrosion Resistance and Adhesive Strength

After anodization, thickness of the protective oxide layer increases and it could lead to less ion release in the human body. The oxide barrier layer (the relatively thin, non-porous oxide layer under the porous oxide structures) is considered to contribute to the improvement of corrosion resistance. However, it was suggested that the implants' mechanical properties could be impaired with increased spark coating thickness.[18]

The interface between the anodic oxide layer and the titanium substrate usually does not show any discontinuity.[18] Besides, the HA crystals on the hydrothermally treated titanium are gradually grown consuming Ca and P in the anodic film. As a result, the interface between the substrate, the anodic film, and the HA film are considerably strong. The adhesive strength between the anodic oxide films and the titanium substrates was reported to be 25 MPa,[43] and the adhesive strength between the oxide/HA coating and the substrate after a hydrothermal treatment was reported to be up to 40 MPa by Ishizawa et al. when using less concentrated electrolytes.[40, 44] These values are equivalent or higher than those of plasma sprayed HA onto titanium surfaces, which were reported between 15 to 30 MPa depending on different processing parameters.[45, 46] Moreover, the HA produced from hydrothermal treatment after anodization (AH-HA) seemed to be more stable than plasma sprayed HA (PS-HA). It was reported that the shear strength of PS-HA in SBF decreased from 28.1 MPa to 20.4 MPa after 4 weeks[47]; meanwhile, Ishizawa et al. found that AH-HA retained high durability after 300 days in SBF.[40] So from a mechanical point-of-view, hydrothermally treated anodic titanium would be a better choice than HA plasma sprayed titanium.

2.3.3 Biological Properties of Anodized Titanium

2.3.3.1 In Vitro Studies

Clearly, coating strength, mechanical and other properties are not the only concern for orthopedic implants. Cytocompatibility leading to promoted bone growth needs to be assessed. Most studies have been reported in vitro bone cell responses to anodized and anodized/hydrothermally treated titanium surface. Fini et al. reported that the adhesion, spreading, proliferation, and differentiation of osteoblast-like cells (HOS-TE85, human osteosarcoma line) were similar on unanodized titanium, titanium anodized enriched with Ca/P, and titanium anodized and hydrothermally treated.[27] An unexpected increase of unattached cells in the latter two substrates was observed. However, the percentage of unattached cells was in the range of 10-20% which is considered a normal range for cytocompatible materials. On contrast, Rodriguez et al. reported increased osteocalcin production on the anodized and hydrothermally treated titanium surfaces but the highest alkaline phosphate (ALP) activity on control titanium throughout an 8 day study using an osteoblast precursor cell line (ATCC, CRL-1468).[48] Both osteocalcin and ALP are markers of osteoblast differentiation to deposit calcium. They explained that a decrease in ALP activity was in part attributed to the maturation of osteoblast precursor cells and in part attributed to the increased production of mineralized matrix. Also using Ca-GP and CA as an electrolyte, Li et al. reported decreased osteoblastic MG63 cell proliferation when anodization voltage increased above 190 V; however, increased ALP activity of human osteosarcoma cell line was reported with voltages above 300 V.[24]

Zhu et al. studied the effects of topography and composition of anodized titanium surfaces on osteoblast (SaOS-2 derived from human osteosarcoma) responses [22]. Their cell experiments showed an absence of cytotoxicity and an increase of cell attachment and proliferation after anodization in an electrolyte composed of Ca-GP and CA. The cells on the surfaces with micro-pores showed an irregular and polygonal growth and more lamellipodia while cells on the titanium control showed many thick stress fibers and intense focal contacts. However, they didn't find any significant difference for ALP activity. Suh et al. studied the effects of hydrothermally treated anodic films similar to the Zhu formulations and they observed no statistical differences in cell viability using the MTT assay when osteoblasts (ROS 17/2.8, a rat osteosarcoma cell line) were cultured for 4 days on untreated, anodized, and anodized/hydrothermally treated surfaces.[25] In contrast, they found that hydrothermal treatment had an effect on early osteoblast attachment, resulting in a more well-spread shape compared to the cellular rounded shape observed on anodized and control titanium after 6 h (Fig. 2.15).

The different observations in the above in vitro studies could be attributed to the use of different anodization parameters and different cell lines. The optimal anodization conditions are still under investigation.

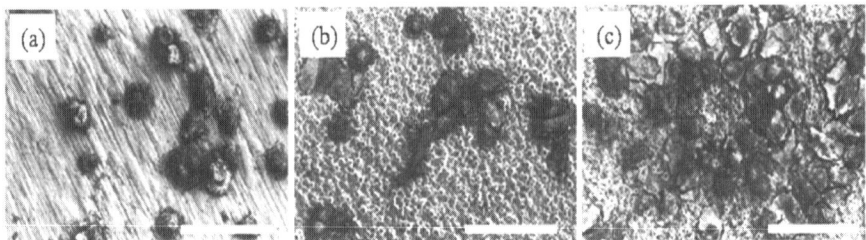

Fig. 2.15. SEM of cells after 6 h culturing on (a) control surface, (b) anodized surface and (c) anodized surface followed by hydrothermal treatment for 4 h. Scale bars =120 μm. (Adapted from Ref. 25)

Since the nano-tubular structure is relatively new, few cytocom-patibility studies have been completed to examine its potential for use as a novel titanium bone implant surface. However, because of the size and order of the titania tubular structure (which somewhat mimics the natural environment of bone) it is very interesting to deter-mine whether there is any morphological or size advantage of using nano-tubular structures compared to a conventional anodized titanium porous structure for enhancing bone cell functions.

Currently, our research has focused on osteoblast functions on such anodized titanium with nano-tubular titania structures. These structures are similar to those formed by Gong et al.[30] (0.5% HF, 20 V, 20 min). After anodization, the tubular structures had increased

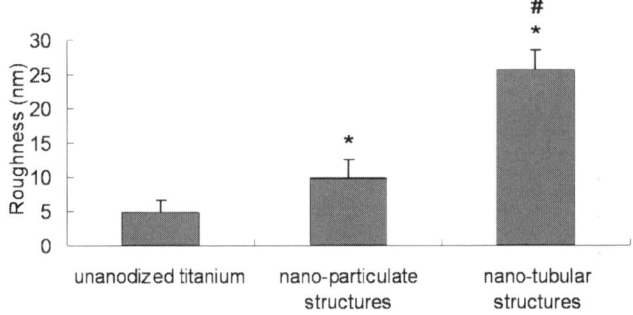

Fig. 2.16. Surface roughness of (a) unanodized titanium, (b) anodized titanium with nano-particulate structures, and (c) anodized titanium with nano-tubular structures. Data = mean ± SEM; n = 3; *p < 0.01 (compared to unanodized titanium) and #p < 0.01. (compared to nano-particulate structure)

surface roughness (Fig. 2.16). The inner diameters of the nano-tubes were about 70 nm and the depth of them was about 200 nm. To study the effects of nanoroughness and morphology, intermediate samples that possessed a nano-particulate structure and a medium roughness in between the unanodized control and the nano-tubular structure were created (0.5% HF, 10 V, 20 min) (Fig. 2.16).

The experiments showed increased osteoblast adhesion after 4 hour of culture with greater anodized titanium roughness (Fig. 2.17). The difference in osteoblast morphology was obvious between nano-tubular structures and unanodized titanium. Most cells were well-spread on anodized titanium with nano-tubular structures while they mostly looked round on the control (Fig. 2.18). After 4 weeks of culture, the

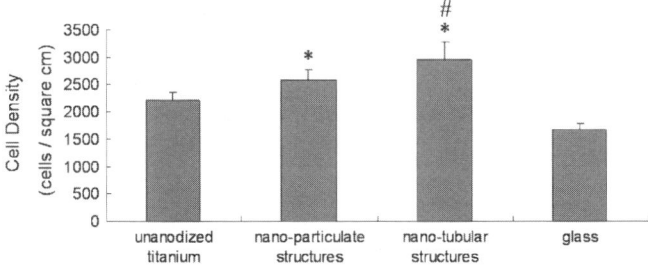

Fig. 2.17. Osteoblast adhesion on unanodized titanium, anodized titanium with nano-particulate structures (10 V, 0.5 % HF, 20 min), anodized titanium with nano-tubular structures (20 V, 0.5 % HF, 20 min), and glass (reference). Values are mean $\pm$ SEM; n = 3; *p < 0.1 (compared to unanodized titanium) and #p < 0.1. (compared to anodized titanium with nano-particulate structures)

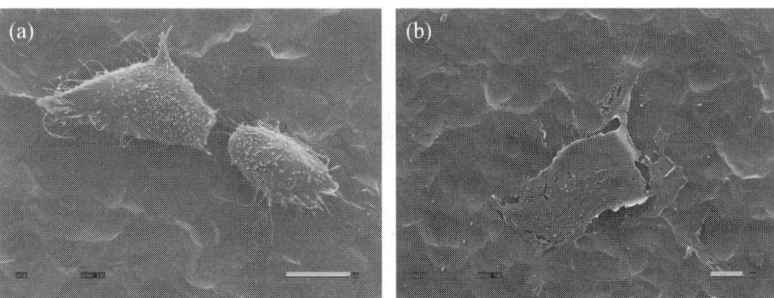

Fig. 2.18. Typical osteoblast morphology on (a) unanodized titanium and (b) anodized titanium surface with nano-tubular structures after 4 h culture. Scale bars = 10 μm.

anodized titanium with nano-tubular structures promoted the highest calcium deposition by osteoblasts among all the samples. These results indicated that the special nano-tubular structures anodized onto the titanium surface may have provided an optimal surface roughness for promoting bone cell function.

2.3.3.2 Mechanisms of Increased Osteoblast Function

Moreover, protein (fibronectin and vitronectin) adsorption on nano-tubular samples has been examined to explore the mechanism of enhanced osteoblast adhesion. Fibronectin and vitronectin are two major proteins that involved in osteoblast adhesion.[49-51] Results showed significantly increased both fibronectin (15%) and vitronectin (18%) adsorption on nano-tubular structures compared to unanodized titanium samples (Fig. 2.19). Because the cells adhered to the titanium surface via pre-adsorbed proteins, increased fibronectin and vitronectin adsorption on anodized titanium substrates with nano-tubular structures may explain the observed enhanced osteoblast functions.

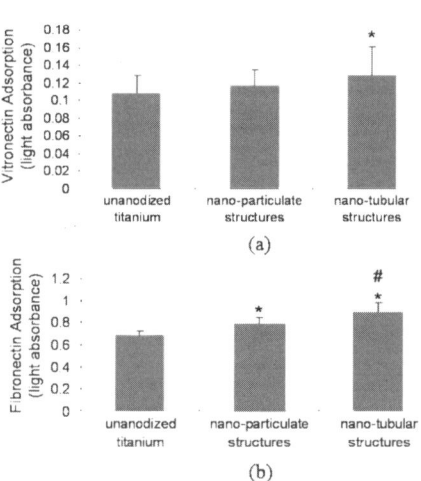

Fig. 2.19. (a) Fibronectin and (b) vitronectin adsorption on unanodized titanium, anodized titanium possessing nano-particulate structures (0.5 % HF, 10 V and 20 min), and anodized titanium possessing nano-tubular structures (0.5 % HF, 20 V and 20 min). Values are mean $\pm$ SEM; n = 3; *p < 0.1 (compared to unanodized titanium) and #p < 0.1. (compared to nano-particulate structures)

2.3.3.3 In Vivo Studies

While in vitro assays may generate a quick assessment of cytocompatibility, in vivo studies are necessary to fully evaluate new bone growth. A survey of in vivo investigations of bone tissue reactions to anodized titanium implants is listed in Table 2.3. As with in vitro analysis, the varied oxide properties not only include thickness, but also morphology, chemical composition, crystallinity, and surface roughness.

Some in vivo studies were mainly interested in the effects of thick, porous oxide coating on new bone growth. Less than 200 nm thick oxide film anodized in acetic acid showed no significant difference compared to unanodized samples after implanted into a rabbit 6 weeks.[21] In contrast, a H_3PO_4/H_2SO_4 electrolyte was usually used to form thick anodic films up to tens of microns. Enhanced bone bonding was found for micron-thick porous anodic oxide films formed in a H_3PO_4/H_2SO_4 electrolyte solution in a rabbit model.[52-54]

More importantly, changes of surface chemistry could play a more important role in inducing new bone growth. Several in vivo studies focused on Ca-P enriched anodized titanium with and without hydrothermal treatment.[24, 27, 55-58] Ishizawa et al. produced 1-2 μm HA films on anodic oxide layer and compared bone growth on them with unanodized titanium.[55] They found strong bone bonding via push-out tests with anodized titanium after 8 weeks of implantation into rabbits. Following their in vitro work, Fini et al. found the lowest affinity index on anodized titanium while the highest was found on the anodized and hydrothermally treated titanium.[27] The low bone contact on anodized titanium was probably due to the in vivo reduction and degradation of the amorphous titania layer while HA crystals aided bone opposition. Giavaresi et al. also supported the positive role of HA produced from hydrothermal treatment in accelerating bone ingrowth and bone mineralization.[56-57] Son et al. reported no significant difference for the percent bone contact for all samples but did find significantly increased removal torque strength for anodized implants after 6 weeks of implantation into a rabbit.[58]

The dissolution of AH-HA and PS-HA in vivo was studied by Ishizawa et al.[59] Basically, the AH-HA was much more durable than PS-HA because of their relatively high crystallinity and their relatively

Table 2.3. Survey of in vivo investigations of bone tissue reactions to anodized titanium implants.

Implant	Treatment	Chemical composition	Oxide thickness (μm)	Oxide/HA morphology (pore size, μm)	Oxide/HA crystallinity	Roughness (μm)	Test	Animal and time (week)	Reference
Cylinder	AO and HT	TiO_2, HA	<10	Porous (1-3)	A+R	ND	Push-out	Rabbits 8	55
Screw	E/M and AO	Mainly TiO_2	0.18-0.2	Pores and pits	N	32.3/40.8 nm (rms)	contact ratio	Rabbits 1, 3, 6	21
Screw	AO	Ti, O, C, P/S	1/10	Porous (1-10)	A+R	1.2 (Sa)	RTQ, RFA	Rabbits 3, 6	52
Rod	Ca-P AO w/ or w/o HT	Ti, O, Ca/P, HA	5	porous (1-3)/ columnar	N/C	1.97 (Ra)	AI	Rats 4 , 8	27
Screw	Ca-P AO and HT	Ti, O, Ca/P, HA	ND	porous	ND	1.97 (Ra)	AI, push-out strength	Sheep 4, 8, 12	56, 57
Screw	Ca-P AO w/ or w/o HT	Ti, O, Ca/P, HA	ND	Porous/needle-like	A+R/C	ND	percent bone contact, RTQ	Rabbits 6,12	58
Screw	AO	Primary TiO_2	0.2-1	Porous (1.27-2.10)	N, A, A+R	0.96-1.04 (Sa)	RTQ, RFA	Rabbits 6	53, 54
Screw	AO	Ti, O, S, P, Ca	1.1-1.3	Porous (<1.5)	A or N	0.82-1.04 (Sa)	RTQ, BMC	Rabbits 6	60

Abbreviations: AO-anodization, HT-hydrothermal treatment, E-electropolished, M-machined. C-crystalline, N-noncrystalline, A-anatase, R-rutile, RTQ-removal torque values, RFA-resonance frequency analysis, AI-affinity index, BMC-bone to metal, ND-not determined.

low thickness. Ishizawa et al. also found these two HA had different bone responses.[59] Specifically, new bone thinly spread over the whole AH-HA area while new bone formed from surrounding bones to the PS-HA area. This is probably due to their different degradation properties. One drawback of most of the above studies is that both chemical composition and surface morphology changed after titanium anodization so that it is hard to verify the role of one material property. However, Sul et al. indirectly verified a chemical bonding in vivo by maintaining surface morphology and roughness and changing chemical characteristics (specifically, S, P and Ca enriched implants via anodization).[60]

The removal torque value (RTQ) showed significant differences between Ca-containing anodized titanium implants and unanodized titanium implants as well as S-containing anodized titanium implants and unanodized titanium implants (Fig. 2.20). The bone to metal contact was 186%, 232%, and 272% higher in S, P, and Ca implants, respectively, when compared to the control groups. These results confirmed that ions incorporated into the titanium oxide layer during anodization could have important roles in enhancing bone juxtaposition.

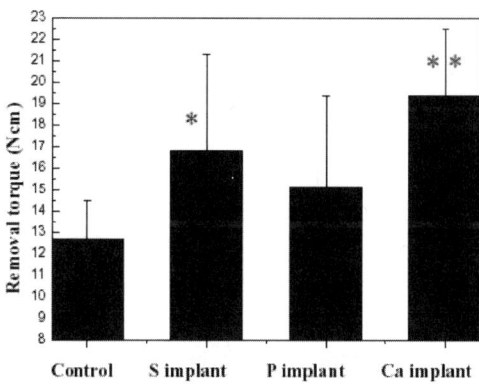

Fig. 2.20. Mean removal torque values after 6 weeks of healing time. *$p < 0.05$, **$p < 0.001$. (Adapted from Ref. 60)

2.4 Future Directions

As a surface modification method, anodization can lead to desired chemistry and/or topography changes and could be used with other treatments (e.g., hydrothermal) together.

First, anodization provides a controlled way to create nano-roughness or even nano-features. Generally, there are two mechanisms that are responsible for osseointegration of bone: biomechanical interlocking and biological interactions. For biomechanical interlocking, it depends on the roughness, and surface irregularity. Current femoral stems made of titanium alloys are usually macro-textured to provide such surface features for bone to mechanically interlock. For biological interactions, it involves complex systems. Considering roughness in different scales, it is reported that increased micro/submicron-roughness could enhance bone cell function, such as ALP activity.[61, 62] while some other studies have revealed the enhanced cell-implant interactions on nanoporous or nanophase materials.[14, 15, 63, 64] Ideally, the future titanium implant should possess roughness in all three scales: macro, micro, and nano. One possible approach to accomplish this is by subjecting implants to techniques like polishing and mechanical grinding that promote micro-roughness, and then to induce nano-tubular structures by a quick anodization process.

Second, micro/nano HA films produced using anodized titanium shows some advantages over conventional ones. Although plasma spray is still widely used for HA coatings on titanium, anodization has a strong role to play to incorporate Ca and P into Ti coatings. For example, anodization has the ability to form uniform and thin HA or calcium phosphate layers on titanium implants of various shapes. Moreover, HA deposited onto the anodized titanium could be nano-scale in dimension. One problem that still needs to be more fully investigated is the bonding strength between apatite crystals and the anodic oxide.

Furthermore, anodization can be used to incorporate drug delivery into titanium-based implants to enhance new bone formation. Porous ASD surfaces could be used as matrices for drug storage and release[65]; similarly, the nano-tubular structures could serve as reservoirs of chemical mediators, such as bone morphogenetic protein-2 (BMP-2) and osteogenic protein-1 (OP-1, BMP-7).[66] In the future, studies should concentrate on embedding these growth factors into the unique porosity that can be well controlled on titanium for orthopedic applications.

In a word, anodization as a quick and efficient modification method of titanium based implants shows significant promise for enhancing their 10 to 15 year lifetime.

Acknowledgements

The authors would like to thank National Science Foundation Nanoscale Exploratory Research Grant for financial assistance.

References

1. D. M. Brunette, P. Tengvall, M. Textor, and P. Thomsen, in *Titanium in medicine*, Springer, p.171 (2001)
2. J. F. Shackelford, vol. 1 *Bioceramics*, Netherlands: Gordon and Breach Science Publishers, p.17 (1999)
3. C. G. Moran, T. C. Horton, BMJ, 320, p.820 (2000)
4. D. M. Brunette, P. Tengvall, M. Textor, and P. Thomsen, in *Titanium in medicine*, Springer, p.232 (2001)
5. C. Larsson, P. Thomsen, B. O. Aronsson, M. Rodahl, J. Lausmaa, B. Kasemo and L. E. Ericson, Biomaterials 17, p.605 (1996)
6. H. M. Kim, F. Miyaji, .T. Kokubo, T. Nakamura, J. Mater. Sci.: Mater. Med. 8, p.341 (1997)
7. T. Kokubo, H. M. Kim, M. Kawashita, and T. Nakamura, J. Mater. Sci.: Mater. Med. 15, p.899 (2004)
8. C. Sittig, M. Textor, N. D. Spencer, M. Wieland, and P. H. Vallotton, J. Mater. Sci.: Mater. Med.10, p.35 (1999)
9. K. Bordji, J. Y. Jouzeau, D. Mainard, E. Payan, P. Netter, K. T. Rie, T. Stucky and M. Hage-Ali, Biomaterials 17, p.929 (1996)
10. R. Furlong, J. F. Osborn, J. Bone Joint Surg. 73B, p.741 (2001)
11. S.-S. Kim, M. S. Park, O. Jeon, C. Y. Choi and B.-S. Kim, Biomaterials, In Press
12. K. C. Baker, M. A. Anderson, S. A. Oehlke, A. I. Astashkina, D. C. Haikio, J. Drelich and S. W. Donahue Growth, Materials Science and Engineering: C, In Press
13. M. Sato, E. B. Slamovich and T. J. Webster, Biomaterials 26, p.349 (2005)
14. J. D. Bronzino, Biomedical Engineering Handbook. CRC Press, p. 274 (1995)
15. T. J. Webster, J. U. Ejiofor, Biomaterials 25, p.4731 (2004)

16. Y. T. Sul, C. B. Johansson, Y. Jeong, T. Albrektsson, Medical Engineering & Physics 23, p.329 (2001)
17. J. Choi, R. B. wehrspohn, J. Lee, U. Gosele, Electrochimica Acta 49, p.2645 (2004)
18. R. Chiesa, E. Sandrini, M. Santin, G. Rondelli, A. Cigada, J. Applied Biomaterials & Biomechanics 1, p.91 (2003)
19. O. Zinger, P. F. Chauvy, D. Landolt, J. of the electrochemical society 150, p.495 (2003)
20. B. Yang, M. Uchida, H.-M. Kim, X. Zhang and T. Kokubo, Biomaterials 25, p.1003 (2004)
21. C. Larsson, P. Thomsen, B.-O. Aronsson, M. Rodahl, J. Lausmaa, B. Kasemo and L. E. Ericson, Biomaterials 17, p.605 (1996)
22. X. Zhu, J. Chen, L. Scheideler, R. Reichl and J. Geis-Gerstorfer, Biomaterials rials 25, p. 4087 (2004)
23. H.-H. Huang, S.-J. Pan, Y.-L. Lai, T.-H. Lee, C.-C. Chen and F.-H. Lu, Scripta Materialia 51, p.1017 (2004)
24. L. H Li, Y. M Kong, H. W Kim, Y. W Kim, H. E Kim, S. J Heo and J. Y Koak, Biomaterials 25, p.2867 (2004)
25. J. Y Suh, B. C Jang, X. Zhu, J. L. Ong, and K. Kim, Biomaterials 24, p.347 (2003)
26. W. W. Son, X. Zhu, H. I. Shin, J. L. Ong, K. H. Kim, J. Biomed. Mater. Res. Part B Appl. Biomater. 66B, p.520 (2003)
27. M. Fini, A. Cigada, G. Rondelli, R. Chiesa, R. Giardino, G. Giavaresi, N. N Aldini, P. Torricelli, B. Vicentini, Biomaterials 20, p.1587 (1999)
28. W.L. Baun, Surf. Technol. 11, p.421 (1980)
29. V. Zwilling, E. Darque-Ceretti, A. Boutry-Forveille, D. David, M. Y. Perrin, M. Aucouturier, Surf. Interface Anal. 27, p.629 (1999)
30. D. Gong, C. A. Grimes, O. K. Varghese, W. Hu, R. S. Singh, Z. Chen, and E. C. Dickey, J. Mater. Res. 16, p.3331 (2001)
31. G. K. Mor, O. K. Varghese, M. Paulose, N. Mukherjee and C. A. Grimesa, J. Mater. Res. 18, p.2588 (2003)
32. R. Beranek, H. Hildebrand, and P. Schmuki, Electrochemical and solid-state letters 6, p.B12 (2003)
33. H. Tsuchiya, J. M. Macak, L. Taveira, E. Balaur, A. Ghicov, K. Sirotna, P. Schmuki, Electrochemistry communications 7, p.576 (2005)
34. Q. Cai, M. Paulose, O. K. Varghese and C. A. Grimes, J. Mater. Res. 20, p.230 (2005)
35. C. Ruan, M. Paulose, O. K. Varghese, G. K. Mor, and C. A. Grimes, J. Phys. Chen. B 109, p.15754 (2005)
36. J. M. Macak, H. Tsuchiya, P. Schmuki, Angew. Chem. Int. Ed. 44, p.2100 (2005)
37. Ghicov, H. Tsuchiya, J. M. Macak, P. Schmuki, Electrochemistry communications 7, p.505 (2005)
38. K. S. Raja, M. Misra, K. Paramguru, Electrochimica Acta 51, p.154 (2005)

39. P. Kurze, W. Krysmann, H. G. Schneider, Cryst. Res. Technol. 21, p.1603 (1986)
40. H. Ishizawa, M. Ogino, J. Biomed. Mater. Res. 29, p.1071 (1995)
41. S.-H. Oh, R. R. Finõnes, C. Daraio, L.-H. Chen and S. Jin, Biomaterials 26, p.4938 (2005)
42. J. L. Delplancke, R. Winand, Electrochim Acta 33, p.1539 (1973)
43. J. F. Schreckenback, G. Marx, F. Schlottig, M. Textor, N. D. Spencer, Journal of Surface Science, Materials in Medicine 10, p.453 (1999)
44. H. Ishizawa, M. Ogino, J. Biomed. Mater. Res. 29, p.65 (1995)
45. Y. P Lu, R. F. Zhu, S. T. Li, Y. J. Song, M. S. Li, T. Q. Lei, Materials Science and Technology 19, p.260 (2003)
46. Y. Yang, J. L. Ong, J. Biomed. Mater. Res. A 64, p.509 (2003)
47. Y. C. Yang, E. Chang, S. Y. Lee, J. Biomed. Mater. Res. A 67, p.886 (2003)
48. R. Rodriguez, K. Kim, J. L. Ong, J. Biomed. Mater. Res. A 65, p.352 (2003)
49. K. Anselme, Biomaterials 21, p.667 (2000)
50. E. G. Hayman, M. D Pierschbacher, S. Suzuki, E. Ruoslahti, Exp. Cell Res. 160, p.245 (1985)
51. H. Thomas, C. D. McFarland, M. L. Jenkins, A. Rezania, J. C. Steel, K. E. Healy, J. Biomed. Mater. Res. 37, p.81 (1997)
52. P. Henry, A. E. Tan, B. P. Allan, APPL Osseointegration Res. 1, p.15 (2000)
53. Y. T. Sul, C. B. Johansson, Y. Jeong, A. Wennerberg, T. Albrektsson, Clin. Oral Implants Res. 13, p.252 (2002)
54. Y. T. Sul, C. B. Johansson, K. Roser, T. Albrektsson, Biomaterials 23, p.1809 (2002)
55. H. Ishizawa, M. Fugino, M. Ogino, J. Biomed. Mater. Res. 29, p.1459 (1995)
56. G. Giavaresi, M. Fini, A. Cigada, Biomaterials 24, p.1583 (2003)
57. G. Giavaresi, M. Fini, A. Cigada, R. Chiesa, G. Rondelli, L. Rimondini, N. Nicoli Aldini, L. Martini, R. Giardino, J. Biomed. Mater. Res. A 67, p.112 (2003)
58. W. W. Son, X. Zhu, H. I. Shin, J. L. Ong, K. H. Kim, J. Biomed. Mater. Res. B Appl. Biomater. 66B, p.520 (2003)
59. H. Ishizawa, M. Fujino, and M. Ogino, J. Biomed. Mater. Res. 35, p.199 (1997)
60. Y. T. Sul, Biomaterials 24, p.3893 (2003)
61. Feng, J. Wang, B. C. Yang, S. X. Qu, X. D. Zhang, Biomaterials 24, p.4664 (2003)
62. D. Boyan, R. Batzer, K. Kiesewetter, Y. Lie, D. L. Cochran, S. Szmuckler-Moncler, D. D. Dean, Z. Schwartz, J. Biomed. Mater. Res. 39, p.77 (1998)
63. M. Karlsson, E. Palsgard, P. R. Wilshaw, L. D. Silvio, Biomaterials 24, p.3039 (2003)
64. T. J. Webster, C. Ergun, R. H. Doremus, R. W. Siegel and R. Bizios, Biomaterials 22, p.1327 (2001)
65. D. S. Dunn, S. Raghaven, R. G. Volz, J. Appl. Biomater. 5, p.325 (1994)
66. M. Varkey, S. A. Gittens, H. Uludag, Expert Opin. Drug Deliv. 1, p.19 (2004)

3. Titanium Dioxide Coatings in Medical Device Applications

3.1 Introduction

Titanium dioxide (TiO_2, titania) is a widely abundant and inexpensive material. In bulk form it is produced as a white powder and it is the most widely used white pigment because of its brightness and very high refractive index ($n=2.4$). Applications include filler pigment in paints, cosmetics, pharmaceuticals, food products (such as E171, e.g., white lettering on M&Ms) and toothpaste. When deposited as a thin film, its refractive index and color make it an excellent reflective optical coating for dielectric mirrors. It is also widely used in sun block creams due to its photostability, high refractive index and UV absorption properties. TiO_2 is chemically and photo-chemically stable, non-toxic and insoluble under normal pH conditions. The corrosion resistance of titanium metal is due to the formation of a native oxide passivation layer.

TiO_2 occurs in three crystalline forms; brookite, anatase and rutile, the latter two being the more common. Rutile is the thermodynamically stable form. In the crystal lattice of TiO_2 each Ti atom is bonded to six O atoms and each O atom is bonded to three Ti atoms to form a tetragonal crystal lattice. Anatase differs from rutile by the number of common edges of the TiO_6 octahedra i.e., 4 for anatase and 2 for rutile. TiO_2 is effectively an insulator at normal temperatures, however, the band gap (3.0 eV for rutile and 3.2 eV for anatase) is such that it will absorb ultra violet light at wavelengths just under 400 nm and it is referred to as a wide band gap semiconductor.

TiO_2 powder is prepared on an industrial scale either by the sulphate process or by the vapour phase oxidation of titanium tetrachloride ($TiCl_4$). In the sulphate process the ilmenite ore ($FeTiO_2$) is dissolved in sulphuric acid, iron is removed and the solution is hydrolysed. The hydrated TiO_2 is calcined to remove water. Anatase is the main crystal form produced from this process because sulphate ions, inherently present in the product, stabilise this phase. In vapour phase oxidation (also know as flame hydrolysis) titanium tetrachloride is sprayed into a high temperature flame to produce nanometer-sized particles. The main contaminant is chloride and the phase purity is not good. Powder preparations can be used to make coatings by methods including plasma spray, dip coating, and electrophoretic coating. This chapter reports on advances made in this important field of medical research [1-25].

Other methods may be used to produce coatings by the formation of TiO_2 from precursors directly on a substrate surface. If the substrate is Ti metal or alloy, the simplest method of producing a thin film of TiO_2 is to oxidise the surface. This may be achieved by simply leaving the titanium sample in the open atmosphere where a natural oxide layer will form with time. Alternatively one may increase the rate of oxidation and control the oxide film thickness by simple thermal treatment in an oxygen atmosphere, exposure to an oxidising solution or atmosphere, or by electrochemical oxidation (anodisation). Methods of producing thin films of TiO_2 on other supporting substrates have been developed including physical, chemical and physicochemical routes. These methods may also be used to coat Ti metal and it's alloys.

The main physical route to TiO_2 films is sputter deposition. This involves the formation of a plasma e.g., argon in a vacuum system. A high electric field is produced between the TiO_2 target and the substrate to be coated. The plasma contains positively charged argon ions, which are accelerated towards the TiO_2. The argon ions impact with a billiard ball effect, dislodging titanium and oxygen atoms, or clusters of TiO_x, with high kinetic energy. These species move out into the plasma and upon collision with the substrate adhere to the surface and form a film of TiO_2. Under optimal conditions one can achieve a uniform coating of TiO_2 with the desired crystal structure and film thickness. The sputter deposition route lends itself to the coating of a wide range of substrate materials, including polymers, and to different substrate conformations. A disadvantage is the requirement for a vacuum plasma system that can be expensive.

An alternative wet chemical route to TiO_2 thin films is sol gel processing, which has been used in the production of ceramic and glass coatings for many years. For TiO_2 sol gels the most common precursors are alkoxides, e.g., titanium (IV) propoxide and titanium (IV) butoxide, although inorganic compounds may also be used. Alkoxides are metal organic compounds where the metal is bonded to a hydrocarbon via a bridging oxygen i.e., the titanium atom replaces the proton in the alcohol. The metal alkoxides are liquid at room temperature and can be produced with high purity. Alkoxides are very reactive with water and the controlled hydrolysis of the alkoxide is followed by a condensation and polymerisation step whereby the titanium hydroxide sol gives up water to form a polymeric TiO_x gel. The sol gel can be coated onto a wide range of substrates using techniques such as dip coating, spin coating, roll coating, etc. In the case of TiO_2 sol gel processing, a high temperature thermal treatment stage is normally required to yield a crystalline film. High temperature treatment is obviously not applicable for thermolabile materials such as polymers. However, other methods such as hydrothermal annealing have been used to produce crystalline films at much lower temperatures (100–150°C).

Another commonly employed route to TiO_2 films is chemical vapour deposition (CVD). In this technique the substrate to be coated is heated and exposed to a volatile metal organic precursor in a carrier gas. Upon contact with the surface the reactive precursor decomposes to yield an oxide film. Other physicochemical approaches may involve a mixture of techniques e.g., reactive sputter deposition or plasma enhanced CVD.

Titanium metal and its alloys are important as biomedical and dental implant materials because of their relatively high corrosion resistance and good biocompatibility. The passivating oxide layer is responsible for these properties. However, despite widely reported low rates of corrosion for titanium *in vitro*, there is evidence to suggest that corrosion rates may be enhanced *in vivo*, leading to the release of titanium and accumulation in adjacent tissues or transport to other areas of the body [22]. Indeed, the biological environment may be aggressive towards titanium or the native oxide film. Furthermore, oxide film growth and ion incorporation into the film have been noted following implant into humans. Different methods of surface modification have been attempted in order to improve the corrosion resistance and biocompatibility of titanium, titanium alloys and stainless steel. It was reported [23] on the corrosion resistance for biomaterial applications of TiO_2 films deposited on titanium and stainless steel by ion-beam-assisted sputter deposition (IBASD). In that approach, a pure titanium target is sputtered by an argon ion beam, and the sputtered atoms are deposited onto the substrate while a flow of neutral oxygen gas is introduced on the substrate (normal reactive sputter deposition). An additional oxygen ion beam is used as the assisting beam. They reported a two-layer model of the oxide film deposited using this method and that the IBASD films exhibited improved corrosion resistance as compared to a native oxide layer on titanium or stainless steel.

Hemocompatability is an important parameter for implant materials that come into contact with blood. Almost any medical device introduced into the human body will initially come into contact with blood. Furthermore, thrombogenicity of artificial implant devices such as artificial heart valves is a serious problem as the implant induces blood clotting and patients with such implant devices must be given anticoagulant drugs as ongoing therapy. Surface modification of implant devices is therefore an important approach to improving hemocompatability. Different materials have been investigated for coating implant devices e.g., diamond like carbon, silicon carbide, titanium nitride, and aluminium oxide, and low temperature isotropic pyrolytic carbon (LTIC). LTIC is widely regarded as the best hemocompatible coating. It has been reported that albumin can passivate a surface, that complement activation can result in the neutrophil recruitment to surfaces, and that fibrinogen initiates the acute inflammatory response. Platelets and inflammatory cells are likely to respond to the layer of adsorbed proteins, not to the material surface itself. However, the composition of the layer of adsorbed proteins is dependent on the properties of the material. Therefore, the initial reactions that take place on the material surface upon exposure to blood will determine the conditions for subsequent reactions. Nygrean, Tengvall and Lundstrom [20] investigated the initial reactions that take place on exposure of TiO_2 surfaces to blood. They compared Ti metal that was passivated in nitric acid to Ti metal which had been passivated by annealing in air at 700°C. In order to study the initial interactions they used capillary blood from human donors (without addition of anticoagulants), which was exposed to the surface for only 5 seconds. Fluorescent immuno-assay was used to determine the presence of platelet cells, fibrinogen,

Cl_q, and prothrombin/thrombin. They reported that the serine proteases of the coagulation and complement systems were initiated within 5 seconds of the blood exposure to the TiO_2 surface. They also found that platelets were adhered to the surface in the initial 5 seconds' exposure. Both plasma proteins and cells were found at the blood-surface interface after only 5 seconds and this implied that a complex "biofilm" was formed within a very short contact time. The interaction of plasma proteins will differ for hydrophilic and hydrophobic surfaces. Both annealed and acid treated samples were macroscopically hydrophilic and there was no significant difference in contact angle, however, there was a significant difference between the levels of prothrombin/thrombin and platelets adhered to the different surfaces. The authors suggest that this may be due to different levels of carbon impurity in the two films. Platelet adhesion to a surface, although probably a pre-requisite to activation, does not in itself mean activation.

Huang et al. [13] reported on the hemocompatibility of titanium oxide and tantalum doped titanium oxide films prepared by plasma immersion ion implantation and deposition (PIIID) and sputtering. The first event to occur following implant's surface contact with blood is the adsorption of a protein layer. If the surface characteristics of the material result in a change in the configuration of the adsorbed protein e.g., fibrinogen or globulin, it may enhance coagulation and/or platelet activation. Alternatively, adsorption of albumin on the surface with maintenance of the native configuration discourages coagulation. Therefore, a reduction in protein adsorption and denaturation is a key strategy to enhancing anticoagulation properties of surfaces. The anticoagulation nature of a surface depends on multiple characteristics of the material including surface energy, surface charge and surface topography and on the surface effects imposed at different stages of the blood-surface interaction process. Huang et al. [13] reported that their TiO_2 films exhibited lower interface energy compared to LTIC, leading to less fibrinogen adsorption on the TiO_2. Furthermore, changing the structure of the TiO_2 film from amorphous to crystalline, and doping of the films with tantalum also can affect the anticoagulation properties. They postulated that the semiconducting nature of the TiO_2 films is an important contributing factor where n-type semiconductor properties helps to prevent protein denaturation on the surface by inhibiting charge transfer from the protein into the

TiO$_2$. Indeed they found improved behaviour of platelet adhesion on crystalline rutile as compared to amorphous TiO$_2$ films.

The surfaces of implants used for dental and orthopaedic applications also become coated with a proteinacious film. The nature of this protein layer depends on the surface of the implant and may affect the biological response to the implant, including cell attachment. Fibronectin is one of the first extracellular matrix proteins produced by odontoblasts and osteoblasts, and therefore, is a useful model to investigate protein surface interactions *in vitro*. Fibronection is composed of two similar polypeptide chains whose subunits are linked by disulphide bonds and this protein is reported to play a major role in interaction of the implant material and the body. Yang et al. [25] investigated fibronectin adsorption on titanium surfaces and its effect on osteoblast precursor cell attachment. They used Ti metal that was pretreated by wet grinding followed by passivation in 40% v/v nitric acid (HNO$_3$) for 30 minutes. Each sample was sterilised by UV irradiation for at least 24 hours, a step that may have had other consequences. X-ray photoelectron spectroscopy analysis confirmed an amorphous oxide layer on the surface of the Ti following the treatment stages. The researchers observed a significant difference in the fibronectin adsorbed after 15 and 180 minutes exposure to protein containing solution. The amphoteric charactistics of TiO$_2$ mean that the surface charge changes with pH. The isoelectric point of TiO$_2$ has been reported to be between pH 4.0 and 6.2, and therefore, at a pH 7.4, the oxide will be mainly anionic (net –ve charge) in character and will electrostatically bind proteins that are cationic (net +ve charge) at pH 7.4. Binding of proteins to surfaces may also involve hydrophobic bonding the extent of which will be affected by the wettability of the implant surface. TiO$_2$ surfaces have been reported to have both hydrophilic and hydrophobic components. Therefore the extent of specific protein attachment to TiO$_2$ surfaces will be dependent on a complex interplay between hydrophobicity/hydrophillicity and electrostatic interaction. Yang et al. [25] also reported that the pre-adsorption of fibronectin to the TiO$_2$ surface enhanced the attachment of osteoblast cells as compared to control Ti samples without pre-adsorption of fibronectin. It has been suggested that the presence of fibronectin promotes cell attachment by binding through cell surface receptors and mediating adhesive interactions. Furthermore, adsorption of low concentrations of fibronectin on surfaces causes unfolding of the protein into an inactive conformation, but a high adsorption concentrations unfolding is prevented by steric hindrance due to molecule packing.

More recently it has been reported that nanostructure control of TiO_2 films can not only improve biocompatibility, but can improve the bioactivity of the surfaces for bone adhesion. It is hypothesised that tissue-biomaterial interactions occur within 1 nm of the material surface, and therefore the ability to engineer surfaces on the nanometre scale will have major impact on the production of materials with improved biocompatibility and bioactivity.

Due to the increased life expectancy of the population there has been an enormous increase in the incidence of bone fracture and the need for bone implant surgery. The improvement of implant-bone interface is a real problem. Titanium and its alloys are the most commonly employed metals used in the manufacture of orthopaedic prostheses on account of their excellent mechanical properties, corrosion resistance and biocompatibility. Even so, osseointegration results have not always been satisfactory under altered metabolic bone conditions. Research and development into improvements in osseointegration have focussed on implant surface properties such as morphology, roughness and chemical composition. All of these factors may affect cell and biochemical responses of host bone and the ability to control these may allow one to promote bone apposition through the acceleration of the chemical bonding between the new bone and the implant surface. For titanium implants, the oxide layer grows slowly on the surface following implant and contributes to the formation of apatite and bone-like tissue. Current passivation methods used in the pretreatment of titanium prosthesis (machining, ultrasonic cleaning, sterilisation, and anodisation) still present limits. For example, low-level contamination with impurities may have a deleterious effect on the osseointegration process, the TiO_2 film thickness is linearly dependent on the anodic potential employed in electrochemical oxidation and the anodically grown oxide may present significant porosity. Giavaresi et al. [10] reported on the osseointegration of a nanostructured titanium oxide coating produced by chemical vapour deposition (CVD). The aim of the study was to compare the *in vivo* (implanted in rabbits) osseointegration of Ti implants coated with TiO_2 CVD thin films with that of Ti machined implants. They reported that the affinity index (AI: the interface contact between bone and implant as calculated as the length of the bone profile directly opposed to the implant divided by the total length of the bone-implant interface) of the Ti/CVD implants were significantly higher than that of the machined implants. SEM analysis confirmed a high level of osteintegration for the Ti/CVD implants in cortical bone and enhanced osseointegration in cancellous bone, as compared to that observed for the machined implants. They concluded that their histomorphometric, ultra-structural and microhardness findings demonstrated that the nanostructured TiO_2 coating positively affected the osseointegration rate of commercially pure Ti implants in terms of bone mineralization in both cortical and cancellous bone. Further studies were planned on mechanical bonding with bone and bone remodelling around implants.

In 2001 Ramires et al. [23] reported on the influence of titania-hydroxyapatite (TiO_2-HA) composite coatings, obtained via a sol gel route, on *in vitro* osteoblast behaviour. They found that these materials have no toxic effects (at least *in vitro*). Cell growth and morphology were similar on TiO_2-HA coatings and TiO_2 coatings. However, alkaline-phosphatase-specific activity and collagen production of osteoblasts cultured on TiO_2-HA coatings were significantly higher than uncoated titanium or polystyrene culture plates. They concluded that the TiO_2-HA coatings were bioactive owing to the presence of hydroxyl groups on the surface that promote calcium and phosphate precipitation and improve interactions with osteoblastic cells.

Other workers have also investigated bone cell proliferation on hybrid TiO_2-HA coatings on Ti implants. For example, Lee et al. [15] produced TiO_2 films by a method known as micro-arc oxidation (MAO) in which a DC pulsed potential is applied to the Ti substrate in an electrochemical cell. In this case the Ti metal was pre-coated with HA (using e-beam evaporation). The logic behind this approach was to introduce Ca and P into the TiO_2 films to improve osseointegration and cellular activity. The rough, porous TiO_2 films produced by MAO should enhance mechanical interlocking of tissue and implant. A CaP layer is thought to enhance the initial cellular response, due to high osteoconductivity and bioactivity. When the MAO treatment was carried out at 230 V the coating surface was observed by SEM to become rough and porous (only isolated areas were affected at lower potentials) and caused the dissolution of the HA layer, however, a large amount of Ca and P were incorporated into the TiO_2 layer and some of the CaP layer remained after MAO treatment. The proliferation of the human osteosarcoma cells was decreased only slightly on the HA surface treated by MAO, similar to Ti surface treated by MAO, and was attributed to the increase in surface roughness. However, there was a marked increase in the alkaline phosphatase activity of the cells on the MAO treated TiO_2-HA compared to the MAO treated Ti. They concluded that the use of hybrid coatings obtained by pre-coating of HA on Ti followed by MAO treatment is a possible route to enhanced cell responses for bone implant materials.

Secondary infections are a cause of implant failure, particularly with percutaneous (skin penetrating) implants. However, infections at skin-penetrating titanium implants anchored in the temporal bone can often be cured by local treatment. As mentioned previously, host protein adsorption to the implant material occurs within seconds of implant. Protein adsorption to the surface depends on a number of parameters. Furthermore, microbial adhesion will depend on the nature of the surface and indeed on the nature of the surface bound proteins. Fibronectin has been proposed to mediate adhesion of staphylococci to implant materials in blood. Staphylococcus epidermidis is a common etiological agent with infections involving polymeric implants whereas Staphylococcus aureus is the most common etiological agent with infections involving metal implants. Understanding the mechanism of bacterial attachment to surfaces is crucial for enabling the engineering of surfaces for reduced biofilm recruitment and implant failure due to secondary infection. Holgers and Ljungh [12] reported a study into the cell surface characteristics of microbial isolates from human percutaneous titanium implants. They found that no isolates expressed a hydrophobic cell surface, however, isolates from infected implants were less hydrophilic than those from non-inflamed tissue. The degree of hydrophillicity of an implant surface will influence the recruitment of biofilm. This leads to the possible exploitation of other, perhaps more exciting, properties of TiO_2.

The important biocompatible and bioactive properties of TiO_2 have been addressed above. TiO_2, has additional properties which bring added value as a material for use in biomedical applications. In 1972 Japanese researchers, Honda and Fujishima [8], published a

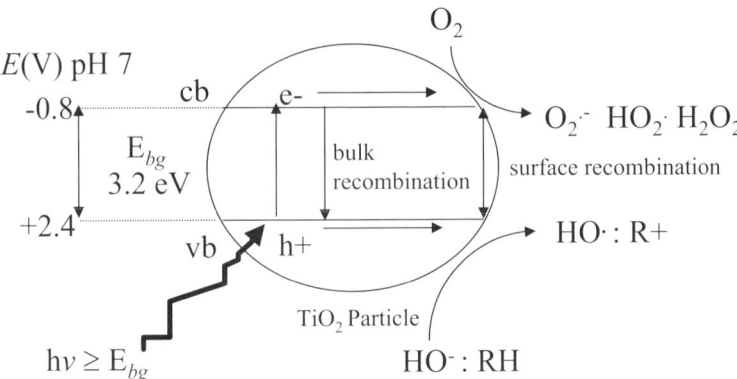

Fig. 3.1. Mechanism of TiO photocatalysis (potentials vs SCE)

paper in Nature reporting the photo-splitting of water using a single crystal of rutile TiO_2 under ultra violet irradiation (Honda and Fujishima, [8]). If one could achieve efficient water splitting into hydrogen and oxygen with solar energy then the world's energy problems might be solved.

Since then, the ability of TiO_2 to absorb UV electromagnetic irradiation and use that energy to drive electrochemical reactions on it's surface, has been investigated for a wide variety of applications, including water and air treatment and purification [1,2,5,11,17-19,24] and "self-cleaning" surfaces [9]. Indeed, Pilkington's are now selling "self cleaning" glass called *Activ*, with a 15 nm layer of TiO_2 on the surface which, under the action of solar UV, can destroy and remove organic contamination. The process has been aptly named photocatalysis, meaning the use of a catalyst to speed up a photochemical reaction. The mechanism of photocatalysis is shown in Figure 3.1.

TiO_2 is a wide band gap semiconductor material. The valence band is filled with electrons and is separated from an empty conduction band by band gap energy (E_{bg}). If illuminated with UV light (E = hv where E_{bg} of 3.2 eV is equivalent to 387 nm) a photon is absorbed and an electron is promoted from the valence band to the conduction band, leaving a hole in the valence band. These charge carriers can recombine in the bulk, or they can move to the surface of the particle. An electron in the conduction band has a negative electrochemical reduction potential and can reduce an electron acceptor species at the interface, e.g., molecular oxygen producing superoxide radical anion, perhydroxyl radical, and hydrogen peroxide. The hole has a very positive electrochemical reduction potential, positive enough to oxidise water or hydroxyl ions to yield hydroxyl radicals. Hydroxyl radicals are powerful and indiscriminate oxidising species. The redox reactions at the surface of the particle lead to the generation of active oxygen species which can attack organic and inorganic species at or near the surface. There are literally thousands of papers in the literature reporting photocatalysis for the destruction of a wide range of organic pollutants, including microorganisms such as viruses, bacteria, and fungi, and even tumour cells [1,7,9,14].

The potential for this "self sterilising" property has been identified for use in medical devices. Ohko et al. [21] reported on self-sterilising and self-cleaning of silicone catheters coated with TiO_2 photocatalyst thin films. They described a sol gel method for coating silicone catheters with TiO_2 to produce a photoactive film. The self-cleaning effect was demonstrated using the bleaching of methylene blue dye and the self-sterilising effect was demonstrated by the killing of *E. coli*. The application proposed was that the catheter could be irradiated prior to insertion thus helping to prevent catheter related bacterial infection. The authors suggested practical use as an intermittent self-sterilising catheter for neurogenic bladder patients and or for self-sterilising suction tubes for frequent draining of sputa and oral fluid. They also suggested that a dark bactericidal action could be incorporated by surface doping of the TiO_2 with silver.

Another property of TiO_2 that has created a great deal of excitement is the phenomenon of photo-induced superhydrophilicity. UV excitation of the TiO_2 generates electrons and holes. The electrons tend to reduce the Ti(IV) cations to the Ti(III) state, and the holes oxidize the O_2^- anions. This results in the ejection of oxygen atoms creating oxygen vacancies. Water molecules can then occupy these oxygen vacancies, producing adsorbed OH groups, which tend to make the surface more polar or hydrophilic. After about 30 minutes or so under a moderate intensity UV source, the contact angle for water approaches zero. It is a combination of this superhydrophilic effect and the photocatalytic effect that is responsible for the 'self-cleaning' nature of these coatings. Even more interesting is that TiO_2 surfaces have been reported to be amphiphilic in nature i.e., displays both hydrophilic and hydrophobic properties. Fujishima, Rao and Tryk [9] reported light induced reduction in contact angle for water approaching zero and light induced reduction in contact angle for glycerol trioleate (a component of vegetable oils) approaching zero.

The wettability of a surface will have an effect on the interactions of the surface with proteins. Therefore the ability to induce changes in contact angle of the surface would have important implications for implant materials. Liu et al. [16] reported that UV irradiation of TiO_2 coatings prior to immersion in simulated body fluid enhanced the formation of bonelike apatite. They used plasma sprayed nano-particle TiO_2 to form coatings on the surface of Ti metal. Following coating, one set of samples was irradiated for 24 h using a 125 W medium pressure Hg lamp (main output 365 nm) and a non-irradiated sample set was used as the control. Following four weeks immersion in simulated body fluid they used energy dispersive x-ray spectro-scopy, XRD and FTIR to analyse the surfaces. They found that the samples that had been irradiated with UV prior to immersion had a newly formed layer on the surface that was carbonate containing hydroxyapatite (bonelike apatite). Without UV irradiation prior to immersion no new surface precipitates were detected. They reasoned that oxygen vacancies were created and that Ti^{3+} sites were more favourable for the dissociation of water to form an abundance of surface Ti-OH groups. OH groups on ceramic surfaces are suggested to be effective for inducing the formation of an apatite layer. The mechanism is thought to involve the Ti-OH surface groups reacting with hydroxyl ion in the simulated body fluid to produce negatively charged Ti-O$^-$.

$$Ti\text{-}OH \ + \ OH^- \rightarrow Ti\text{-}O^- + H_2O \qquad (3.1)$$

The Ca^{2+} ions in the solution are electrostatically attracted to the negatively charged surface, followed by reaction with HPO_4^{2-} to form calcium hydrogen phosphate. The CaP continues to grow, also incor-porating carbonate anions from solution, and crystallises to form carbonate-containing hydroxyapatite (bonelike apatite). Liu et al. [16] reported that the photo-induced bioactive surface remained for at least one week following UV irradiation.

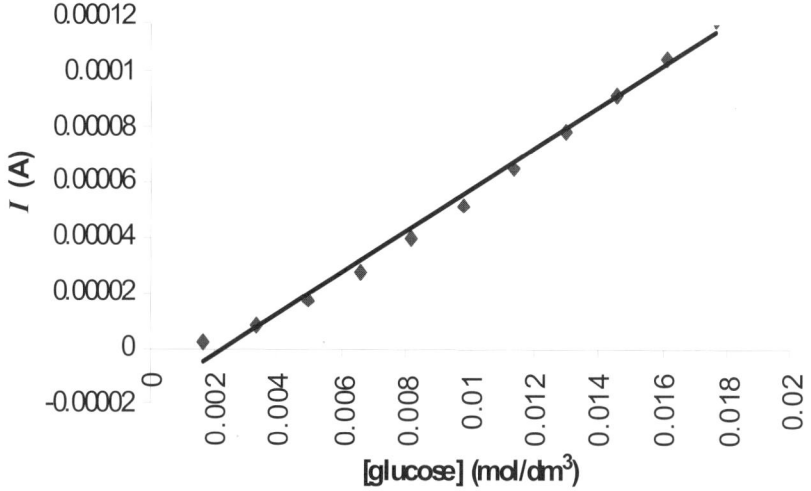

Fig. 3.2. Nanocrystalline TiO_2 electrode's amperometric response to standard additions of glucose. 10 mg GOD was present as free enzyme in 30 cm^3 pH 6 phosphate buffer. The TiO_2 electrode was held at a fixed potential of –0.4 V (SCE) [4].

Cosnier et al. [6] reported the use of nanocrystalline TiO_2 films for use in biomedical sensor applications i.e. an electrochemical transducer material for the detection of hydrogen peroxide in the presence of oxygen. There is a wide range of oxidase enzymes which play different roles in the body. The best known is glucose oxidase (GOD), which acts to oxidise glucose to gluconic acid and in the process, reduces molecular oxygen to hydrogen peroxide. GOD is commonly employed in commercial glucose biosensors in which the electron transfer to the electrode is mediated by a free redox couple. However, the Holy Grail of glucose sensing is to develop an implantable sensor for feedback control of an insulin delivery system (artificial pancreas) and remove the need for routine blood sampling.

Mediated biosensors are not suitable for implant as free mediators will simply diffuse away into the blood stream and immobilised mediators are not as efficient or the mediator may be toxic. An alternative is to use non-mediated electrochemical sensing. TiO_2 has a lower overpotential for the electrochemical reduction of hydrogen peroxide than for the reduction of dissolved oxygen. Therefore, as hydrogen peroxide is a by product of the action of GOD, a TiO_2

sensor incorporating GOD could be used to detect glucose by the electrochemical reduction of hydrogen peroxide without interference from oxygen reduction. Therefore, there is no requirement for a mediator and TiO_2 is biocompatible. Figure 3.2 shows electrochemical reduction current measured for nanocrystalline TiO_2 electrode as a function of glucose concentration in a solution containing free GOD. The response is linear in the concentration range for glucose found in physiological blood.

In conclusion, TiO_2 coatings are important for improving the biocompatibility and bioactivity of implant materials. The additional properties of photocataltyic sterilisation and photo-induced super-hydrophilicity may bring added value. This remains a vibrant field of research.

References

1. Blake, D.M., Maness, P.C., (1999), Separation and Purification Methods 28(1): 1-50.
2. Byrne J.A., Eggins B.R., Brown N.M.D., McKinney B., Rouse M., (1998) Applied Catalysis B: Environmental, 1998, 17, 25-36.
3. Byrne, J.A., Davidson, A., Dunlop, P.S.M., Eggins, B.R., (2002) Journal of Photochemistry and Photobiology A: Chemistry, 148, 365-374.
4. Byrne, J.A., Hamilton, J.W.J., McMurray, T.A., Dunlop, P.S.M., Jackson, V., Donaldson, A., Rankin, J., Dale, G., Alrousan, D., (2006) Abstracts of the NSTI conference, Boston.
5. Coleman, H.M., Routledge, E.J., Sumpter, J.P., Eggins, B.R., Byrne, J.A., (2004) Water Research, 38, 3233-3240.
6. Cosnier, S., Gondran, C., Senillou, A., Gratzel, M., Vlachopoulos, N., (1997) Electroanalysis, 9 (18), 1387-1392.
7. Dunlop, P.S.M., Byrne, J.A., Manga, N., Eggins, B.R., (2002) Journal of Photochemistry and Photobiology A: Chemistry, 148, pp 355-363.
8. Fujishima, A., Honda, K., (1972) Nature, 238, 37-38.
9. Fujishima, A., Rao, T.N., Tryk, D.A., (2000) Journal of Photochemistry and Photobiology C: Photochemistry Reviews, 1, 1-21.
10. Giavaresi, G., Ambrosio, L., Battiston, G.A., Casellato, U., Gerbasi, R., Finia, M., Aldini, N.N., Martini, L., Rimondini, L., Giardino, R., (2004) Biomaterials, 25, 5583-5591.
11. Hoffman, M.R., Martin, S.T., Choi, W., Bahnemann, D.W., (1995) Chem. Rev., 95, 69-96.
12. Holgers, K.M., Ljungh, A., (1999) Biomaterials, 20, 1319-1326.

13. Huang, N., Yang, P., Leng, Y.X., Chen, J.Y., Sun, H., Wang, J., Wang, G.J., Ding, P.D., Xi, T.F., Leng, Y., (2003) Biomaterials, 24, 2177-2187.

14. Kuhn, K.P., Chaberny, I.F., Massholder, K., Stickler, M., Benz, V.W., Sonntag, H.-G., Erdinger, L., (2003) Chemosphere, 53, 71-77.

15. Lee, S.-H., Kim, H.-W., Lee, E.-J., Li, L.-H., Kim, H.-E., (2006) Journal of Biomaterials Applications, 20, 195-208.

16. Liu, X., Zhao, X., Ding, C., Chu, P.K., (2006) Applied Physics Letters, 88, 13905.

17. McMurray, T.A., Byrne, J.A., Dunlop, P.S.M., Winkelman, J.G.M., Eggins, B.R., McAdams, E.T., (2004) Applied Catalysis A: General, 262, 1, 105-110.

18. McMurray, T.A., Byrne, J.A., Dunlop, P.S.M., McAdams, E.T., (2005) Journal of Applied Electrochemistry, 35, 723-731.

19. Mills, A., Le Hunte, S., (1997) Journal of Photochemistry and Photobiology A: Chemistry, 108, 1-35.

20. Nygren, H., Tengvall, P., Lundstrom, I., (1997) Journal of Biomedical Materials Research, 34, 487-492.

21. Ohko, Y., Utsumi, Y., Niwa, C., Tatsuma, T., Kobayakawa, K., Satoh, Y., Kubota, Y., Fujishima, A., (2001) Journal of Biomedical Materials Research (Applied Biomaterials) 58, 97-101.

22. Pan, J., Leygraf, C., Thierry, D., Ektessabi, A.M., (1997) Journal of Biomedical Materials Research, 35, 309-318.

23. Ramires, P.A., Romito, A., Cosentino, F., Milella, E., (2001) Biomaterials, 22, 1467-1474.

24. Shani Sekler, M., Levi, Y., Polyak, B., Dunlop, P.S.M., Byrne, J.A., Marks, R.S., (2004) Journal of Applied Toxicology, 24, 395-400.

25. Yang, Y., Glover, R., Ong, J.L., (2003) Colliods and Surfaces B: Biointerfaces, 30, 291-297.

.

4. The Effect of Shape and Surface Modification on the Corrosion of Biomedical Nitinol Alloy Wires Exposed to Saline Solution

4.1 Introduction

The Corrosion behaviour of Nitinol wire that has been chemically etched and mechanically polished was studied in a corrosive 0.9% Saline solution. The electrochemical corrosion tests conducted on the as-received straight and curved wires of nitinol included open circuit potential measurement, polarisation resistance and Tafel plots. The chemically etched looped wires exhibited the highest recorded corrosion potential Ecorr and the lowest values of corrosion current icorr. The results of the open circuit potential (OCP) measurements and polarisation resistance, combined with scanning electron microscopy (SEM) indicated the presence of a protective passive corrosion resistant film on the chemically etched wires.

One of the most popular shape-memory and super-elastic alloys used for biomedical stents is called Nitinol. It exhibits a number of favourable material properties that makes it well suited for use as a stent [1]. Its most important properties are the super elasticity and shape memory abilities. Nitinol's ability to be deformed by more than 10% strain and elastically recover its original shape is described as super elasticity. The material is able to withstand a stress, which induces an elastic deformation, thus causing the material to undergo a phase transformation. Superelasticity results from a stress-induced transformation while shape memory results from a thermal phase transformation. Both of these properties are employed in Nitinol stents [2].

Biocompatibility is always an issue whenever implanting a foreign material into the human body [3]. Implanting metals brings its own set of problems, such as corrosion.

When a metal is introduced into the body, a wide variety of processes and interactions with the biological environment can take place. Such processes include metal ion release [4]; oxide formation, either as a apassive oxide film [5] or as a particulate oxide; and corresponding reduction reactions that typically involve oxygen reduction but may also include biochemical species such as proteins [6]. These phenomena vary microscopically over the alloy surface and depend on the local environment, local alloy composition and structure as well as local mechanical factors [7].

The aim of the current investigation is to study the corrosion and pitting resistance behaviour of nitinol wire with different surface finishes i.e. chemical etching and mechanical polishing. The corrosion test was conducted in 0.9% saline solution at 37°C by means of electrochemical measurements. OCP, pitting potentials, corrosion potentials and corrosion currents were analysed in each of the materials studied.

4.2 Experimental Methods

The materials under investigation were Nitinol wire that had been mechanically polished with a diameter of 0.2mm and nitinol wire that had been chemically etched with a diameter of 0.2mm provided by Vascutek of Renfrewshire, Scotland, U.K. The material compositions are shown in Table 4.1, from EDX analysis.

The solution used was a commercial NaCl solution, from Parkfields Pharmaceutical limited which contains sodium chloride B.P. 0.9% w/v in purified water B.P. The experimental arrangement used consists of the electrochemical unit including the Voltalab 40 (PGZ301) potentiostat that is able to perform conventional electrochemical measurements dedicated to corrosion techniques. The overall system was controlled using a PC-compatible microcomputer with Voltamaster 4 electrochemical software and a basic corrosion cell whose body is made of pyrex with a double jacketed device allowing thermostatic experiments to be performed (ISO 10993-15).

The experiments were carried out at 37 +/– °C by circulating the water through the double wall of the cell. The working electrode with a surface area of about 1cm^2 was inserted into the cell through a rubber sealer. The apparatus and the electrochemical cell conformed

Table 4.1. Materials composition: (e) Elemental percentage, (a) Atomic weight percentage

Material	Ni(e)	Ti(e)	Total(e)	Ni(a)	Ti(a)	Total(a)
Mechanical Polished	54.4	45.5	100.0	49.5	50.5	100.0
Chemical Etched	53.8	46.2	100.0	48.8	51.2	100.0

to ATM G5-94. The potential of the working electrode was measured against the reference calomel electrode, which contained a porous pin liquid junction and a saturated KCl salt bridge solution from Radiometer analytical. A platinum wire metal electrode was used as the auxiliary counter electrode. The electrolyte was continuously purged with purified nitrogen gas to de-aerate the solution for 1hr before OCP and polarization experiments.

4.2.1 Electrochemical Testing Procedure

OCP and potentiodynamic anodic polarization measurements were conducted according to ASTM G5 test generated using a potentiostat. The OCP was determined after 22hr for 0.9% saline solution, the value obtained after stabilisation of the curve being called the rest potential. Potentiodynamic anodic polarization test was scanned at −1000mV to 2500mV at a scan rate of 0.167 mV/sec in order to determine corrosion and breakdown potentials. Current corrosion density (*i*corr), corrosion potential (Ecorr) and breakdown potential (E$_{bp}$) were obtained from polarization curves using the Tafel slopes.

4.2.2 Experimental Results

Open Circuit Potential (OCP)

OCP is the thermodynamic measurement of how likely the surfaces are to corrode. The more positive an OCP the more stable is the oxide film on the surface and therefore the less reactive the surface. The OCP was measured to determine the stable corrosion potential (Ecorr) of the nitinol alloy samples, i.e., the values obtained when the potential becomes constant (Fig. 4.1).

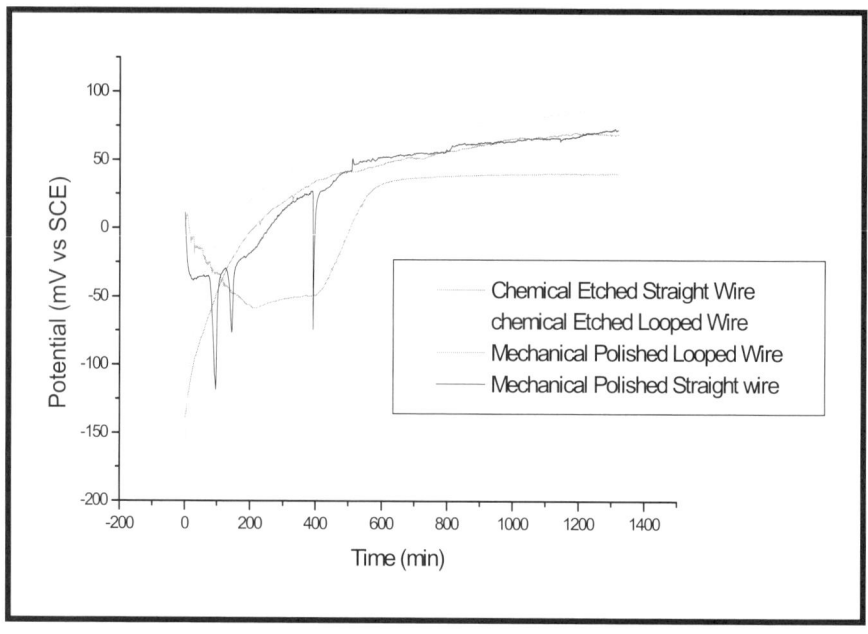

Fig. 4.1. OCP vs. time curves for the stable potential determination of nitinol in 0.9% saline solution

The OCP value for chemical etched wire in looped geometry proved to be the most positive value of OCP, the wire progressively becomes more anodic with time, the oxide layer film remains intact and the material stays stable in the passive state, and is therefore the least reactive surface in 0.9% saline. The Mechanically polished wire initially shifted towards more negative values before stabilising in more positive values after 22hrs. Table 4.2 shows the OCP values table of the nitinol wires after 22hrs.

Tafel Plot

Tafel plots of the different surface treatments Fig. 4.2 are used to measure the corrosion current i_{corr}, which is then used to calculate the corrosion rate. The Tafel plot yields the Tafel constants β_A and β_C which is used with polarisation resistance (Rp) [8,9] data to calculate corrosion rate as shown in Table 4.3. Results show chemical etched looped sample with the highest corrosion potential and the chemical etched straight wire with the lowest corrosion potential. The results

Table 4.2. OCP values after 22hrs in 0.9% saline

Sample	OCP (mV)
Chemical Etched	67
Mechanical Polished	32
Chemical Etched Looped	87
Mechanical Polished Looped	40

indicate that the chemically etched looped wire is the most stable of the wires, with the highest corrosion potential, a low corrosion current and low corrosion rate as shown in Fig. 4.3.

Potentiodynamic Curves

The potentiodynamic curves are as shown in Fig. 4.4. The potentiodynamic curves indicate the passivation tendencies of the wires, in the passive region. Results from Fig. 4.4 show that the chemically etched looped wire has the largest passivation region, i.e., (–0.78mV to +448mV) [9].

The breakdown potentials (E_{bd}) correspond to the breakdown of the passive surface film, which is identified by the sudden increase in the current flowing in the solution, as shown in Table 4.4.

Table 4.3. Tafel parameters for NiTi wires in 0.9% saline solution

Sample	E_{corr} (mV/)	i_{corr} (μA/cm^2)	βA (mV/)	βC (mV/)	Rp (mV/)	CR (μm/yr)
Chemical Etched	–317.8	0.14	68.8	–162.0	0.13	3.74
Mechanical Polished	–235.6	0.13	116	–69.6	0.14	3.60
Chemical Etched Looped	–164	0.18	97.4	–75.5	0.10	4.89
Mechanical Polished Looped	–175	0.11	72.8	–56.9	0.12	3.07

Table 4.4. Breakdown potential of the nitinol wires in 0.9% saline

Sample	E_{corr} (mV/)	E_{brk} (mV/)	i_{brk} (μA/cm^2)
Chemical Etched	–317.8	331	0.66
Mechanical Polished	–235.6	248	0.8
Chemical Etched Looped	–164	445	1.2
Mechanical Polished Looped	–175	172	0.81

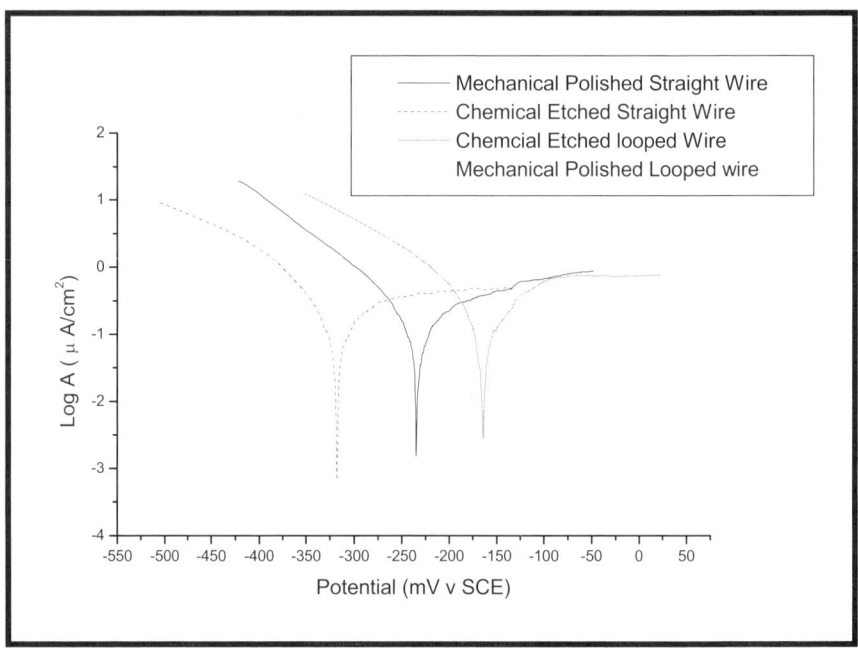

Fig. 4.2. Tafel plot of wires in 0.9% Saline solution

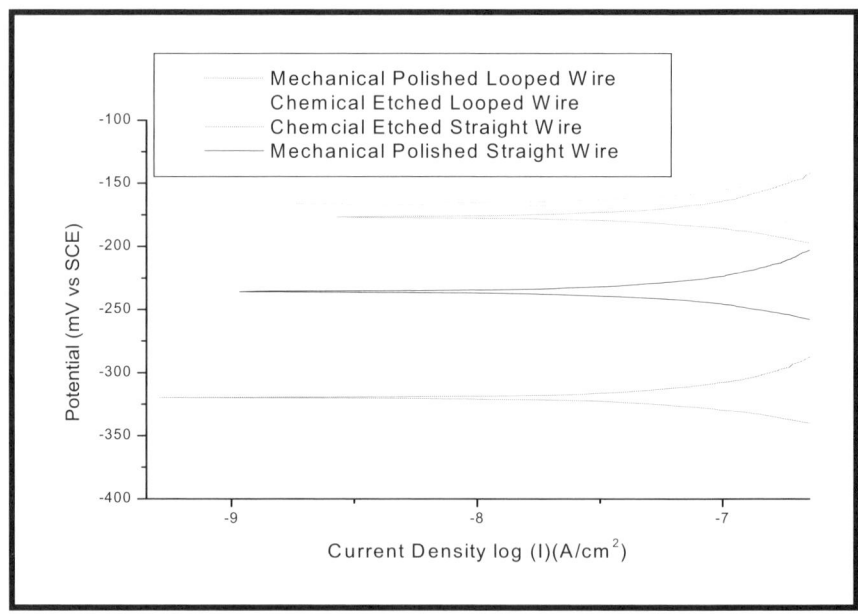

Fig. 4.3. Tafel curve looking at Ecorr and icorr saline solution

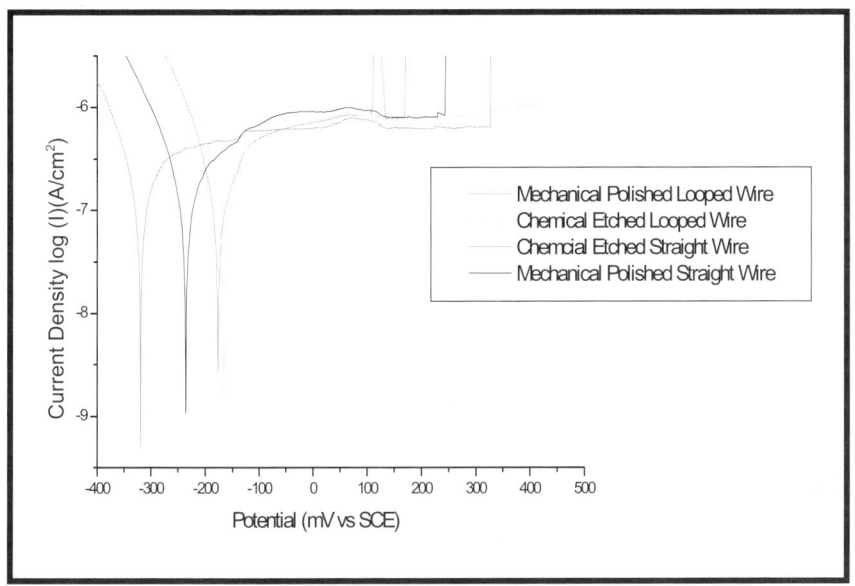

Fig. 4.4. Passivation region of potentiodynamic curve in 0.9% saline

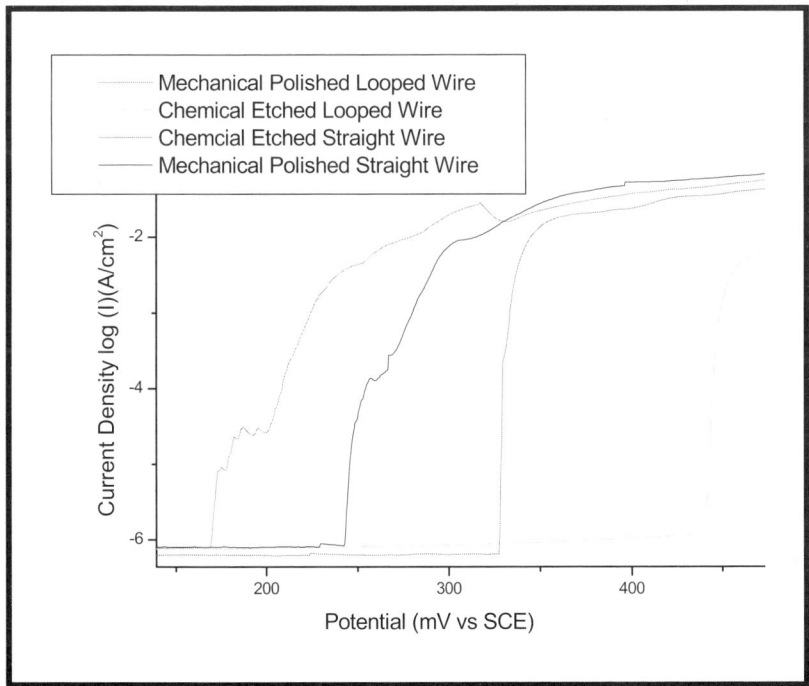

Fig. 4.5. Corrosive region (transpassive region) of the potentiodynamic curve in 0.9% saline

Fig. 4.5 shows the corrosive region of the different wires from the potentiodynamic curve, where there is a sudden increase in the current.

Anodic Polarisation (Localised Pitting Test)

Anodic polarisation behaviour of passive metals examines pitting and crevice corrosion susceptibility. Pitting corrosion occurs when the oxide passive film layer breaks down locally. After local breakdown of the film an anode forms where the film has broken down whilst the unbroken film acts as the cathode. This then accelerates localised attack and pits develop at the unbroken spots. The "pitting potential" corresponds to the potential at which the current starts to increase on the anodic scan. The "repassivation potential" corresponds to the potential at which the current becomes negligible on the reverse (cathodic) scan. A "repassivation potential" close to the "pitting potential" indicates that the sample is capable of re-protecting itself easily after pitting. Figures 4.6–4.7 are graphs of anodic polarisation. From these curves it is possible to determine how prone to pitting a metal is in a particular environment.

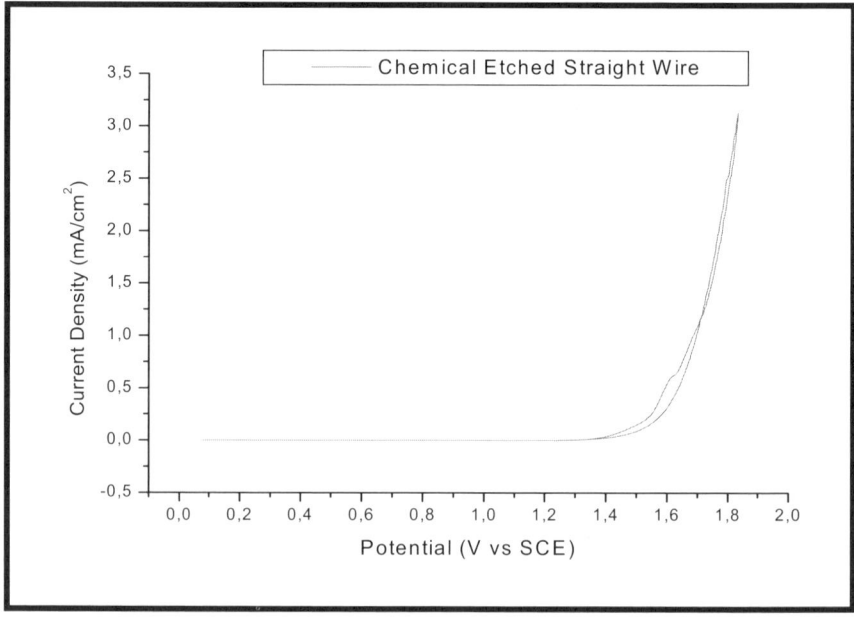

Fig. 4.6. Anodic polarisation of chemical etched straight wire in 0.9% saline

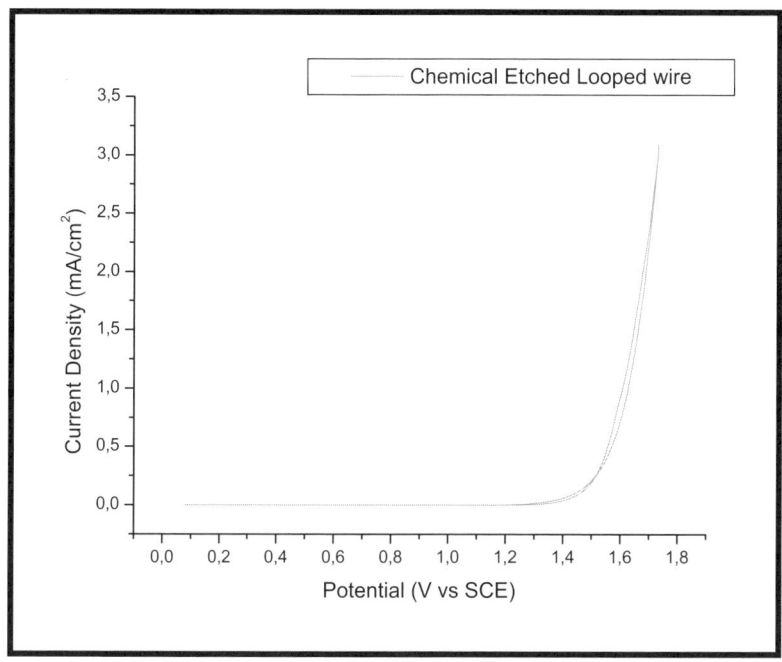

Fig. 4.7. Anodic polarisation of chemically etched looped wire in saline solution

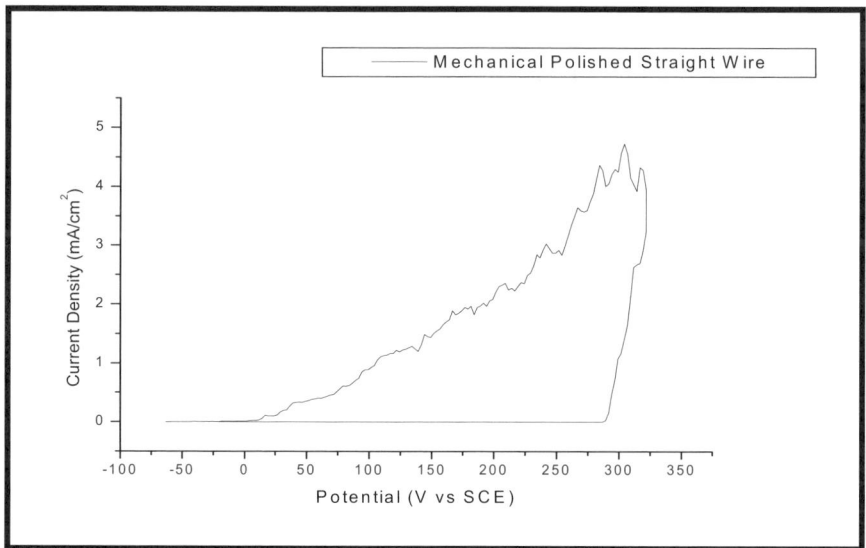

Fig. 4.8. Anodic polarisation of mechanically polished straight wire in 0.9% saline

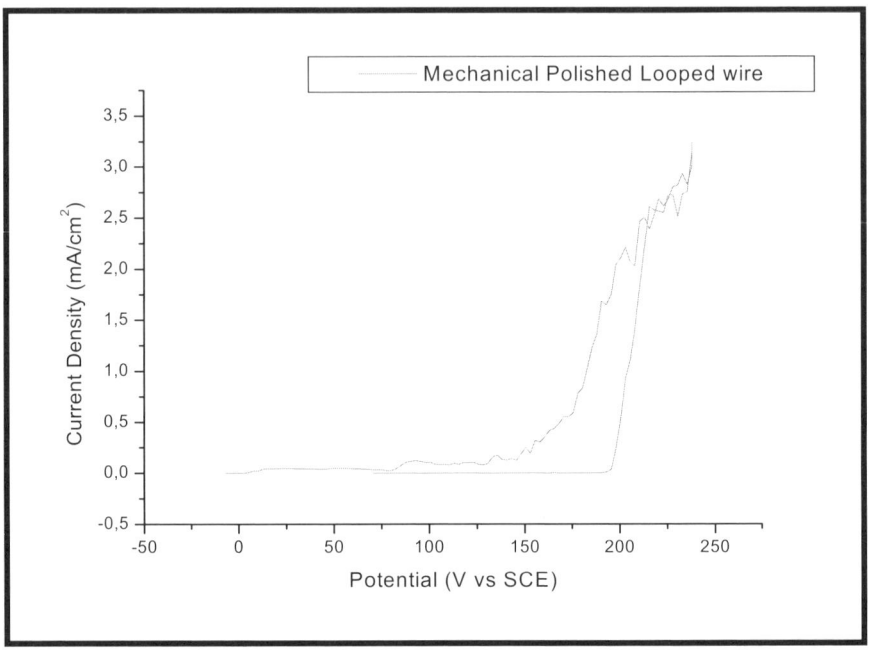

Fig. 4.9. Mechanically polished looped wire in Saline solution

Anodic polarisation curves reveal the pitting corrosion potential of the chemical etched straight wire at 1415mV and the chemical etched looped wire at 1451mV as shown in Figs. 4.6–4.7 whilst the pitting in the mechanically polished straight wire is at 289mV and the mechanical polished looped wire is at 233mV as shown in Figs. 4.8–4.9. The protection potential also known as the repassivation potential is the potential at which the current becomes negligible on the reverse scan. For the mechanical polished wire the repassivation potential is observed at 13.3mV, whilst the repassivation for the chemical etched wire is observed to 1382mV. The mechanical polished surfaces undergo early pitting below 300mV as shown in Figs. 4.8–4.9 whilst the chemical etched surfaces remain stable for a while before pitting takes place above 1400mV as shown in Figs. 4.6–4.7. Results also show pitting potential (Epitt) and repassivation potentials (Erep) are rather close for the chemically etched surface samples. For the mechanically polished surfaces the Epitt and Erep potentials are very far apart. Pitting potential and protection potential calculated from the pitting corrosion curves are shown in Table 4.5.

Table 4.5. Pitting potential and repassivation potential of material in 0.9% saline

Sample	E_{Pitt} (mV)	E_{rep} (mV)	E_{Pitt}- E_{rep} (mV)
Chemical Etched	1415	1382	33
Mechanical Polished	289	13.3	275
Chemical Etched Looped	1451	1451	42
Mechanical Polished Looped	233	200	33

A repassivation potential close to the pitting potential indicates that the sample is capable of re-protecting itself easily after pitting.

Surface Studies

SEM images of chemical etched and mechanical polished wires before and after corrosion in 0.9% saline are as shown in Fig. 4.10 (a-f)

The SEM images in Figs. 4.10b-c show the chemical etched wire after the pitting test. It is evident that no pitting or corrosion took place. The surface of the wire is not as smooth as the as-received wire. There was no significant difference between the looped and the straight geometry. Both pitting and corrosion would seem to have taken place on the surface of the mechanically polished wires in Figs. 4.10e-f. The mechanically polished looped wire in Fig. 4.10f appears to have undergone excessive pitting. In certain areas on the

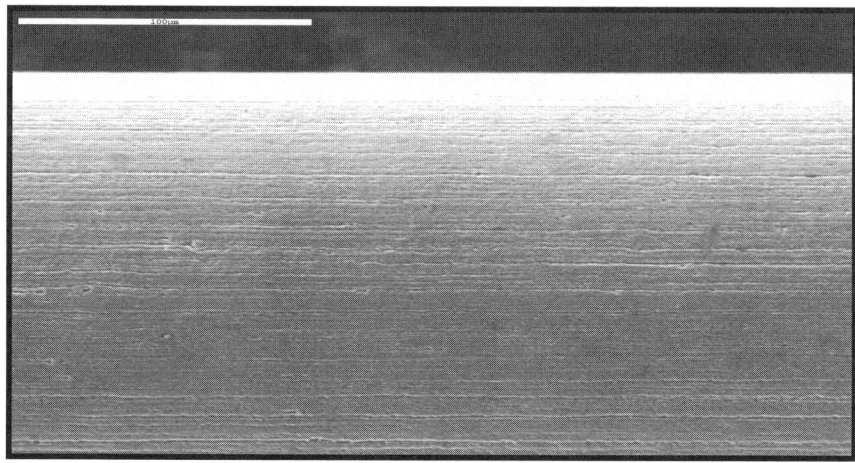

Fig. 4.10a. SEM image of chemical etched surface before corrosion (scale bar 100μm)

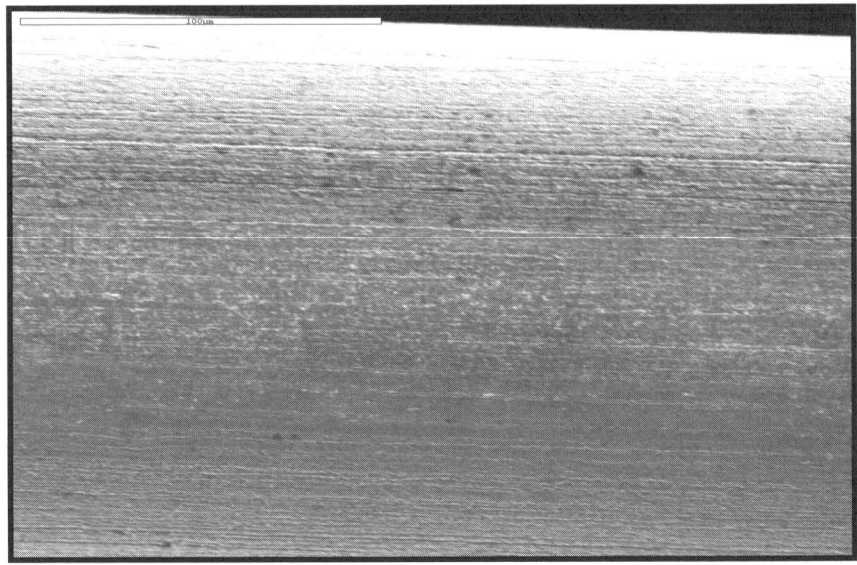

Fig. 4.10b. SEM image of chemically etched straight surface after the potentiodynamic test (Magnification x500)

Fig. 4.10c. SEM image of chemically etched looped surface after potentiodynamic test (Magnification x500)

Fig. 4.10d. SEM image of mechanically polished wire before corrosion (scale bar 100μm)

Fig. 4.10e. SEM image Mechanically Polished Straight wire after potentiodynamic test (+3500mv) (scale bar magnification x500)

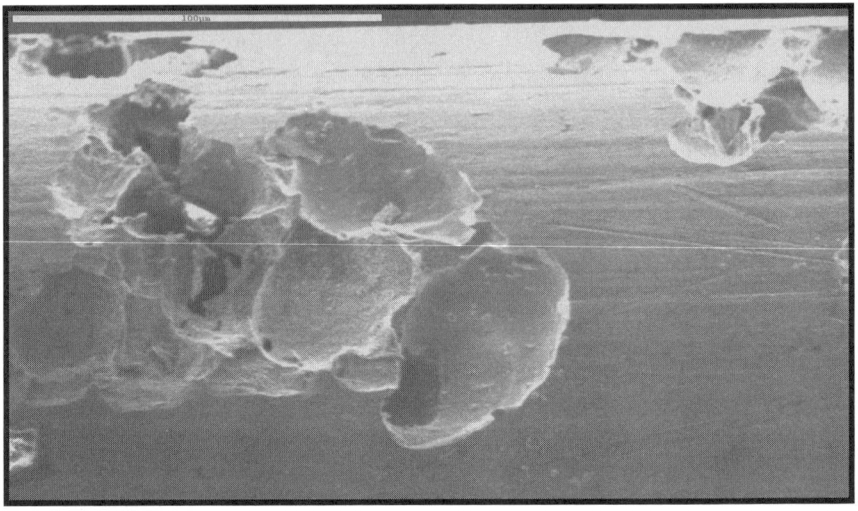

Fig. 4.10f. SEM image Mechanically Polished looped wire after potentiodynamic test (+3500mv) (scale bar magnification x500)

surface the pits have turned into cracks. Pitting is only observed on certain areas of the surface and other areas have been left untouched. The crumbling effect of the oxide layer on the mechanically polished straight wire can be observed in Fig. 4.10e. Also it appears that the oxide layer has been broken into smaller parts so that another layer on the mechanical polished wire can be observed. EDX analysis of the chemical etched wire after potentiodynamic test of +3500mV. Fig. 4.11a showed reduced nickel content of less than 10cps on the surface; whilst nickel content of the mechanical polished wire had remained the same at around 70cps as shown in Fig 4.11b.

4.3 Summary

The effect of a saline corrosive environment on a typical composition of nitinol alloy used as an implant material in the human body has been investigated. The investigation considered the effect of two surface treatments, namely, mechanical polishing and chemical etching using a proprietary technique, on the resistance of nitinol wires to a corrosive saline environment. It is also considered that the effect of

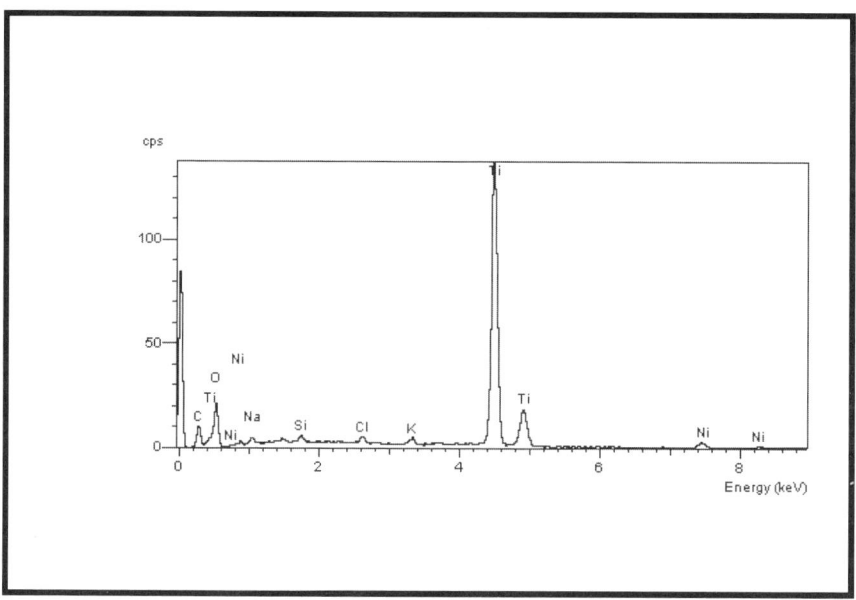

Fig. 4.11a. EDX analysis of chemically etched wire PD +3500mV having very little nickel on its surface

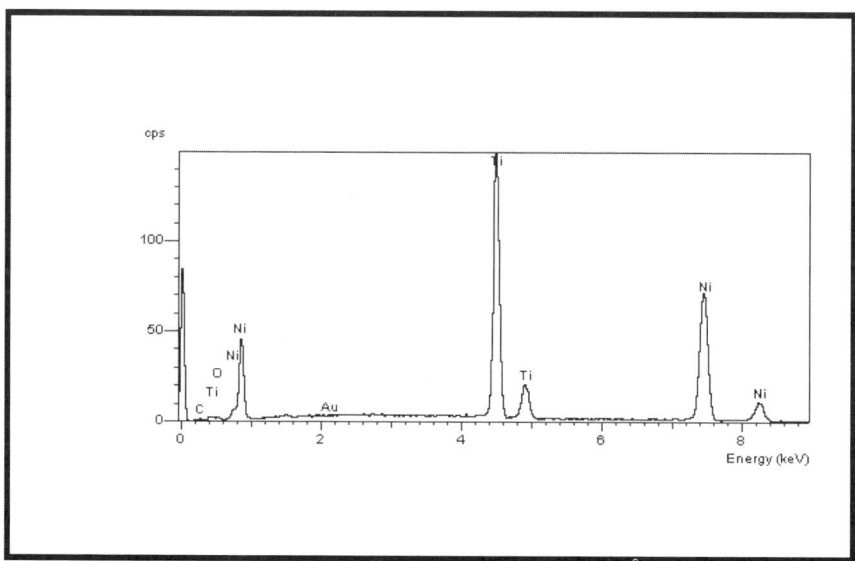

Fig. 4.11b. EDX analysis of mechanically polished wire PD +3500mV

an imposed geometry may change the corrosion that the effect of an imposed geometry may change the corrosion conditions, i.e., whether the wires are exposed to the saline solution as straight or looped wires on their resistance to corrosion in a saline environment. It is observed the most positive open circuit potential (OCP) after 22 hours on the chemically etched wire. Tafel plot results also indicated that the chemically etched looped wire had the largest passive region and the re-passivation and pitting potentials were found to be quite close to each other for the chemically polished wire. There was a more substantial difference between the re-passivation and the pitting potential of the mechanically polished nitinol wires, indicating a higher susceptibility to pitting corrosion. This was confirmed by scanning electron microscopy (SEM) that showed substantial pitting on the mechanically polished wire, whilst the chemically etched wire appears to be almost unaffected by the electrochemical test in saline solution. The proprietary chemical treatment on the nitinol wire seems to be necessary for corrosion protection in applications where nitinol wire will be exposed to a corrosive saline environment.

Acknowledgements

We would like to acknowledge support for this investigation by Mr. Tim Ashton, Vice-president of R&D at Vascutek (Terumo), Scotland.

References

1. Wu S. K., Lin H. C., Recent development of TiNi-based shape memory alloys in Taiwan, Mat. Chem & Phys., 2000: 81-92.
2. Stoeckel. D. P. A., Duerig. T., Self-expanding nitinol stent: material and design considerations, Eur Radiol 2004, **14**, 292-301.
3. Williams D. F., Fundamental aspects of biocompatibility, 1981, CRC Press: Boca Raton, Florida.
4. Cisse O., Savadogo O., Wu M., Yahia L. H., Effect of surface treatment of NiTi alloy on its corrosion behavior in Hanks' solution, Journal of Biomedical Materials Research, 2002, 339-345.
5. O'Brien B., Carroll W. M., Kelly M. J., Passivation of nitinol wire for vascular implants - a demonstration of the benefits, Biomaterials, 2002, 1739-1748.

6. Khan M. A., Williams R. L., Williams D. F., The corrosion behaviour of Ti-6Al-4V, Ti-6Al-7Nb and Ti-13Nb-13Zr in protein solution, Biomaterials 1999, 631-637.
7. Shabalovskaya S. A., Anderegg J., Laab F., Thiel P. A., Rondelli G., Surface conditions of nitinol wires, tubing and as-cast alloys: the effect of chemical etching, aging in boiling water and heat treatment., J. Biomed. Mat. Res., 2003, 339-345.
8. Ademosu O., Ogwu A. A., McLean J., Corrigan M., Placido F., Electro-chemical corrosion of chemically etched and mechanically polished nitinol wires in Hanks' and saline solution, International Conference on Surfaces, Coatings and Nanostructured Materials, nanoSMat2005, 2005a. Aveiro, Portugal.
9. Ademosu O., Ogwu A. A., Placido F., McLean J., Ashton T., Surface char-acterisation of modified biocompatible nitinol alloy wires exposed to saline and Hanks' physiological solution, Institute of Physics (UK) conference on Novel Applications of surface modification. 2005b., Chester College, UK.
10. Ernest B. Yeager, Center for Electrochemical Sciences (YCES) and the Chemical Engineering Department, Case Western Reserve University, Cleveland, Ohio. http://electrochem.cwru.edu/ed/encycl/art-c02-corrosion.htm#passiv. 2004.

5. Cardiovascular Interventional and Implantable Devices

5.1 Introduction

Cardiovascular interventional and implantable devices must be safe and efficacious, as well as biocompatible. Surface treatment is of importance to the design and function of these devices. Lubricity, wear resistance, thrombogenicity, inflammation, and infections can all be affected significantly by surface treatments.

The surfaces of cardiovascular interventional and implantable devices can either be modified with active or passive coating. Devices with active coating such as drug eluting stents (DES) deliver therapeutic agents that can enhance the mechanical function and modulate long-term vascular responses. In some implantables such as vascular grafts, endothelial cell growth is desirable. This is achievable with the addition of a coating or a modification to the surface properties of the device [1].

This chapter reviews some of the most commonly used cardiovascular interventional and implantable devices with an overview of the role that surface treatments have in their functionality and safety.

5.2 Cardiovascular Interventional Tools

Cardiovascular interventional tools are used in the treatments of coronary arterial diseases, heart failure, and peripheral arterial diseases. These devices are not intended for long-term implantation. There are two main interventional procedures that are performed today: percutaneous transluminal coronary angioplasty (PTCA) and stenting. Both are performed via catheterizations.

The earliest known cardiac catheterization was performed by Hales in 1711. More recently, Gruntzing and Myler performed the first human coronary angioplasty in May 1977 [3]. Currently, 1.2 million Americans undergo cardiac catheterization and over one-half million receive a percutaneous coronary intervention such as balloon angioplasty, atherectomy, or stent implantation annually [4].

Percutaneous transluminal coronary angioplasty (PTCA), commonly known as balloon angioplasty, is performed by threading a balloon catheter through the femoral artery, located in the groin, to a trouble spot in an artery of the heart (Fig. 5.1). The balloon is then inflated, compressing the plaque and dilating (widening) the narrowed coronary artery so that blood can flow more easily. This is often accompanied by inserting an expandable metal stent. Stents are wire mesh tubes used to prop open arteries after PTCA.

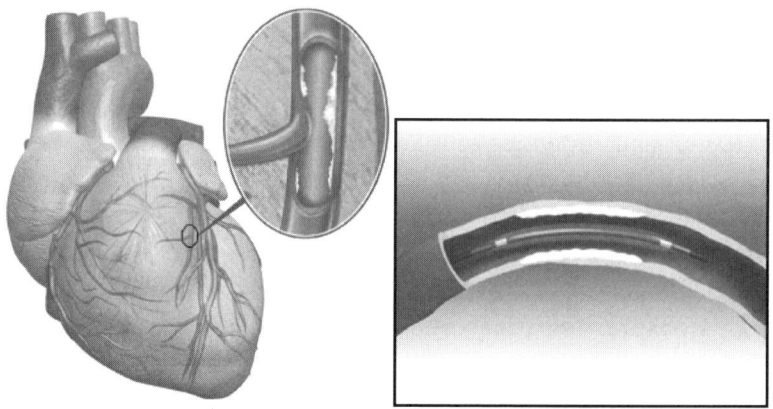

Fig. 5.1. Schematic of Percutaneous Transluminal Coronary Angioplasty (Courtesy: Medtronic. Inc)

Cardiovascular interventional devices commonly used include the following:

Guiding catheter - a long, thin, flexible tube that acts as a conduit for injection of contrast. A guiding catheter delivers other devices, i.e., a balloon catheter or a stent delivery catheter into the coronary vasculature. A *guiding catheter* (Fig. 5.2) often contains metallic braiding to help provide push-ability and torque-ability [5].

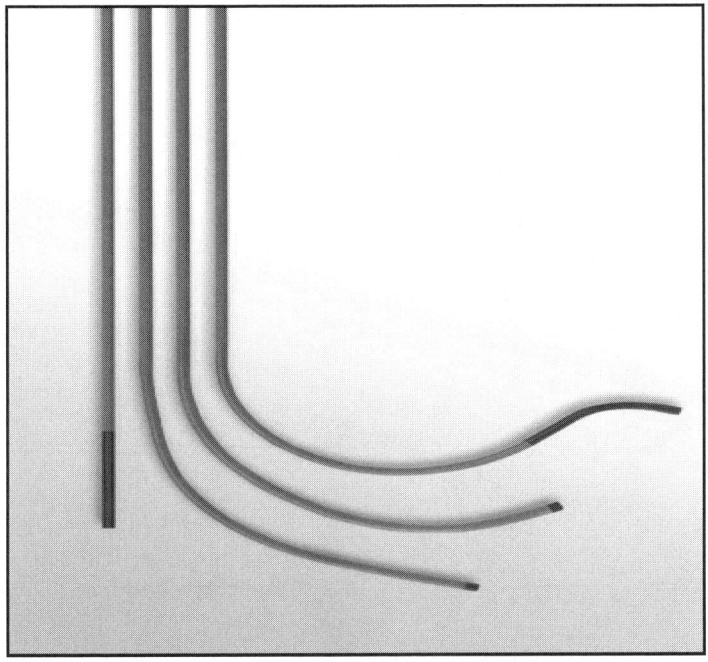

Fig. 5.2. Attain® Fixed Shape Guide Catheters (Courtesy: Medtronic Inc.)

Guide wire - a long, thin, flexible wire that acts as a railing for delivery of a balloon catheter and stent delivery system. It is typically made from stainless steel or nickel titanium (Fig. 5.3).

Track-ability (the ease of tracking a balloon over the guide wire up to the target lesion) and push-ability (the ability to advance the balloon across the lesion) are more important to the practice of interventional cardiology than any other in vitro measure. However, there are no reliable in vitro methods for measuring these two properties. A variety of lubricious coatings exist for the *guidewires*, which reduce friction, facilitate advancement of interventional devices, and enhance lesion crossing. Examples include hydrophilic coatings, silicone (hydrophobic) and PTFE coating [6].

Hydrophilic coating must be hydrated to be effective. However, this might not be appropriate for all devices. Hydrophobic coatings, such as silicone, when applied to the surface, are not as lubricious as hydrophilic coatings but provide more tactile feedback as the physician manipulates the devices.

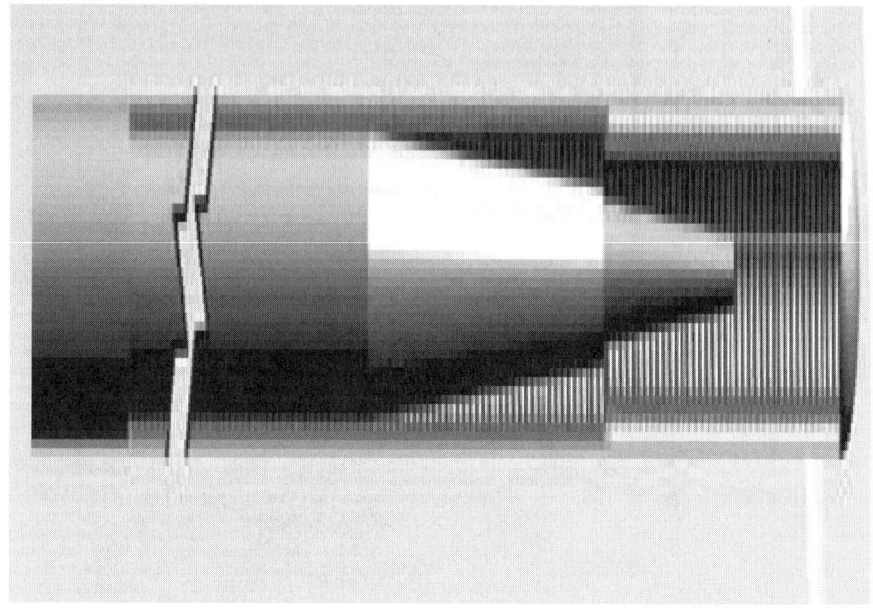

Fig. 5.3. Schematic of Guidewire Crossection (Courtesy: Medtronic Inc.)

5.3 Key Surface Properties for Cardiovascular Interventional Devices

Friction is the primary metric of concern. However where energy is transmitted from the surface of the devices, for example during abla-tion, the insulation material property also becomes a concern. Friction generally dissipates energy in the form of heat. This deficit must be overcome to maintain two objects moving at a constant velocity with respect to one another. The process of wear also accompanies the presence of frictional forces between two surfaces.

Magnitude of friction and wear are generally regulated by modifi-cation of the interface between the two surfaces. The various surface treatment methods, solid, liquid, and gas, all provide lubrication. Not only is the coefficient of friction a measurable of importance but also provides insights pertaining to the mechanism of lubrication. The prin-ciple lubrication mechanisms are hydrodynamic, elastohydrodynamic, mixed and boundary [7].

In cardiovascular applications, there is usually not sufficient film thickness between two surfaces to provide hydrodynamic lubrication, as the typical distance between the two surfaces is not sufficient. In hydrodynamic lubrication the surfaces are at greater distances from one another. Boundary lubrication represents the opposite situation where the distance between the two surfaces is minimal.

The lubrication mechanism is largely a function of surface property modification. An effective boundary lubricant usually interacts with the surface of the interventional or implantable device. Hydrophilic coatings usually provide effective boundary lubrication of cardiovascular interventional and implantable devices. Elastohydrodynamic lubrication provides a film thickness that is slightly less than that found in hydrodynamic lubrication. In this scenario, pressure waves can cause a deformation in the tissue surface opposite of the interventional or implantable device.

As the distance between the surfaces decreases from elastohydrodynamic lubrication, mixed lubrication occurs. Mixed lubrication has a higher coefficient of friction and wear rate than hydrodynamic lubrication. As a result of the dimensional constraints, most cardiovascular interventional and implantable devices provide lubrication via boundary and mixed lubrication.

The surface properties are modified in an attempt to minimize coefficient of friction. The two primary methods to accomplish this are hydrophobic and hydrophilic coating. Hydrophilic coating is activated by an aqueous environment; where a layer of slippery hydrogel is created on the device surface, reducing the friction force [8-10]. Hydrophilic coating might not be appropriate for all devices such as those that are not constantly hydrated [11]. Hydrophobic coatings such as silicone can also be applied to the surface but it is not as lubricious as a hydrophilic coating. It does, however, provide more tactile feedback to the operating physician.

5.4 Cardiovascular Implantable Devices

Unlike interventional tools, cardiovascular implantable devices remain in the cardiovascular system usually until the device must be removed as a result of functional failure. These failures can be classified as either mechanical or surface property failure. The heart, along with the arterial and venous system, can be defined as an electromechanical system. Certain implantable devices focus on treatment of electrical aspects while others focus on mechanical aspects.

5.5 Electrical Implantable Devices

The rhythm management of the heart is regulated by two different types of myocardial cells; pacemaker cells and conducting cells. Pacemaker cells, located in the sinoatrial (SA) node and atrioventricular (AV) node are respectively located where the superior vena cava and right atrium join in the inferior portion of the interatrial septum [4]. Conducting cells, referred to as the Bundles of His and Purkinje Fibers, serve the function of relaying action potentials from the SA node to the AV node and to all parts of the ventricle.

Devices used by clinicians in the regulation of the rhythm management of the heart include: implantable pulse generators (pacemakers), implantable cardioverter defibrillators (ICD), pacemaker and ICD leads, and implantable loop recorders. These devices are generally implanted in the muscle cavity of the pectoral region (Figs. 5.4–5.8).

Pacemakers deliver electrical impulses to the heart via a special conductor called a pacemaker lead (Fig. 5.4 and Fig. 5.5). One end of the lead is connected to the pacemaker and the other end of the lead contains a metallic electrode, which is in contact with the heart tissue, stimulating the heart to contract. The lead also carries information from the heart back to the pulse generator, which the physician accesses via a special external programmer [12,13].

The pacemaker case is made of titanium, a metal that is 10 times as strong as steel, but much lighter. Titanium and its alloy are biocompatible. The titanium casing was developed to enclose the battery and circuitry. Epoxy resin with silicone rubber encased the inner components in previous designs.

Titanium casings with special filters help shield the components and greatly reduce outside electromagnetic interference. Patients with these newly designed pacemakers can now safely use microwave ovens and other appliances and equipment found in the home and office.

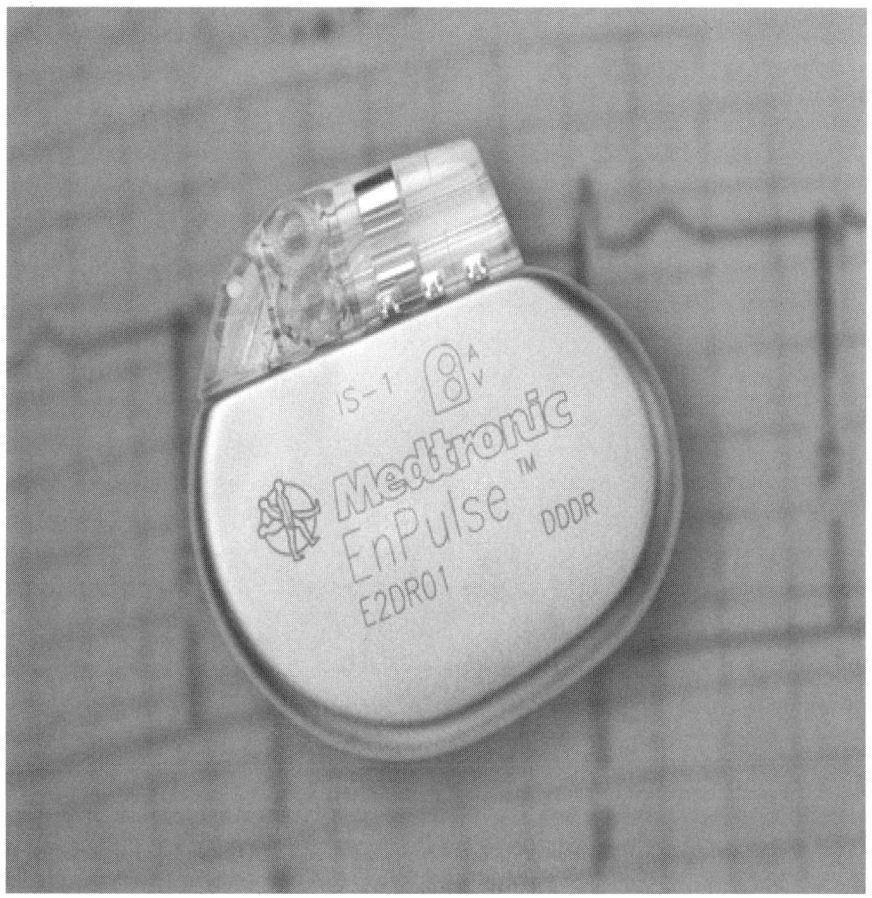

Fig. 5.4. Pacemaker – EnPulse. (Courtesy of Medtronic, Inc.)

A nitride coating can be deposited via sputtering technique onto the surface of the pacemaker metallic electrode on the pacemaker leads to improve the sensing and pacing performances over an electrode with a smooth, unsputtered surface [14].

When the heart beats in asynchronous rhythms, the ICD delivers an electric shock that can "reset" the heart back to its normal rhythm. The standard material construction of the casing is the same as that of the pulse generator. Another implantable electrical device called an implantable loop recorder provides the function of recording an electrocardiogram (ECG) before and after the onset of symptoms (Figure 5.8).

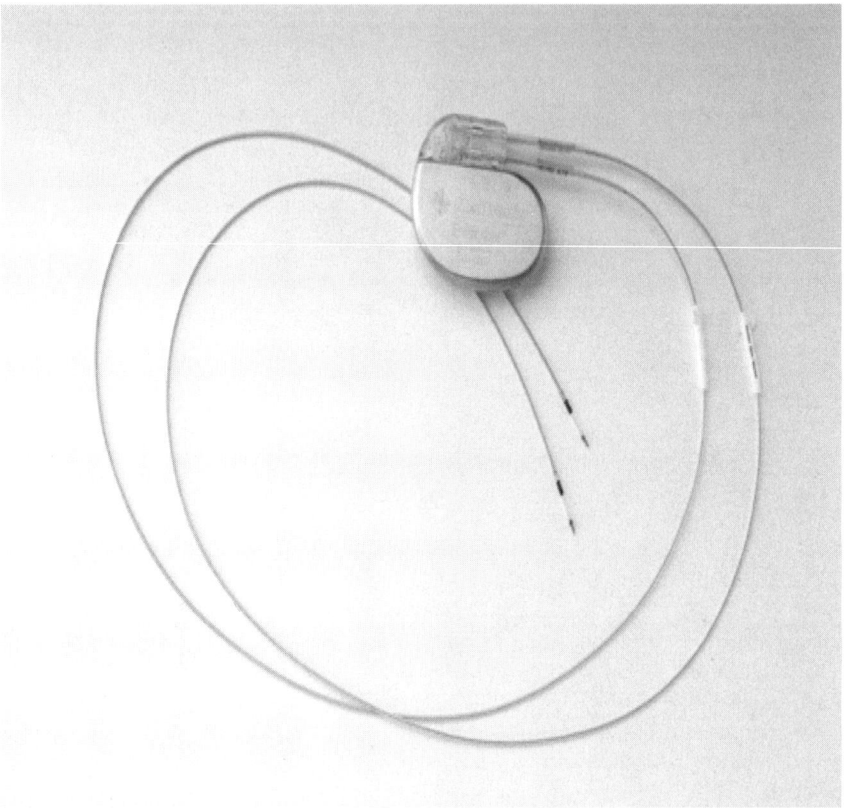

Fig. 5.5. Medtronic EnRhythm pacemaker with leads. (Courtesy of Medtronic, Inc.)

Fig. 5.6. Implantable Cardioverter Defibrillator (Courtesy: Medtronic)

Fig. 5.7. External programmer (Courtesy: Medtronic)

5.6 Mechanical Implantables

The heart, as a mechanical system, acts as a positive displacement pump to circulate blood through the cardiovascular system. Devices used by clinicians in the treatment of cardiovascular disease include: stents, vascular grafts and stent grafts, and heart valves.

A stent is a wire mesh tube used to prop open an artery during angioplasty (Figure 5.9). The stent is collapsed to a small diameter and put over a balloon catheter. It is then moved into the area of the blockage. When the balloon is inflated, the stent expands, locks in place and forms a scaffold. This holds the artery open. The stent stays in the artery permanently, holds it open, improves blood flow to the heart muscle and relieves symptoms (i.e., chest pain) [15].

Stents have virtually eliminated many of the complications that used to accompany "plain old balloon angioplasty" (POBA) such as abrupt and unpredictable closure of the vessel, which can result in emergency bypass surgery. The additional structural strength of the stent can also help keep the artery open while the healing process progresses.

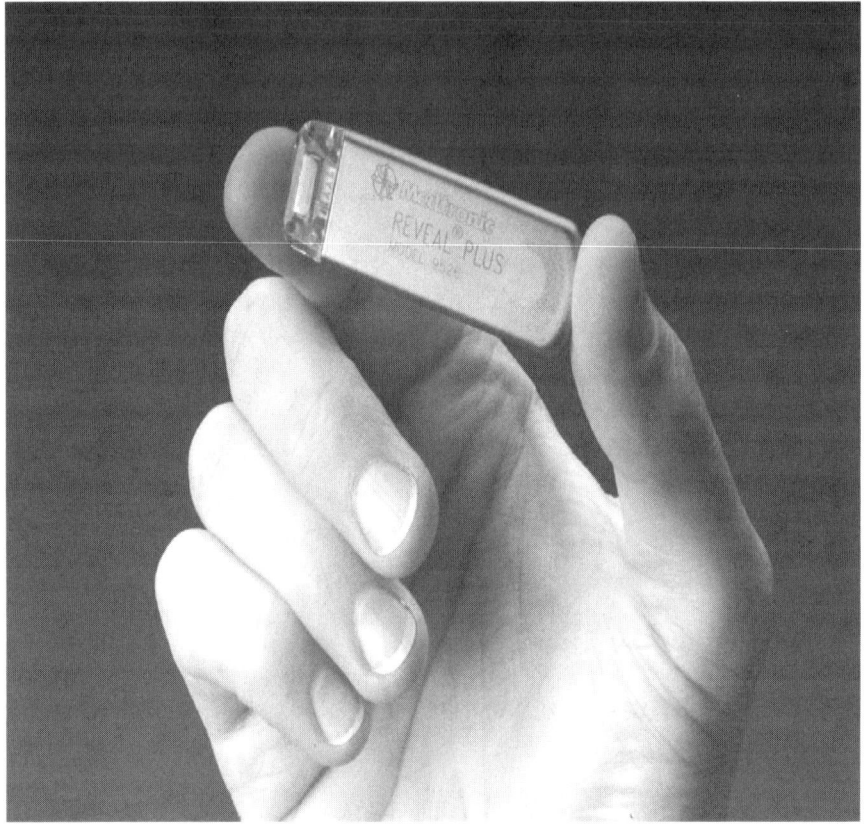

Fig. 5.8. Implantable Loop Recorder (Courtesy: Medtronic)

Vascular grafts are classified into two major categories. Large diameter grafts (>10 mm) primarily made up of Dacron (80%) and PTFE (20%) are primarily used for aortic and iliac artery reconstruction. The small caliber grafts (<10 mm) are primarily used for coronary artery bypass grafts (CABG), lower-extremity bypass procedures, and hemodialysis access. The patency rates for the synthetic graft material are better for the aortic and iliac artery reconstruction than for when small caliber grafts are used. Saphenous vein grafts and internal mammary grafts both have greater patency rates than the small caliber vascular grafts with the internal mammary artery providing the superior patency rates.

Dacron (polyethylene terphthlate) and PTFE (polytetrafluoroethylene) maintain their tensile strength for years after being implanted.

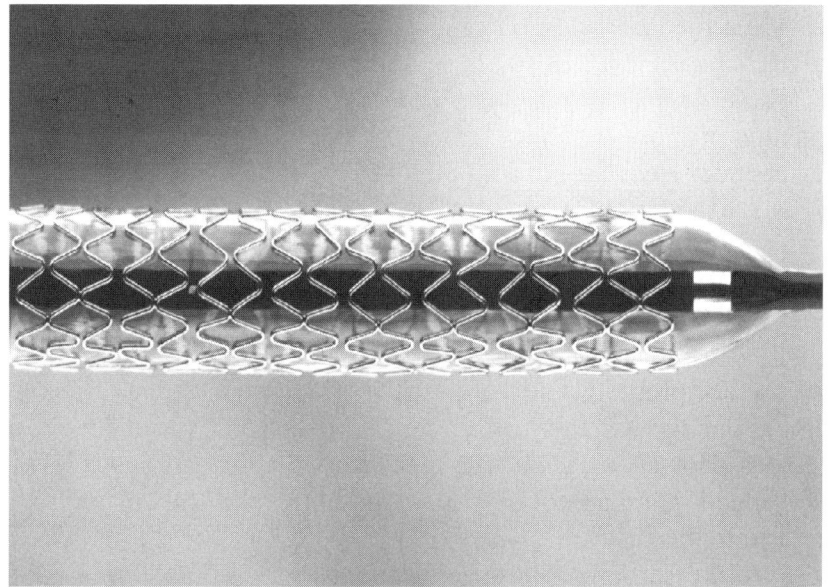

Fig. 5.9. Endeavor Drug Eluting Stent (Courtesy: Medtronic Inc.)

However other materials such as Nylon (polyamide), Ivalon, and Orlon decrease in tensile strength after months of being implanted. Woven Dacron grafts are nonporous with no stretch while knitted Dacron grafts have variable stretch and are porous. As previously mentioned, the patency rates for the use of these grafts is better for large vessel grafts [16].

Neointimal hyperplasia, which occurs in the anastomotic area, is the prime reason of failure of small caliber grafts. The low compliance rates leading to compliance mismatch plays a role as well. The neointimal hyperplasia could be caused by platelet deposition with local release of platelet-derived growth factor (PDGFP), other growth factor stimulation of smooth muscle cell (SMC) proliferation, monocyte recruitment, complement activation, leukocyte deposition, chronic inflammation, and mechanical stimuli such as stress and shear abnormalities [16]. Additionally, the optimal sizing of a graft is important to provide a wall shear rate that will increase the long-term patency of the synthetic grafts [17].

Other devices within this classification include stent grafts and heart valves. Stent grafts combine the mechanical scaffolding function of stents and conduit function of vascular grafts. Heart valves provide a mechanical replacement for failing bicuspid and tricuspid valves.

These devices primary function is to serve either a structural or dynamic primary function. However, the surface treatment of these devices does have a significant effect on the efficacy of these devices.

5.7 Important Surface Properties for Implantable Cardiovascular Devices

The surface properties of materials are modified to reduce adherence to vascular tissue. This can be accomplished via various treatment methods. However, the surface modification is not always solely for mechanical purposes. The surface modification can also provide therapeutic benefit.

Pyrolitic carbon is used in cardiovascular applications in artificial hearts and prosthetic heart valves. The material has similarities to graphite with covalent bonding present between the graphene sheets. The risk of thrombosis is reduced when using pyrolytic carbon, as blood clots are less likely to form when it is used. However, as a precaution, the patient who receives a prosthetic heart valve must be on an anti-coagulant regimen.

In addition to prosthetic heart valves, bioprosthetic heart valves, made from porcine valves or bovine pericardium, can be used to replace the natural valves. The tissue of the valves can be treated to prevent calcification via special processing [18].

The Mosaic® aortic and mitral bioprosthesis, manufactured by Medtronic, uses the AOA method, which has been observed to reduce the buildup of calcium in animal studies. The only stented valve in the U.S. that incorporates this tissue treatment is the Mosaic® bioprosthesis. One likely method of the AOA process function is via the binding of free aldehyde groups of glutaraldehyde to the amino group of AOA molecules. This process, referred to as *capping*, is believed to inhibit the mineralization of tissues. Another viable mechanism by which AOA treatment minimizes calcification is slowing down the diffusion of calcium ions though the treated tissues. (Fig. 5.10)

The surfaces of stents have evolved to include the delivery of drugs that have proven successful in the reduction of restenosis rates.

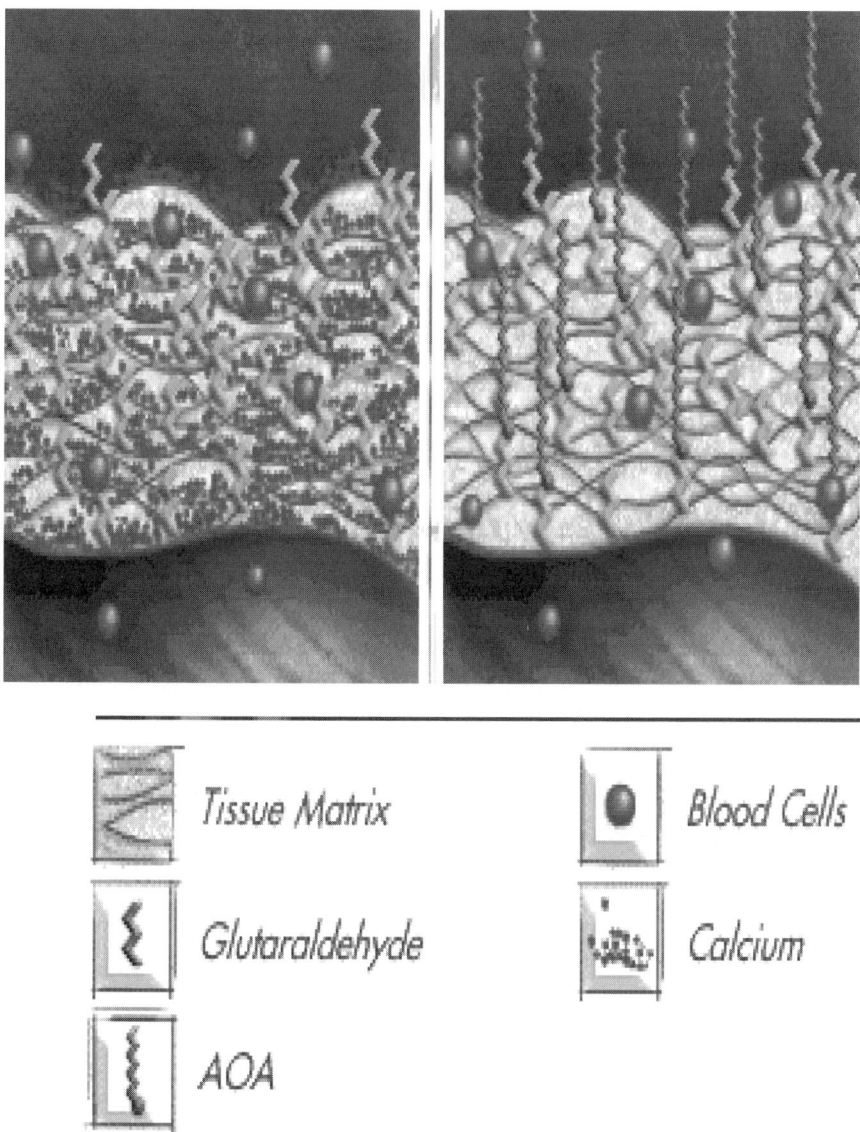

Fig. 5.10. AOA Treatment Process (Courtesy: Medtronic)

Restenosis rates have been observed to reduce significantly when the stent surface is coated such that siromilus and paclitaxel are eluted [19]. Biodegradable materials using the mechanism of autologous cell seeding has shown potential. However, the pretreatment of materials increases the opportunity for infection. A biodegradable graft material made up of polylactic-co-glycolic acid as a biodegradable scaffold material compounded with a collagen microsponge has shown positive results. No thrombus formation was observed, while the polylactic-co-glyclolic acid scaffold was almost completely absorbed over a 6-month period [20]. Polyethylene oxide (PEO) has also been used to modify biomaterial surfaces to minimize or prevent protein absorption and cell adhesion.

Minimization of thrombosis is a key to the reduction or restenosis where stents are implanted. To achieve this goal, usually the mechanisms focused on are minimizing coagulation, platelet adhesion, and platelet activation. However, the role of complement activation prior to platelet activation and the role of leukocytes could play a significant role in the minimization of thrombosis as well [21].

Surface oxide formation has also been observed to be a significant factor in determining the degree of thrombosis. However, amorphous oxide film may provide a solution to the thrombosis caused by metal [22].

Several other surface modifications of the stents to reduce restenosis and thrombogenicity by improved biocompatibility have also been evaluated with varying degrees of success: metallic coatings: gold, titanium, copper; phosphorylcholine (PC), a synthetic mimetic outer wall of red blood cells; heparin; silicon carbide; and carbon and fluorinated diamond like carbon (F-DLC) [23-25].

References

1. Blake, D. M., Manes, P. C., (1999), Separation and Purification Methods 28 (1): 1-50.
2. Byrne, J.A., Eggins, B.R., Brown, N.M.D., McKinney, B., Rouse, M., (1998) Applied Catalysis B: Environmental, 1998, 17, 25-36.
3. Byrne, J.A., Davidson, A., Dunlop, P.S.M., Eggins, B.R., (2002) Journal of Photochemistry and Photobiology A: Chemistry, 148, 365-374.
4. Byrne, J.A., Hamilton, J.W.J., McMurray, T.A., Dunlop, P.S.M., Jackson, V., Donaldson, A., Rankin, J., Dale, G., Alrousan, D., (2006) Abstracts of the NSTI conference, Boston.
5. Coleman, H.M., Routledge, E.J., Sumpter, J.P., Eggins, B.R., Byrne, J.A., (2004) Water Research, 38, 3233-3240.

6. Cosnier, S., Gondran, C., Senillou, A., Gratzel, M., Vlachopoulos, N., (1997) Electroanalysis, 9(18), 1387-1392.
7. Dunlop, P.S.M., Byrne, J.A., Manga, N., Eggins, B.R., (2002) Journal of Photochemistry and Photobiology A: Chemistry, 148, pp 355-363.
8. Fujishima, A., Honda, K., (1972) Nature, 238, 37-38.
9. Fujishima, A., Rao, T.N., Tryk, D.A., (2000) Journal of Photochemistry and Photobiology C: Photochemistry Reviews, 1, 1-21.
10. Giavaresi, G., Ambrosio, L., Battiston, G.A., Casellato, U., Gerbasi, R., Finia, M., Aldini, N.N., Martini, L., Rimondini, L., Giardino, R., (2004) Biomaterials, 25, 5583-5591.
11. Hoffman, M.R., Martin, S.T., Choi, W., Bahnemann, D.W., (1995) Chem. Rev., 95, 69-96.
12. Holgers, K.M., Ljungh, A., (1999) Biomaterials, 20, 1319-1326.
13. Huang, N., Yang, P., Leng, Y.X., Chen, J.Y., Sun, H., Wang, J., Wang, G.J., Ding, P.D., Xi, T.F., Leng, Y., (2003) Biomaterials, 24, 2177-2187.
14. Kuhn, K.P., Chaberny, I.F., Massholder, K., Stickler, M., Benz, V.W., Sonntag, H.-G., Erdinger, L., (2003) Chemosphere, 53, 71-77.
15. Lee, S.-H., Kim, H.-W., Lee, E.-J, Li, L.-H., Kim, H.-E., (2006) Journal of Biomaterials Applications, 20, 195-208.
16. Liu, X., Zhao, X., Ding, C., Chu, P.K., (2006) Applied Physics Letters, 88, 13905.
17. McMurray, T.A., Byrne, J.A., Dunlop, P.S.M., Winkelman, J.G.M., Eggins, B.R., McAdams, E.T., (2004) Applied Catalysis A: General, 262, 1, 105-110.
18. McMurray, T.A., Byrne, J.A., Dunlop, P.S.M., McAdams, E.T., (2005) Journal of Applied Electrochemistry, 35, 723-731.
19. Mills, A., Le Hunte, S., (1997) Journal of Photochemistry and Photobiology A: Chemistry, 108, 1-35.
20. Nygren, H., Tengvall, P., Lundstrom, I., (1997) Journal of Biomedical Materials Research, 34, 487-492.
21. Ohko, Y., Utsumi, Y., Niwa, C., Tatsuma, T., Kobayakawa, K., Satoh, Y., Kubota, Y., Fujishima, A., (2001) Journal of Biomedical Materials Research (Applied Biomaterials) 58, 97-101.
22. Pan, J., Leygraf, C., Thierry, D., Ektessabi, A.M., (1997) Journal of Biomedical Materials Research, 35, 309-318.
23. Ramires, P.A., Romito, A., Cosentino, F., Milella, E., (2001) Biomaterials, 22, 1467-1474.
24. Shani Sekler, M., Levi, Y., Polyak, B., Dunlop, P.S.M., Byrne, J.A., Marks, R.S., (2004) Journal of Applied Toxicology, 24, 395-400.
25. Yang, Y., Glover, R., Ong, J.L., (2003) Colliods and Surfaces B: Biointerfaces, 30, 291-297.

6. Surface Engineering Artificial Heart Valves to Improve Quality of Life and Lifetime using Modified Diamond-like Coatings

6.1 Introduction

There are two types of artificial heart valves, namely, (i) biological valves and (ii) mechanical valves. Biological heart valves are made from tissue taken from animals or human cadavers. They are treated with preservatives and sterilized for human implantation. On the other hand, mechanical heart valves are made of man-made materials. The advantage of mechanical valves over biological valves is that they normally last for a comparatively longer lifetime. The biological valves exhibit a shorter lifetime and tend to wear out with time in service. This chapter discusses mechanical heart valves and highlights the underlying problems faced with biomaterials used in the manufacture of such valves. The history relating to mechanical heart valves will be reviewed, which dates back from 1952. We describe the principal cause of concern facing modern biomaterials used in the manufacture of heart valve components, which is thrombus formation (thrombosis). The hemocompatibility of biomaterials used in human implants will be discussed. Further, we describe the endothelium layer and discuss endothelial cell seeding as a tool for developing a heart valve that overcomes the common problems faced by modern valve materials. Diamond-like carbon (DLC) and its hemocompatibility and/or biological properties have been discussed and reviewed. Finally, we present results of endothelial cell seeding on chromium and silicon modified DLC films. A range of traditional and sophisticated techniques have been used to characterize the physical and biological properties of DLC and modified DLC films for applications in mechanical heart valves.

Heart disease is one of the most common causes of death in the world today, particularly in the western countries. There are various causes of heart disease, related most commonly to diet and exercise. The failure of heart valves accounts for about 25–30% of heart problems that occur today. Faulty heart valves need to be replaced by artificial ones using sophisticated and sometimes risky surgery. However, once a heart valve has been replaced with an artificial one there should be no need to replace it again and it should last at least as long as the life of the patient. Therefore, any technique that can increase the operating life of heart valves is highly desirable and valuable. Currently, pyrolytic-carbon (PyC) is used for the manufacture of mechanical heart valves. PyC belongs to the family of "turbostratic carbons", which have a similar structure to graphite.

Graphite consists of carbon atoms that are covalently bonded in hexagonal arrays. These arrays are stacked and held together by weak interlayer binding. PyC differs from graphite in that the layers are disordered, thus resulting in wrinkles or distortions within layers. This feature gives PyC improved durability compared to graphite. Although, PyC is widely used for heart valve purposes, it is not the ideal material. In its processed form, PyC is a ceramic-like material and like ceramics, it is subject to brittleness. Therefore, if a crack appears, the material, like glass, has very little resistance to the growth/propagation of the crack and may fail under loads. In addition, its blood compatibility is not ideal for prolonged clinical use. As a result, thrombosis often occurs in patients who must continue to take anti-coagulation drugs on a regular basis [1]. The anti-coagulation therapy can give rise to some serious side effects, such as birth defects. It is therefore extremely urgent that new materials, which have better surface characteristics, blood compatibility, improved wear properties, better availability and higher resistance towards breaking are developed.

6.2 History of Mechanical Heart Valves

Charles Hufnagel [2] was the first person, in 1952, to implant an artificial heart valve, in the form of Lucite tube and methacrylate ball, in the descending aorta of the heart with clamps that facilitated rapid insertion of the valve without arresting the heart. Subsequently, the methacrylate ball was replaced with a hollow nylon ball coated with silicone rubber that was designed to reduce noise. More than 200 Hufnagel valves were implanted in patients with aortic insufficiency and remarkably some of these valves functioned for up to 30 years without any significant wear [3]. Later on Murray [4] employed a similar technique to insert a human homograph in the descending aorta.

Initially flexible polyurethanes and silicon rubber materials were employed for use in heart valve components. The reason for using such materials was that the valves would not only reside in the exact position anatomically but would effectively show mechanical characteristics similar to that of natural heart valves. Valves made from such materials were implanted in humans, in different ways. For example, in some cases, individual leaflet was made from flexible material and in others the entire valve was constructed from elastic materials. However, these elastic valves could not last longer than approx. 2–6 years due to a number of reasons, such as tearing, material fatigue and calcification [5-9].

In 1960, Dwight Harken [3,10] implanted double cage ball valves in seven patients. Unfortunately, only two of the seven patients survived the surgery. Later the same year, Albert Starr [3,10] was more fortunate than Dwight Harken, as he implanted prosthetic valve(s), comprising of metal cage and silicon ball, in a number of patients. The success rate increased as the number of survivals was six out of eight patients. These devices though proved durable but continued suffering from several problems like high profile, high rates of thromboembolism (blocking of blood vessel by a blood clot dislodged from its site of origin) and stenotic central position of the ball. Soon after in 1962, Vincent Gott and Ron Daggett [11] addressed two of the problems by introducing a low profile reinforced silicon flap valve that played a significant role in the reduction of central obstruction and the resulting valvular stenois in patients. In 1963, Antolio Cruz [12] devised a heart valve with a free-floating disc tilting on the edge of an orifice ring whose excursion was retained by a cage. Although it had good hemBODYnamic qualities, it was as stenotic as the ball valves and remained suffering from thromboembolic problems [10]. Wada-Cutter valve [13] was designed which significantly reduced valvular stenosis and was thus first implanted in 1965. This valve enabled the disc to rotate or tilt about an axis slightly offset from the centre opening. With the Wada-Cutter valve, pivot points were fixed so that the stress and wear of the repeated openings was focused on two pivots on the disc. Later in 1974, its production was halted due to excessive wear of the Teflon occluder [10]. Subsequently, Don Shiley and Viking Bjork [3,10] designed a tilting disk valve known as the Bjork-Shiley valve. This provided the same features, as that of Wada-Cutter valves only with the difference that unlike it, the disk of the Bjork-Shiley valve was not fixed at pivots points [14]. In the original Starr-Edwards, Wada, Shiley prosthetic valves designs, instead of using the metal occluders, polymeric occluders were employed because of their lower density made for more light and more responsive moving parts. Owing to some key characteristics of plastics, plastic occluders were employed in heart valves. The elimination of strain flexure, considered essential for the elastic valves, increased the life of the more rigid plastic parts. But eventually all the polymeric occluders suffered from degenerative failure. Silicone balls would absorb lipids from the blood stream and crack if not cured appropriately [15]; Tefflon was subject to failure through cold flow from repeated stress in a focused location [16]; and Derlin would warp/distort in typical steam sterilization cycles as Derlin absorbs water and changes the configuration of the disc [3,17].

Gott [3,10] not just designed the first low profile heart valve but he contributed significantly in the field of biomaterials by performing a series of detailed investigations. Gott's experiments were performed on more than 200 dogs in which he implanted small tubes of different materials in vena cava in order to address thromboembolism through improvement in material selection [18]. The tubes made from polymer and stainless steel materials were used in the construction of heart valves in 1962 and it was found that they uniformly occlude within two hours. Later on it was discovered that an application of a coating made from graphite mixed with Heparin bonded with benzalkonium chloride removed clot formation in five animals for a period of up to two weeks. Lillehei and Nakib Ahmad [10] devised a non-tilting disk valve in 1967 that was entirely made of machined titanium, including the poppet and was called Lillehei-Nakib Toroidal Valve. A special division within a nuclear research company called

General Atomics was formed after Gott's publication on the bio-compatibility of carbon and working here Dr Jack Bokros [3] developed a new carbon material called Pyrolite (also known as pyrolytic carbon). Eventually, in 1969, after more than six years of development, plastic and metal materials used for heart valves started to become replaced by Pyrolite. Michael DeBakey [19] implanted a valve that was very similar to Starr aortic valve with a hollow pyrolite ball as occluder. Taking advantage of the improved biocompatibility and durability of pyrolytic carbon, translating disk valves like the Beall-Suritool and Cooley Cutter valve were redesigned [120]. Later in 1974, Shiley valve became the most frequently implanted mechanical valve just after when the Derlin disk in the Shiley valve was changed to pyrolytic carbon [21].

The Lillehei-Kaster (LK) [22-23] tilting disk was introduced in 1970 with a pyrolytic carbon disk and solid titanium orifice and it has a remarkable record of 30 years with only one case reported of structural failure. In 1976, the valve was redesigned by replacing flat disk by a curved disk, which in turn, improved the catastrophic thrombosis [3,21,24]. Further improvement was achieved later in 1984 when the orifice was changed from titanium to Pyrolite carbon [3, 25].

Bob Kaster [26] redesigned the LK valve that was introduced seven years ago, by replacing the outflow struts with a single wire running through the centre of the disk. Although the efficiency of this valve was reduced due to the hole through the centre of the disk allowing increased blood leakage through it yet this device maintained the excellent reliability of the original LK design. The only reported structural failure for this design was combated by changing the flat disk to a rounded surface in a D-16 design in order to avoid the observed impacts of disk with the large flat stop when it closes [27]. Basically this change was implemented in order to reduce haemolysis but this change resulted in three reported cases of disk fracture when combined with Pyrolite disk attached to the extremes of the allowable tolerances.

Today, heart valves employing two leaflets in their design are implanted but the idea of bileaflet valve was documented by John W. Holter in 1958 the time when valves were still being implanted in the descending aorta. In 1963, Vincent Gott and Dr. Ronald Daggett developed a valve called "Gott-Daggett valve", which was a poly-carbonate ring containing a disc of Teflon fabric impregnated with silicone rubber [28]. Then Dr Lillehei implanted an all titanium bileaflet valve in 1968 that was developed by Kalke [29] but the patient died within 48 hour so the idea was abandoned. After seven years, the founder of cardiac pacemakers designed a valve very similar to the Kalk Valve (with little modifications) and requested Carbomedics to build it from Pyrolite carbon. This design known as "St Jude Medical heart valve" was the first that was entirely made up of pyrolytic carbon [10, 21] and quickly captured a leading market position. Although this valve had design and material improvement as major factors in its growth but its market acceptance was also pushed up due to strut failure problems in the Shiley valve [30].

The combination of Pyrolyte carbon with a bileaflet valve proved significant but it failed to provide a cure for the patients with prosthe-tic valves. Soon after five years of the introduction of St Jude's valve, a new valve Baxter-Duromedics [31] entered the market with an improvement in pivot design and with a seating lip that the occluders rest against in the closed position. The seating lip proved effective in reducing the regurgitation but the desired gain in the efficiency was counterbalanced by the loss of orifice area. Further the seating lip enhanced the formation of cavitation bubbles that resulted in signi-ficant structural damage and thereby the design was withdrawn after 3 years with only 20,000 valves due to 12 reported cases of structural failure [32].

Sulzer Carbomedics [33] introduced a new bileaflet valve called the Carbomedics Prosthetic Heart Valve (CPHV) in 1986, which improved the pivot design with a titanium stiffening ring to protect the valve. The implant of over 325,000 of these valves in 14 years with no post-operative structural failure demonstrated clearly the effectiveness of titanium stiffening ring. Sorin Biomedics [34] who had been manufacturing tilting disc valves similar to Shiley valves started offering a bileaflet valve as well in 1990. Sorin combined a curved pyrolitic carbon leaflet, similar to Baxter Duromedics design with a titanium orifice. The titanium orifice was basically used both to avoid expense and manufacturing complexities associated with Pyrolitic carbon. The Sorin valve was coated with a thin layer of vapour deposited carbon on the titanium. Later Edward Life Sciences

[35] obtained the rights of marketing this valve in U.S with little modification in the new sewing ring design with oversized silicon filler but otherwise left the design unchanged. Several independent investigators came up with the conclusion that the carbon layer is subjected to rapid erosion both in vivo and in vitro explaining the limited acceptance of the Sorin bileaflet and Edwards Mira valve [35]. Sorin [36-37] accepted the need for further studies after knowing that insufficient safety and effectiveness data had been collected for this valve.

In cooperation with the Carbon implants, the Medtronic [21] company began a clinical study and designed Medtronic Parallel Valve to open parallel to the flow of blood with a 90° opening angle rather than the 78-80° angles used by other manufactures. Pivot in this valve was quite complex, intended to decrease the leakage when the valve was fully closed, and it has been shown to be an important risk factor for valvular thrombosis [38]. Hence it was withdrawn from the market within one year of clinical trials.

Certain employees of Carbon Implants, Inc. formed a spin-out company called as Medical Carbon Research Institute (MCRI) after completing the design of the Medtronic Parallel Valve and developed another valve similar to this. This new valve was called the On-X^{TM} [21] with an elongated orifice, 90° opening angle and 40° closed angle featuring four sizes of carbon orifices labelled 19, 21, 23, and 25 mm. Finally in 1999, Medtronic began another bileaflet valve clinical trail in Europe with an internally developed bileaflet valve called the Medtronic Advantage [34].

Two of the most recent valves are the ATS Open Pivot® bileaflet valve and the On-X® prosthetic valve [39]. ATS open pivot valve (made by ATS Medical, Inc. in 2000) is a mechanical heart valve with two leaflets (flap like structures) in the shape of a circle, each leaflet is half the circle, surrounded by a ring made of polyester fabric. The leaflets are made of carbon.

The On-X® valve (made by Medical Carbon Research Institute, LC in 2001) is a mechanical heart valve with two movable half-discs (bileaflets), contained within a housing surrounded by a man-made fabric-covered ring. The leaflets are made of graphite and tungsten, with a carbon coating.

Table 6.1. Summary of the material composition of nine key designs of mechanical valves developed since 1959

Year	Valve Name	Type	Poppet	Material
1959	Hfnagel	Ball	Polypropylene	Methacrylate
1964	Starr-Edwards 1000	Ball	Silastic	Stellite
1968	Wada-Cutter	Tilting Disc	Teflon	Titanium
1969	Bjork-Shiley	Tilting Disc	Derlin	Stellite
1970	Lillehei-Kaster	Tilting Disc	Pyrolitic carbon	Titanium
1971	Bjork-shiley	Tilting Disc	Pyrolitic carbon	Stellite
1977	Medtronic-Hall	Tilting Disc	Pyrolitic carbon	Titanium
1977	St Jude Medical	Bi-leaflet	Pyrolitic carbon	Pyrolitic C
1991	Jyros	Bi-leaflet	Vitreous carbon	Vitreous C
1999	Medtronic Advantage	Bi-leaflet	Pyrolitic carbon	Pyrolitic C

Mechanical heart valves can be classified into three primary types: (i) ball and cage, (ii) tilting disk, and (iii) bileaflet valves. Table 6.1 shows the types of heart valves used for implantation since 1959. All mechanical heart valves consist of an occluder working as an element to retard the flow of blood and of a mechanism to restrain the motion of the occluder. The occluder functions relatively in response to the pressure across the valve during the cardiac cycle. The occluder opens to permit the blood flow through the valve in case if the upstream pressure is greater than the downstream pressure and closes when the downstream pressure is greater than the upstream pressure to stop the blood flow through the valve. Figure 6.1 shows a typical mechanical heart valve made by Medtronic, USA.

Fig. 6.1. A typical PyC leaflet heart valve (Medtronic)

6.3 Thrombosis

As mentioned earlier, the principal concern with mechanical heart valves is thrombus formation. The process of a blood clot (thrombus) formation inside a blood vessel to stop bleeding is termed as thrombosis. Excessive blood loss is undesirable and physiological feedback systems exist for protection. Therefore, the body has ways of protecting itself in cases when the blood escapes the body. When the human body loses a small amount of blood through a minor wound, the platelets cause the blood to clot in order to stop the bleeding. Because new blood is always being made by hemopoetic stem cells located in bone marrow, the body can replace the lost blood effectively and efficiently. However, when the human body loses a large amount of blood through a major wound, then in such cases blood needs to be replaced through a blood transfusion from other people.

Clotting involves thrombus formation that prevents further blood loss from damaged tissues, blood vessels or organs. Figure 6.2 shows the schematic of the thrombosis process. The first step in the clotting process is the adsorption of blood proteins on the skin surface. There are three types of proteins in human blood: (i) albumin, (ii) fibrinogen and (iii) globulin. Albumin, fibrinogen and globulin are present in the blood in the following percentage (%) ratio: 60:4:35, respectively. Soon after vessel injury occurs, blood clotting is initiated resulting in platelets and tissues to secrete a clotting factor called prothrombin activator. Prothrombin activator and calcium ions catalyse the conversion of prothrombin to thrombin (activated form), which subsequently catalyses the conversion of fibrinogen to fibrin threads. Fibrin threads are sticky and trap more platelets thus further sealing the point of blood leak.

Platelets are irregularly shaped, colourless bodies that are present in the blood. Their sticky surface enables them, along with other substances, form clots at the blood outlet point(s) to stop bleeding. The mineral calcium, vitamin K and fibrinogen help the platelets form a clot. If blood is lacking these nutrients, it will take longer than normal for the blood to clot. If these nutrients are missing, one could bleed to death. Some clots can be extremely dangerous. A blood clot that forms inside of a blood vessel can be deadly because it blocks the flow of blood, cutting off the supply of oxygen to the body.

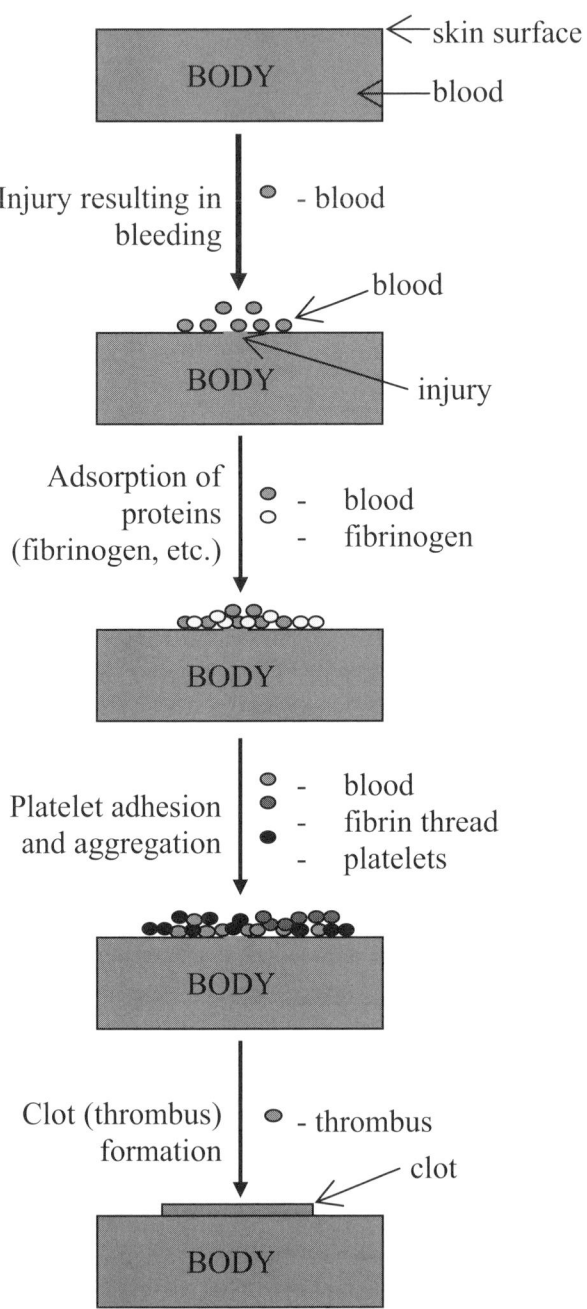

Fig. 6.2. Schematic diagram depicting the thrombosis process in the human body

There are two pathways that effectively lead to the formation of a blood clot, namely, the intrinsic and extrinsic pathways. Although they are initiated by distinct mechanisms, the two converge on a common pathway that leads to clot formation. The formation of a blood clot in response to an abnormal vessel wall in the absence of tissue injury is the result of the intrinsic pathway. Thrombus formation in response to tissue injury is the result of the extrinsic pathway. Both pathways are complex and involve numerous different proteins termed clotting factors.

There are many circumstances that may lead to thrombosis, however, the principal three situations include: (i) damage to endothelial cells, which line the interior surface of blood vessels; (ii) slow blood flow; and (iii) changes in the composition of blood. There are many ways endothelial cells can be damaged. For example, increased blood pressure (hypertension) is one common cause of endothelial damage. Diseases of blood vessel walls [40] (e.g., atherosclerosis) can damage endothelial cells directly or roughen the endothelial lining producing turbulent blood flow that, in turn, damages endothelial cells. If blood flow becomes abnormally slow, platelets have the opportunity to contact the endothelial lining. Once the platelets are in contact they may stick and initiate the intrinsic blood coagulation pathway. Similarly, other conditions may indirectly initiate blood clotting. For example, certain alterations in red blood cells (e.g., some anemias) can make them stickier than normal resulting in clumping of RBCs (erythrocytes) and reduced blood flow. Similarly, overproduction of RBCs (polycythemia) can cause the blood flow to slow.

6.4 Hemocompatibility

Hemocompatibility can be defined as the implanted material's compatibility with the blood. This definition takes into account the following [41]:

- The activation of the blood coagulation system at the blood-material interface
- The response of the immune system induced after the blood-material contact
- The other tissue responses which appear as consequence of the blood-material contact

The hemocompatibility of the implant material is closely related to the reactions between the surface of the biomaterial and the inflammatory host response [42]. There are several factors that contribute to this and these may depend on individual patient characteristics, such as general health, age, tissue perfusion, immunological factors, or implant characteristics, such as surface roughness and porosity, chemical reactions at the surface, corrosion properties of the material, and the toxicity of the individual metals present in the alloy [43].

The human body is able to resist against any form of attack from unknown foreign material(s). When a material (foreign body) is implanted in the human body, it comes in direct contact with blood. This contact between the blood and material leads to an inflammatory reaction directed against the material. The following are some of the interactions that can be considered either as good or bad depending on the circumstances [44]:

1. Adsorption of proteins, lipids, or calcium from the blood onto the surface of the device. At a later step, migration of the surface adsorbed material into the bulk may occur.
2. Adhesion of platelets, Leukocytes or erythrocytes onto the surface of the device.
3. Formation of "capsules" on the outer surface of the device, or "pseudo-intima" on the inner surface of a device, e.g., for a vascular graft.

The interactions that are generally regarded as undesirable are as follows [45]:

1. Platelet activation and aggregation.
2. Formation of thrombi (blood clot) on the device surface.
3. Transport of thrombotic or other material from the surface of the device to another site via the circulatory system.
4. Injury to cells or tissues adjacent to the device.
5. Injury to circulating blood cells, e.g., Hemolysis (Hemolysis is the breakage of the red blood cell's (RBC's) membrane, causing the release of the hemoglobin and other internal components into the surrounding fluid).

6. Overenthusiastic cell proliferation (increase in cell number by division) on or adjacent to the device resulting in reduced or turbulent blood flow of the blood at the device location.

7. Adsorption of proteins, lipids, or calcium from the blood onto the surface of the device.

The surface of the biomedical device implanted in the body is critically important because it comes into direct contact with the surroundings (i.e., tissues, blood, etc). It is the response of the host to the material that defines the character of the biomaterial that is used in medical surgery. Surface roughness/smoothness is another important parameter, which can have a noticeable influence on the hemocompatibility of a biomaterial carrying the surface. For example in artificial heart valve applications, a smooth surface is essential as surface roughness causes turbulence in the blood, which leads to the integrity of the red cells being damaged causing bacteria to adhere, and blood coagulation and clots.

A relationship has been proposed between surface charges on materials and protein deposition levels on such surfaces [46]. It has also been stated that the biocompatibility of materials can be influenced by factors such as hydrophobicity and topography [47, 48]. Reports from Ahluwalia et al. [49] and Bowlin and Rittger [50] based on surface potential measurements using the vibrating Kelvin probe method suggest that positively charged surfaces enhance cell adhesion in comparison to neutral or negatively charged surfaces. The hydrophylic or hydrophobic nature of a surface has also been associated with the extent of cell interactions with the surface [49,51]. There has also been suggestions in the literature that the electron conduction of materials might have an effect on their biocompatibility. Bruck [52, 53, 54] reported, based on his studies on pyrolytic polymers, that intrinsic electronic conduction and semi-conduction may be involved in the extent of hemocompatibility observed in the materials. Bruck [52, 53, 54] observed clotting times, six to nine times longer than those observed with non-conducting polymers and also observed little or no platelet aggregation in electro-conducting polymers, when compared to non-conducting control samples.

Boldz and Schaldach [55] and Chen et al. [56] have also reported that there is a relationship between the electronic structure of a surface and it's hemocompatibility. According to these authors [55, 56], the denaturing of fibrinogen at a surface involves an electron transfer from the inactive site of the protein to the surface of the solid, i.e., an oxidation process takes place (a redox reaction). They suggested that fibrinogen decomposes after the electron transfer and transforms into fibrin-monomer and fibrin-peptides, with a cross-linking to form thrombus. Fibrinogen has been reported to have an electronic structure similar to a semiconductor [55] and a band-gap of ~1.8ev. Boldz and Schaldach [55] have suggested that haemocompatible surfaces will be expected to have a low density of unoccupied states in the energy range of the transfer level with an energy band-gap of greater than 1.8eV. Their reasoning was that if the charge transfer energy range lies in the band-gap energy range of the artificial surface, the oxidation process involving electron transfer is inhibited. It is well known from the theory of hetero-junctions, that when the band-gap of a semiconductor is contained within the band-gap of another semiconductor (i.e., straddling hetero-junction), electrons and holes need energy (ΔEc and ΔEV) respectively to move from the semi-conductor with the small band-gap to the one with the larger band-gap. ΔEc and ΔEV represent the differences of the conduction and valence band edges of the two semiconductors.

6.5 Endothelium and Endothelial Cell Seeding

Endothelium is the layer of thin, flat cells that line the interior surface of blood vessels, forming an interface between circulating blood inside the vessel and rest of the vessel wall. All blood vessels and lymph vessels (a network of vessels throughout the body that drains fluid from tissues and returns it to the bloodstream, managing fluid levels in the body, filtering out bacteria, and housing types of white blood cells.) are lined with endothelial cells; the layer being called the *endothelium.*

Under normal circumstances the endothelial surface prevents blood clotting and allows smooth flowing of the blood thus controlling blood pressure [57]. Surface molecules of endothelial cells act as receptors and interaction sites for a whole host of important molecules, especially those that attract or repel WBCs (leucocytes*).* Leucocyte adhesion molecules are important in inflammation and are normally repelled by endothelium in order to allow the free flow of blood cells over the surface. But in inflammatory states the WBCs are actually attracted to the endothelium by adhesion molecules. In addition platelets start sticking to the surface and later to each other making a network that end up making a clot [58].

Endothelium is known as "nature's haemocompatible surface", and the performance of any biomaterial's hemocompatibility must be compared with that of the endothelium [59].

A promising and logical approach to reduce thrombogencity is endothelial cell seeding to synthetic vascular grafts. Endothelial cell function on synthetic grafts can be improved by inserting genes to reduce or inhibit thrombus formation. The graft can be modified to release agents, which inhibit smooth muscle cell growth and promote endothelial cell adhesion and function. For synthetic grafts with diameters less than 6 mm, vascular graft acceptance requires an endothelial monolayer on the graft's luminal surface to reduce thrombosis. Grafts are seeded with endothelial cells prior to insertion in the body in order to encourage the development of a monolayer of endothelial cells. Seeding and implantation must occur rapidly in order to minimize the chance of infection [60]. Otherwise, a confluent endothelial monolayer will not form and a subconfluent layer of endothelium will be exposed to fluid shear stresses, resulting in complications. Endothelial cell adhesion to a synthetic surface involves a defined set of molecular interactions that influence subsequent cell division and/or protein synthesis. Endothelial cell adhesion to vascular grafts involves:

1. attachment and spreading on a rough synthetic surface;
2. maintenance of attachment following exposure to blood flow;
3. normal function.

Experimentally, cell adhesion to foreign biomaterials, such as polymers is influenced by manipulation of the following variables:

1. The surface properties which influence non-specific and specific interactions (e.g., hydrophobicity, surface charge, presence of oxygen and amine groups);

2. The density and affinity of adhesion molecules on the surface;
3. Covalent and non-covalent interactions between the cell and surface molecules;
4. The time of interaction between the cell and the surface;
5. Signalling events within the cell to promote or inhibit cell spreading.

Cell attachment and spreading on artificial surfaces is mediated by bonds formed between cell adhesion proteins at the substrate surface and protein receptors embedded in the cell membrane. The role of adsorbed proteins in cell adhesion has been reviewed elsewhere [61].

6.6 Surface Engineering Artificial Heart Valves

In the following section(s) we focus on biocompatible coatings used to improve the surface characteristics of blood contacting devices, such as mechanical heart valves in order to overcome thrombus formation.

6.6.1 Biological Properties of Diamond-like Carbon

Diamond like carbon (DLC) is a very promising coating because it is chemically inert, extremely hard, wear resistant and biocompatible [62, 63]. Biological properties of DLC can be altered by alloying [64, 65, 66, 67]. DLC, as a biocompatible base material, can be easily alloyed with other biocompatible materials such as titanium or silicon, or toxic materials such as silver, copper and vanadium [68].

In the investigation of hydrogen free DLC, test results show that the higher bonding ratio of sp^3/sp^2 may contribute to the blood compatibility of DLC. However, as the DLC films doped with nitrogen of certain concentration, even if the bonding ratio of sp^3/sp^2 is decreased, the behavior of platelet adhesion can be improved significantly [69]. A similar phenomenon has been observed in hydrogen free DLC film doped with argon [70]. As argon atoms are introduced into DLC to a certain content, the bonding ratio sp^3/sp^2 of the DLC films changes from 78/22 to 56/44, and the platelets adhesion behavior of the films is modified significantly. It is also found that above phenomena may be related to changes in surface tension on the films. If DLC films are annealed at the temperature of 600°C, the DLC films present a p-type characteristic. The blood compatibility becomes deteriorated [71,72]. The DLC film, which has a more hydrophilic nature, seems to have better blood compatibility than that of the DLC film with more hydrophobic nature [73].

The platelet adhesion is reduced on DLC when compared with Ti surface. Jones et al. [74,75] showed by In vitro experiments that the DLC surfaces expressed a decreased area coverage of platelets compared to titanium, TiN and TiC. Whereas on the Ti containing surface platelet activation, clotting of platelets and thrombus formation was observed, no such reaction took place on the DLC surface. High albumin to fibrinogen ratio was also observed in DLC as compare to Ti, TiN and Tic coatings which is an indication of the ability of DLC to prevent thrombus formation [75]. Dion et al. [76] also observed higher albumin/fibrinogen ratio in DLC as compare to silicone (a polymer widely used for implants). Cui and Li [77] state from in vitro experiment that DLC and CN films show good tissue- and blood compatibility.

Gutebsohn et al. [78] analysed the intensity of the platelet activation antigens CD62p and CD63 and showed that DLC coating of a stainless steel coronary artery stent resulted in a decrease of these antigens. It indicates a low platelet activation on DLC and, therefore, a low tendency for thrombus formation. In this in vitro experiment, they showed that DLC coating suppressed metal ion release from the stainless steel stent which may influence negatively the hemocompatibility of the surface. Alanazi et al. [79] conducted an in vitro experiment using a flow chamber and whole human blood, which showed a low percentage of DLC to adhere to the surface. However, the results differed with the deposition conditions used to produce DLC:

Thomson et al. [80] have studied the inflammatory potential of DLC-coated tissue culture plates by measuring the levels of the lysosomal enzymes released from macrophages cells (cells play major role in inflammation and the response to foreign bodies). They showed that no inflammatory response was elicited in vitro. Similarly Fibroblast is another cell used to test biocompatibility. Preliminary result [81] showed no statistically significant differences between fibroblasts grown on in control polystyrene plate and those grown in DLC coating.

Some work has been done showing change in bioreaction on DLC due to alloying [82, 83, 84, 85, 86, 87, 88]. A reduced inflammatory reaction was observed on implanted stents when SiO_x was added to DLC [82]. It was noticed that proteins adsorption could be altered as a function of Ti content [85, 87] in the DLC Osteoclasts (osteoclast is a large multinucleated cell that plays an active role in bone resorption) was inhibited by the addition of Ti into DLC [83, 84]. The addition of Ca-O to a DLC film resulted in decreased wetting angle, improved cell morphology and viability of mouse fibroblast cells [88].

6.6.2 Other Biocompatible Coatings

It is worth considering that titanium and its alloys have been used in biomedical implants for many years now and therefore it is sensible to look at the surface treatment of titanium alloys for producing superior surface characteristics, which could be ideal for heart valves. It should be noted that titanium alloys are not brittle like PyC. A large number of research scientists have deposited Ti-based coatings, using energised vapour-assisted deposition methods, and studied their potentials for use in biomedical areas, such as heart valves or stents. Yang et al. [89] deposited Ti-O thin films using plasma immersion ion implantation technique and characterised the anti-coagulant property employing in-vivo methods. They found that the Ti-O film coatings exhibited better thrombo-resistant properties than low-temperature isotropic carbon (LTIC), in long-term implantation. Chen et al. [90] deposited TiO coatings doped with Ta, using magnetron sputtering and thermal oxidation procedures, and studied the antithrombogenic and hemocompatibility of $Ti(Ta^{+5})O_2$ thin films. The blood compatibility was measured in-vitro using blood clotting and platelet adhesion measurements. The films were found to exhibit attractive blood compatibility exceeding that of LTIC. Leng et al. [91] investigated the biomedical properties of tantalum nitride (TaN) thin films. They demonstrated that the blood compatibility of TaN films was superior to other common biocompatible coatings, such as TiN, Ta and LTIC. Potential heart valve duplex coatings, consisting of layers of Ti-O and Ti-N, have been deposited onto biomedical Ti-alloy by Leng et al. [92] and their blood compatibility and mechanical properties have been characterised. The TiO layer was designed to improve the blood compatibility, whereas, TiN was deposited to improve the mechanical properties of the TiO/TiN duplex coatings. They found that the duplex coatings displayed (i) better blood compatibility than LTIC; (ii) greater microhardness; and (iii) improved wear resistant than Ti6Al4V alloys. It has been reported that the TiO coatings display superior blood compatibility to LTIC [93].

6.6.3 Chromium Modified DLC

Chromium modified DLC (Cr-DLC) films with varying %Cr content were deposited on 50 mm circular silicon wafers using magnetron sputtering utilising the intensified plasma assisted processing (IPAP) process [94]. Table 6.2 shows the conditions employed to deposit the Cr-DLC samples with varying %Cr contents. The as-deposited Cr-DLC films were nanocomposite and consisted of nano-sized chromium carbide particles embedded in an amorphous DLC matrix.

Figure 6.3 shows the XRD spectrum for a typical Cr-DLC film, with 10%Cr, in the 2-theta range 15-115 degrees. It was found from HRTEM studies that the nano-structured Cr-DLC film consisted of nano-sized (5 nm) chromium carbide particles. The presence of chromium carbide particles was evident from the XRD peak centered at around 33°2-theta value. The other major intense peak centered at around 68°2-theta value corresponds to silicon, which was from the silicon wafer substrate material. We found that the Cr-carbide nanoparticles were embedded deep into the amorphous DLC matrix and were not present on the DLC surface. This enabled the nano-particles to be protected by the amorphous DLC film.

Table 6.2. Deposition parameters employed in depositing Cr-DLC films using the IPAP process

Sample	Deposition rate (nm/min)	Magnetron Current (mA)
Cr(1%)DLC	6.6	155
Cr(5%)DLC	7.6	220
Cr(10%)DLC	15.2	310

Bias Voltage: -1000 V; flow rate (sccm): CH_4:Ar 7.4:40;
Chamber pressure: 2.66 Pa; processing time: 2 hrs;
sputter cleaning: Ar^+: 3.3 Pa, -1500 V, ~20 minutes.

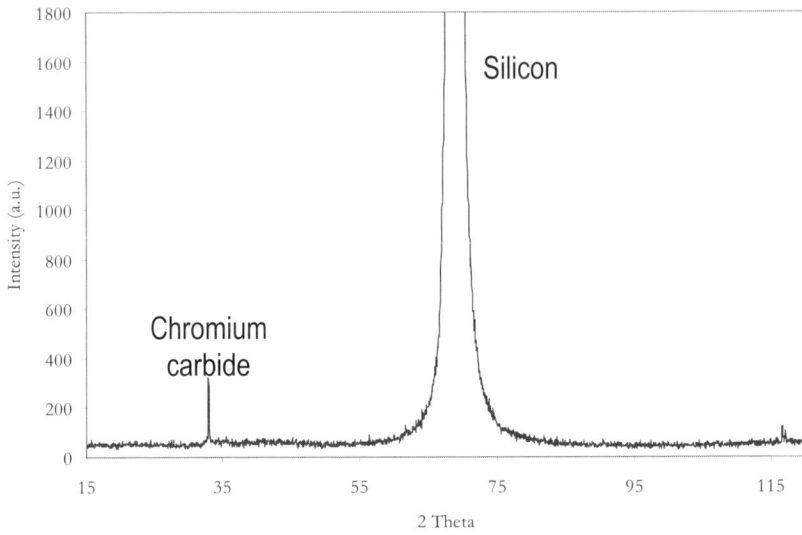

Fig. 6.3. XRD spectrum for sample Cr(10%)-DLC showing the silicon substrate and chromium carbide peaks

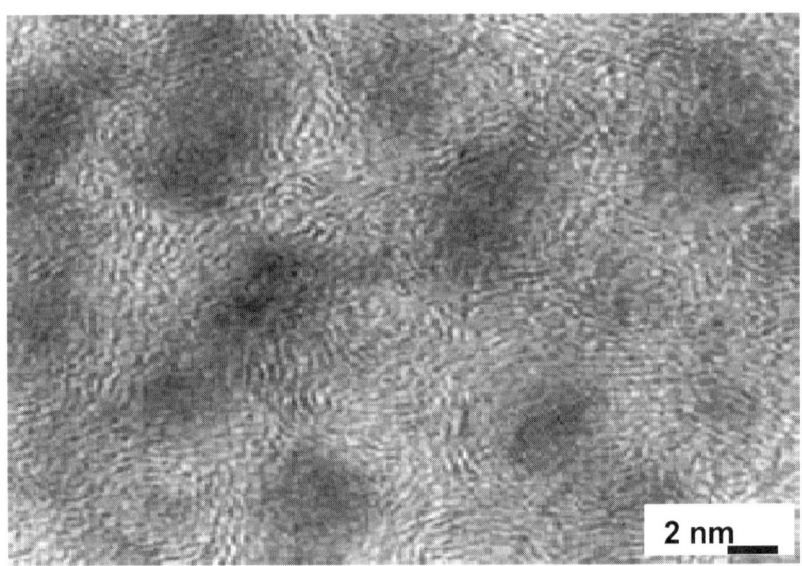

Fig. 6.4. High Resolution TEM micrograph of the 5%Cr-DLC sample showing the microstructure exhibited by the thin film coating

Figure 6.4 Displays the high-resolution (HRTEM) micrograph representing 5%Cr-DLC and showing the microstructure exhibited by the same as-deposited film. The micrograph shows the presence of nanoclusters (NCs) around 5 nm in diameter surrounded by a ~2 nm thick amorphous matrix. Electron diffraction showed that the dark contrast NCs correspond to Cr carbide, encapsulated by an amorphous matrix.

The contact angle measurements obtained using the optical method are as shown in Figure 6.5. Chromium doping leads to a gradual increase in the contact angles as shown in Figure 6.5. The increase begins to level out with Cr content above 5% where it begins to reach saturation point. The average contact angles displayed by samples 1%Cr-, 5%Cr- and 10%Cr- DLC samples were calculated to be 80.285, 97.23 and 99.274 degrees, respectively. Raman spectroscopy was used to characterise the quality of the as-deposited Cr-DLC films with different Cr contents, in terms of diamond carbon-phase purity. The data from the Raman spectroscopy studies, including intensities of D (I_D) and G (I_G) bands, Full width at half maximum (FWHM) of I_D and I_G bands, I_D/I_G ratio and the positioning of the D and G band peaks can all be found in Table 6.3. From the Raman investigations, it was found that the Cr-DLC films displayed the two D and G bands of graphite. The G and D bands are usually assigned to zone centre of phonons of E_{2g} symmetry and K-point phonons of A_{1g} symmetry, respectively. The D band peaks for the three Cr-DLC films were positioned at 1401.6, 1408.4 and 1306.8 cm^{-1}, whereas, the G band peaks were centered at 1538.8, 1540.6 and 1511.8 cm^{-1}. From both D and G bands, 5%Cr-DLC sample displayed the smallest values for FWHM. The I_D/I_G ratio was the least for 10%Cr-DLC and the highest out of the three samples for the 5%Cr-DLC film. This suggests that there are more disordered graphitic phases in sample 5%Cr-DLC and the least similar disorder in 10%Cr-DLC film.

The results of the cell count analysis shown in Figure 6.6, give an indication of the influence of Cr content in Cr-DLC films on the adherent cell population of the three samples, 5%Cr-DLC provided the best conditions for HMV-EC seeding, while, 10%Cr-DLC film resulted in the least population of adherent human endothelial cells onto its surface. It should be noted that sample 1%Cr-DLC was a better base material for seeding endothelial cells than 10%Cr-DLC. Figure 6.7 displays the SEM micrographs showing the population of endothelial cell attachment onto 1%Cr-, 5%Cr- and 10%Cr- DLC film surfaces. All three films displayed smooth surface profiles, which is a key requirement in artificial heart valve applications.

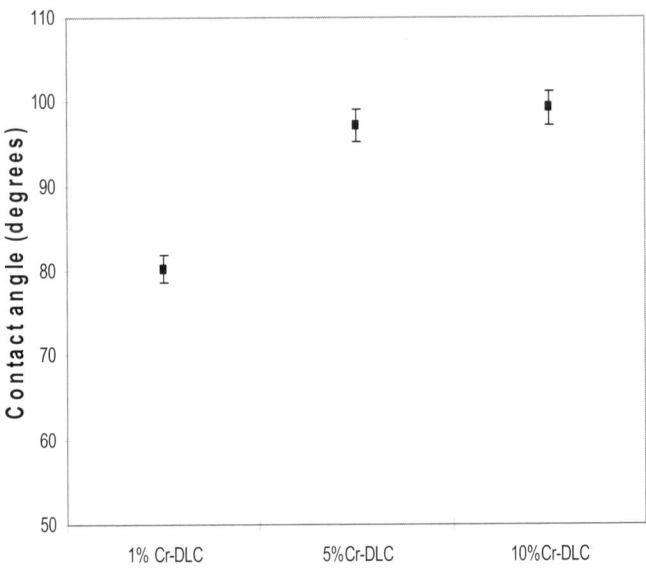

Fig. 6.5. Graph showing the contact angle measurements as obtained using the optical method for the three Cr-DLC samples

Table 6.3. Raman data obtained for the three Cr-DLC samples. The data includes I_D and I_G peak intensities; Full-Width at Half Maximum (FWHM) values for D band; and D and G band intensities

Sample	%Cr	I_D (a.u.)	I_G (a.u.)	FWHM (D)	D-Peak (cm^{-1})	G-Peak (cm^{-1})
CrDLC1	1	21.833	21.043	363.68	1401.6	1538.8
CrDLC2	5	21.628	20.415	361.08	1408.4	1540.6
CrDLC3	10	45.588	49.748	389.90	1306.8	1511.8

It was noted that there was a direct correlation between the I_D/I_G ratio and the population of endothelial cells attaching to the three Cr-DLC films with different Cr contents. We noted that from the three Cr-DLC films, the highest value displayed for the I_D/I_G ratio was by 5%Cr-DLC film, which also gave the highest adherent cell population onto its surface. The lowest I_D/I_G ratio value was for 10%Cr-DLC, which showed the least, from the three samples investigated in this study, population of cell attachment to its surface after conducting the cell seeding procedures. Furthermore, the density of nano-sized

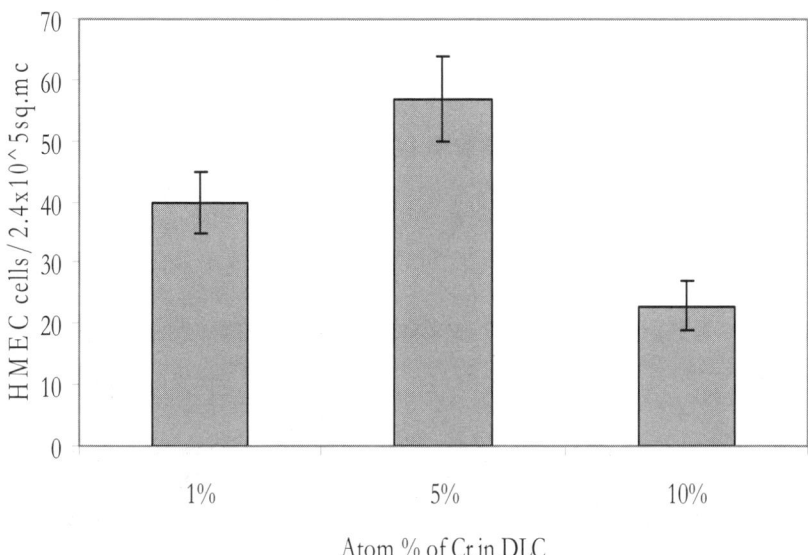

Fig. 6.6. Graph showing the cell count results obtained after seeding the human microvascular endothelial cells onto the Cr-DLC surfaces

Cr-carbide particles produced during film growth is expected to be the highest in 10%Cr-DLC and the least in 1%Cr-DLC sample. This difference in the Cr-carbide content in the three films is sure to influence the surface chemistry of the DLC films.

It was noted that there was a direct correlation between the I_D/I_G ratio and the population of endothelial cells attaching to the three Cr-DLC films with different Cr contents. We noted that from the three Cr-DLC films, the highest value displayed for the I_D/I_G ratio was by 5%Cr-DLC film, which also gave the highest adherent cell population onto its surface. The lowest I_D/I_G ratio value was for 10%Cr-DLC, which showed the least, from the three samples investigated in this study, population of cell attachment to its surface after conducting the cell seeding procedures. It is apparent that increased Cr content into the growing DLC films alters the microstructure of the deposited films. Furthermore, the density of nano-sized Cr-carbide particles produced during film growth is expected to be the highest in 10%Cr-DLC and the least in 1%Cr-DLC sample. This difference in the Cr-carbide content in the three films is sure to influence the surface chemistry of the DLC films. It was difficult to correlate the water contact angle results with the cell seeding efficiency of the films.

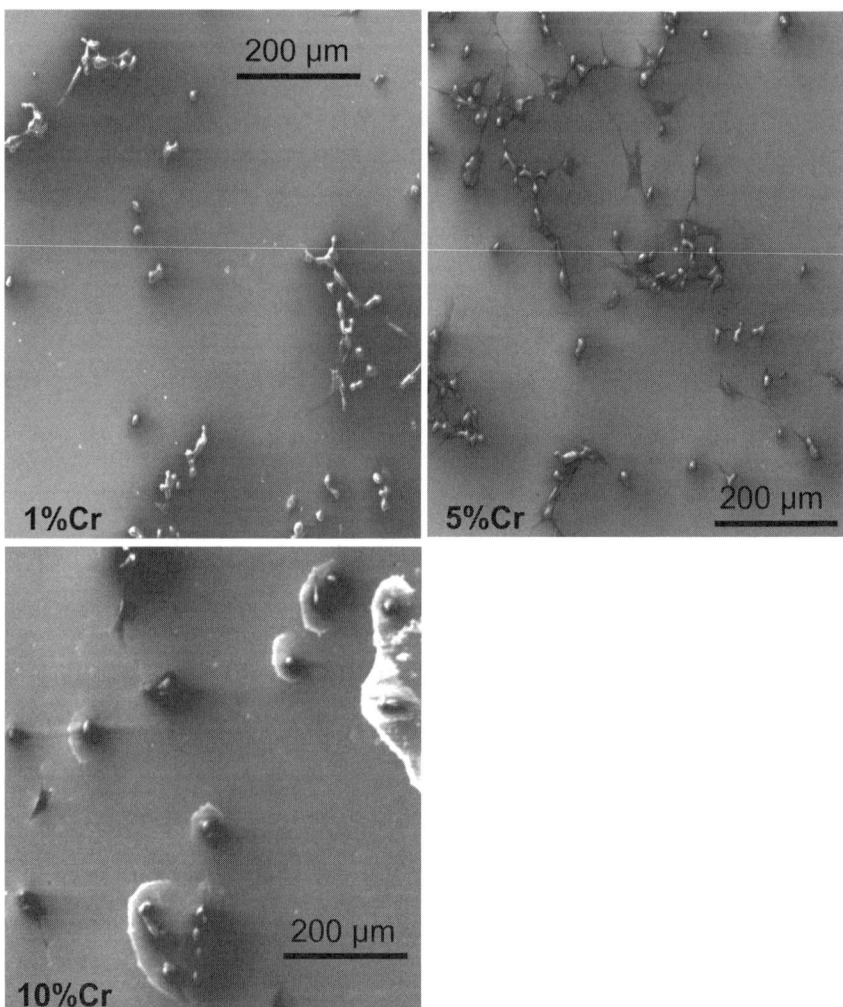

Fig. 6.7. SEM micrographs showing endothelial cell seeding on the three types of Cr-DLC film surfaces

However, Grinell [95] reported that wettable (hydrophilic) surfaces tend to be more conducive to cell adhesion than similar hydrophobic surfaces.

6.6.4 Silicon Modified DLC

We investigate to evaluate the potential of a more readily available synthetic silicon modified carbon film (Si-DLC) for thrombo-resistant applications. We study the adhesion and spreading of micro-vascular endothelial cells on the surface of Si-DLC and thermally annealed Si-DLC films. Table 6.4 shows the deposition conditions employed for preparing DLC and Si-DLC film samples.

Raman spectroscopy was used to characterise both the as deposited and the thermally annealed DLC and Si-DLC films. The thermal annealing was conducted at temperatures between 200°C and 600°C under a flowing nitrogen atmosphere for 2 hours. The information collected from the Raman spectroscopy investigation included the Raman-peak intensities for the D and G-peaks, that is I_D and I_G peaks, the I_D/I_G ratios and the full-width at half maximum (FWHM) for both peaks as shown in Table 6.5.

Table 6.4. deposition conditions for the DLC and Si-DLC film (The numbers in SD5, SD10, SD15 and SD20 represents the flow rates of TMS in sccm used during deposition)

Sample	DLC	SD5	SD10	SD15	SD20
DC-Volts (volts)	400	400	400	400	400
RF-power (watts)	~150	~160	~165	~175	~210
Temperature (°C)	ambient	ambient	ambient	ambient	ambient
Pressure (base)/torr	6×10^{-6}	6×10^{-6}	6×10^{-6}	6×10^{-6}	6×10^{-6}
Gas ratio (sccm) Ar : C_2H_2: TMS	10:20: 0	10:20:5	10:20:10	10:20:15	10:20:20
Pressure (deposition)/mtorr	~1.3×10^{-2}	~1.9×10^{-2}	~2.7×10^{-2}	~3.2×10^{-2}	~4.5×10^{-2}
Time (deposition)/mins	5	5	5	5	5

Table 6.5. Raman features of a-C:H and a-C:H:Si thin films

Bias Volts (V)	TMS sccm	I_D (a.u)	I_G (a.u)	FWHM (D)	FWHM (G)	I_D/I_G	D-Peak (cm^{-1})	G-Peak (cm^{-1})
400	5	1231.7	4397.6	211	104.6	0.28	1336.48	1504.99
400	10	1383.2	6396.5	192.8	103.5	0.21	1310.1	1490.07
400	15	2206.7	9687.6	198.8	103.4	0.22	1296.36	1483.18
400	20	1584.4	8423.5	205.5	102.7	0.18	1278.99	1477.16
400	0	1775.9	4501.8	194.8	108.5	0.39	1349.5	1541.96

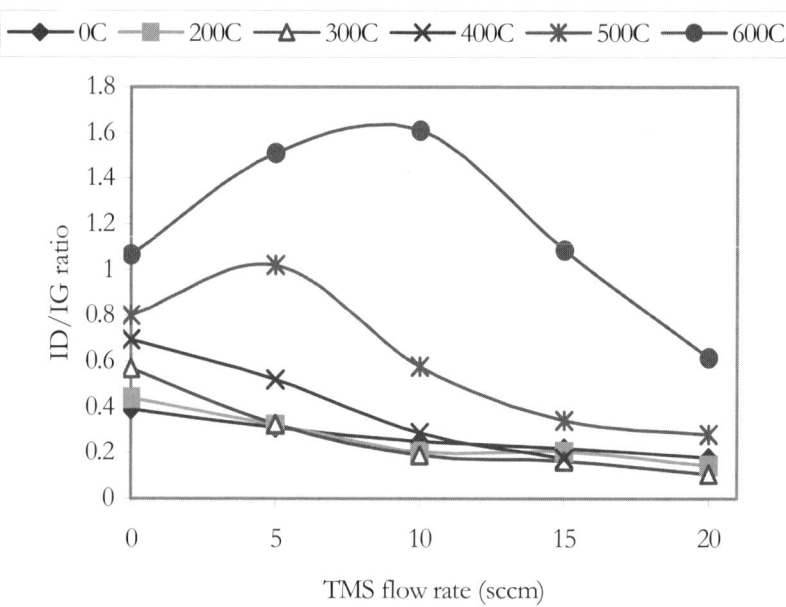

Fig. 6.8. Raman I_D/I_G ratios for the DLC and Si-DLC films both non-annealed and thermally annealed (200-600°C) under flowing nitrogen for 2 hours versus the TMS flow rate during deposition

    The results of the Raman spectroscopy show an increase in the I_D/I_G ratios (Fig.6.8) with annealing temperature for the DLC and Si-DLC films in agreement with earlier reports in the literature [96]. The increase in I_D/I_G ratio on the annealing of DLC has been associated with the growth of crystallite structures in the DLC thin film. In the DLC films the increase in the I_D/I_G ratio with thermal annealing is linear, but for the Si-DLC films the I_D/I_G ratio increase occurred only at relatively higher annealing temperatures of 300°C and above. The I_D/I_G ratio also decreases with increasing amount of silicon in the films as shown in Fig.6.8.
    Physical and chemical changes in carbon materials resulting in graphitisation occurs during thermal annealing. Graphitisation is associated with an increased sp^2 content, while silicon doping tends to increase the number of sp^3 sites [96]. Silicon does not form π-bonds and it therefore increases the number of sp^3 sites in the Si-DLC films.

Typical x-ray photoelectron spectroscopy (XPS) chemical analysis for the as deposited and annealed DLC and Si-DLC films are as indicated in Table 6.6. The peak binding energies of the films are

Table 6.6. XPS chemical analysis and binding energies (B.E) of elements in the as-deposited DLC and Si-DLC thin films

Sample	C(%)	Peak B.E/eV	O(%)	Peak B.E/eV	Si(%)	Peak B.E/eV
DLC/undoped/400V	89.58	285.11	10.07	532.51	0.36	102.41
DLC(SD5)/400V	82.83	285.01	12.21	533.31	4.96	101.51
DLC(SD10)/400V	79.07	285.09	13.32	533.09	7.61	101.29
DLC(SD15)/400V	76.57	285.10	13.80	533.10	9.63	101.30
DLC(SD20)/400V	77.09	285.10	12.29	533.10	10.62	101.30

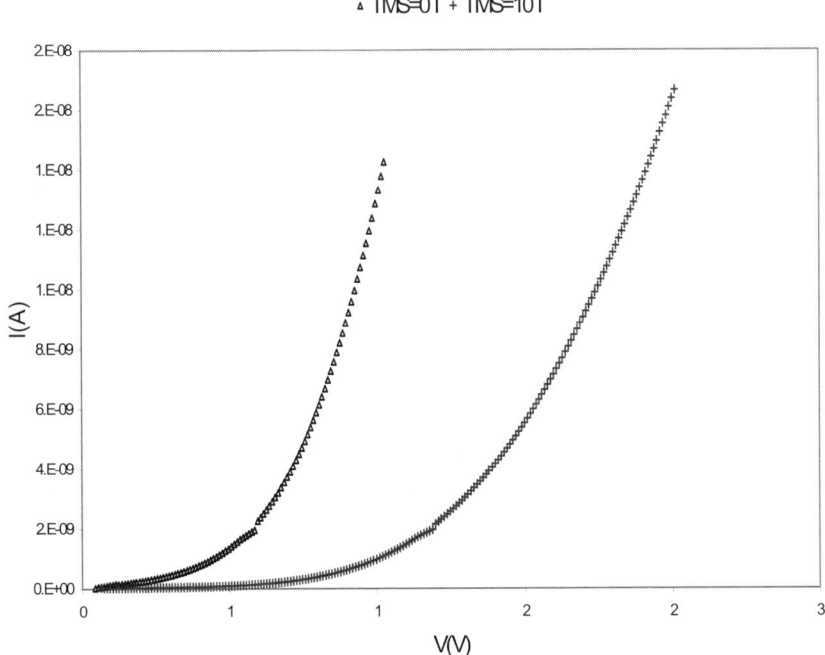

Fig. 6.9. Typical current-voltage (I-V) curves of DLC and Si-DLC (TMS flow rate of 10sccm)

consistent with those reported in the literature by Dementjev [97], Grill [98], Constant and Le Normand [99] and Baker and Hammer [100]. There was only a slight change in the values of the binding energies for the silicon-modified films even after annealing to 600°C. We also observed an increase in sp^3/sp^2 ratios after peak de-convolution.

Typical I-V curves of the metal-semiconductor-metal (MSM) sandwich (Fig. 6.9) shows that the electrical conduction mechanism is not simple ohmic but semi-conducting. We observed that silicon addition to DLC lowers it's resistivity. Thermal annealing of both DLC and Si-DLC leads to a decrease in resistivity (Fig. 6.10), which is likely to be associated with microstructural changes as indicated by the sp^3/sp^2 ratio changes observed for the annealed films by Raman spectroscopy.

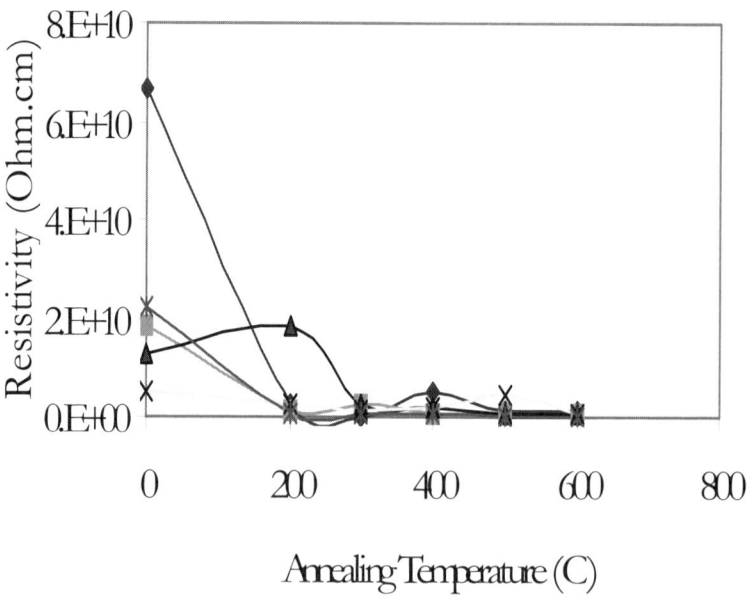

Fig. 6.10. Resistivity versus annealing temperature of DLC and Si-DLC films thermally annealed at 200°C-600°C (on x-axis 0 indicates room temperature)

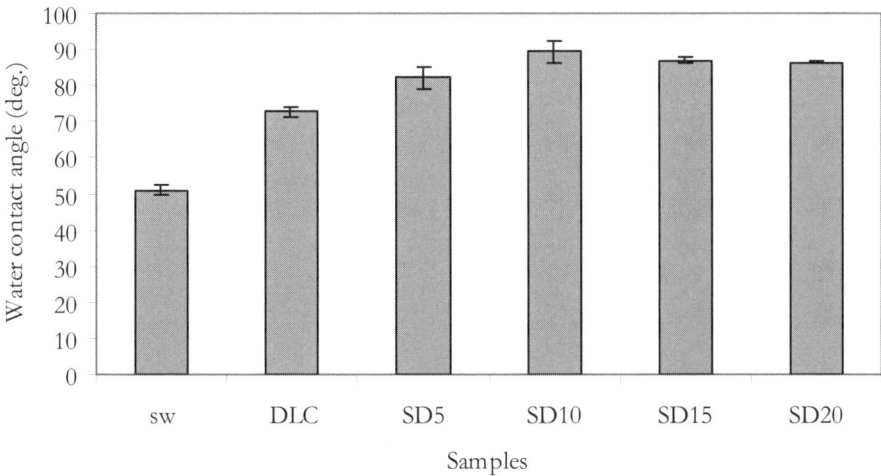

Fig. 6.11. Contact angle in water using direct optical technique

When thermal annealing at 600°C, the conductivity of both DLC and Si-DLC becomes simple Ohmic. This result is consistent with the Raman spectroscopy investigation, which revealed graphitisation at this annealing temperature. The increasing graphitic content of the films at an annealing temperature of 600°C increases the proportion of de-localised π-bonded electrons and therefore increases the electrical conductivity of the films at this annealing temperature, which results in the ohmic-behaviour.

The contact angle measurement results obtained using the optical method are as shown in Fig. 6.11 and the results of the surface energy measured by the Wilhemy plate technique for films deposited on silicon substrates are as shown in Table 6.6. Silicon doping leads to an increase in the contact angles as shown in Fig. 6.11. Silicon doping also results in a slight reduction of the surface energy values as shown in Table 6.7.

As shown in Fig. 6.12a, there seems to be a little cell adhesion on the non-doped DLC seeded with endothelial cells in comparison to the silicon modified DLC films Fig. 6.12b. The measured contact angle for the non-doped DLC film is 72° (which is lower than those of Si-DLC, Fig. 6.11). The general trend observed in our experiments based on scanning electron microscopy (SEM) and cell counting

Table 6.7. Surface energy and contact angle measurements for the films deposited on silicon substrates (γ_P = polar components, γ_D = dispersive components, and γ_S = total surface energy)

TMS/sccm	Bias volt (V)	θ_{adv} water (deg.)	θ_{rec} water (deg.)	γ_P mN/m	γ_D nM/m	γ_S mN/m
(DLC)	400	88.01	55.32	1.17	41.05	42.22
(SD5)	400	82.12	49.27	2.5	41.53	44.03
(SD10)	400	89.54	27.7	1.5	35.73	37.24
(SD15)	400	90.76	41.48	1.24	35.88	37.13

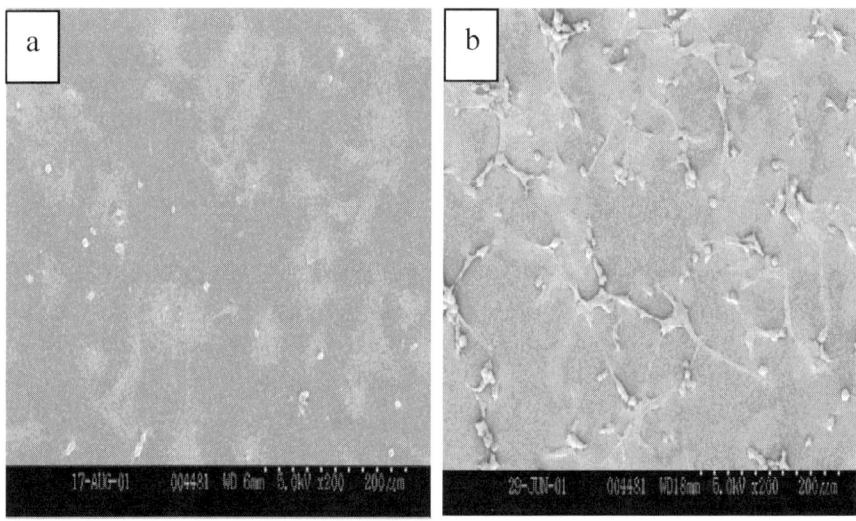

Fig. 6.12. (a) SEM image of DLC seeded with endothelial cells for ~6 hours; x200: (b) Si-DLC (TMS flow rate of 5sccm) seeded with endothelial cells for ~6 hours; x200

using SEM images indicated an increase in cell adhesion with silicon addition in the Si-DLC films. Contrary to earlier reports in the literature by Grinell [101] that wettable (hydrophilic) surfaces tend to be more conducive to cell adhesion, we have observed that more hydrophobic surfaces containing increasing amounts of silicon in the Si-DLC films and increasing contact angle with water, tends to promote human endothelial cell growth and adhesion on the films. The results of the measurements of the surface energy of the DLC and Si-DLC films indicates a substantial contribution to the total surface energy by the dispersive component term, which might play a part in the cell adhesion process.

The contact potential difference (CPD) measures the surface potential difference between the surface of the films and the vibrating reference electrode made of brass (Cu-Zn). The relationship between the work function of the films $\Phi_{(film)}$, the work function of the reference electrode $\Phi_{(brass)}$ and the CPD is given by:

$$CPD = \Phi_{probe\ (brass)} - \Phi_{sample} = \Phi_{probe\ (brass)} - [\chi + (E_C - E_F)_{bulk}] - \Phi_{ss}...$$

$$(6.1)$$

The changes in the CPD are related to changes in the electron affinity, χ, band bending due to surface states Φ_{ss}, or a shift of the bulk Fermi level $(E_C-E_f)_b$. If χ remains constant, then the changes in the CPD are directly related to the shift of the Fermi level in the bulk material and band bending in the surface states [102-104].

The result of the work-function measurements is as shown in Fig. 6.13. The work function decreased with silicon addition. The decrease in work function values has also been associated with a reduction in the net surface dipole [105]. Bolz and Schaldach [106] have already

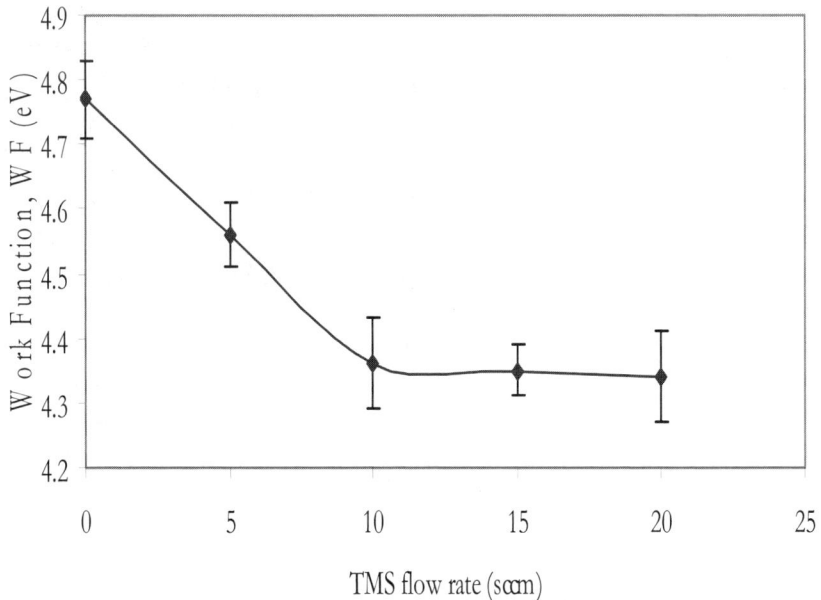

Fig. 6.13. Work function results of DLC and Si-DLC

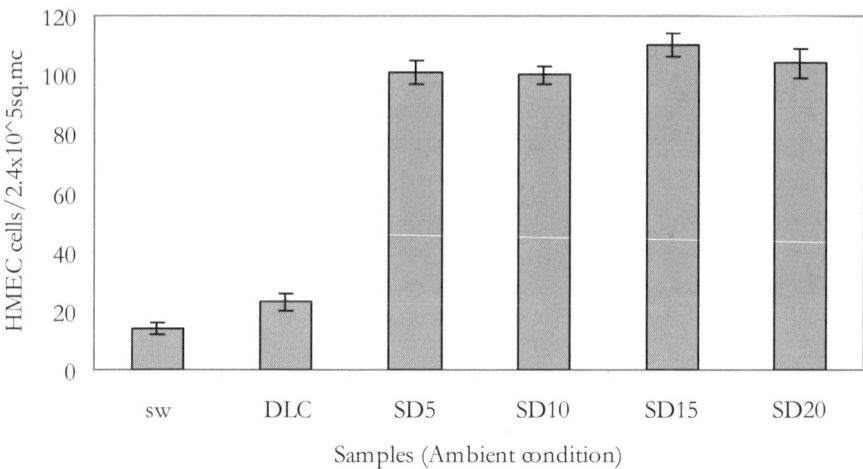

Fig. 6.14. Human micro-vascular endothelial cells (HMEC-1) adhesion on DLC and Si-DLC as obtained (ambient condition, un-annealed samples)

proposed that the early stages of thrombus formation is preceded by an electron transfer oxidation process, resulting in the transformation of fibrinogen to a fibrin polymer. We observed a substantial increase in the number of adherent endothelial cells in the Si-DLC compared to the DLC films as shown in Fig. 6.14, but further silicon addition did not lead to any substantial change in the number of adherent cells on the Si-DLC film surface. It has been suggested that the cell adhesion process depends on the sign of the charge carried by the adherent cell. Positively charged surfaces will attract cells with a negative charge or dipole and vice-versa [106].

The results of the cell count analysis, giving an indication of the adherent cell population against the annealing temperatures are as shown in Figs. 6.15a-c. There is a considerable drop in the population of adherent human endothelial cells for DLC thermally annealed at 600°C compared to the film annealed at 200°C as shown in Fig. 6.15a.The Raman spectrum of the film annealed at 600°C indicates a substantial graphitisation, and this is also the case for the film annealed at 500°C. Annealing DLC at 500°C and 600°C leads to a substantial graphitisation and our current observation is that this does not appear favourable for human endothelial cell adhesion and growth, although the graphite phase is associated with an improvement in electrical

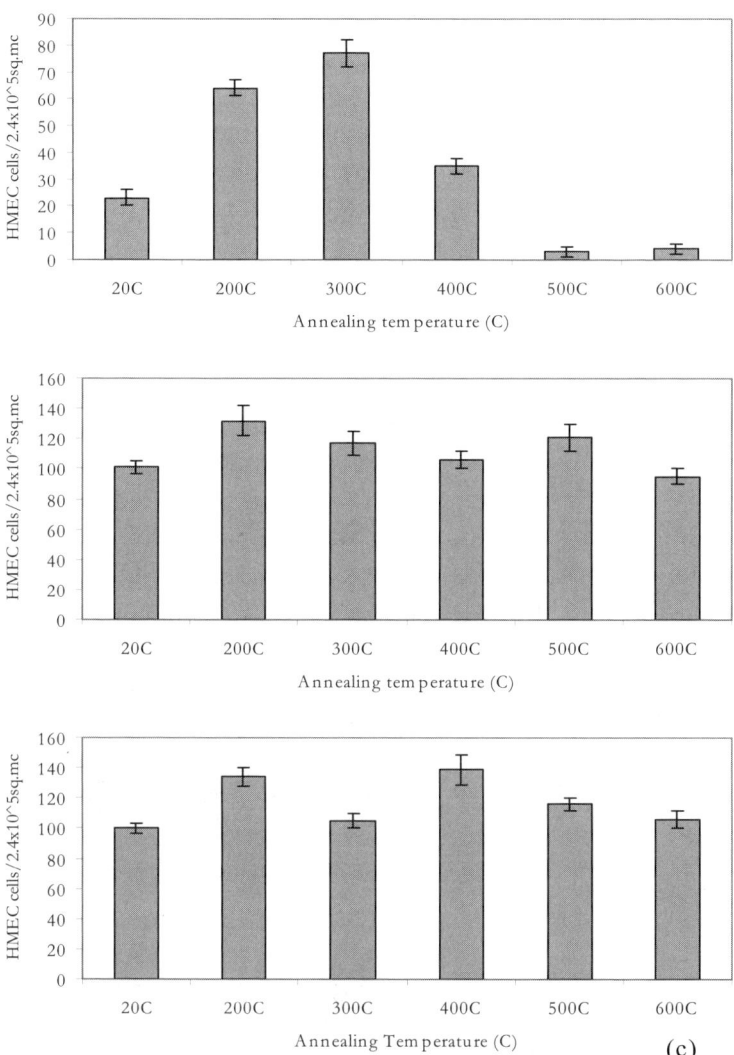

Fig. 6.15. (a) Human endothelial cell adhesion on DLC (as deposited and thermally annealed, 200°C-600°C; 20C = ambient temperature): (b) Si-DLC (TMS flow rate, 5sccm, as deposited and thermally annealed, 200°C-600°C, 20C = ambient temperature); (c) Si-DLC (TMS flow rate, 10sccm, as deposited and thermally annealed, 200°C-600°C, 20C = ambient temperature)

conduction. For the Si-DLC samples annealed between 200°C and 600°C, there was a fairly consistent population of endothelial cells for the 5 and 10sccm TMS flow rate films as shown in Fig. 6.15b and 6.15c. In all the cases investigated for the DLC and Si-DLC films, we observed a direct correlation between electronic conduction in the films and the population of adherent human endothelial cells whenever there was no significant graphitisation detectable by Raman spectroscopy. However, even when electronic conduction was relatively high in the films, graphitisation in such films always led to a substantial reduction in the population of adherent endothelial cells on it's surface. There is a significant statistical difference in endothelial cell count (at 95% confidence interval level, with $p < 0.05$, paired t-test and Tukey test) between the samples DLC and Si-DLC at ambient temperature.

The results of the MTT-assay shown in the histogram in Fig. 6.16, give the optical density of the active cells in culture after fifty-six (56) hours. The averaged-blanks values were subtracted from the total reading in order to take into account the optical density contribution due to the colour of the as-deposited DLC and Si-DLC thin films. This was confirmed by using the reader to read the optical density of the empty culture dishes, both coated and non-coated, and it was observed that there were some readings. The assay detects metabolism in the living cells and the intensity obtained is dependent on the degree of activity of the cells [107]. Further details of the statistical analysis of the MTT assay of HMECs on DLC and Si-DLC films

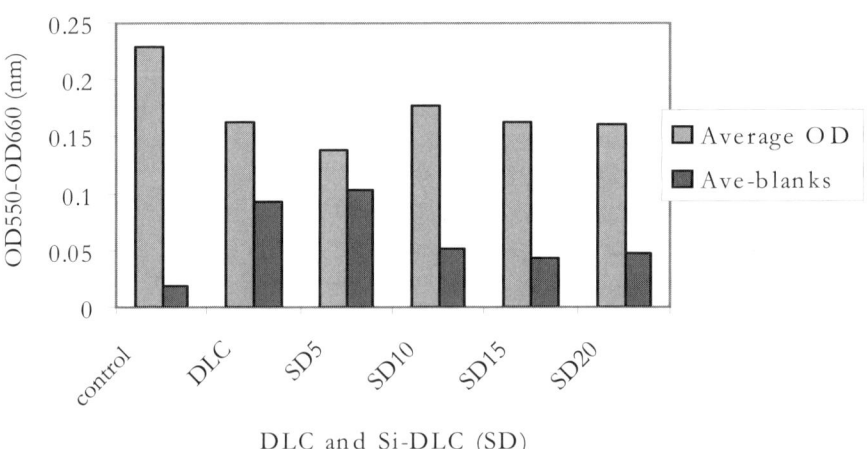

Fig. 6.16. Histogram of MTT-assay of HMEC seeded on 96 well culture plates that were coated with DLC and Si-DLC (TMS flow-rate, 5-20sccm, SD5-20)

using the ANOVA test has been reported elsewhere by some of the authors of this chapter [108].

6.7 Summary

The contents of the current chapter can be summarised as follows:

- Mechanical heart valves were introduced and the principal concerns relating to such valves were put forward.
- The history of mechanical heart valves has been reviewed dating back from 1952.
- The process of thrombosis has been defined, described and the principals of thrombus formation have been outlined.
- The chapter also discusses the hemocompatibility of biomaterials. Further, we describe the endothelium, its functions and endothelial cell seeding in modern biomaterials engineering.
- The hemocompatibility of DLC films has been reviewed and in particular results of investigations on Cr-DLC and Si-DLC films have been presented and discussed.
- The diamond-like films were deposited containing different levels of Cr and Si contents.
- The contact angle measurements showed that the surface contact angle gradually increased with %Cr content in the films films and began to level off at around 10%Cr content.
- Raman spectroscopy was used to characterise the D and G bands present in all the Cr-DLC films.
- The I_D/I_G ratios were correlated to the population of endothelial cell attachments onto Cr-DLC surfaces. High values of I_D/I_G ratio correlate with better HMC-EC attachment to 5%Cr-DLC film. This finding suggests that disordered graphitic phases in the DLC film leads to enhancement in the seeding of endothelial cells.
- With silicon modified DLC films, we observed a direct correlation between adherent cell population on the film surfaces and electronic conduction.

- This correlation does not hold when graphitisation, which is detected by Raman spectroscopy has occurred in the Si-DLC films regardless of their electronic conductivity.
- The micro-structural variation in the films was achieved by varying silicon content and by thermal annealing.
- A direct relationship between increasing contact angle of water (i.e., hydrophobicity) and adherent endothelial cell population was observed. This trend is the opposite of those reported in the literature for other cells, where a hydrophilic interaction was considered necessary for cell attachment on surfaces.
- We suspect that the interaction between endothelial cells and the Si-DLC films could be mediated by dispersive forces.
- Although the work function decreased with increasing silicon addition, and silicon addition lead to an initial increase in the population of adherent endothelial cells on the film surface, further silicon addition did not lead to any substantial changes in the number of adherent endothelial cells on the Si-DLC film surface.
- The cytotoxicity of the endothelial cells on the Si-DLC film was measured by the MTT-assay test and the cells are established to be viable after adhesion to the Si-DLC films.
- The biological response of human microvascular endothelial cells seeded on chromium and silicon modified diamond-like carbon films and on control surfaces was evaluated in terms of initial cell enhancement, growth and cytotoxicity.
- Endothelial cell adhesion and growth was found to be affected by changes in the microstructure of the films induced by chromium/silicon doping and thermal annealing of silicon modified diamond-like carbon films.
- We observed a significant statistical difference in endothelial cell count between DLC and Si-DLC films using the paired t-test.
- MTT-assay tests showed the endothelial cells to be viable when seeded on Si-DLC films before and after thermal treatment based on the ANOVA statistical test.

References

1. K. Barton, A. Campbell, J.A. Chinn, C.D. Griffin, D.H. Anderson, K. Klein, M.A. Moore and C. Zapanta, *Biomedical Engineering Society (BMES) Bulletin*, Vol. 25, No. 1, (2001) 3.
2. C.A. Hufnagel and W.P. Harvey, *Bull. Georgetown Univ. Med. Cent.*, 6: 60-63 1953).
3. L. Vincent, M.D. Gott, E. Diane Alejo, and E. Duke, M.D. Cameron, *Ann Thorac Surg* 2003; 76: S2230-S2239.
4. G. Murray, *Angiology*, 7: 466 (1956).
5. N.S. Braunwald, T. Cooper and A.G. Morow, *J. Thorac. Cardiovasc. Surg.*, 40: 1-11 (1960).
6. H.T. Bahnson, F.C. Spencer, E.F.G. Busse and F.W. Davis, Jr., *Ann. Surg.*, 152: 494 (1960).
7. B.B. Roe, J.W. Owsley and P.C. Boudoures, *J. Thorac. Cardiovasc. Surg.*, 36: 563-570 (1958).
8. B.B. Roe, *J. Thorac. Cardiovasc. Surg.*, 58: 59-61 (1969).
9. N.S. Braunwald and A.G. Morrow, *J. Thorac. Cardiovasc. Surg.*, 49: 485-496 (1965).
10. A. Richard, M.D. DeWall, M.D. Naureen Qasim and Liz Carr, *Ann Thorac Surg* 2000; 69: 1612-21.
11. V.L. Gott, R.L. Daggett, J.D. Whiffen et al., *J. Thorac. Cardiovasc. Surg.*, 48: 713-725 (1964).
12. A.B. Cruz, R.L. Kaster, R.L. Simmons, C.W. Lillehei, *Surgery* 1965; 58: 995-8.
13. J. Wada, S. Lomatsu, K. Ikeda et al., A new hingeless valve. In: *Prosthetic Heart Valves* (Brewer, L.A., editor), Springfield Thomas, pp. 304-314 (1969).
14. V.O. Björk, *J. Thorac. Cardiovasc. Surg.*, 3: 1-10 (1969).
15. R. Carmen and S.C. Mutha, *J. Biomed. Mater. Res.*, 6: 327-346 (1972).
16. V.O. Björk, *The Surgical Treatment of Aortic Valve Disease*. Ingelhelm and Rhein, C.H. Boehringer Sohn, (1974).
17. B.J. Messmer, M. Rothlin and A. Senning, *J. Thorac. Cardiovasc. Surg.*, 65, 386-390 (1973).
18. V.L. Gott, J.D. Whiffen and S.M. Valiathan, *Ann. NY. Acad. Sci.,* 146: 21-29 (1968).
19. S.M. Scott, G.K. Sethi, A.H. Bridgman and T. Takaro, *Ann. Thorac. Surg.*, 21: 483-486 (1976).
20. A.C. Beall, Jr., G.C. Morris, Jr., G.P. Noon et al., *Ann. Thorac. Surg.*, 15: 25-34 (1973).
21. Cedars-Sinai Medical center Prosthetic heart valve information, Division of Cardiology: http://www.csmc.edu/pdf/Heart_Valves.pdf
22. C.W. Lillehei, R.L. Kaster, M. Coleman and J.H. Bloch, NY *State Med. J.*,74: 1426-1438 (1974).
23. FDA panel meeting for approval of the OmniCarbon®valve,PMA# P830039, 1998.
24. di Summa, M. Poletti, Breno, L. et al., *J Heart valve Dis* 2002; 11: 517-23.

25. P.J.K. Starek, L.P. McLaurin, B.R. Wilcox, G.F. Murry, *Ann. Thorac. Surg.*, 22: 362-368 (1976).

26. P.J.K. Starek, R.L. Beaudet and V.K. Hall, The Medtronic-Hall valve: Development and clinical expe-rience. In: *Cardiac Surgery Current Heart Valve Prosthesis* (Crawford, F.A., ed.), Hanley & Belfus,Philadelphia, pp. 223-236,1987.

27. *Food and Drug Administration Enforcement Report*, September 7, (1988).

28. W.P. Young, R.L. Daggett, VL. Gott, Long-term follow-up of patients with a hinged leaflet prosthetic heart valve. In: Brewer L.A., ed. *Prosthetic heart valves*. Springfield, IL: Charles C. Thomas, (1969) 622-32.

29. B.R. Kalke, *Evaluation of a double-leaflet prosthetic heart valve of a new design for clinical use*, Ph.D. Thesis, University of Minnesota, (1973).

30. *Lessons of Björk-Shiley Heart Valve Failure.* www.me.utexas.edu/~uer/heartvalves/shiley.html.

31. J.J. Klawitter, Design and in vitro testing of the Duromedics bileaflet valve. *First International Hemex Symposium on the Duromedics Bileaflet Valves* (1985).

32. R. Richard, A. Beavan and P. Strzepa, *J. Heart Valve Dis.*, 3: S94-S101 (1994).

33. J. Craver, Carbomedics Prosthetic Heart Valve(tm). *Euro Jl Cardio-Thorac. Surg.* 15 (Suppl.1) S3-Sll (1999).

34. Prosthetic heart valves: History of mechanical heart valve replacement: *BMES BULLETIN*; No4: vol. 24, (2000).

35. A. Campbell, T. Baldwin, G. Peterson, J. Bryant and K. Ryder, *J. Heart Valve Dis.*, 5: S124-S132 (1996).

36. B. Lung, T. Haghigat, E. Garbaz et al., Incidence and predictors of prosthetic thrombosis on mitralbileaflet prostheses during the postoperative period. *Congress of the European Society of Cardiology*, August 1999.

37. E. Bodnar, P. Arru, E.G. Butchard et al., Panel discussion, *J. Heart Valve Dis.*, 5: S148 (1996).

38. J. Gross, M. Shu, F. Dai, J. Ellis and A. Yoganathan, *J. Heart Valve Dis.*, 5: 581-590 (1996).

39. http://www.accessdata.fda.gov/scripts/cdrh/cfdocs/cfTopic/mda/mda-cardio.cfm?topic=421>

40. M.H. Henri Spronk, Danielle van der Voort and Hugo ten Cate, *Thrombosis Journal* 2004, 2: 12 doi: 10.1186/1477-9560-2-12.

41. William DF (ed.) *Definitions in Biomaterials*, Elsevier, (1987)b.

42. S.H. Teoh, *International Journal of Fatigue* 22 (2000) 825-837.

43. A. Klinger D. Steinberg, D. Kohavi & M.N. Sela, (1997), *J. Biomed. Mater. Res.* 36: 387-392.

44. BMEn 5001-"cardiovascular" *Applications of Biomaterials*; November 25, 1998-W.Gleason.

45. http://www.biomed.metu.edu.tr/courses/term_papers/artif-heart-valves_erol.htm

46. C.E. Soltys-Robitaille, D.M. Ammon, Jr. , P.L. Valint Jr. and G.L. robe III, *Biomaterials* 2001; 22: 3257-3260.

47. V. Pesakova, Z. Klezl, K. Balik, M. Adam. *Journal of Material Science: Materials in Medicine* 2000; 11(12): 793-798.
48. G.M. Bruinsma, H.C. Van der Mei, H.J. Busscher. *Biomaterials* 2001: 22; 3217-3224.
49. A. Ahluwalia, G. Basta, F. Chiellini, D. Ricci, G. Vozzi, *Journal of Material Science: Materials in Medicine* 2001; 12(7): 613-619.
50. G.L. Bowlin and S.E. Rittger. *Cell Transplantation* 1997; 6(6): 623-629.
51. P.B. Van Wachem, J.M. Schakenraad, J. Feijen, T. Beugeling, W.G. Van Aken, E.H. Blaauw, P. Nieuwenhius, I. Molenaar. *Biomaterials* 1989; 10: 532-539.
52. S.D. Bruck, *Polymer* 1975; 16: 25.
53. S.D. Bruck, *Nature* 1973; 243: 416-417.
54. S.D. Bruck, *J. Poly Sc* 1967; C17: 169-185.
55. A. Boldz, M. Schaldach, *Artificial Organs* 1990; 14(4): 260-269.
56. J.Y. Chen, L.P. Wang, K.Y. Fu, N. Huang, Y. Leng, Y.X. Leng, P. Yang, J. Wang, G.J. Wan, H. Sun, X.B. Tian, P.K. Chu, *Surface and Coatings Technology* 2002; 156: 289-294.
57. http://greenfield.fortunecity.com/rattler/46/endothelium.htm
58. http://teaching.anhb.uwa.edu.au/mb140/MoreAbout/Endothel.htm
59. J.L. Gordon, 1986 in *Blood-Surface Interactions: Biological Principles Underlying Hemocompatibility with Artificial Materials* edited by Cazenave J.P, Davies J.A, Kazatchkine M.D, and W.G van Aken; Elsevier Science Publishers (Biomedical Division) 1986; pp. 5.
60. S. Williams, *Cell Transplantation* 4: 401-410, 1994.
61. T. Horbett, *Colloids Surf. B: Biointerfaces* 2: 225-240, 1994.
62. Veli-Matti Tianen, *Diamond and Related Materials* 10 (2001) 153-160.
63. A. Grill, B.S. Meyerson, Development and status of diamondlike carbon, in K.E. Spear, J.P Dismukes (eds), *Synthetic Diamond; emerging CVD science and technology*, Wiley, New York, 1994, Chapter 5.
64. A. Schroeder, G. Francz, A. Bruinink, R. Hauert, J. Mayer, E. Wintermantel, *Biomaterials* 21 (5) (2000) 449-456.
65. R. Hauert, L. Knoblauch-Meyer, G. Francz, A. Schroeder, E. Wintermantel, *Surf. Coat. Technol.* 120-121 (1999) 291-296.
66. R. Hauert, U. Muller, G. Francz, et al., *Thin Solid Films* 308-309 (1997) 191-194.
67. A. Dorner-Reisel, C. Schurer, C. Nischan, O. Seidel, E. Muller, *Thin Solid Films* 420-421 (2002) 263-268.
68. R. Hauert, U. Muller, *Diamond and Related Materials* 12 (2003) 171-177.
69. N. Huang, P. Yang, Y.X. Leng, J. Wang, J.Y. Chen, H. Sun, G.J. Wan, A.S. Zhao and P.D. Ding, "Surface Modification for Controlling the Blood-Materials Interface": invited report on *6th Asia Symposium on Biomedical Materials*, July 20-23, 2004, Chengdu, China and published in *Key Engineering Materials* http://www.paper.edu.cn/scholar/download.jsp?file=huangnan-6
70. Y.X. Leng, N. Huang et al., *Surface Science*, Vol. 531 (2003), p. 177.

71. P. Yang, J.Y. Chen, Y.X. Leng, H. Sun, N. Huang, P.K. Chu, 7th International Workshop on Plasma Based Ion Implantation, Sept. 16-20, 2003, San Antonio, USA, 2003.
72. N. Huang, P. Yang, Y.X. Leng, J. Wang, H. Sun, J.Y. Chen, G.J. Wan, *Surface & Coatings Technology* 186 (2004) 218- 226.
73. P. Yang, S.C.H. Kwok, P.K. Chu, Y.X. Leng, J.Y. Chen, J. Wang, N. Huang, *Nuclear Instruments and Methods in Physics Research*. Section B, Beam Interact. Mater. Atoms 206 (2003) 721.
74. M.I. Jones, I.R. McColl, D.M. Grant, K.G. Parker, T.L. Parker, *Diamond and and Related Materials* 8 (1999) 457-462.
75. M.I. Jones, I.R. McColl, D.M. Grant, K.G. Parker, T.L. Parker, *J. Biomed. Mater. Res.* 52/2 (2000) 413-421.
76. I. Dion, X. Roques, C. Baquey, E. Baudet, B. Basse Cathalinat, N. More, *Bio-Med. Mater. Eng.* 3y1 (1993) 51-55.
77. F.Z. Cui and D.J. Li., *Surf. Coat. Technol.* 131 (2000), pp. 481-487.
78. K. Gutensohn, C. Beythien, J. Bau, T. Fenner, P. Grewe, R. Koester, K. Padmanaban and P. Kuehnl, *Thrombosis Research* 99 (2000) 577-585.
79. A. Alanazi, C. Nojiri, T. Noguchi, T. Kido, Y. Komatsu, K. Kirakuri, et al., *ASAIO J.* 46 (2000) 440-443.
80. L.A. Thomson, F.C. Law, N. Rushton, J. Franks, *Biomaterials* 12 (1991) 37.
81. M. Allen, F. Law, N. Rushton, *Clin. Mater.* 17 (1994) 1.
82. I. De Scheerder, M. Szilard, H. Yanming, et al., *J. Invasive Cardiol.* 12 (8) (2000) 389-394.
83. A. Schroeder, G. Francz, A. Bruinink, R. Hauert, J. Mayer, E. Wintermantel, *Biomaterials* 21 (5) (2000) 449-456.
84. G. Francz, A. Schroeder, R. Hauert, *Surf. Interface Anal.* 28 (1999) 3.
85. R. Hauert, L. Knoblauch-Meyer, G. Francz, A. Schroeder, E. Wintermantel, *Surf. Coat. Technol.* 120-121 (1999) 291-296.
86. R. Hauert, U. Muller, G. Francz, et al., *Thin Solid Films* 308-309 (1997) 191-194.
87. A. Schroeder, *Ph.D. Thesis*, Dissertation Nr. 13079, ETH Zurich (1999).
88. A. Dorner-Reisel, C. Schurer, C. Nischan, O. Seidel, E. Muller, *Thin Solid Films* 420-421 (2002) 263-268.
89. P. Yang, N. Huang, Y.X. Leng, J.Y. Chen, H. Sun, J. Wang, F. Chen and P.K. Chu, *Surface and Coatings Technology*, 156 (2002) 284-288.
90. J.Y. Chen, Y.X. Leng, X.B. Tian, L.P. Wang, N. Huang, P.K. Chu and P. Yang, *Biomaterials*, 23 (2002) 2545-2552.
91. Y.X. Leng, H. Sun, P. Yang, J.Y. Chen, J. Wang, G.J. Wan, N. Huang, X.B. Tian, L.P. Wang and P.K. Chu, *Thin Solid Films*, 398-399, (2001) 471-475.
92. Y.X. Leng, P. Yang, J.Y. Chen, H. Sun, J. Wang, G.J. Wang, N. Huang, X.B. Tian and P.K. Chu, *Surface and Coatings Technology*, 138 (2001) 296-300.
93. J. Li, *Biomaterials*, 14 (1993) 229.
94. A.A. Adjaottor, E. Ma and E.I. Meletis, *Surface and Coatings Technology*, 89 (1997) (3), pp. 197-203.

95. F. Grinnell, *Int. Rev. Cytol.* 1978; 53: 65-144.
96. A.A. Ogwu, R.W. Lamberton, S. Morley, P. Maguire, J. McLaughlin. *Physica B* 1999; 269: 335-344.
97. A.P. Dementjev, M.N. Petukhov, A.M. Baranov, *Diamond and Related Materials* 1998; 7: 1534-1538.
98. A. Grill, B. Meyerson, V. Patel, J.A. Reimer and M.A. Petrich, *J. Applied Physics*, 1987; 6: 2874.
99. L. Constant, Le Normand, *Diamond and Related Materials* 1997; 6: 664-7.
100. M.A. Baker and P. Hammer, *Surface and Interface Analysis* 1997; 25: 629-642.
101. F. Grinnell, *Int. Rev. Cytol.* 1978; 53: 65-144.
102. D.P. Magill, A.A. Ogwu, J.A.D. McLaughlin, P.D. Maguire, *J. Vac. Sci. Technol. A* 2001; 19(5): 2456-2462.
103. A. Hadjaj, R.E. Cabarrocas, B. Equar, *Philosophical Magazine* 1997; B76:941.
104. A. Hadjaj, M. Favre, B. Equer, R.I. Cabaroccas, *Solar Energy Materials and Solar Cells* 1998; 51: 145-153.
105. G. Attard and C. Barnes, *Surfaces* 1998; Oxford University Press: 64-65.
106. A. Boldz, M. Schaldach, *Artificial Organs*, 1990; 14(4): 260-269.
107. H. Wan, R.L. Williams, P.J. Doherty, D.F. Williams, *Journal of Materials Science: Materials in Medicine* 1994; 5: 441-445.
108. T.I.T. Okpalugo, E. McKenna, A.C. Magee, J.A. McLaughlin, N.M.D. Brown, *Journal of Biomedical Materials Research Part A*, 2004; 71A (2): 201-208.

7. Diamond Surgical Tools

7.1 Introduction

Deposition technology has played a major part in the creation of today's scientific devices. Computers, electronic equipment, biomedical implants, cutting tools, optical components, and automotive parts are all based on material structures created by thin film deposition processes. There are many coating processing ranging from the traditional electroplating to the more advanced laser or ion-assisted deposition. However, the choice of deposition technology depends upon many factors including substrates properties, component dimensions and geometry, production requirements, and the exact coating specification needed for the application of interest. For complex geometry components, small feature sizes, good reproducibility and high product throughput chemical vapor deposition (CVD) is a highly effective technology. For example, low-pressure and plasma assisted CVD is a well-established technology for semiconductor devices, which has very small feature sizes and complex geometrical arrangements on the surface.

In order to understand both physical vapor deposition (PVD) and CVD processes, one has to model them in terms of several steps. These processes can be divided into the following stages:

- **Generation of Vapor Phase Species**

  The precursor materials are converted into a convenient form so that transport to the substrates is efficient. A vapor is generated in the reactor. Hot filaments, lasers, microwave, ion beams, electrons guns, etc., can be used to activate the source materials, enabling deposition to be carried out.

- **Transport of Source Materials to the Substrate Region**

  The vapor species are transported from the source to the substrate with or without collisions between the atoms and molecules. During transport, some of the species can be ionized by creating plasma in this space. This is normally carried out in a vacuum system; however, atmospheric CVD systems are also employed.

- **Adsorption of Active Species on the Substrate Surface**

  For deposition to take place the active species must first be adsorbed onto the active sites on the surfaces. Initially this occurs via physisorption where the species adhere to the surface with weak van der Waals forces and then strong covalent bonds are formed between the species and the surface known as chemisorption.

- **Decomposition Adsorbed Species on the Substrate Surface**

  Once the gaseous species are adsorbed onto surface site and the energy of the species is sufficient then decomposition of the precursors can take place resulting in the creation of nucleation center.

- **Nucleation and Film Formation**

  The process involves the subsequent formation of the film via nucleation and growth processes. These can be strongly influenced by process parameters resulting in a change in the microstructure, composition, impurities, and residual stress of the films. The final film properties are highly dependent on the microstructural and interfacial characteristics of the deposited coating.

Independent control of these stages is critical and determines the versatility or flexibility of deposition process. For example, PVD process parameters can be independently and precisely monitored and controlled. Thus allowing microstructure, properties, and deposition rates to be tailored specifically to the performance requirements of the product. Generally, CVD processes have the advantage of good throwing power enabling complex geometry substrates to be coated, while the deposition rates in PVD processes are much higher than those in CVD processes at lower deposition temperatures.

Although CVD and PVD processes are simple in principle, one must be well versed in vacuum technology, physics, chemistry, materials science, mechanical and electrical engineering as well as in elements of thermodynamics, chemical kinetics, surface mobility and condensation phenomena in order to obtain a detailed fundamental understanding of these processes. In this chapter we restrict our attention to the deposition of diamond thin films for use in cutting tools.

7.2 Properties of Diamond

Diamond is an advanced material with an excellent combination of physical and chemical properties. If high-quality diamond films with comparable properties to natural diamond can be formed with low surface roughness numeral potential applications will emerge in the near future particularly in the emerging field of nanotechnology.

Diamond as a material possesses a remarkable range of physical attributes, which make it a promising material for a large range of applications. Selections of these are given in Table 7.1. However, owing to the cost and availability of large natural diamonds, most of these applications have not been developed to their full potential.

7.3 History of Diamond

7.3.1 Early History of Diamond Synthesis

Diamond is one of the most technologically and scientifically valuable crystalline solids found in nature. Their unique blends of properties are effectively incomparable to any other known material. Sir Isaac Newton was the first to characterize diamond and determine it to be of organic origin while in 1772, the French chemist Antoine L. Lavoisier established that the product of diamond combustion was limited to carbon dioxide.

Table 7.1. Properties of diamond

Properties	Applications
High wear resistance	Cutting tools
Chemical inertness	Electrochemical Sensors
High thermal conductivity	Heat spreaders
Biological inertness	In vitro applications
Semiconducting when doped	Electronic devices
High resistivity (insulator)	Electronic devices
Negative electron affinity	Cold cathode electron sources

English chemist Smithson Tennant showed that diamond combustion products were no different than those of coal or graphite and resulted in "bound air". Later, the discovery of x-rays enabled Sir William Henry Bragg and his son Sir William Lawrence Bragg to determine that carbon allotropes were cubic (diamond), hexagonal (graphite) and amorphous. With this information, early attempts to synthesize diamond began in France in 1832 with C.C. de la Tour and later in England by J.B. Hanney and H. Moisson. The results of their work are disputed to this day.

Synthesis of diamond has attracted widespread attention ever since it was established that diamond is a crystalline form of carbon. Since diamond is the densest carbon phase, it became immediately plausible that pressure, which produces a smaller volume and therefore a higher density, may convert other forms of carbon into diamond. As understanding of chemical thermodynamics developed throughout the 19th and 20th centuries, the pressure-temperature range of diamond stability was explored. In 1955, these efforts culminated in the development of a high pressure-high temperature (HPHT) process of diamond synthesis with a molten transition metal solvent-catalyst at pressures where diamond is the thermodynamically stable phase [1]. Three major problems can be isolated for emphasizing the difficulty of making diamond in the laboratory. First, there is difficulty in achieving the compact, strongly bonded structure of diamond, which requires extreme pressure. Secondly, even when such a high pressure has been achieved, a very high temperature is required to make the conversion from other forms of carbon to diamond proceed at a useful rate. Finally, when diamond is thus obtained, it is in the form of very small grains and to achieve large single crystal diamond requires yet another set of constraints. However, less well known has been a parallel effort directed toward the growth of diamond at low pressures where it is metastable. Metastable phases can form from precursors with high chemical potential if the activation barriers to more stable phases are sufficiently high. As the precursors fall in energy, they can be trapped in a metastable configuration. Formation of a metastable phase depends on selecting conditions in which rates of competing processes to undesired products are low [2]. In the case of diamond, achieving the appropriate conditions has taken decades of research [3]. The processes competing with diamond growth are spontaneous graphitization of the diamond surface as well as nucleation and growth of graphitic deposits.

The most successful process for low-pressure growth of diamond has been chemical vapor deposition (CVD) from energetically activated hydrocarbon/hydrogen gas mixtures. CVD is a process whereby a thin solid film, by definition, is synthesized from the gaseous phase via a chemical reaction. The development of CVD in common with many technologies, has been closely linked to the practical needs of society. The oldest example of a material deposited by CVD is probably that of pyrolytic carbon, since as Ashfold et al. [4] pointed out, some prehistoric art was done on cave walls with soot condensed from the incomplete oxidation of firewood. A similar procedure formed the basis of one of the earliest patents and commercial exploitation of a CVD process, which was issued for the preparation of carbon black as a pigment. The emerging electric lamp industry provided the next major application of CVD with a patent issued for improvements to fragile carbon filaments [4]. Since these improved filaments were far from robust, the future for a pyrolytic carbon CVD industry was limited and a few years later, processes for the deposition of metals to improve the quality of lamp filaments were described [5]. From the turn of the century through to the late 1930s a variety of techniques appeared for the preparation of refractory metals for a number of applications. It was also during this period that silicon was first deposited by hydrogen reduction of silicon tetrachloride [5] and the use of that material for electronic applications was foreseen by the development of silicon-based photo cells [6] in 1946 as well as rectifiers [7]. The preparation of high-purity metals, various coatings, and electronic materials has developed significantly in the last 45 years or so, but it is undoubtedly the demands and requirements of the semiconductor and microelectronic industries that have been the main driving force in the development of CVD techniques as well as the greater efforts for understanding the basics of CVD processes. Consequently, a large body of literature and reviews now exists on CVD.

Indeed, it was the chemical vapor deposition from carbon-containing gases that enabled W.G. Eversole, referred to in reference [8], at the Union Carbide Corporation to be the first to grow diamond successfully at low pressures in 1952, after which conclusive proof and repetition of the experiments took place. In the initial experiments, carbon monoxide was used as a source gas to precipitate diamond on a diamond seed crystal. However, in subsequent experiments, methane and other carbon-containing gases were used as well as a cyclic growth etches procedure to remove co-deposited graphite. In all of his studies, it was necessary to use diamond seeds in order to initiate diamond growth. The deposits were identified as diamond by density measurements, chemical analysis, and diffraction techniques. The synthesis by Eversole preceded the successful diamond synthesis at high pressure by workers at the General Electric Company [1], which was accomplished in 1954. However, the important difference was that Eversole grew diamond on pre-existing diamond nuclei whereas the General Electric syntheses did not initiate growth on diamond seed crystals. Deryagin [9,10] in the former Soviet Union began work on low-pressure diamond synthesis in 1956, in which many approaches were taken, which started with the growth of diamond whiskers by a metal-catalyzed vapor-liquid-solid process. Subsequently, epitaxial growth from hydrocarbons and hydrocarbon/hydrogen mixtures was investigated as well as different forms of vapor transport reactions. In addition, theoretical investigations of the relative nucleation rates of diamond and graphite were also performed. Angus and co-workers at Case Western Reserve University concentrated primarily on diamond CVD on diamond seed crystals from hydrocarbons and hydrocarbon/hydrogen mixtures [11,12]. They grew p-type semiconducting diamond from methane/diborane gas mixtures and studied the rates of diamond and graphite growth in methane/hydrogen gas mixtures and ethylene. They were the first to report on the preferential etching of graphite compared to diamond by atomic hydrogen and noted that boron had an unusual catalytic effect on metastable diamond growth.

The role of hydrogen in permitting metastable diamond growth was also recognised by some early workers. The low energy electron diffraction (LEED) study of Lander and Morrison [13,14] showed that a {111}-diamond surface saturated with hydrogen gave an unrecon-structed (1×1) LEED pattern. The unsatisfied dangling bonds normal to the surface are terminated with hydrogen atoms, which maintain the bulk terminated diamond lattice to the outermost surface layer of carbon atoms. When hydrogen is absent, the surface reconstructs into more complex structures. They also showed that carbon atoms are very mobile on the diamond surface at temperatures above 1200 K and stated that these conditions should permit epitaxial growth. Other work [15,16] suggested that the presence of hydrogen enhanced diamond growth. Chauhan et al. [17,18] as well as Deryagin et al. [19] showed that addition of hydrogen to the hydrocarbon gas-phase sup-pressed the growth-rate of graphite relative to diamond thus resulting in higher diamond yields. Eventually, however, graphitic carbons nucleated on the surface and suppressed further diamond growth. It was then necessary to remove the graphitic deposits preferentially with atomic hydrogen [20] or oxygen [21], and to repeat the sequence. By the mid-1970s diamond growth at low pressures had been achieved by several groups. The beneficial role of hydrogen was known to some extent and growth rates of 0.1 μm hr^{-1} had been achieved. Although the growth-rates were too low to be of any commercial importance, the results provided the experimental foundation for much of the work that followed.

7.3.2 Modern Era of Metastable Diamond Growth

Japanese researchers associated with the National Institute for Research in Inorganic Materials (NIRIM) made the first disclosures of methods for rapid diamond growth at low pressures. Research on metastable diamond growth was initiated at NIRIM in 1974. In 1982, they des-cribed techniques for synthesizing diamond at rates of several microns per hour from gases decomposed by a hot filament as well as micro-wave or DC discharges [22–25]. These processes produced individual faceted crystals without the use of a diamond seed crystal. The current worldwide interest in new diamond technology can in fact be directly

traced to the NIRIM effort. Although Deryagin, as reported in reference [26], had reported high rate diamond growth earlier, process
details were not disclosed [26]. All of the techniques are based on
the generation of atomic hydrogen in the vicinity of the growth surface during deposition. Although the chemical vapor deposition of
diamond from hydrogen rich/hydrocarbon-containing gases has been
the most successful method of diamond synthesis, numerous other
methods have been attempted with varying degrees of success, with
ion beam methods being the most successful [27]. In 1971 hard carbon
films were first deposited using a beam of carbon ions. As the films
had many of the properties of diamond, they were called diamond-
like carbon because definitive diffraction identification was not possible. In 1976, Spencer et al. [28] formed finely divided polycrystalline
diamond using a beam of carbon ions with energies between 50 and
100 eV, and subsequently Freeman [29] grew diamond via ion implantation.

With further research and additional technological progress in
improving and devising new methods for synthesis and fabrication,
it becomes increasingly likely that new applications will be discovered.
In order to be able to take full advantage of the unique characteristics of diamond as a material for the construction of solid-state devices,
basic scientific understanding of the various experimental process
techniques and in particular the introduction and activation of dopants
must be obtained. Attention also needs to be paid to proper design of
devices incorporating novel features utilizing concepts and practices
established in silicon and gallium arsenide device technology. The
potential of diamond as a material for solid-state devices has been
the subject of a few reviews [30–36] that have discussed the electronic
material parameters of diamond and the simulated characteristics
that can be obtained. Simple devices incorporating diamond have
been demonstrated primarily incorporating natural or HPHT diamond.
Photodetectors, light-emitting diodes, nuclear radiation detectors,
thermistors, varistors, and negative resistance devices in synthetic
crystals have been demonstrated. Several groups [37–41] have also
demonstrated basic field effect transistor device operation in epitaxial
diamond films and boron-doped layers on single crystal diamond
substrates. However, for wide application of diamond solid-state

devices, high-quality films on more commonly available substrates are essential as well as studies on the device potential of polycrystalline films. So far only thermistors [42] and Schottky diodes [43] have been produced and characterized in the polycrystalline material. This is due to material problems, in that the polycrystalline nature of the films results in grain boundaries, twins, stacking faults, and other defects, which have restricted exploitation in the electronic industries. To date there has been no confirmed observations of a means of achieving heteroepitaxy, that is, single-crystal diamond grown on a non-diamond substrate, and therefore no means of achieving diamond devices for practical applications. Indeed, achieving heteroepitaxy stands as the single most prominent technological hurdle for diamond-based electronics. However, CVD synthesis is a very active area that is improving with experience. In the near future, in situ probes may be used to optimize various diamond CVD processes by providing a maximization of the flow of diamond precursors to the surface while simultaneously minimizing the competing deposition of non-diamond carbon forms. The wide variety of means by which diamond is being routinely formed as a film will enhance its deployment and the potential for active electronic exploitation. Indeed, diamond coatings in general are expected to make so large an impact in the future that many people believe that that the future age will be known as the diamond age, going chronologically from the stone age, bronze age, to the iron age of the past and the silicon age of the present.

7.4 CVD Diamond Technology

The reactor system (comprising the reaction chamber and all associated equipment) for carrying out CVD processes must provide several basic functions common to all type of systems (Fig. 7.1). It must allow transport of the reactant and diluents gases to the reaction site, provide activation energy to the reactants (heat, radiation, plasma), maintain a specific system pressure and temperature, allow the chemical processes for film deposition to proceed optimally, and remove the by-product gases and vapors. These functions must be implemented with adequate control, maximal effectiveness, and complete safety.

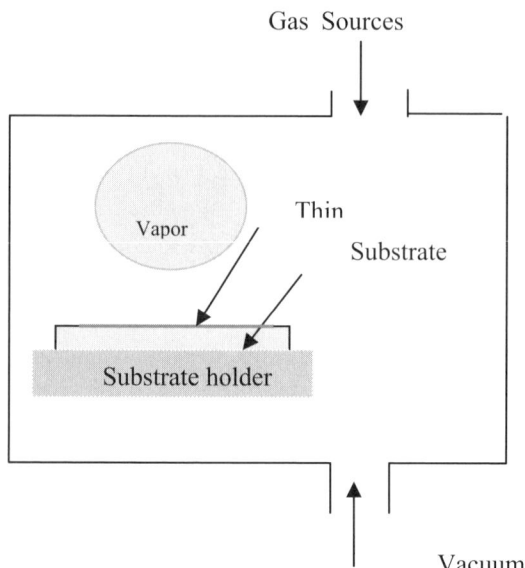

Fig. 7.1. A simple schematic of a vapor deposition process

Chemical vapor deposition is a crystal growth process used not only for diamond but also for a range of different semiconductor and other crystalline materials such as silicon or gallium arsenide. These industrial fields are diverse and range from gas turbines to gas cookers and from coinage to nuclear power plants.

The CVD process relies first on the generation of a species that is produced by the reaction of the element that is to be deposited with another element that results in the substantial increase in the depositing elements vapor pressure. Second, this volatile species is then passed over or allowed to come into contact with the substrate being coated. This substrate is held at an elevated temperature, typically from 800 –1150°C. Finally, the deposition reaction usually occurs in the presence of a reducing atmosphere, such as hydrogen. The film properties can be controlled and modified by varying the problem parameters associated with the substrate, the reactor, and gas composition.

7.5 CVD Diamond Processes

Several different approaches to the deposition of diamond have been investigated and these include the ones described in the following sections.

7.5.1 Plasma-Enhanced CVD

Plasmas generated by various forms of electrical discharges or induction heating have been employed in the growth of diamond. The role of the plasma is to generate atomic hydrogen and to produce the necessary carbon precursors for diamond growth. The efficiencies of the different plasma processes vary from method to method. Three plasma frequency regimes will be discussed. These are microwave plasma CVD, which typically uses excitation frequencies of 2.45 GHz; Radio-frequency (RF) plasma excitation, which employs frequencies of usually 13.56 MHz, and direct-current (DC) plasmas, which can be run at low electric powers (a "cold" plasma) or at high electric powers (which create an *arc* or a *thermal* plasma).

RF Plasma-Enhanced CVD

Generally, radio-frequency power can be applied to create plasma in two electrode configurations, namely, in an inductively coupled or a capacitively coupled parallel plate arrangement. A number of workers have reported the growth of diamond crystals and thin films using inductively coupled RF plasma methods [21–24] as well as capacitively coupled methods [25–26]. A high power in the discharge leading to greater electron densities was found to be necessary for efficient diamond growth. However, the use of higher power results in physical and chemical sputtering from the reactor walls, leading to contamination of the diamond films [21]. The advantage of RF plasmas is that they can easily be generated over much larger areas than microwave plasmas but the method is not routinely applied for the deposition of diamond films.

DC Plasma-Enhanced CVD

In this method, plasma in a H_2-hydrocarbon mixture is excited by applying a DC bias across two parallel plates, one of which is the substrate [27–29]. DC plasma-enhanced CVD has the advantage of being able to coat large areas as the diamond deposition area is limited by the electrodes and the DC power supply. In addition, the technique has the potential for very high growth rates. However, diamond films produced by DC plasmas were reported to be under high stress and to contain high concentrations of hydrogen as well as impurities resulting from plasma erosion of the electrodes.

Microwave Plasma-Enhanced CVD

Microwave-plasma assisted CVD has been used more extensively than any other method for the growth of diamond films [14–20]. Microwave plasmas are different from other plasmas in that the microwave frequency can oscillate electrons. Collision of electrons with gaseous atoms and molecules generates high ionization fractions. This method of diamond film growth has a number of distinct advantages over the other methods of diamond film growth. Microwave deposition, being an electrodeless process, avoids contamination of the films due to electrode erosion. Furthermore, the microwave discharge at 2.45 GHz, being a higher frequency process than the RF discharges at typically 13.56 MHz, produces a higher plasma density with higher energy electrons, which effectively results in higher concentrations of atomic hydrogen and hydrocarbon radicals leading to efficient diamond growth. In addition, as the plasma is confined to the center of the deposition chamber as a ball, carbon deposition onto the walls of the chamber is prevented.

7.5.2 Hot Filament CVD (HFCVD)

In the early 1970s it was suggested that the simultaneous production of atomic hydrogen during hydrocarbon pyrolysis may enhance the deposition of diamond. Soviet researchers who generated H by dissociating H_2 using an electric discharge or a hot filament tested this suggestion [8]. It was observed that atomic hydrogen could easily be produced by the passage of H_2 over a refractory metal filament, such as tungsten, heated to temperatures between 2000 and 2500 K. When atomic hydrogen was added to the hydrocarbon typically with a C/H ratio of ~0.01, it was observed that diamond could be deposited while graphite formation was suppressed. The generation of atomic hydrogen during diamond CVD enabled (a) a dramatic increase in the diamond deposition rate to approximately 1 μm hr^{-1} and (b) the nucleation and growth of diamond on non-diamond substrates [8–13]. Because of its inherent simplicity and comparatively low operating cost, HFCVD has become very popular in industry. Table 7.2 outlines typical deposition parameters used in the growth of diamond films by this technique.

A wide variety of refractory materials have been used as filaments including tungsten, tantalum, and rhenium due to their high electron emissivity. Refractory metals, which form carbides (e.g., tungsten and tantalum) typically must carburize their surface before supporting the deposition of diamond films. The process of filament carburization results in the consumption of carbon from the CH_4, and thus a specific incubation time is needed for the nucleation of diamond films. Therefore, this process may affect the early stages of film growth, although it is insignificant over longer periods. Furthermore, the volume expansion due to carbon incorporation leads to cracks along the length of the wire. The development of these cracks is undesirable, as it reduces the lifetime of the filament.

Table 7.2. Typical deposition parameters used in the growth of diamond films by HFCVD

Gas mixture	Total pressure (Torr)	Substrate temperature (K)	Filament temperature (K)
CH_4 (0.5 – 2.0%)/H_2	10 – 50	1000 – 1400	2200 – 2500

It is believed that thermodynamic near-equilibrium is established in the gas phase at the filament surface. At temperatures around 2300 K, molecular hydrogen dissociates into atomic hydrogen and methane transforms into methyl radicals, acetylene species, and other hydrocarbons stable at these elevated temperatures. Atomic hydrogen and the high-temperature hydrocarbons then diffuse from the filament to the substrate surface. Although the gaseous species generated at the filament are in equilibrium at the filament temperature, the species are at a super equilibrium concentration when they arrive at the much cooler substrate. The reactions that generate these high-temperature species (e.g., C_2H_2) at the surface of the filament or anywhere where there are hydrogen atoms, proceed faster than any reactions that decompose these species during the transit time from the filament to the substrate. Consider the equilibrium between methane and acetylene:

$$2CH_4 \rightleftharpoons C_2H_2 + 3H_2$$

At the filament surface, the reaction is immediately driven to the right, creating acetylene. After acetylene diffuses to the substrate, thermodynamic equilibrium at a substrate temperature of ~1100 K calls for the formation of methane, but the reverse reaction proceeds much slower. Solid carbon precipitates on the substrate in order to reduce the superequilibrium concentration of species such as acetylene in the gas phase. The diamond allotrope of carbon is "stabilised" by a concurrent super equilibrium concentration of atomic hydrogen. This simple explanation emphasizes the importance of reaction kinetics in diamond synthesis by HFCVD.

Advantages of the CVD Process

The process is gas phase in nature, and therefore given a uniform temperature within the coating retort and likewise uniform concentrations of the depositing species the deposition rate will be similar on all surfaces. Therefore, variable and complex shaped surfaces, given reasonable access to the coating powders or gases, such as screw threads, blind holes, or channels or recesses, can be coated evenly without build-up on edges.

Disadvantages of the CVD Process

The CVD process is carried out at relatively high temperatures and therefore limitations due to dimensional tolerances are an important consideration. Components that have tight dimensional tolerances will not be amenable to CVD. However, the reduction of distortion during coating can sometimes be controlled by careful stress relieving after rough machining of the component during fabrication.

7.6 Treatment of Substrate

7.6.1 Selection of Substrate Material

Deposition of adherent high-quality diamond films onto substrates such as cemented carbides, stainless steel, and various metal alloys containing transition element has proved to be problematic. In general, the adhesion of the diamond films to the substrates is poor and the nucleation density is very low [44–51]. Mainly refractory materials such as W (WC-Co), Mo, and Si have been used as substrate materials. Materials that form carbide, are found to support diamond growth. However, materials such as Fe and steel possess a high mutual solubility with carbon, and only graphitic deposits or iron carbide result during CVD growth on these materials. For applications in which the substrate needs to remain attached to the CVD diamond film, it is necessary to choose a substrate that has a similar thermal expansion coefficient to that of diamond. If this is not done the stress caused by the different rates of contraction on cooling after deposition will cause the film to delaminate from the substrate. The influence of different metallic substrates on the diamond deposition process has been examined. Interactions between substrate materials and carbon species in the gas phase are found to be particularly important and lead to either carbide formation or carbon dissolution. Carbides are formed in the presence of carbon-containing gases on metals such as molybdenum, tungsten, niobium, hafnium, tantalum, and titanium. The carbide layer formed allows diamond to form on it since the minimum carbon surface concentration required for diamond nucleation cannot be reached on pure metals. As the carbide layer increases in thickness, the carbon transport rate to the substrate decreases until a critical level is reached where diamond is formed [52–58]. Substrates made from metals of the first transition group such as iron, cobalt, and nickel, are characterized by high dissolution and diffusion rates of carbon into those substrates (Table 7.3) [59]. Owing to the absence of a stable carbide layer, the incubation time required to form diamond is higher and depends on substrate thickness. In addition, these metals catalyze the formation of graphitic phases, which is reflected in the graphite-diamond ratio of during the deposition process, yielding a low diamond. The importance of this mechanism in relation to

Table 7.3. Solubility and diffusion rates of carbon atoms in different metals at 900 °C

	α–Fe	γ–Fe	Co	Ni
Solubility of carbon (wt.%)	1.3	1.3	0.1	0.2
Carbon diffusion rate (cm/s)	2.35×10^{-6}	1.75×10^{-8}	2.46×10^{-8}	1.4×10^{-8}

diamond deposition decreases from iron to nickel, corresponding to a gradual filling of the 3d-orbital [59]. This effect occurs whenever the metal atoms come into contact with the carbon species, which can take place on the substrate or in the gas phase [60].

7.6.2 Substrate Pre-Treatment

In order for continuous film growth to occur, a sufficient density of crystallites must be formed during the early stages of growth. In general, the substrate must undergo a nucleation enhancing pre-treatment to allow this

This is particularly true for Si wafer substrates that have been specially polished to be smooth enough for micro-electronic applications. Substrates may be pre-treated by a variety of methods including:

- Abrasion with small (~nm/μm size) hard grits (e.g., diamond, silicon carbide).
- Ultrasonication of samples in slurry of hard grit (e.g., diamond).
- Chemical treatment (acid etching and Murakami agent)
- Bias enhanced nucleation (BEN) (negative/positive substrate biasing)
- Deposition of hydrocarbon/oil coatings

The basis for most of these methods is to produce scratches, which provide many sites for nucleation diamond of crystallites. It is also possible that small (~nm size) flakes of diamond, produced during abrasion with diamond grit, become embedded in the substrate, and that CVD diamond grows on this material [61].

It could be desirable to produce nucleation sites without damage to the underlying substrate. This is particularly important for some applications such as diamond electronics and optical components. One method for encouraging nucleation without damaging the substrate material has been developed: bias enhanced nucleation (BEN).

(a) Pre-Treatment on Mo/Si Substrate

Prior to pre-treatment Si/Mo substrate are ultrasonically cleaned in acetone for 10 minutes to remove any unwanted residue on the surface. Abrasion with 1 μm sizes of diamond powder is performed for 5 minutes. Alternatively, substrate was immersed in diamond solution containing 1–3 μm of diamond particles and water for I hr in ultrasonic bath. These methods produce scratches on the surface, which create many nucleation sites. The substrates are then washed with acetone in the ultrasonic bath for 10 minutes. SEM and energy–dispersive x-ray spectroscopy (EDX), characterized abraded surface of substrates.

(b) Pre-Treatment on WC-Co Substrate

The application of diamond coatings on cemented tungsten carbide (WC-Co) tools has attracted much attention in recent years in order to improve cutting performance and tool life. However, deposition of adherent high-quality diamond films onto substrates such as cemented carbides, stainless steel, and various metal alloys containing transition element has proved to be problematic. In general, the adhesion of the diamond films to the substrates is poor and the nucleation density is very low [44–50]. WC-Co tools contain 6% Co and 94% WC substrate with grain size 1–3 micron is desirable for diamond coatings.

In order to improve the adhesion between diamond and WC substrates it is necessary to etch away the surface Co and prepare the surface for subsequent diamond growth. In particular, the cobalt (Co) binder, which provides additional toughness to the tool but is hostile to the diamond adhesion. The adhesion strength to diamond films is relatively poor, and can lead to catastrophic failure of coating in metal cutting [59]. The Co binder can also suppress diamond growth, favoring the formation of non-diamond carbon phases resulting in poor adhesion between the diamond coating and the substrate [62]. Most importantly, it is difficult to deposit adherent diamond onto untreated WC-Co substrates. Figure 7.2 (a) shows sub-micron size Co crystals on diamond films that were deposited on an untreated substrate (without removal of surface Co). EDX spectra show that trace amount of Co elements on the surface. It also appeared poor adhesion of diamond film and shows delaminated film on the surface (Fig. 7.2 (b)).

Poor adhesion can be related to the cobalt binder that is present to increase the toughness of the tool; however, it suppresses diamond nucleation and causes deterioration of diamond film adhesion. To eliminate this problem, it is usual to pre-treat the WC-Co surface prior to CVD diamond deposition. Various approaches have been used to

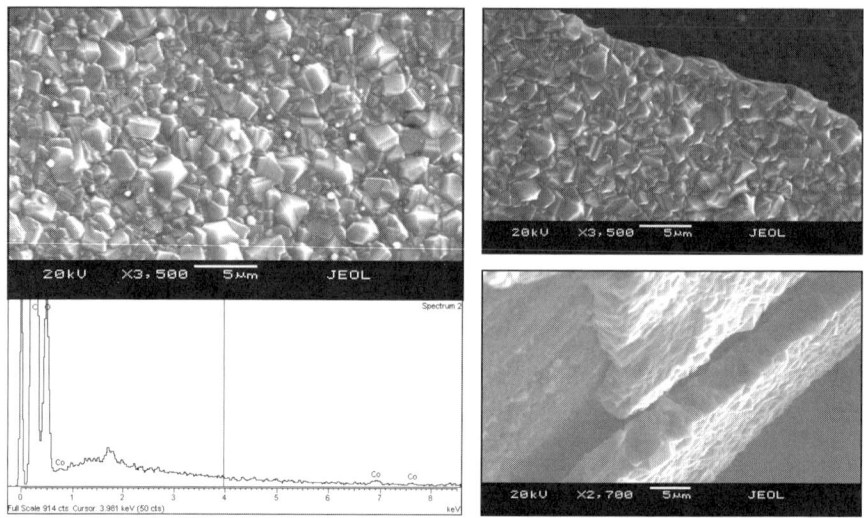

Fig. 7.2. (a) Co trace on diamond film, (b) delaminated diamond film

suppress the influence of Co and to improve adhesion. Therefore, a substrate pre-treatment, for reducing the surface Co concentration and achieving a proper interface roughness, will enhance the surface readily available for coating process [62]. For example, chemical treatment using Murakami agent and acid etching has been used successfully for removal of the Co binder from the substrate surface [63].

The WC-Co substrates (Flat) used were 10 × 10 mm by 3 mm in thickness. The hard metal substrates used were WC-6wt% Co with WC average grain size of 0.5 μm (fine grain) and 6 μm (coarse grain). Table 7.4 show that the data for substrate, which consist of the chemical composition, density, and hardness of samples, used for diamond deposition. Figure 7.3 (a) and (b) show that coarse and fine grain of etched WC-Co insert surface.

Table 7.4. WC-Co insert chemical composition

WC grain size (μm)	WC	Co	TaC	Density (g/cm³)	Hardness (HRA)
WC fine grain 0.5 μm	94.2	5.8	0.2	14.92	93.40
WC coarse grain 6 μm	94.0	6.0		14.95	88.50

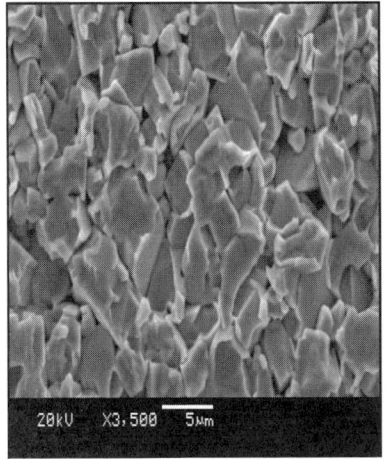 

Fig. 7.3. (a) Etched coarse grain; (b) etched fine grain

The Co cemented tungsten carbide (WC-Co) rotary tools (surgical tools, surgical tools), 20 mm in length including the bur head (WC-Co) & shaft (Fe/Cr) and ~1 mm in diameter, were also used. Prior to pre-treatment both set of substrates are ultrasonically cleaned in acetone for 10 mins to remove any loose residues. The following two-step chemical pre-treatment procedure is used. A first step etching, using Murakami's reagent [10 g $K_3Fe(CN)_6$ + 10 g KOH + 100 ml water] is carried out for 10 min in ultrasonic bath to etch WC substrate, followed by a rinse with distilled water. The second step etching is performed using an acid solution of hydrogen peroxide [3 ml (96% wt.) H_2SO_4 + 88 ml (30% w/v) H_2O_2], for 10 s, to remove Co from the surface. The substrates are then washed again with distilled water in an ultrasonic bath. After wet treatment the surgical tool is abraded with synthetic diamond powder (1 μm grain size) for 5 mins and followed by ultrasonic treatment with acetone for 20 mins. Etched surface of substrates can be characterised by SEM and energy–dispersive x-ray spectroscopy (EDX).

7.7 Modification of HFCVD Process

7.7.1 Modification of Filament Assembly

The filament material and its geometrical arrangement are important factors to consider in order to have improved coatings using the

CVD method. Therefore, in order to optimize both the filament wire diameter and the filament assembly/geometry it is necessary to understand the temperature distributions of the filament. Research by the author indicated that the best thermal distribution and diamond growth uniformity is obtained using tantalum wires of 0.5 mm in diameter. To ensure uniform coating around the cylindrical shape samples (surgical tools or surgical tools), tools were positioned centrally and co-axially within the coils of the filament, the six-spiral (coil) filament was made with 1.5 mm spacing between the coils (Fig. 7.4a and b).

Tantalum wire of 0.5 mm in diameter and 12–14 cm in length is used as the hot filament. The filament is mounted vertically with the surgical tool held in between the filament coils, as opposed to the horizontal position used in the conventional HFCVD system. To ensure uniform coating the surgical tool is positioned centrally and coaxially within the coils of the filament. A schematic diagram of the modified HFCVD system is presented in Fig. 7.5 and has been designed for surgical tool or wire or surgical tool with similar diameter. The new vertical filament arrangement used in the modified HFCVD system enhances the thermal distribution, ensuring uniform coating, increased growth rates, and higher nucleation densities.

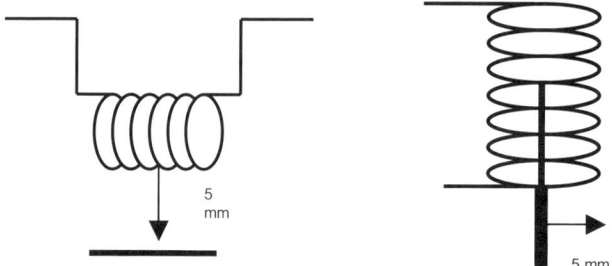

Fig. 7.4. (a) Conventional filament arrangement, (b) modified filament arrangement.

7.7.2 Process Conditions

The CVD reactor is a cylindrical stainless steel chamber measuring 20 cm in diameter and 30 cm in length. Diamond films were deposited onto the cutting edge of the tools at 5 mm distances from the filament. The gas source used during the deposition process is composed of a mixture containing 1% methane with an excess of hydrogen; the volume flow rate for hydrogen is 100 sccm, while the volume flow rate for methane is 1 sccm. The deposition time and pressure in the vacuum chamber were 5–15 hours and 20 Torr (2.66 k Pa) employed.

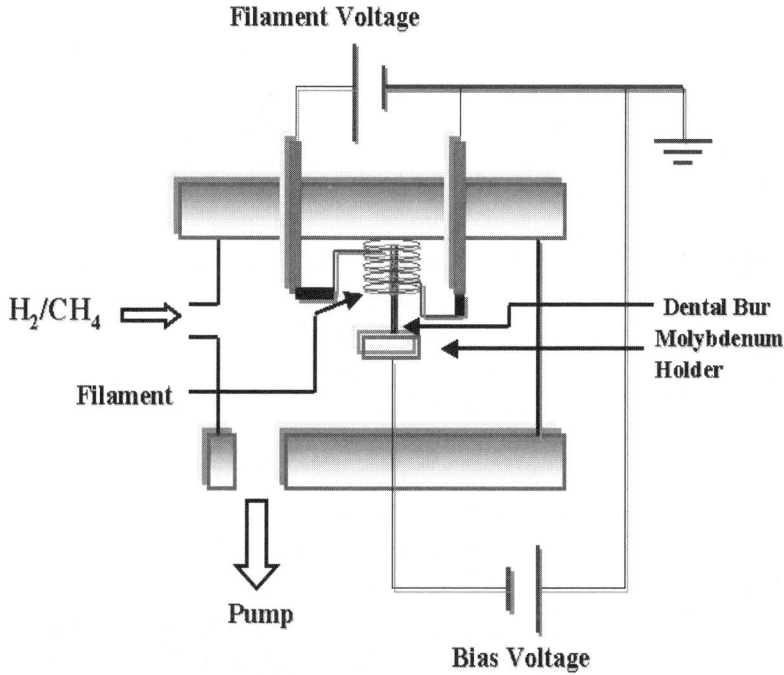

Fig. 7.5. Schematic diagram of HFCVD apparatus

Table 7.5. Process conditions used for diamond film deposition on surgical tools

Process variables	Operating parameters
Tantalum filament diameter (mm)	0.5
Deposition time (hours)	5–15
Gas mixture	1% CH_4 in excess H_2
Gas pressure (Torr)	20 (2.66 k Pa)
Substrate temperature ($^\circ$C)	800–1000
Filament temperature ($^\circ$C)	1800–2100
Substrate (WC-Co/Mo/Ti) diameter (mm)	Wire/drill/surgical tool (approx. 1 mm)
Distance between filament and substrate (mm)	5
Pre-treatment (Murakami etching and acid etching)	20 minutes plus 10 seconds

The substrate temperature was measured by a K-type thermocouple mounted on a molybdenum substrate holder. The depositions are carried out between 800 and 1000°C. The filament temperature is measured using an optical pyrometer and found to be between 1800 and 2100°C depending upon the filament position. A summary of the process conditions is shown in Table 7.5.

7.8 Nucleation and Growth

The growth of diamond thin films at low pressures, at which diamond is metastable, is one of the most exciting developments in materials science of the last two decades. However, low growth rates and poor quality currently limit applications. Diamond growth is achieved by a variety of processes using very different means of gas activation and transport. Generalized models coupled with experiments show how process variables, such as gas activation temperature, pressure, characteristic diffusion length, and source gas composition, influence diamond growth rates and diamond quality. The modeling is sufficiently general to permit comparison between growth methods. The models indicate that typical processes, e.g., hot-filament, microwave and thermal plasma reactors, operate at pressures where concentrations of atomic hydrogen, [H], and methyl radicals, [CH_3], reach maxima. The results strongly suggest that the growth rate maxima with pressure arise from changes in the gas phase concentrations rather than changes in substrate temperature.

The results also suggest that, at 1 atmosphere pressure using only hydrocarbon chemistry, growth rates saturate at gas activation temperatures above 4000 K. Models of defect incorporation indicate that the amount of sp^2, non-diamond material incorporated in the diamond is proportional to [CH_3]/[H] and therefore can be correlated with the controllable process parameters. The unusual and interesting connection between diamond nucleation and growth with the process of the vapor synthesis of diamond is essentially quite simple. Carbon containing precursor molecules (like CH_4) are excited and/or dissociated and subsequently condensing via a free dangling bond of the radical in diamond configuration on a surface [64–65] (Fig. 7.6).

A nucleation pathway occurs through a stepwise process includ-ing the formation of extrinsic (pre-treatment) or intrinsic (in situ) nucleation sites, followed by formation of carbon-based precursors. It is believed that nucleation sites could be either grooves of scratching lines or protrusions produced by etching-re-deposition.

The gas activation is done either by hot filaments, microwave, or radio frequency plasmas. The most crucial parameter in all this proc-esses is besides a carbon source the presence of large amounts of atomic hydrogen. The role of atomic hydrogen in the process is:

- Creation of active growth sites on the surface
- Creation of reactive growth species in the gas-phase
- Etching of non-diamond carbon (like graphite) graphitic, sp^2, precursors will be explored.

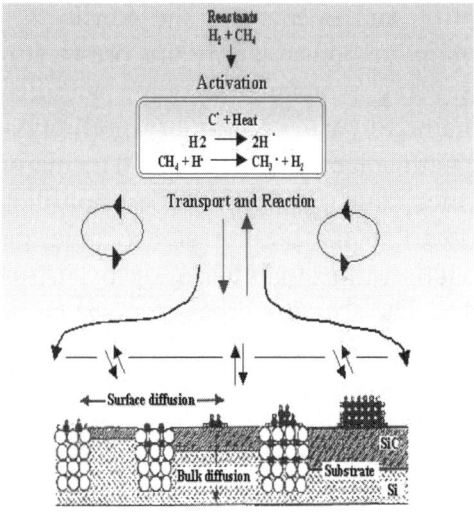

Fig. 7.6. Schematic diagram of diamond nucleation and growth

7.8.1 Nucleation Stage

Nucleation of diamond is a critical and necessary step in the growth of diamond thin films, because it strongly influences diamond growth, film quality and morphology [66]. Growth of diamond begins when individual carbon atoms nucleate onto the surface to initiate the beginnings of an sp^3 tetrahedral lattice. There are two types of diamond growth:

Homo-epitaxial growth: It is an application of diamond substrates, the template for the required tetrahedral structure is already there, and the diamond lattice is just extended atom-by-atom as deposition proceeds.

Hetero-epitaxial growth: It uses the non-diamond substrates, there is no such template for the C atoms to follow, and those C atoms that deposit in non-diamond forms are immediately etched back to the gas phase by reaction with atomic [H].

To deal with the problem of the initial induction period before which diamond starts to grow, the substrate surface often undergoes pre-treatment prior to deposition in order to reduce the induction time for nucleation and to increase the density of nucleation sites. There are two main methods to apply this pre-treatment:

Generally, seeding or manual abrading with diamond powder or immersing in diamond paste containing small crystallites processed in an ultrasonic bath enhances nucleation. The major consideration is the nucleation mechanism of diamond on non-diamond substrates. It has been shown that the pre-abrasion of non-diamond substrates reduces the induction time for nucleation by increasing the density of nucleation sites. The abrasion process can be carried out by mechanically polishing the substrate with abrasive grit, usually diamond power of 0.1–10 μm particle size, although other nucleation methods do exist such as bias-enhanced nucleation which is used in heteroepitaxial growth. The most promising in-situ method for diamond nucleation enhancement is substrate biasing. In recent years, more controlled techniques, such as bias-enhanced nucleation and nanoparticle seeding, have been used to deposit smoother films [67–68]. In this method, the substrate is biased negatively during the initial stage of deposition [69]. Before CVD diamond deposition the filament was pre-carburised for 30 minutes in 3% methane with excess hydrogen to enhance the formation of tantalum carbide layer on the filament surface in order to reduce the tantalum evaporation during diamond deposition [70].

7.8.2 Bias-Enhanced Nucleation (BEN)

The substrate can be biased both negatively and positively; however, there is much research work and a large volume of literature on negative biasing. Negative substrate biasing is attractive because it can be controlled precisely; it is carried out in-situ, gives good homogeneity, and results in improved adhesion. On flat substrates, such as copper and silicon, biasing has been shown to give better adhesion, improved crystallinity and smooth surfaces.

A negative bias voltage up to −300 V has been applied to the substrate relative to the filament. This produced emission currents up to 200 mA. The nucleation times used were between 10 and 30 minutes. In the activated deposition chamber CH_4 and H_2 were decomposed into various chemical radicals species CH_3, C_2H_2, CH_2, CH, C and atomic hydrogen H by the hot tantalum filament. The methyl radicals and atomic hydrogen are known to play important roles in diamond growth. In the biasing process electrons were emitted from the diamond-coated molybdenum substrate holder and moved to the filament after they gained energy from the electrical field. When the negative bias was applied to the anode the voltage was gradually increased until a stable emission current was established and a luminous glow discharge was formed near the substrate [71]. The nucleation density of diamond has been calculated from the SEM micrographs. Figure 7.7 (a) and (b) shows the effects of bias time on the nucleation density at bias voltage of −300 V. As bias time is increased the nucleation density also increases. The highest nucleation density was calculated to be 0.9×10^{10} cm^{-2} for a bias time of 30 minutes. At a bias time of 10 minutes the nucleation density obtained was 2.7×10^8 cm^{-2}.

Wang et al. [72] also reported that an increase in the emission current produced higher nucleation densities [73]. Since the bias voltage and emission current are related, the enhancement of the nucleation density cannot be attributed to solely ion bombardment or electron emission of the diamond-coated molybdenum substrate holder, but may be a combination of these mechanisms [74]. Results were based on negatively bias enhanced nucleation related to the grounded filament. However, it was reported that very low electric biasing current values (μA) were detected for applied substrate biases voltages either positive or negative. Furthermore, when increasing negative biases of up to –200 V resulted in a value of nucleation density similar to that obtained with positively bias enhanced nucleation related to the filament. In contrast, an application of negative bias applied to the substrate at –250 V resulted in (10^{10} cm^{-2}) maximum values of nucleation density. The enhancement in the nucleation density can be attributed to the electron current from the filament by increasing the decomposition of H_2 and CH_4. The increase in the nucleation density is expected since negatively biasing the substrate increases the rate of ion bombardment into the surface creating greater numbers and density of nucleation sites. Therefore, the greater the density of nucleation sites the higher the nucleation density. Kamiya et al. reported that reproducibility of the experiment was poor and that no definite trend in the nucleation density could be found with respect to different bias conditions [74].

7.8.3 Influence of Temperature

The temperature is a major factor in influencing the deposition rate, crystallite size, and controlling the surface roughness. Variation in the average crystallite size of diamond along the length of the substrate (surgical tool) can be attributed to the variation in the variation in the substrate temperature. The substrate temperature from the end to the center of the filament is more accentuated for molybdenum wire with a smaller diameter [9]. Figure 7.8 demonstrates the ability of this CVD process to coat 3-D shaped components, illustrating that the process is in the kinetic control regime rather than transport control regime. Most physical vapor deposition type of processes operates at conditions where the rate-determining step of the deposition process is the diffusion of precursor gases to the substrate surface. Generally, this results in poor film uniformity in grooves and at the sharp edges. By operating under kinetic control regime film uniformity is much enhanced.

Deposition temperature can also influence the diamond film thickness in terms of substrate and filament position. Analysis of temperature distribution along the coiled filament showed that the temperature is highest at the center of the filament with a rapid decrease toward the edges (Fig. 7.8). This suggests that position A is the hottest followed by position B and C on the bur. Generally, higher substrate temperatures increase diamond film growth rate and the crystallite size. At the bottom of the filament coil temperature is lower; therefore, the part of the bur parallel to the coil at this temperature will be coated with the diamond film at a lower growth rate. Therefore, it can be

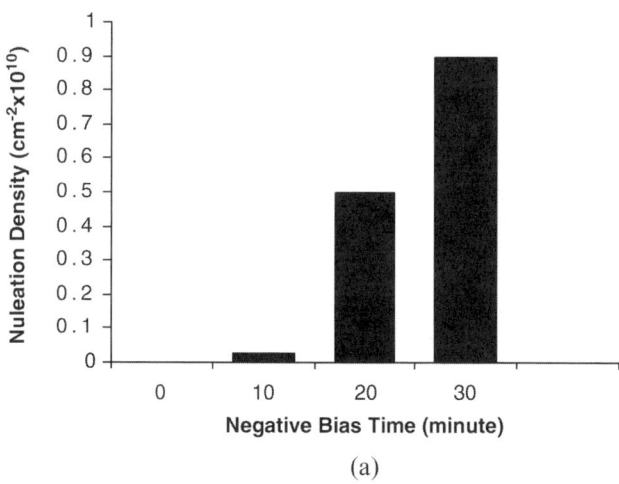

(a)

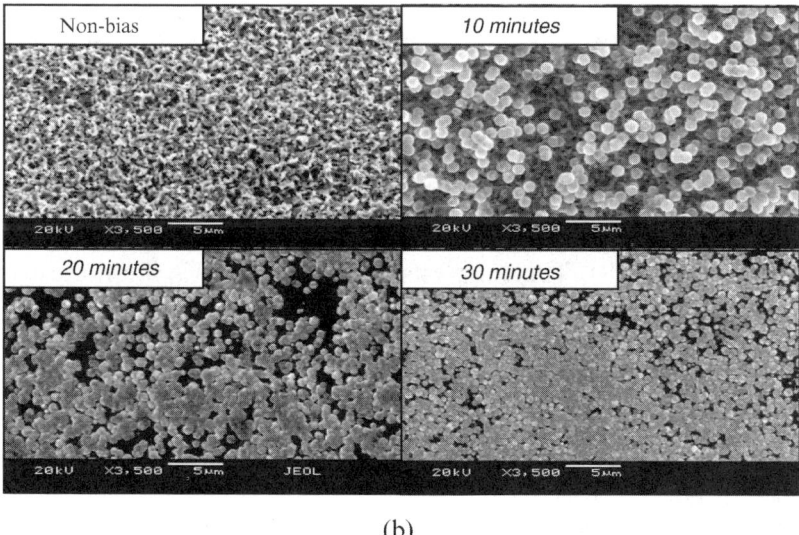

(b)

Fig. 7.7. (a) nucleation density of diamond by BEN; (b) SEM of nucleation density on substrate

expected that at these regions the film will be thinner. The thermal gradient gives variations in the film thickness and crystal sizes as evident from figure. Generally, with columnar growth, the average crystallite size increases as the films become thicker. The films were thicker and the crystallite size was larger at position A compared to position C.

Analysis of temperature distribution along the coiled filament showed that the temperature is highest at the center of the filament with a rapid decrease toward the edges. The bur substrate and filament temperature have been measured parallel to the positions A, B, and C respectively (Fig. 7.9). This suggests that position A is the hottest followed by position B and C on the bur. Variations in the film thickness and crystal sizes are mainly due to thermal gradients at various positions on the bur.

Figure 7.10 indicates that the coated surgical tool was cut in order to study the cross section of the tool. It was found that the coating is thicker at the cutting teeth with average thickness of about 43 μm due to the slightly higher temperature at the bur tip because cutting teeth is closer to the filament coil. At the base of the bur the heat is carried away faster and therefore it is at a lower temperature giving rise to lower growth rates and hence thinner films, at about 23 μm in thickness. Thicker coating at the tip is expected to give the tool longer life. Further work is required to study the effects of film thickness at the tooth tip and at the base on tool performance and lifetime.

Deposition temperature can also influence the diamond film thickness in terms of substrate and filament position. Analysis of temperature distribution along the coiled filament showed that the temperature is highest at the center of the filament with a rapid decrease toward the edges (Fig. 7.8). This suggests that position A is the hottest followed by position B and C on the bur. Generally, higher substrate temperatures increase diamond film growth rate and the crystallite size. At the bottom of the filament coil temperature is lower; therefore, the part of the bur parallel to the coil at this temperature will be coated with the diamond film at a lower growth rate. Therefore, it can be expected that at these regions the film will be thinner. The thermal gradient gives variations in the film thickness and crystal sizes as evident from figure. Generally, with columnar growth, the average

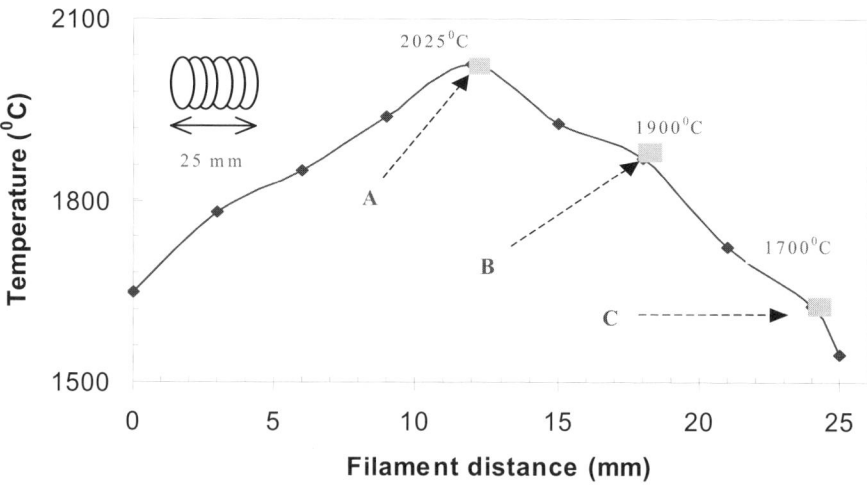

Fig. 7.8. Deposition temperature against filament position

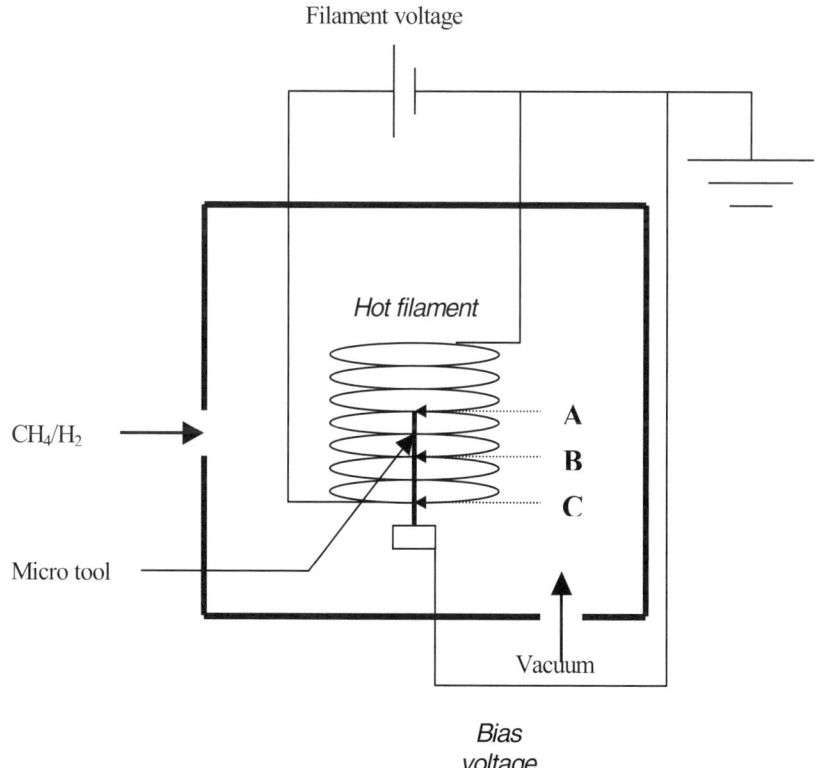

Fig. 7.9. Schematic diagram of a surgical tool assembled with filament

crystallite size increases as the films become thicker. The films were thicker and the crystallite size was larger at position A compared to position C.

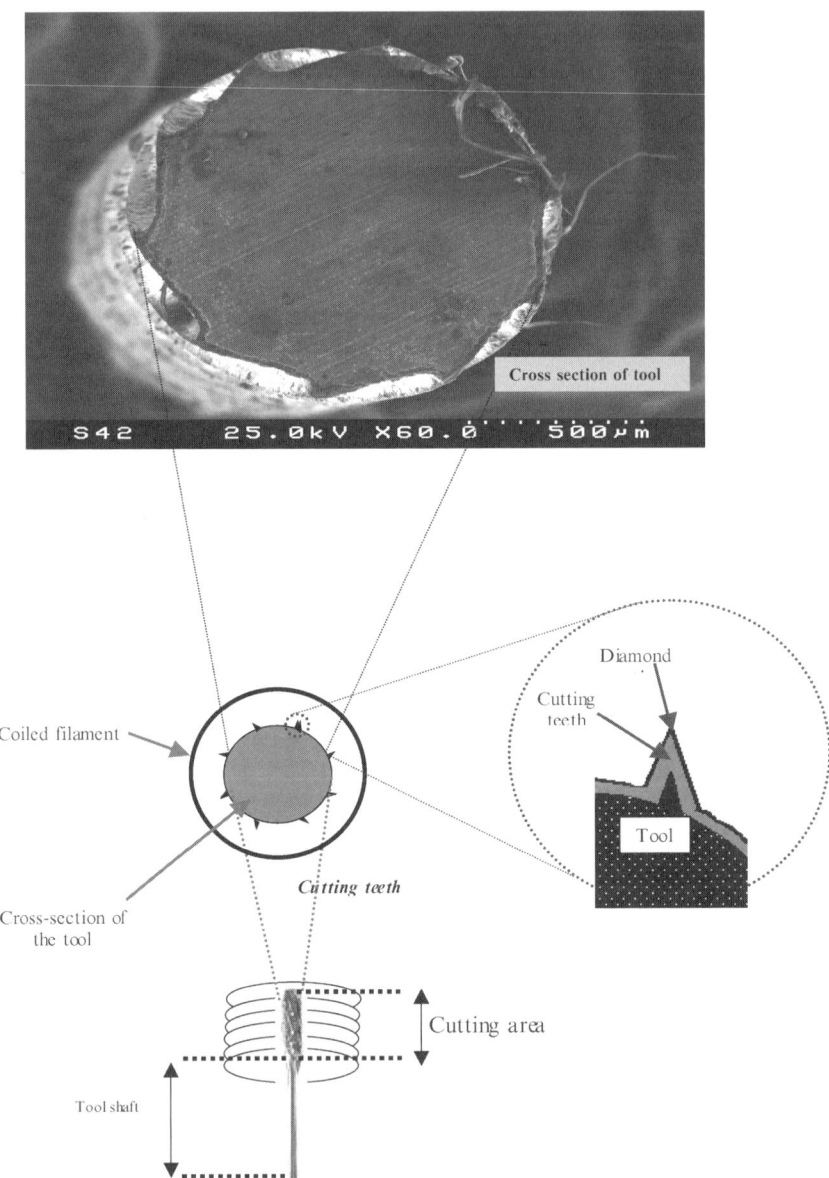

Fig. 7.10. Schematic diagram of surgical tool cutting edges

Analysis of temperature distribution along the coiled filament showed that the temperature is highest at the center of the filament with a rapid decrease toward the edges. The bur substrate and filament temperature have been measured parallel to the positions A, B, and C respectively (Fig. 7.9). This suggests that position A is the hottest followed by position B and C on the bur. Variations in the film thickness and crystal sizes are mainly due to thermal gradients at various positions on the bur.

7.9   Deposition on Three-Dimensional Substrates

7.9.1   Diamond Deposition on Metallic (Molybdenum) Wire

It is difficult to deposit CVD diamond onto cutting tools, which generally have a 3-D shape and possess complex geometry and sharp edges, using a single step growth process [76]. The cylindrical shape wire, which has complex geometry, can be used as a model application for deposition of diamond on cutting tools such as surgical tools and

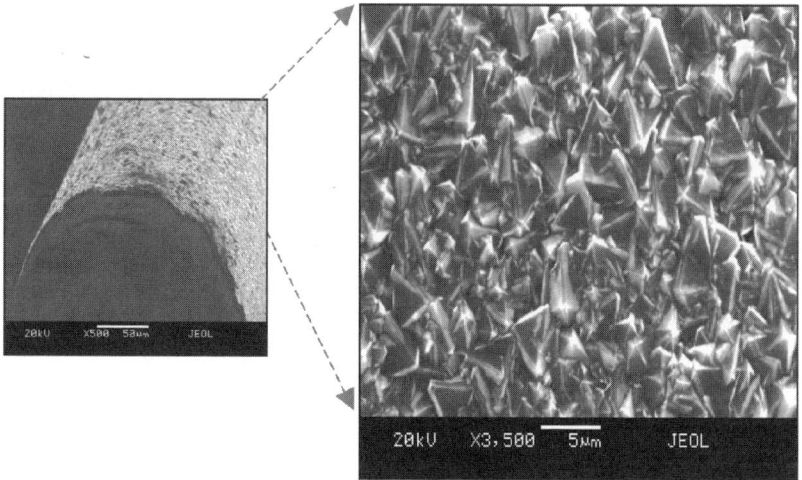

Uniform growth of (111) faceted
octahedral diamond film on

Fig. 7.11. Diamond film on molybdenum

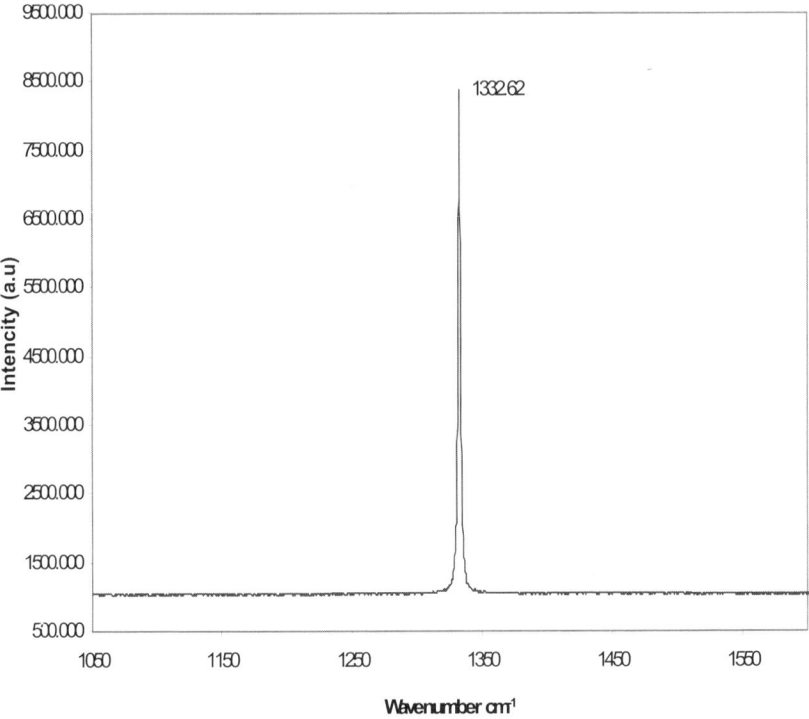

Fig. 7.12. Raman spectra of CVD diamond on molybdenum wire

surgical tools. The molybdenum (Mo) wires are deposited with CVD diamond by modified vertical filament approach. After deposition time of 5 hours is continuous films of 5 μm thick CVD diamond were obtained (Fig. 7.11). The film morphology showed that it has good uniformity and high purity of diamond. The Raman spectroscopy confirmed that sp^3 diamond peak at wave number 1332.6 cm^{-1} as shown on Fig. 7.12.

7.9.2 Deposition on WC-Co Surgical tool

Deposition of diamond on wires can be readily extended to surgical tool used for machining tool, NEMS, and MEMS devices. The uniform and adherent coating are essential in order to obtain an improved performance. Figure 7.13 (a) shows an SEM micrograph of an uncoated surgical tool. The WC-Co cutting edges are welded onto

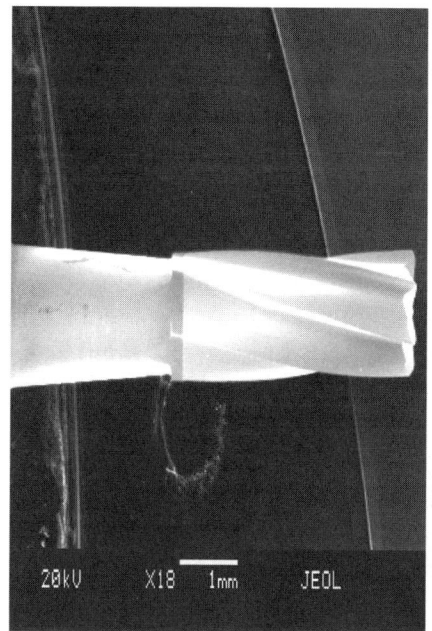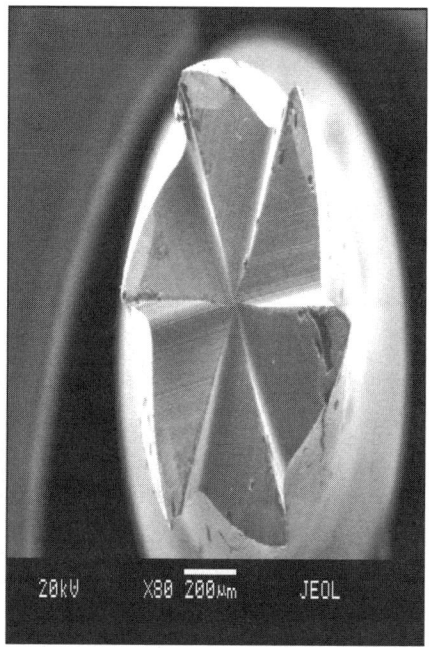

Fig. 7.13. Tip and cutting edge of surgical tool

the steel shaft (Fe-Cr). The cutting tip is 4 mm in length and 0.8 mm in diameter. The surgical tool has six sharp cutting edges, which is clearly visible in Fig. 7.13 (b).

Figure 7.14 shows the SEM micrographs and the corresponding EDX spectra of the WC-Co surgical tool before and after the chemical etching process. Before etching, the EDS spectrum (Figure 7.14 (a)) show the peaks for cobalt (Co), carbon (C), and tungsten (W). High cobalt content inhibits diamond deposition, resulting generally in graphitic phases, which degrade the coating adhesion. The Co diffuses to the surface regions, preventing effective bonding between the substrate surface and the film coating. To improve the coating adhesion of diamond on WC-Co tools, several approaches can be employed. For example, first, the use of interlayer material such as chromium can act as a barrier against cobalt diffusion during diamond CVD. Second, the cobalt from the tool surface can be etched using either chemical or plasma methods. Third, the cobalt can be converted into stable intermediate interlayer cobalt compounds.

These can act as a barrier to cobalt diffusion from the substrate during film growth [77]. Murakami solution followed by H_2SO_4/H_2O_2 etch can be used to chemically remove the cobalt from the bur surface. The EDX spectrum shows that the Co peak has disappeared after etching. This will prove to be beneficial in enhancing the coating adhesion. Comparison of the SEM micrograph in Figs. 7.14 (a) and (b) shows that the surface topography is significantly altered after etching in Murakami and H_2SO_4/H_2O_2 solutions. The etching process makes the surface much rougher with a significant amount of etch pits, which act as low-energy nucleation sites for diamond crystal growth.

An SEM micrograph of a diamond-coated WC surgical tool is shown in Fig. 7.15. Six cutting edges of the surgical tool tip were coated with a polycrystalline diamond film using the modified vertical HFCVD method. Analysis of the SEM picture shows that the coating uniformly covered the cutting edges as well as the nearby regions in which the placement of the surgical tool within the coils of the filament, ensuring uniform deposition. The diamond crystal structure and morphology are uniform and adherent, as shown in Figs. 7.15 (a) and (b). It is also shows a close-up view of the diamond-coated region of the surgical tool in Fig. 7.15(c).

Typically the crystallite sizes are of the order of 5–8 μm. The visibly adherent diamond coatings on the WC-Co surgical tools consist of mainly (111) faceted diamond crystals. The design of the filament and substrate in the reactor offer the possibility of uniformly coating even larger diameter cylindrical substrates.

Raman analysis was performed in order to evaluate the diamond carbon-phase quality and film stress in the deposited films. The Raman spectrum in Fig. 7.16 shows a single peak at 1335 cm^{-1} for the tip of the diamond-coated surgical tool. The Raman spectrum also gives information about the stress in the diamond coatings. The diamond peak is shifted to a higher wave number of 1335 cm^{-1} than that of natural diamond peak 1332 cm^{-1} indicating that stress, which is compressive in nature, exists in the resultant coatings [78].

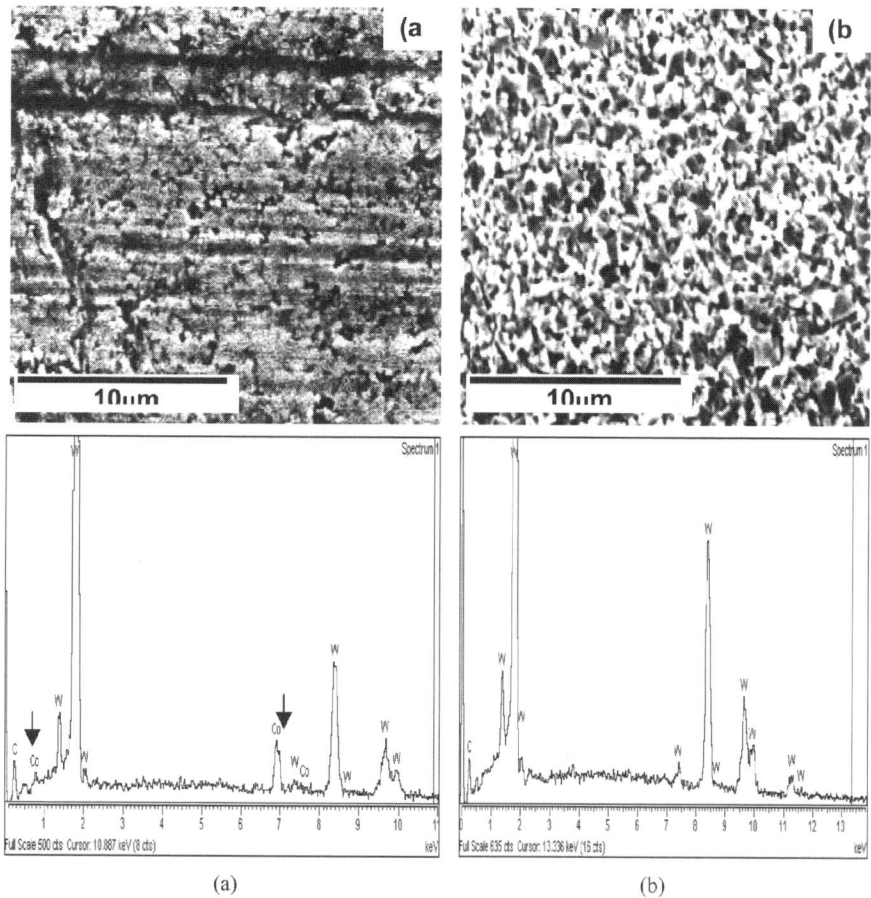

Fig. 7.14. (a) SEM and EDX of WC-Co surgical tool before etching; (b) WC-Co surgical tool after etching

7.9.3 Diamond Deposition on Tungsten Carbide (WC-Co) Surgical tool

Laboratory grade tungsten carbide (WC-Co) surgical tools are shown in Fig. 7.17(a) and (b) (AT23 LR) with fine WC grain sizes (1 μm) 20–30 mm in length and 1.0–1.5 mm in diameter [supplied by Metrodent Ltd, UK] that are used for the CVD diamond deposition process.

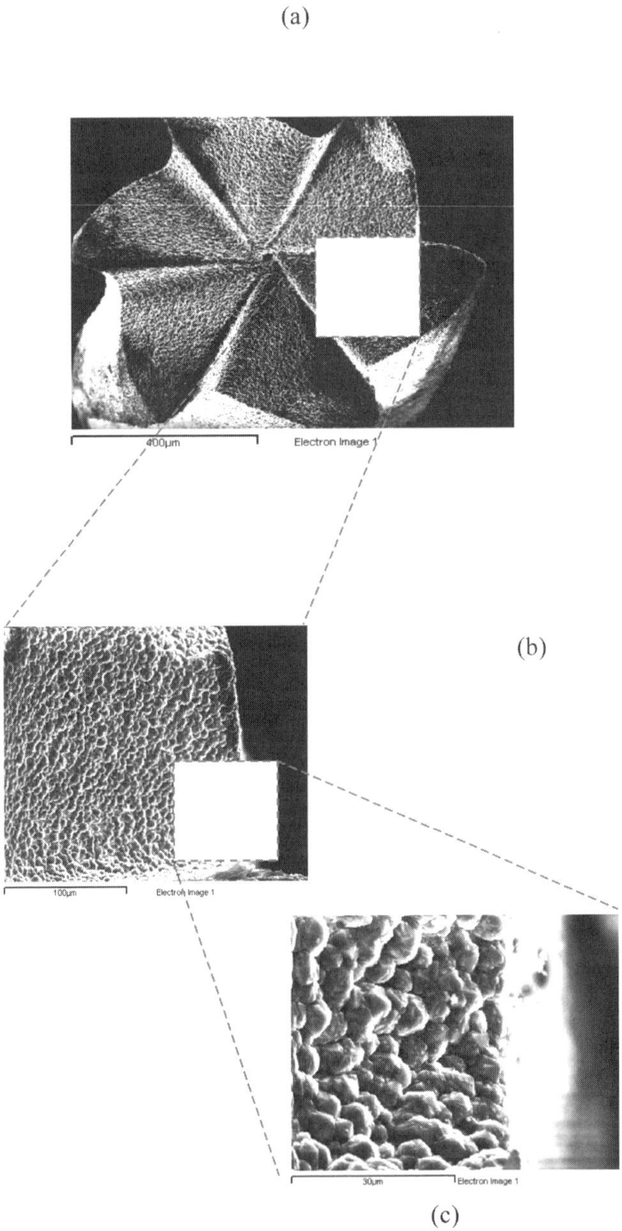

(a)

(b)

(c)

Fig. 7.15. (a) cutting edge of surgical tool after depositing with CVD diamond; (b) cutting edge of surgical tool uniformly coated with diamond; (c) SEM of surgical tool after coating with diamond

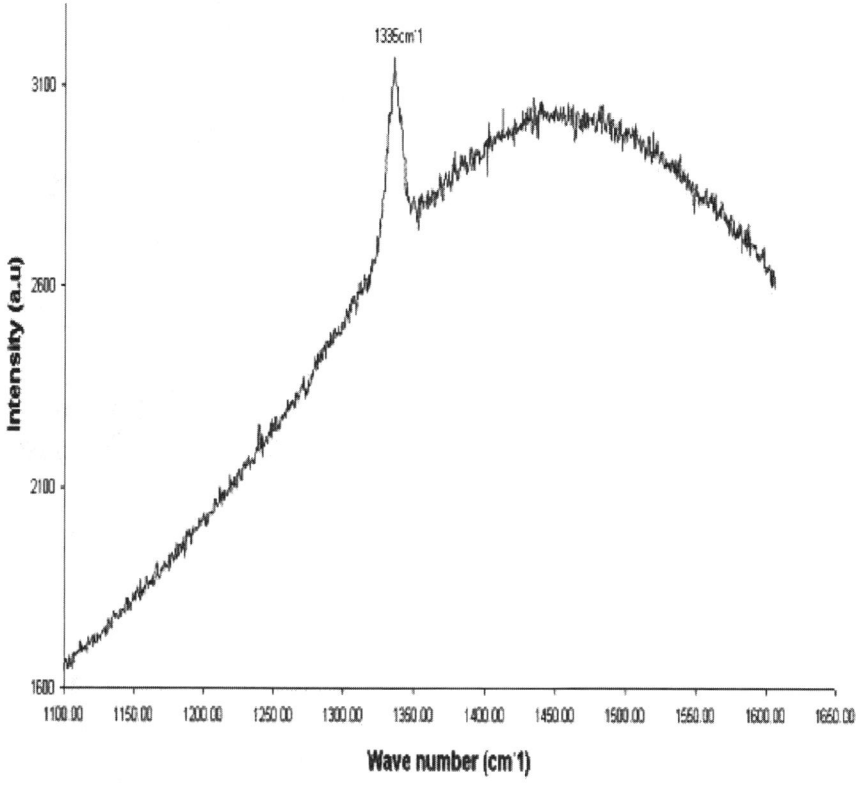

Fig. 7.16. Raman spectra of diamond-coated WC-Co surgical tool

The WC surface has etched away with Murakami solution and surface Co has been removed by acid etching followed by ultrasonically washing in distilled water. The EDX results confirmed that there is no indication of Co left on the surface of the etched surgical tool (Fig. 7.18). Diamond films have been deposited onto the cutting edge of the tools at a 5 mm distance from the tantalum wire filament.

Surface morphology of predominantly (111) faceted octahedral shape diamond films was obtained. The film thickness was measured to be 15–17 μm after diamond deposition for 15 hours. Figure 7.19 shows the SEM micrograph of a CVD diamond-coated surgical tool (AT 23LR) at the cutting edge. The film is homogeneous with uniform diamond crystal sizes, typically in the range of 6–10 μm. As expected the surface morphology is rough making the surgical tools extremely desirable for abrasive applications.

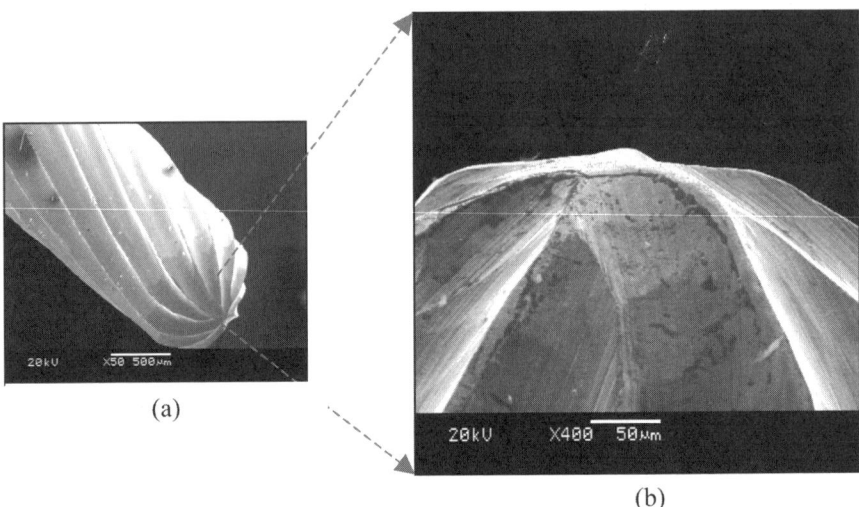

Fig. 7.17. (a) laboratory used WC-Co surgical tool; (b) laboratory used WC-Co surgical tool before surface treatment

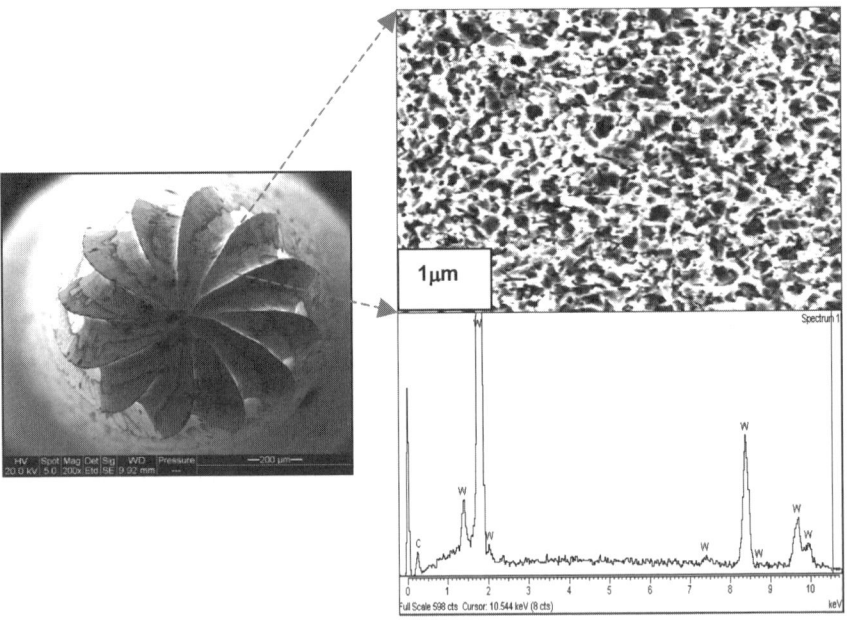

Fig. 7.18. Etched surgical tool surface after chemical treatment

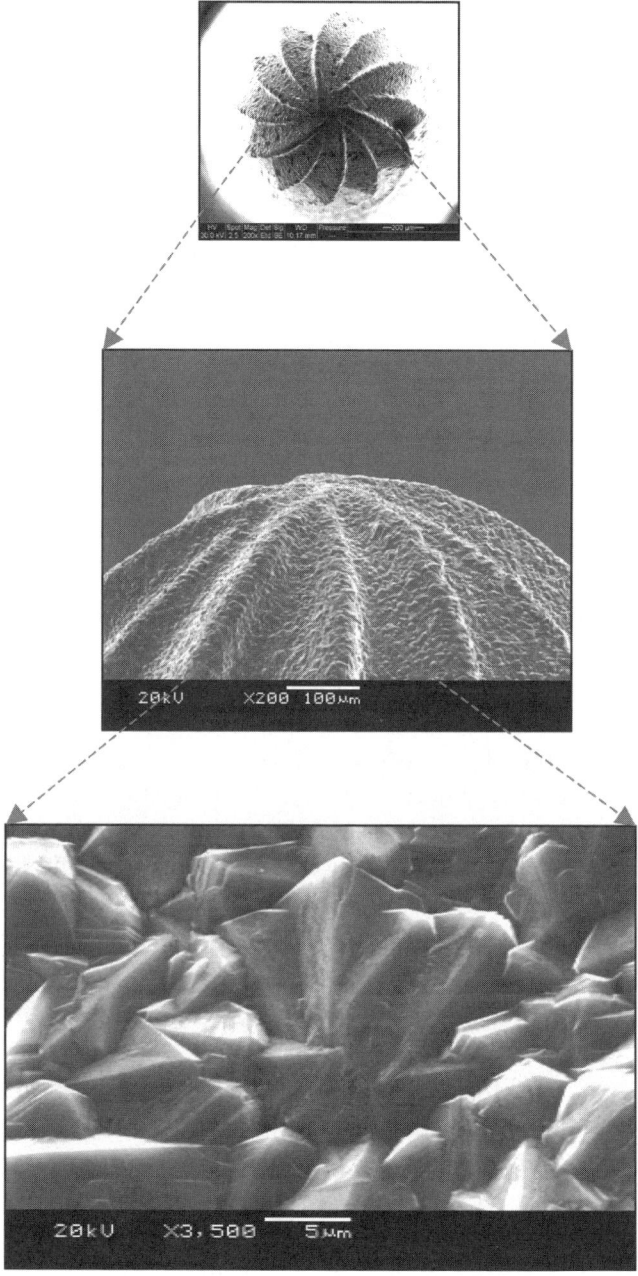

Fig. 7.19. (111) faceted octahedral shape diamond films on a surgical tool

The Raman spectra shown show that at the tip, center, and end of cutting tool single sharps peak at 1336, 1336 and 1337 cm^{-1} respectively was observed for different positions Fig. 7.20. The Raman spectrum also gives an indication about the stress in the diamond coating. The diamond peak is shifted to a higher wave number of magnitude such as 1336, 1337 cm^{-1} than that normally experienced in an unstressed coating where the natural diamond peak occurs at 1332 cm^{-1} thus indicating that the stress is compressive.

7.10 Wear of Diamond

The quality and economy of industrial production processes are to a great extent determined by the selection and the design of appropriate manufacturing operations. For many machining operations, especially for the technologically relevant processing of metallic materials, machining with geometrically specified cutting edge is applicable. Enhancing the performance of machining operations is therefore an economically important goal; to achieving this, coating technology can contribute in varying ways. The cutting tool is the component that is most stressed, and therefore limits the performance in NEMS and MEMS operations. Thermal and mechanical loading affects the cutting tool edges in a continuous or intermittant way. As a result, in addition to good wear resistance, high thermal stability and high mechanical strength are required for cutting materials. Opposing this objective of an ideal cutting material is the fundamental contradiction of properties hardness, strength at elevated temperature, and wear resistance on one hand, and bending strength and bending elasticity on the other hand.

Cutting materials for extreme requirements (for example, interrupted cuts or machining of high strength materials) consequently cannot be made from one single material, but may be realized by employing composite materials.

Surface coatings may improve the tribological properties of cutting tools in an ideal way and therefore allow the application of tough or ductile substrate materials, respectively.

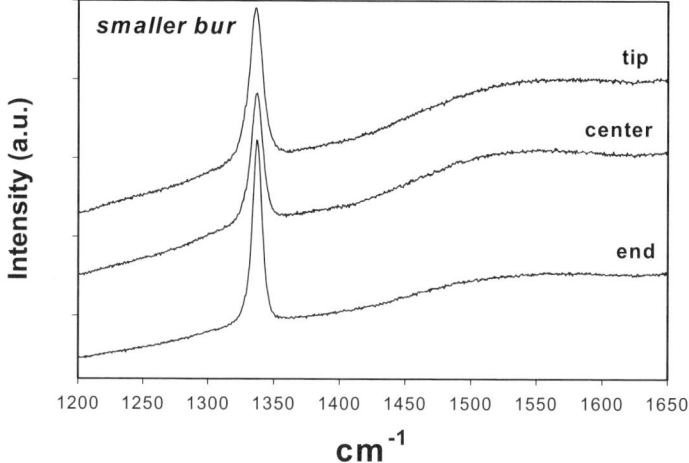

Fig. 7.20. Raman spectra of diamond film on surgical tool

The coated surgical tools have been used to machine a number of materials including copper, aluminum, and iron alloys. The coated tools were compared with uncoated surgical tools to distinguish them in terms of their machining behavior. A micro-machining unit was specifically constructed at Purdue University for such a purpose with a maximum spindle speed of 500,000 revolutions per minute, feed rates of between 5 and 20 µm per revolution, and cutting speeds in the range 100 to 200 meters per minute [79]. The micro-machining unit is shown in Fig. 7.21. The machining center is constructed using three principal axes each controlled using a D.C motor connected to a Motionmaster^{TM} controller. A laser light source is focused onto the rotating spindle in order to measure the speed of the cutting tool during machining. Post machining analysis was performed using a scanning electron microscope to detect wear on the flanks of the cutting edges.

7.10.1 Performance of Diamond-coated Surgical tool

After machining an aluminum alloy material, very low roughness and chipping of the diamond-coated surgical tool were detected. Figure 7.22 shows a typical machined surface in aluminum alloy.

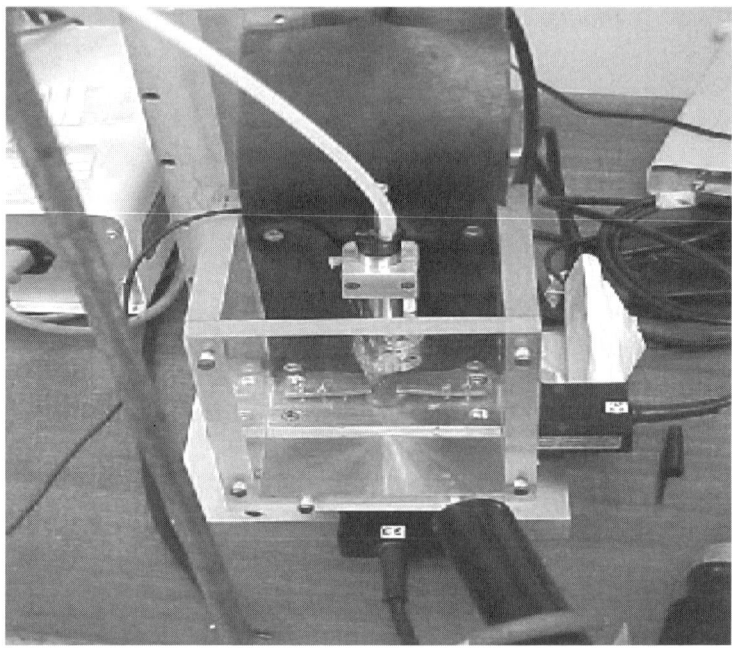

Fig. 7.21. Micro-machining center

A metal chip created from this machining operation is shown in Fig. 7.23. The chip clearly shows shear fronts separated by lamellae caused by plastic instabilities within the material generated at such high speeds. Diamond-coated tools and uncoated were compared by drilling a series of holes in the aluminum alloy. The wear of each tool was determined by examining the extent of flank wear. Uncoated tools appeared to chip at the flank face, and diamond-coated tools tended to lose individual diamonds at the flank face. Uncoated tools drilled an average of 8,000 holes before breakdown occurred, and the diamond-coated tools drilled an average of 24,000 holes [80].

7.10.2 Performance of Diamond-Coated Surgical tool

In order to examine the cutting performance of the diamond-coated surgical tools machining materials such as borosilicate glass, acrylic

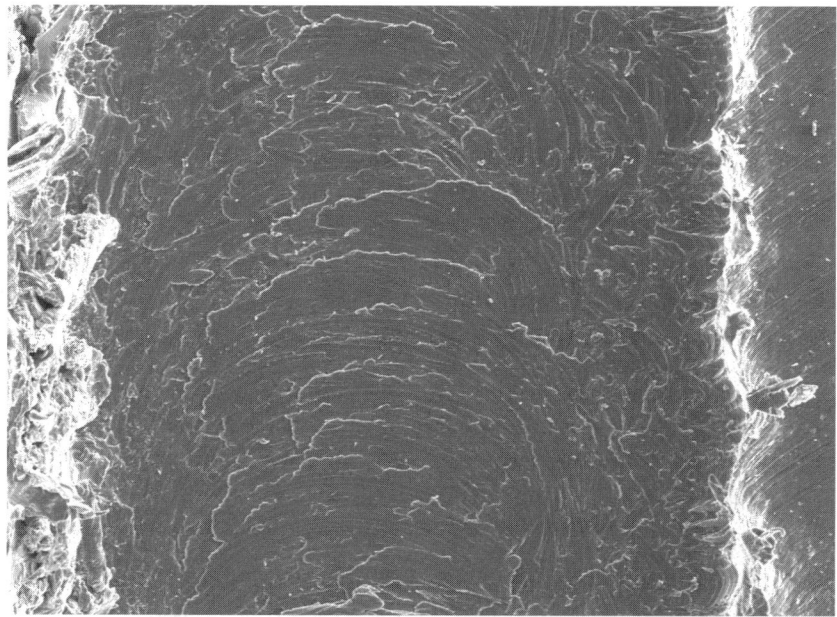

Fig. 7.22. Aluminum alloy material showing a machined track produced by a diamond-coated surgical tool

teeth, and natural human teeth were used. Machining unit was set up for the laboratory bur (AT23LR supplied by Metrodent, UK), which used to operate at 20,000–30,000 rpm with a feed rate of 0.2–0.5 mm/rev without water-cooling.

The flank wear of the burs were estimated by SEM analysis at selected time interval of 1 and 3 min. Prior to SEM analysis diamond-coated burs were ultrasonically washed with 6M Sulfuric acid solution to remove any unwanted machining material, which eroded onto surface of CVD diamond-coated bur. For comparison, the commonly used conventional PCD (polycrystalline diamond) sintered burs with different geometry were also tested on the same machining materials. These burs are made by imbedding synthetic diamond particles into a nickel matrix material to bond the particles at the cutting surfaces.

The morphology of a sintered diamond bur after being tested on borosilicate glass at a cutting speed of 30,000 rpm for 5 minutes with an interval at every 30 sec is shown in Fig. 7.24. It is clearly evident

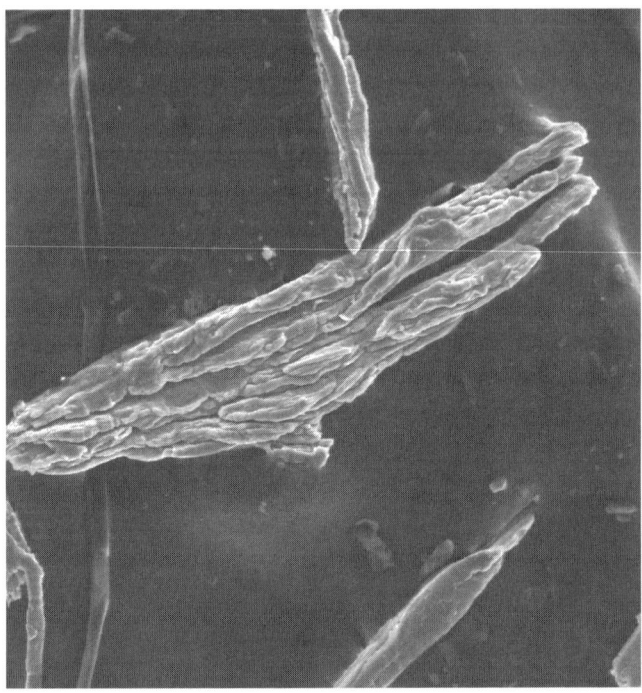

Fig. 7.23. Aluminum alloy chips generated during high-speed micro-machining operations using a diamond-coated surgical tool

that there is significant removal of diamond particles from the surface of the tool after 500 holes. As expected, there is the deterioration of the abrasive performance of the PCD sintered diamond surgical tools.

SEM images of sintered diamond surgical tool tested on borosilicate glass (Fig. 7.24) and CVD diamond-coated laboratory bur after machining tests on borosilicate glass and acrylic/ porcelain teeth respectively for 5 min at a cutting speed of 30,000 rpm are shown in Figs. 7.25 and 7.26. After machining the diamond films are still intact on the pre-treated WC substrate and diamond coating displayed good adhesion. There is no indication of diffusive wears after the initial test for 500 holes.

However, it was observed that the machining of materials such as glass pieces are eroded to cutting edge of the diamond surgical tool as adhesive wear (Fig. 7.25). After testing on acrylic teeth the

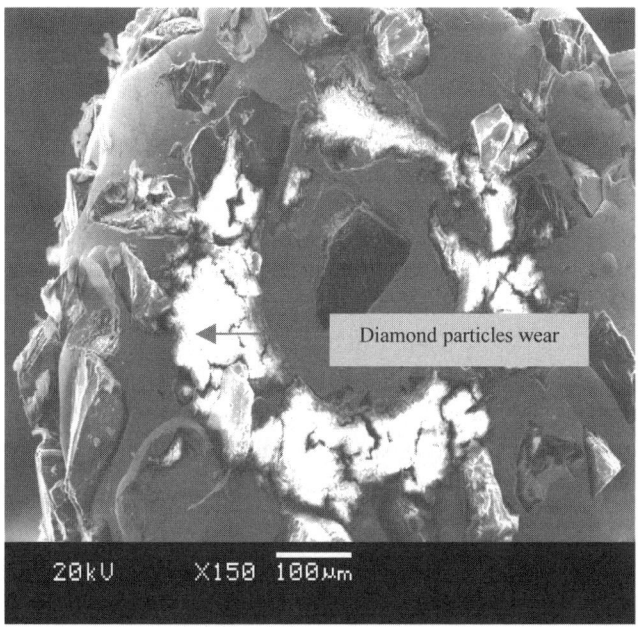

Fig. 7.24. Morphology of sintered diamond surgical tool after machining glass

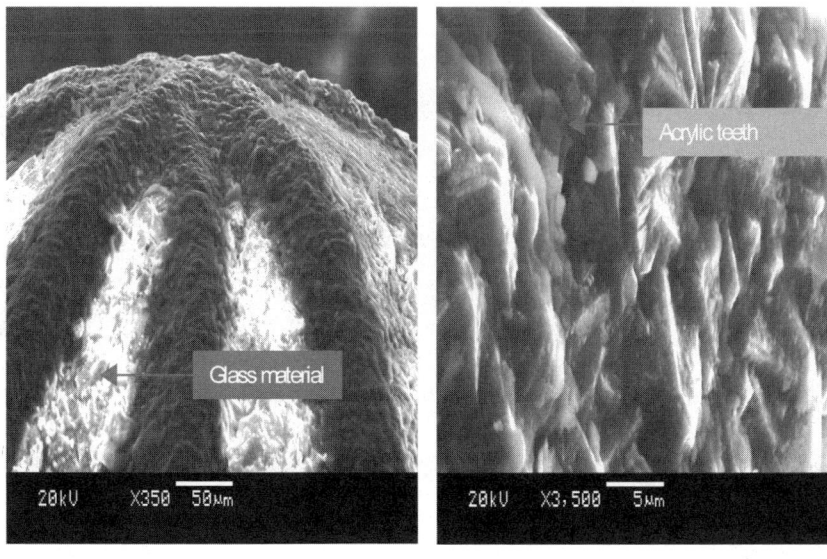

Fig. 7.25. Diamond-coated surgical tool after machining glass

Fig. 7.26. Diamond-coated surgical tool after machining acrylic

mechanism of wear probably involves adhesion as well as abrasion. Figure 7.26 shows that inorganic fillers from acrylic teeth adhered to the cutting tool surface in localized areas when a higher rate of abrasion was used [81].

A micrograph of an uncoated WC-Co surgical tool tested on the borosilicate glass using the same machining conditions are shown in Fig. 7.27(a) and (b). The uncoated WC-Co bur displayed flank wear along the cutting edge of the bur. The areas of flank wear were investigated at the cutting edge of the surgical tool. A series of machining experiments have been conducted using uncoated, diamond-coated surgical tools, and sintered diamond burs when machining extracted human tooth, acrylic tooth, and borosilicate glass. The life of the burs in the machining sense was compared by using the amount of flank wear exhibited by each type of surgical tool. The flank wear was measured at time intervals of 2, 3, 4, 5, 6, and 7 minutes' machining duration. Again, the surgical tools were examined using optical and scanning electron microscopic techniques and observed similar trends with burs that were used in drilling experiments.

The measurements of flanks wear for each bur that machined different dental materials are shown in Figs 7.28–30. It is evident that a longer duty cycle of machining could cause higher rate of flank wear

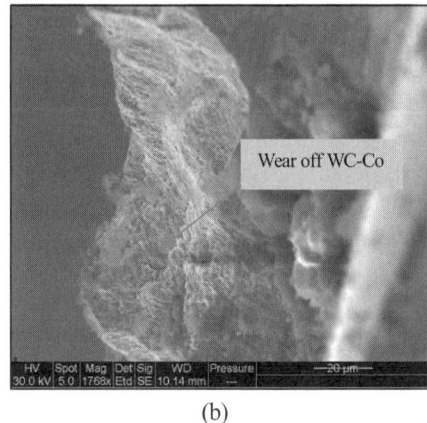

(a)   (b)

Fig. 7.27. (a) cutting edge of WC-Co surgical tool after machining glass; (b) close-up view of WC-Co surface of the surgical tool.

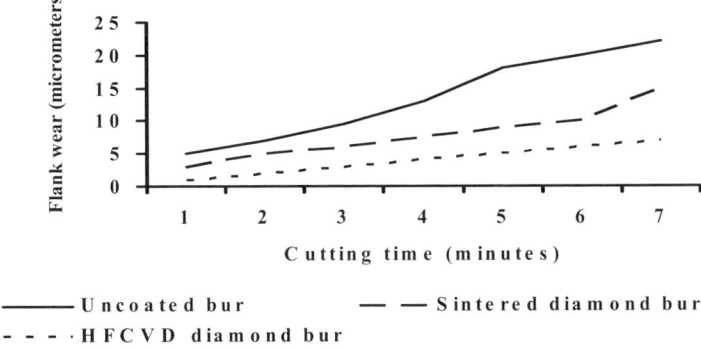

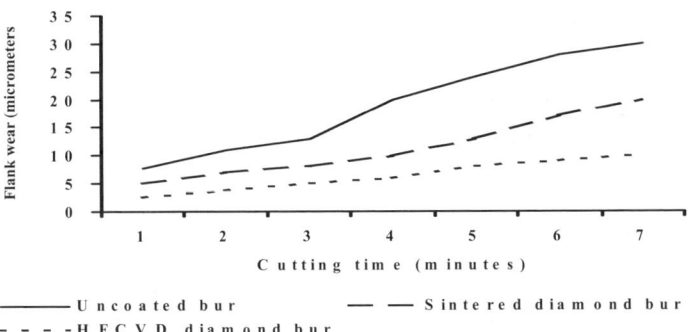

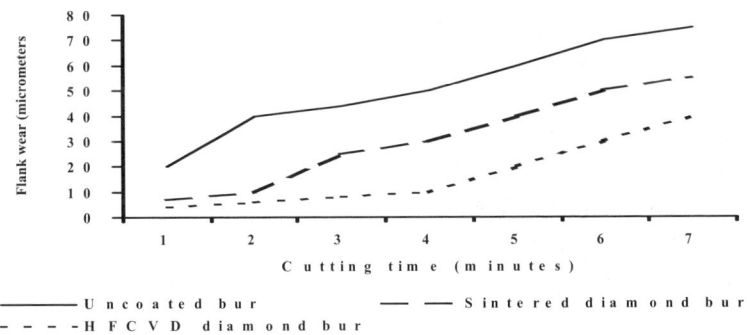

Figs. 7.28–7.30. Flank wear of cutting tools when machining borosilicate glass, acrylic material, and dentine/enamel

on the cutting edge of tool. Therefore, the cutting edge of WC-Co surgical tool should have significant thickness of CVD diamond, which will enhance not only quality of cutting but will also prolong the tool life.

7.11 Time-Modulated CVD Diamond

In the conventional HFCVD method, using the relatively short bias-voltage pulses in bias-enhanced nucleation in which the CH_4 concentration is increased, the flow of CH_4 during film growth is kept constant. Diamond growth in a CVD vacuum reactor is conventionally performed under constant CH_4 flow, while the excess flow of hydrogen is kept constant throughout the growth process. In developing the new TMCVD process it was considered that diamond deposition using CVD consists of two stages: (1) the diamond nucleation stage and (2) crystal growth stage.

Diamond grains nucleate more efficiently at higher CH_4 concentrations. However, prolonged film growth performed under higher CH_4 concentration leads to the incorporation of non-diamond carbon phases, such as graphitic and amorphous. The TMCVD process combines the attributes of both growth stages. This technique has the potential to replicate the benefits obtained by using pulsed power supplies, which are relatively more expensive to employ.

The key feature of the new process that differentiates it from other conventional CVD processes is that it pulses CH_4, at different concentrations, throughout the growth process, whereas in conventional CVD, the CH_4 concentration is kept constant, for the full growth process. In TMCVD, it is expected that secondary nucleation processes occur during the stages of higher CH_4 concentration pulses. This can effectively result in the formation of a diamond film involving nucleation stage, diamond growth, ad secondary nucleation and the cycle is repeated. The secondary nucleation phase can inhibit further growth of diamond crystallites. The nuclei grow to a critical level and then are inhibited when secondary nuclei form on top of the growing crystals and thus fill up any surface irregularities. This type of film growth can potentially result in the formation of a multi-layer type film coating. In such coating systems, the quality and the surface roughness of the film coatings is dependent not on the overall thickness of the film but instead on the thickness of the individual layer of the film coating.

In demonstrating the CH_4 flow regimes, typically employed in conventional diamond CVD and TMCVD processes, Fig. 7.31 shows, as an example, the variations in CH_4 flow rates during film deposition. In the CH_4 pulse cycle employed in Fig. 7.31, the CH_4 flow rate remained constant throughout the conventional CVD process at 25 sccm. It is important to note that the hydrogen flow rate remains constant under both growth modes. CH_4 modulations at 12 and 40 sccm for 15 and 2 minutes, respectively, were performed during the TMCVD process using the microwave CVD system. Since higher CH_4 content in the vacuum chamber results in the incorporation of non-diamond carbon phases in the film, such as graphitic and amorphous, and degrades the global quality of the deposited film, the higher CH_4 pulse duration was kept relatively short. The final stage of the time-modulated process ends with a lower CH_4 pulse. This implies that hydrogen ions will be present in relatively larger amount in the plasma and these will be responsible for etching the non-diamond phases to produce a good quality film.

Figure 7.32 (a) and (b) displays the close-up cross-sectional SEM images of diamond films grown using TMCVD and conventional CVD processes. The conventional MCD film displays a columnar growth structure. However, the time-modulated film displayed a somewhat different growth mode. Instead, the cross section consisted of many coarse diamond grains that were closely packed together. A pictorial model of the mechanism for the TMCVD process is depicted in Fig. 7.32 (c) as compared with the conventional CVD process (Fig. 7.32 d). Primarily, diamond nucleation occurs first in both the TMCVD and conventional CVD processes. However, in TMCVD, diamond nucleates more rapidly as a result of the high CH_4 pulse at the beginning. The high CH_4 pulse effectively ensures the diamond grains to nucleate quicker to form the first diamond layer. The second stage, in which CH_4 content is reduced to a lower concentration, the diamond crystallites are allowed to grow for a relatively longer period. This step enables the crystals to grow with columnar growth characteristics. The surface profile of the depositing film becomes rough, as expected. The third stage involves increasing the CH_4 flow back to the higher pulse. This enables further secondary nucleation of NCD to occur in between the existing diamond crystals, where the surface energy is lower. As a comparison, much less secondary nucleation occurs when the CH_4 flow is kept constant throughout the growth process. The distinctive feature of the TMCVD process is that it promotes secondary diamond-particle nucleation to occur on top of the existing grains in order to fill up any surface irregularities. Figure 7.33 displays the SEM micrograph showing secondary nucleation occurring after a high CH_4 pulse. This result justifies the proposition of the mechanism for the TMCVD process, as shown in Fig. 7.32. It can be expected that at high CH_4 bursts, carbon-containing radicals are present in the CVD reactor in greater amount, which favor the growth process by initiating diamond nucleation. The average secondary nucleation crystallite size was in the nanometer range (≤ 100 nm). It is evident that the generation of secondary nano-sized diamond crystallites has lead to the successful filling of the surface irregularities found on the film profile, in between the mainly (111) crystals. Figure 7.34 shows some randomly selected SEM images of as-deposited diamond films deposited using the TMCVD process at different CH_4 pulse duty cycles.

Figure 7.35 shows the graph displaying the growth rates of conventional and time modulated films grown using hot-filament CVD (HFCVD) and microwave plasma CVD (MPCVD) systems. As expected, the MPCVD system gave much higher growth rates under both growth modes, conventional and time-modulated, compared to films produced using HFCVD. A growth rate of 0.9 μm/hr was obtained using the HFCVD system under constant CH_4 flow. The time-modulated films deposited using HFCVD were grown at a rate of 0.7 μm/hr, whereas with the MPCVD system, films grown using constant CH_4 flow were deposited at a rate of 2.4 μm/hr and using modulated CH_4 flow the films were grown at a rate of 3.3 μm/hr. Although it is

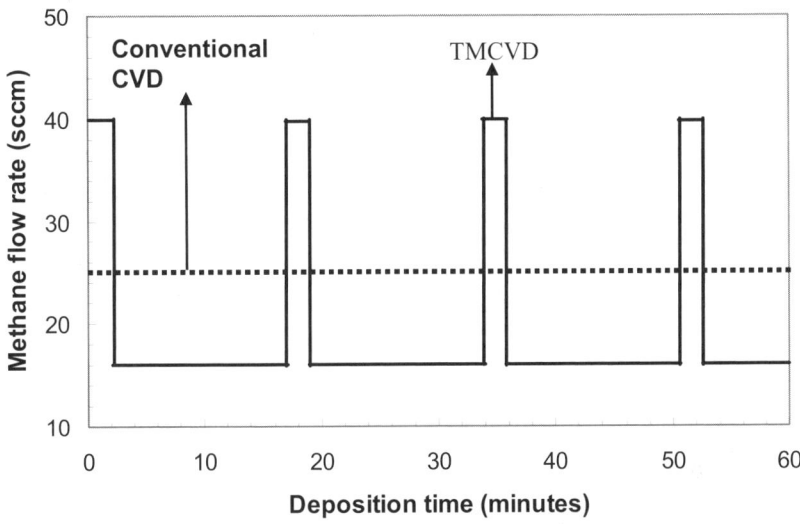

Fig. 7.31. Variations in CH_4 flow rates during film deposition in a typical time-modulated pulse cycle

known that growth rates increase with CH_4 concentration, in the present case using the HFCVD system, the TMCVD process employs greater CH_4 flow than conventional CVD. Our results show that the growth rate of films deposited using constant CH_4 flow is slightly higher than of similar films grown using timed CH_4 modulations.

However, films grown under both modes, conventional and time modulated, using the MPCVD system produced results that were contrary to those obtained using the HFCVD system. Using the MPCVD system, the trend observed was that the films were deposited at a higher growth rate using the TMCVD process than conventional CVD. The substrate temperature is a key parameter that governs the growth rate in diamond CVD. Since the TMCVD process pulsed CH_4 during film growth, it was necessary to monitor the change in the substrate temperature during the pulse cycles. Figure 7.36 shows the graph relating substrate temperature to CH_4 concentration for both HFCVD and MPCVD systems. For the HFCVD system, it was observed that the substrate temperature decreased with CH_4 concentration. In explaining the observed trend, it needs to be considered that the dissociation

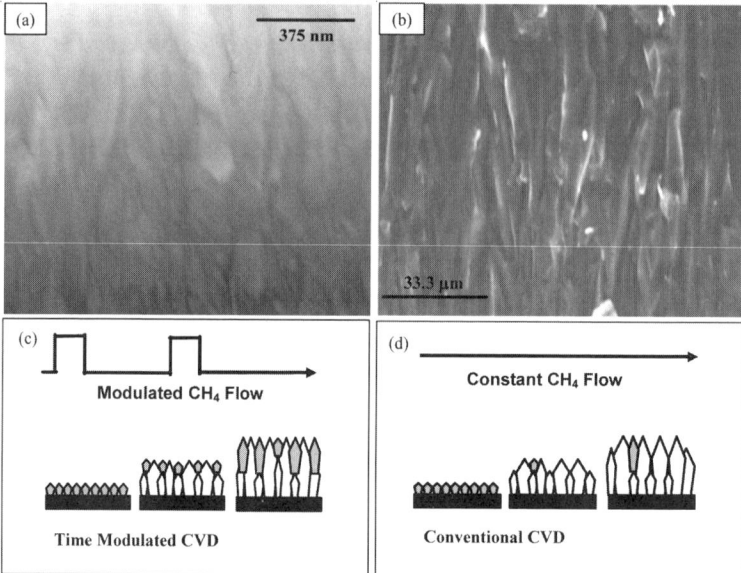

Fig. 7.32. Close-up cross-sectional SEM images of diamond films grown using (a) TMCVD and (b) conventional CVD processes. In addition, the pictorial mechanisms for film growth using TMCVD (c) and conventional CVD (d) processes are also shown

Fig. 7.33. SEM micrograph showing secondary nucleation occurring after a high CH_4 pulse

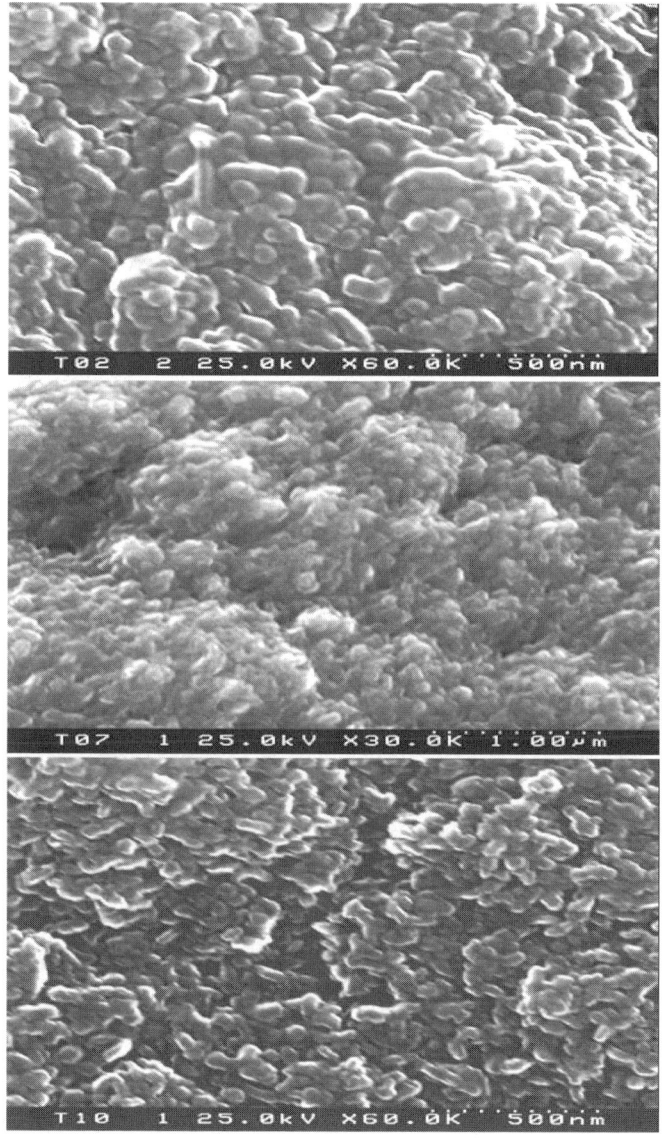

Fig. 7.34. SEM micrographs showing the morphologies of as-deposited diamond films deposited using TMCVD at different CH_4 modulation duty cycles [115]

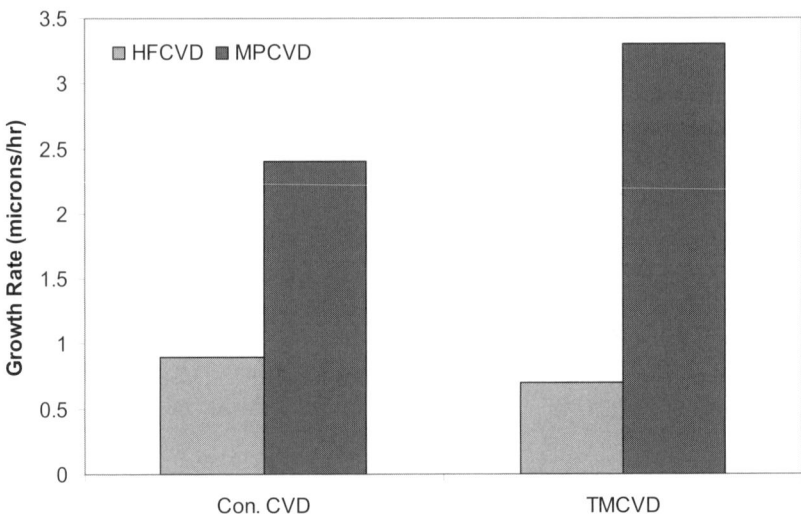

Fig. 7.35. Graph displaying the growth rates of conventional and time-modulated films grown using HFCVD and MPCVD systems

of CH_4 by the hot filament absorbs energy (heat) from the filament and is considered as a cooling process. In our case the filament power was kept constant; therefore, less heat can be expected to radiate to the substrate. In addition, only a small percentage of the thermally dissociated CH species reach the substrate and transfer kinetic energy to the substrate.

It is known that the deposition of diamond films increases with substrate temperature. During the high CH_4 pulse in TMCVD, the lower substrate temperature may be sufficient to lower the growth rate significantly. Generally, in a MPCVD reactor, the substrate temperature increases with CH_4 concentration, as shown in Fig. 7.36.

As a comparison, H_2 is dissociated more extensively in a MPCVD reactor than in a HFCVD reactor to produce atomic hydrogen. Furthermore, in a MPCVD reactor, the plasma power is much greater, 3400 W, than the plasma power in the HFCVD reactor. In fact, the plasma power used in MPCVD for growing diamond films was approximately 10 times greater than the power used in the HFCVD reactor. Therefore, the dissociation of CH_4 absorbs a lower percentage

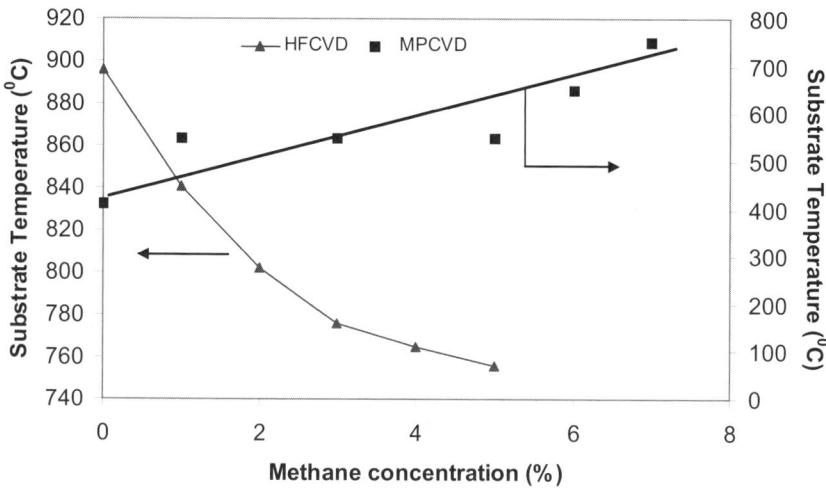

Fig. 7.36. Graph relating substrate temperature to CH_4 concentration for HFCVD and MPCVD systems

of the energy/heat from the plasma in a MPCVD compared to the HFCVD reactor. It is also understood that the reaction between atomic hydrogen and CH species at close proximity to the substrate releases heat. Since in a MPCVD reactor there is a greater intensity of atomic H and CH species, there will be a greater number of reactions between atomic H and CH species. This means that more heat will be released in a MPCVD reactor than in a HFCVD reactor because of such reactions.

This effect contributes to the heating of the substrate. In a HFCVD reactor, the hot filament displays much lower ability to dissociate H_2. Since atomic hydrogen is a critical species that plays an important role in producing a good quality diamond film during CVD, the quality of the films grown using HFCVD is generally lower than of similar films grown using MPCVD. In addition, the hydrogen atoms are required for the effective deposition of diamond onto the substrates. As mentioned earlier, atomic H radicals are present in greater concentration in a MPCVD reactor than in a HFCVD reactor, the MPCVD process gives higher growth rates than the HFCVD process. It is also known that by controlling the temperature during the high/low CH_4

pulse cycles, a greater number of secondary diamond grains were generated and the resultant films displayed (1) smoother surfaces and (2) higher growth rates.

7.12 Conclusions

Thin film deposition technologies, particularly CVD and PVD, have become critical for the manufacture of a wide range of industrial and consumer products. Trends in historical developments in the CVD diamond suggest that the technology is highly likely to yield substantial benefits in emerging technological products in fields of nanotechnology, biomedical engineering, NEMS, and MEMS devices. Several methods including plasma CVD, low pressure CVD, and atmospheric pressure CVD have matured into processes that are routinely used in industry. Microwave and hot filament CVD methods are now commonly used to grow diamond and these can be modified to coat uniformly for tools, NEMS, MEMS and biomedical applications. Diamond coatings examined on tools and biomedical tools showed much enhanced performance compared to uncoated tools.

References

1. Spear KE and Dismukes JP, in Synthetic Diamond: Emerging CVD Science and Technology, The Electrochemical Society, John Wiley and Sons Inc., New York, 1994.
2. Wentorf RH, J. Phys. Chem., 69 (1965) 3063.
3. Butler JE and Woodin RL, Phil. Trans. R. Soc. Lond., A342 (1993) 209.
4. Ashfold MNR, May PW, Rego CA and Everitt NM, Chem. Soc. Rev., 23 (1994) 21.
5. Bachmann PK and Messier R, Chem. Eng. News, 67 (1989) 24.
6. Spear KE, J. Am. Ceram. Sci., 72 (1989) 171.
7. Joffreau PO, Haubner R and Lux B, Mater. Res. Soc. Proc., EA–15 (1988) 15.
8. Spitsyn BV, Bouilov LL and Deryagin BV, J. Cryst. Growth, 52 (1981) 219.
9. Angus JC, Electrochem. Soc. Proc., 89 (1989) 1.
10. Yarbrough WA and Messier R, Science, 247 (1996) 688.
11. Messier R, Badzian AR, Badzian T, Spear KE, Bachmann PK and Roy R, Thin Solid Films, 153 (1987) 1.

12. Angus JC and Hayman CC, Science, 241 (1988) 913.
13. Spear KE, Am J, Ceram. Soc., 72 (1989) 171.
14. Kamo M, Sato U, Matsumoto S and Setaka N, Cryst J. Growth, 62 (1983) 642.
15. Saito Y, Matsuda S and Nagita S, Mater J. Sci. Lett., 5 (1986) 565.
16. Saito Y, Sato K, Tanaka H and Miyadera H, Mater J. Sci., 24 (1989) 293.
17. Williams BE, Glass JT, Davis RF, Kobashi K and Horiuchi T, Vac J Sci. Technol. A, Vac. Surf. Films, 6 (1988) 1819.
18. Kobashi K, Nishimura K, Kawate Y and Horiuchi T, Vac J Sci. Technol. A, Vac. Surf. Films, 6 (1988) 1816.
19. Liou Y, Inspector A, Weimer R and Messier R, Appl. Phys. Lett., 55 (1989) 631.
20. Zhu W, Randale CA, Badzian AR and Messier R, Vac J. Sci. Technol. A, Vac. Surface Films, 7 (1989) 2315.
21. Matsumoto S, Mater J. Sci. Lett., 4 (1985) 600.
22. Matsumoto S, Hino M and Kobayashi T, Appl. Phys. Lett., 51 (1987) 737.
23. Vitkayage DJ, Rudder RA, Fountain GG and Markunas RJ, Vac J. Sci. Technol., A6 (1988) 1812.
24. Meyer DE, Ianno NJ, Woolam JA, Swartzlander AB and Nelson AJ, Mater J. Res., 3 (1988) 1397.
25. Wood P, Wydeyen T and Tsuji O, in Programs and Abstracts of the First International Conference on New Diamond Science and Technology, New Diamond Forum, Tokyo, Japan, 1988.
26. Jackman RB, Beckman J and Foord JS, Appl. Phys. Lett., 66 (1995) 1018.
27. Suzuki K, Sawabe A, Yasuda H and Inuzuka T, Appl. Phys. Lett., 50 (1987) 728.
28. Akatsuka F, Hirose Y and Kamaki K, Jap. J. Appl. Phys., 27 (1988) L1600.
29. Suzuki K, Sawabe A and Inuzuka T, Jap. J. Appl. Phys., 29 (1990) 153.
30. Niu CM, Tsagaropoulos, Baglio J, Dwight K and Wold A, J. Solid State Chem., 91 (1991) 47.
31. Popovici G, Chao CH, Prelas MA, Charlson EJ and Meese JM, Mater J. Res., 10 (1995) 2011.
32. Chao CH, Popovici G, Charlson EJ, Charlson EM, Meese JM and Prelas MA, J. Cryst. Growth, 140 (1994) 454.
33. Postek MT, Howard KS, Johnson AH and Macmichael KL, in Scanning Electron Microscopy, 1980.
34. Spirsyn BV, Bouilov LL and Deryagin BV, J. Cryst. Growth, 52 (1981) 219.
35. Kobashi K, Nishimura K, Kawate Y and Horiuchi T, Phys. Rev. B, 38 (1988) 4067.
36. Pickrell D, Zhu W, Badzian AR, Messier R and Newnham RE, Mater J. Res., 6 (1991) 1264.
37. Oatley CW, in Scanning Electron Microscope, Cambridge University Press, 1972.
38. Tobin MC, in Laser Raman Spectroscopy, Wiley Interscience, New York, 1971.

39. Colthup NB, Daley LH and Wiberley SE, in Introduction to Infrared and Raman Spectroscopy, Academic Press, New York, 1975.
40. Raman CV and Krishnan KS, Nature, 121 (1928) 501.
41. Nemanich RJ, Glass JT, Lucovsky G and Shroder RE, Vac J. Sci. Tech., 6 (1988)1783.
42. Knight DS and White WB, Mater J. Res., 4 (1989) 385.
43. Solin SA and Ramdas AK, Phys. Rev. B, 1 (1970) 1687.
44. Leyendecker T, Lemmer O, Jurgens A, Esser S and Ebberink J, Surf. Coat.Technol, 48 (1991) 253.
45. Murakawa M and Takeuchi S, Surf. Coat. Technol, 49 (1991) 359.
46. Yaskiki T, Nakamura T, Fujimori N and Nakai T, Surf. Coat. Technol, 52 (1992) 81.
47. Reineck J, Soderbery S, Eckholm P and Westergren K, Surf. Coat. Technol, 5 (1993) 47.
48. Wang HZ, Song RH and Tang SP, Diamond and Relat. Mater, 2 (1993) 304.
49. Inspector A, Bauer CE and Oles EJ, Surf. Coat. Technol, 68/69 (1994) 359.
50. Kanda K, Takehana S, Yoshida S, Watanabe R, Takano S, Ando H and Shimakura F; Surf. Coat. Technol, 73 (1995) 115.
51. Luz B and Haubner R, in Diamond and Diamond-like films and coatings, NATO-ISI Series B, Physics, 266, Edited by R.E. Clausing, 579, L.L. Horton, J.C Angus, and P. Koidl, Plenum Press, NY, 1991.
52. Chen X and Narayan J, Journal of Applied Physics, 74 (1993) 1468.
53. Klass W, Haubner R, and Lux B, Diamond and Related Materials, 6 (1997) 240.
54. Zhu W, Yang PC, Glass JT, and Arezzo F, Journal of Materials Research, 10 (1995) 1455.
55. Lux B and Haubner R, Ceramics International, 22 (1996) 347.
56. C.R.C. Handbook of Chemistry and Physics, Edited by R. C. Weast, C.R.C. Press, FL, 1989–1990.
57. Haubner R, Lindlbauer A and Lux B, Diamond and Related Materials, 2, (1993) 1505. 72
58. Chang CP, Flamm DL, Ibbotson DE and Mucha JA, J. Appl. Phys., 63 (1988) 1744.
59. Gusev MB, Babaey VG, Khvostov VV, Lopez-Ludena GM, Yu Brebadze A, Koyashin IY and Alexanko AE, Diamond and Related Materials, 6 (1997) 89–94.
60. Endler I, Barsch K, Leonhardt A, Scheibe HJ, Ziegele H, Fuchs I and Raatz C, Diamond and Related Materials, 8 (1999) 834–839.
61. Kamiya S, Takahashi H, Polini R and Traversa E, Diamond and Relat. Mater, 9 (2000) 191–194.
62. Inspector A, Oles EJ and Bauer CE, Int. J. Refract; Metal Hard Materials, 15 (1997) 49.
63. Itoh H, Osaki T, Iwahara H and Sakamoto H, J. Materials Science, 26 (1991) 370.

64. Liu H and Dandy DS; Diamond Chemical Vapor Deposition, Noyes, 1996.
65. Nazare MH and Neves AJ: Properties, Growth and Application of Diamond, 1998.
66. Zhang GF and Buck V, Surf. Coat. Technol, 132 (2000) 256.
67. Haubner R, Kubelka S, Lux B, Griesser M and Grasserbauer M, J. Physics. 4th Coll, C5, 5 (1995) 753.
68. May P, Rego C, Thomas R, Ashfold MN and Rosser KN, Diamond and Related Materials, 3 (1994) 810.
69. Gouzman I and Hoffmann A, Diamond and Related Materials, 7 (1998) 209.
70. Wang W, Liao K, Wang J, Fang L, Ding P, Esteve J, Polo MC and Sanchez G, Diamond and Related Materials, 8 (1999) 123.
71. Wang BB, Wang W and Liao K, Diamond and Related Materials, 10 (2001) 1622.
72. Kim YK, Han YS and Lee JY. Diamond and Related Materials, 7 (1998) 96.
73. Wang WL, Liao KJ and Gao GC, Surf. Coat. Technol, 126 (2000) 195.
74. Polo MC, Wang W, Sanshez G, Andujar J and Esteve J, Diamond and Related Materials, 6 (1997) 579.
75. Kamiya S, Yoshida N, Tamura Y, Saka M and Abe H, Surf. Coat. Technol, 142–144 (2001) 738.
76. Sein H, Ahmed W, Rego CA, Jones AN, Amar M, Jackson MJ, Polini R, J. Phys: Condes. Matter, 15 (2003) S2961–S2967.
77. May PW, Rego CA, Thomas RM, Ashford MNR and Rosser KN, Diamond and Related Materials, 3 (1994) 810–813.
78. Amirhaghi S, Reehal HS, Plappert E, Bajic Z, Wood RJK, Wheeler DW, Diamond and Related Materials, 8 (1999) 845–849.
79. Jackson MJ, Gill MDH, Ahmed W, Sein H, Proceedings of the Institute of Mechanical Engineers – (Part L): J. Materials, 217 (2003) 77–83.
80. Sein H, Jackson MJ, Ahmed W, Rego CA, New Diamond and Frontier Carbon Technology, 12 (6) (2000) 1–10.
81. Sein H, Ahmed W, Jackson MJ, Woodwards R and Polini R, Thin Solid Films, 447–448 (2004) 455–461.

8. Dental Tool Technology

8.1 Introduction

Dental technology is a discipline of dentistry concerned with the custom manufacture of dental devices to meet the prescription of a dentist. From the earliest times missing teeth have been replaced with dentures or crowns made from a wide variety of materials including gold, human or animal teeth, bone and tusks, and wood. Natural teeth were used for dentures, collected from battlefields, hospitals or by grave diggers, these were mounted in carved dentures of walrus or hippopotamus ivory, or on gold (1). By the late 18th century dentures fused porcelain teeth were introduced, dentures could be carved from blocks of ivory or carved fixed to a gold plate by gold pins. In the mid 19th century the first artificial denture base materials were introduced, vulcanite (or hard rubber) and celluloid, superseded in the 1940s with the introduction of polymethyl methacrylate. During the 20th century base a wide range of new materials and techniques have been introduced to dentistry, including precision lost wax casting for dental alloys, a wide range of precious metal and base metal alloys, and dental ceramics (2).

Dentures were often made by the dentist who extracted the teeth, or their apprentice, sometimes the dentures were made by craftsmen such as jewellers or silversmiths. As clinical dentistry progressed a mechanical assistant specialising in the making of crowns and dentures developed. Dentistry became regulated form the late 19th century onwards and gradually national legislation restricted the practice to qualified dentists only. In 1921 the Dentists' Act, which restricted clinical practice to qualified dentists, stimulated the British Army to begin training dental mechanics, the first formal courses in dental technology were offered in London in 1936 (3). Initially known as the dental mechanic the term dental technician was first used in the 1930s. Dental technology qualifications were generally craft based and gradually progressed to more technically and scientifically based programmes of study from the 1970s onwards, the first-degree programmes were approved in a number of universities in the 1990s. This educational basis of dental technology has resulted in few publications in the scientific literature documenting or evaluating materials and techniques and currently a limited research base.

Dental technology has evolved over the last 100 years from the mechanical assistant to a professional discipline, with an estimated global turnover in the billions of dollars it is however still largely unregulated in many countries. Emerging markets such as China are providing competitive challenges to dental laboratories in the west, particularly in the USA. Internationally the levels of regulation vary, across Europe all dental laboratories must be registered with national medical devices agencies. In the UK statutory registration with the General Dental Council is expected to commence in the summer 2006. There are estimated to be up to 250,000 dental technicians across Europe. In the UK there are approximately 2700 dental laboratories and 8–10,000 people working in dental technology. In the USA there are about 12,000 dental laboratories, which employ about 46,000 technicians. About 40 percent, of the laboratories, are single-handed. It is estimated that the laboratory industry in the USA is responsible for about $6 billion to $8 billion of productivity annually with growth expected to increase by about 6 percent per year for the next several years (4).

Devices are custom-made, largely by hand, and require individual machining with small burs. Before the introduction of lost wax casting, metal crowns and components were made using wire or swaged plate and soldered, filed, polished and buffed using hand files and rotary tools that would be found in jewellery making. The rotary tools will have included bowstring drills using hand cut tools. Power drills for clinical dentistry were slow and difficult to work with were developed from spinning wheels, carpenters drills, jewellers' drills and clockwork mechanisms (5,6). James Beal Morrison, who patented the dental chair in 1867, patented the foot-treadle drill in 1871 (possibly influenced by the Singer sewing machine introduced in the 1850s) and in 1875 added a flexible shaft (7). Later developments included electrically powered motors directly powering the drill, or via belts or flexible drives to the drill handpiece. The flexible shaft invented by James Hall Nasmyth (1808–1890) the Scottish engineer is still used by many dental technicians. The handpiece itself was developed to improve reliability, ease of use and the speed of operation. Air driven turbines running at very high speeds have been tried but the perceived need of dental technicians for sufficient torque when grinding has limited its appeal. Currently the electric micromotor offering speeds of up to 50,000 rpm with high torque are commonly found in most dental laboratories (Figure 8.1).

Fig. 8.1. Dental handpieces and drills

8.2 Burs and Abrasive Points

The history of developments in dental burs is limited (8). The introduction of powered engines stimulated the manufacture and use of dental burs, cutting and grinding tools (6). Originally hand cut and ground the burs were costly and inconsistent, mass production began in the USA in the 1870s made from carbon steel. The SS White "Revelation Bur" was the first to have a continuous drill edge (9). Corundum introduced in 1872, which enabled enamel to be worked, was later superseded by silicon carbide (Carborundum) stones and discs. Diamond burs were initially produced in the 1890s, and in SS

White introduced tungsten carbide (TC) burs to dentistry in 1948 (6). These burs designed for clinical use will have been used in the dental laboratory for fine work. Most trimming and grinding will have been done using hand files and lathe-mounted wheels.

Almost all devices made in dental laboratories involve the use of a dental laboratory handpiece and burs at some stage. A bur of various abrasive materials, sizes, shapes and cutting characteristics is mounted in a handpiece or on the chuck of a fixed lathe. On a fixed lathe the speed is set and the dental device is held in both hands and manipulated onto the rotating bur altering its position to grind or polish the surfaces. It is commonly used to remove sprues from metal castings, where gross grinding is required. However, some technicians find the use of both hands to secure the dental device whilst grinding beneficial and will undertake all machining on the lathe (Figure 8.2).

In many laboratories the moveable handpiece is the tool of choice, this may be driven by a flexible drive or by an electric micromotor (Figure 8.3).

The speed of the handpiece may be fixed by a bench top-speed controller, or more normally, a variable speed controller controlled by foot or knee pressure. The speeds vary depending upon the design of the motor but generally range from 0–30,000 rpm however many

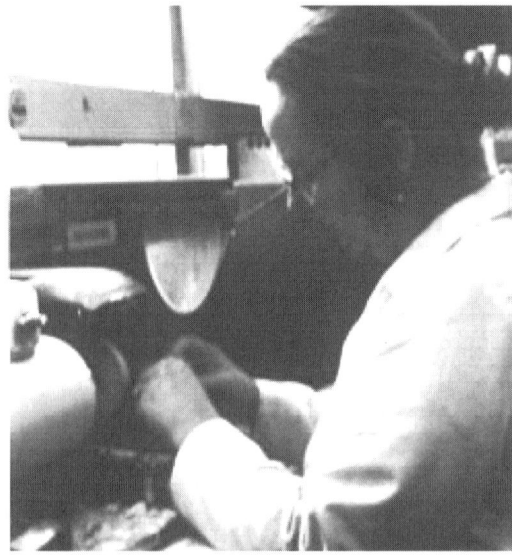

Fig. 8.2. Polishing on the dental lathe

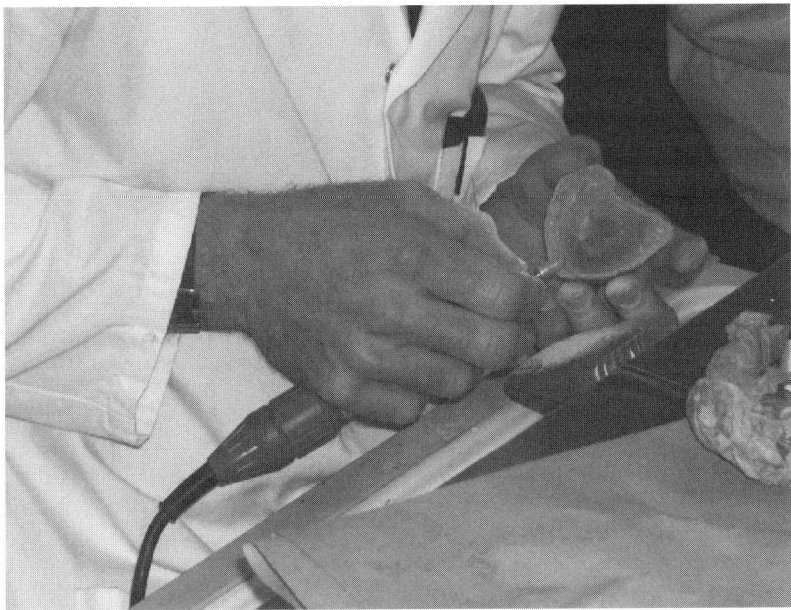

Fig. 8.3. Handpiece and bur being used to grind dental impression tray

machines offer speeds of up to 50,000 rpm. Most machines offer the ability for the operator to set the speed required. They may then operate the motor using a foot or knee controller used either as an on/off switch to the set speed or as a variable speed controller up to the maximum set speed.

When using the handpiece the denture, crown or other device is held in one hand, the handpiece and bur are then brought onto the device varying the position of both hands to enable all surfaces to be trimmed. The pressure used varies during the process, initially for rapid removal of large quantities of material the operator may use heavy pressure and/or high speed. As the device is nearing its required size and shape the operator reduces the pressure and sometimes the speed. The combination of excessive speed and pressure is deprecated in teaching as it generates too much friction which can both damage the tool and the device being trimmed. Acrylic devices can warp and the mechanical properties of metal components can be altered by excessive heat. Pressures in practice together with poor supervision and work-based training can result in misuse of the tools. Complaints of poor performance of burs often relate to use at excessive speed and pressure (10).

The technician uses the drill and bur as a sculptor holding it onto the piece and drawing it along the piece removing the material required. For fine detail such as creating minor texture effects on the surface of a tooth the technician may hold the drill like a pen and gently grind the surface.

The selection of the shape, size and type of bur, cutter or abrasive tool is affected by the material being trimmed, the amount of trimming needed, the nature or form of the trimming being undertaken, the speed of the handpiece or motor and local custom and practice. The range of tool shapes, sizes and materials is very extensive, many of the tools are capable of being used on a wide range of materials and techniques. There are local, regional and national variations in the range of dental rotary tools purchased with no apparent to patterns in usage (10). Each dental laboratory, or even individual technician in the laboratory, select their own preferences. Large dental laboratories may restrict the choices of tools available to its employees for stock control. In a recent oral survey (unpublished) of 10 laboratories, including three dental schools, each laboratory identified using different burs in the finishing of chromium-cobalt denture frameworks. The bur use varied between lathe and handpiece-mounted tools, course diamond barrels, silicon carbide stones (pink or brown), tungsten carbide burs or mandrel mounted abrasive disks. The costs of diamond or tungsten carbide burs can be 10–20 times the cost of silicone carbide stones however the life of the former is greater. One commercial laboratory considered the reduction in production time and longevity as more important that initial cost of the bur. One identified the use of diamond to increase cutting efficiency and reduce friction to avoid overheating of the alloy as the most important (11). The teaching in schools of dental technology is influenced by available funding, although more expensive burs such as diamond are recommended cost implications on the school or student influence bur selection (11,12). The initial pattern of instruction may affect future practice strongly. Responses in the survey from several senior technicians recounted following the instruction they received as an apprentice and had barely changed since as they had adapted their technique to suite the burs they selected. In the teaching of dentists in the USA there is a broad consensus on rotary instrumentation used by dental students with coarser grit burs being used at post- graduate level (13).

8.3 Classification of Dental Burs

The wide range of materials, shapes and cutting surfaces resulted in 20 international standards associated with the manufacture of dental burs these specify aspects of the performance in terms of composition, form, reliability and packaging. BS EN ISO 6360 parts 1, 2, 3, 4, 6, 7 now provides a general numbering system for all types of dental rotary instruments. It was prepared to meet the need for a universal system of classification and establishes a comprehensive coding system. They do not however specify abrasive/cutting efficiency.

The standards are shown in Table 8.1. Although these are current there has been consolidation to create a comprehensive series of standards.

8.4 Coding of Dental Tools

Each item has a 15-digit code, the first 3 numbers identify the materials used in the working part, the next 3 the shank and overall length, the third the shape, the fourth special characteristics, the nominal size of the working part. There are also an optional further 3 numbers for diamond instruments.

The first group is the materials used for the tool, there are 42 materials used for rotary tools of which most are used in the dental laboratory.

8.4.1 Shapes

The shapes are classified in BS EN ISO 6360-2:2004 Dentistry. This standard comprises 5 tables that describe the coding for general shapes and designs of which there are 257 shapes summary information in Table 8.3. In the standard, *table 2* describes 70 disc types, *table 3* special instruments, *table 4* mandrels, and *table 5* root canal instruments (these are not used in dental technology).

Table 8.1. International Standards for Dental Rotary Instruments

BS EN ISO 6360-1:2004	Dentistry. Number coding system for rotary instruments. General characteristics
BS EN ISO 6360-2:2004	Dentistry. Number coding system for rotary instruments. Shapes
BS EN ISO 6360-3:2005	Dentistry. Number coding system for rotary instruments. Specific characteristics of burs and cutters
BS EN ISO 6360-4:2004	Dentistry. Number coding system for rotary instruments. Specific characteristics of diamond instruments
BS EN ISO 6360-6:2004	Dentistry. Number coding system for rotary instruments. Specific characteristics of abrasive instruments
BS 6828-8.1:1987, EN 27787-1:1989, ISO 7787-1:1984	Dental rotary instruments. Cutters. Specification for steel laboratory cutters
BS EN 27787-3:1994, ISO 7787-3:1991	Specification for dental rotary instruments. Cutters. Carbide laboratory cutters for milling machines
BS EN ISO 10323:1996	Dental rotary instruments. Bore diameters for discs and wheels
BS EN ISO 13295:1997	Dental rotary instruments. Mandrels
BS EN ISO 1797-1:1995	Dental rotary instruments. Shanks. Shanks made of metals
BS EN ISO 1797-2:1995	Dental rotary instruments. Shanks. Shanks made of plastic
BS EN ISO 2157:1995	Dental rotary instruments. Nominal diameters and designation code number
BS EN ISO 3823-1:1999	Dental rotary instruments. Burs. Steel and carbide burs
BS EN ISO 3823-2:2003	Dental rotary instruments. Burs. Finishing burs
BS EN ISO 7711-1:1998	Dental rotary instruments. Diamond instruments. Dimensions, requirements, marking and packaging
BS EN ISO 7711-2:1996	Dental rotary instruments. Diamond instruments. Discs
BS EN ISO 7711-3:2004	Dentistry. Diamond rotary instruments. Grit sizes, designation and colour code
BS EN ISO 7786:2001	Dental rotary instruments. Laboratory abrasive instruments
BS EN ISO 7787-2:2001	Dental rotary instruments. Cutters. Carbide laboratory cutters
BS EN ISO 7787-4:2002	Dental rotary instruments. Cutters. Miniature carbide laboratory cutters
BS EN ISO 8325:2004	Dentistry. Test methods for rotary instruments

Table 8.2. Abrasives used in dental technology

Grinding	Polishing
Tungsten carbide	Rubber
Silicon carbide	Natural bristles
Diamond, medium, coarse, very coarse, ultra fine, extra fine, fine,	Synthetic bristles
Ruby	Brass
High speed steel	German silver
Normal grade corundum	
High-grade grade corundum, pink	**Buffing**
High-grade grade corundum white	Felt
Tungsten carbide grit	Leather
Titanium	Flannel
Nickel Titanium	Muslin
Quartz	Felt cloth
Sapphire	Yarn
Cubic boron nitride	Goat hair
Electrocorundum, red	
Free cutting steel	**Other**
Cold worked steel	Plastic
Spring steel	Quill
Stainless steel	Paper
Stainless spring steel	Gutta percha
	Cuttlefish bone

8.4.2 Types of Toothing

Referring to BS EN ISO 6360-3:2005 Dentistry, the number coding system for rotary instruments and specific characteristics of burs and cutters are included in this standard. The standard details the toothing of burs, finishing burs and, very fine, medium, coarse and very coarse cutters. It also includes a *table 8* for surgical tools.

8.4.3 Specific Characteristics of Diamond Instruments

Referring to BS EN ISO 6360-4:2004 Dentistry, the number coding system for rotary instruments and specific characteristics of diamond instruments are included in this standard. The standard describes the additional information that can apply to diamond burs, this is the angle of tapered diamond instruments and the width of diamond-coated discs (Fig. 8.4).

Table 8.3. Rotary shapes and variants

Basic shape	Example variants
Spherical	With collar, hemispherical
Wheel	End cutting, rim cutting, conical, half circle rim.
Cylindrical	side cut, end and side cut, pointed end, hemispherical end, rounded end, distal end hemispherical proximal end hemispherical, plus others
Conical	slender, truncated conical, 30% flatter, side cut only, rounded end
Inverted conical	Side cutting, end cutting, concave collar, rounded conical pointed, others
Bud	slender, rounded, rounded slender, long, flat end rounded edge
Pear, Flame, Bullet	
Egg	long, side cutting
Barrel	
Torpedo	Conical
Lens	

Table 8.4. Summary of coding for toothing burs and cutters

Straight	Right helicoidal
Straight, left crosscut	Right helicoidal left cross cut
Straight, sharp cutting angle	Right helicoidal fine left cross cut
Straight, blunt cutting angle	Right helicoidal, sharp cutting angle left cross cut
Straight, left cutting	Right helicoidal x-cut, transverse blade at the tip
Straight, left cross cut	Right helicoidal left cutting
Straight, x-cut	Right helicoidal left cut; right cross cut
Straight, with grooves	Right helicoidal x-cut
Straight left serpentine cut	Right helicoidal, fine right cross cut
Left helicoidal	Right helicoidal, right cutting x-cut
Left helicoidal left cross cut	Right helicoidal, right cutting x-cut right cross cut
Left helicoidal right cutting	Right helicoidal with grooves
Left cutting x-cut	Right helicoidal, right cutting, 2 cuttings
Left helicoidal right cutting x-cut	Right helicoidal, right and left cutting
Cardia	
Side, finishing bur toothing	
Diamond toothing	
x-cut	

Fig. 8.4. Example of toothing drill

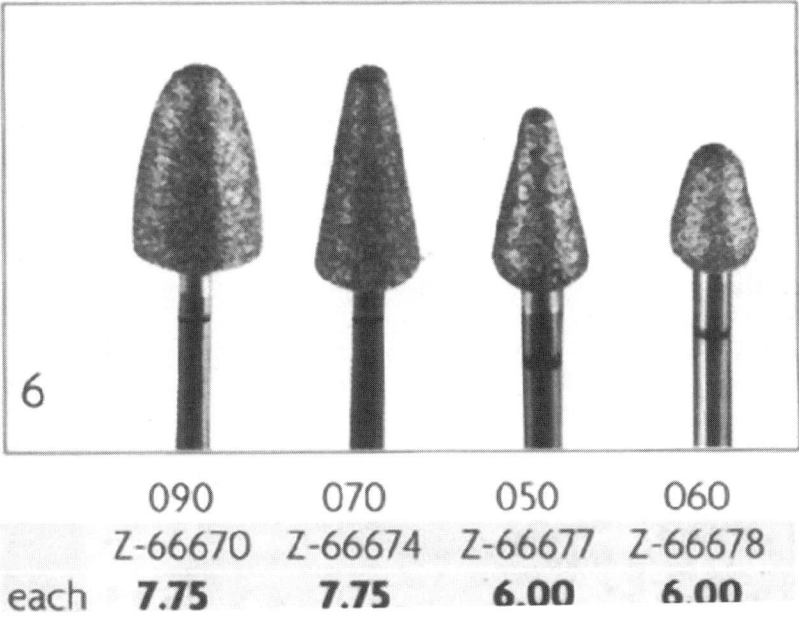

Fig. 8.5. Dental bur coding information

Referring to BS EN ISO 6360-6:2004 Dentistry, the number coding system for rotary instruments and specific characteristics of abrasive instruments, groups abrasive instruments by the fineness of the grit size: ultra-fine, extra-fine, fine, medium, course and very coarse, and within each group it is further subdivided by the hardness of the binding materials very soft, soft, medium, hard, very hard, extra hard. A dental bur will thus be coded as shown in Figure 8.5.

8.5 Dental Devices

Dental devices may be used for a short-term, e.g., removable orthodontic devices, or be in-situ for many years, e.g., dental crowns or bridges. All devices are made largely by hand using a variety of equipment and techniques. The techniques are predominantly based on lost wax techniques at some stage in their manufacture whether the devices are made from acrylic, dental alloys, or even some ceramics. Most newly processed devices normally require trimming, polishing and buffing to make them fit for use in the mouth. As every patient is different, the mouth is a complex three-dimensional structure (Figure 8.6). The dental technician needs to appreciate the structures and functions of the oral cavity, an understanding of the properties

Fig. 8.6. Range of devices made by dental technicians

of the materials and their processing and the dexterity and artistic ability to make the device that will fit, function and appear to be natural. The dental handpiece can often be used as a carver to create the lifelike appearance required.

8.6 Dental Laboratory Materials

8.6.1 Gypsum

Gypsum based products are used for the construction of models or casts of the dental mouth and in refractory investment materials. These may be need to be ground using dental burs to shape the cast for use, for example to expose the edges of a tooth preparation in the making of a crown. A fine crosscut of diamond cut tungsten carbide bur is the tool of choice for many technicians for this task. The gypsum used may be modified during its manufacture to increase hardness. Vacuum mixing can enhance its hardness and the surface may be altered by sealants to improve abrasion resistance, with a hardening solution high strength stones the Knoop hardness of a stone was increased to 79 kg/mm^2 (2). Epoxy resins may also be used for these tasks (2). Gypsum products will be ground dry under extraction at speeds of under 20,000 rpm due to dust Fine burs generate significant amounts of dust, course burs create large particles that can travel

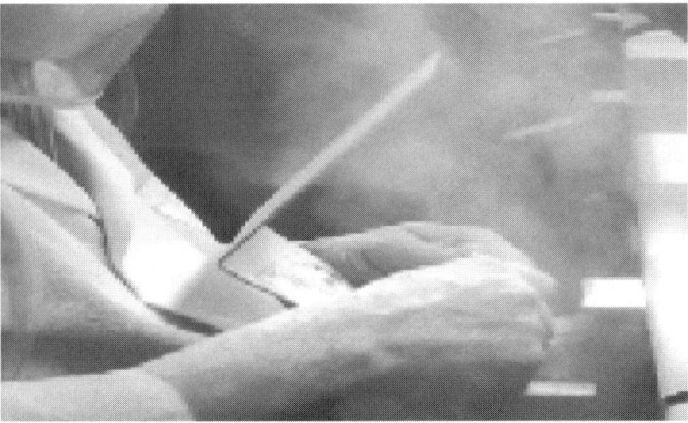

Fig. 8.7. Dust generated grinding dry plaster cast

in all directions across a laboratory at considerable velocities (Figure 8.7). Damp gypsum will adhere to the bur and clog the cutting surface.

8.6.2 Light Activated Dental Impression Tray Materials

To obtain an accurate cast of a patient's mouth, a clinician will take two impressions. The primary impression is taken using a dental impression material in an impression tray available in a limited range of sizes. The impression is disinfected and then poured in dental plaster upon which a custom-made impression tray can be for the individual patient for greater accuracy. There are many materials for these trays, thermosplastic polymers that are pressure formed onto the cast and then trimmed using dental burs. A common material used is a photopolymerised polymer with high inorganic filler content, up to 76% (15). The inorganic filler can be quartz and barium or lithium aluminium silicate glasses, borosilicate glasses, strontium or zinc glasses. These materials are provided as mouldable sheet that can be applied to a cast and cured in about five minutes in curing unit using light at about 500nm (2). Once cured these materials are trimmed using dental burs.

The grinding of these materials creates a very fine dust that many technicians often identified as a nuisance or irritant (Figure 8.8). Although there is no objective data many dental technicians have reported high wear rates grinding this material. Large tungsten carbide cross-cut burs are the tool of choice for many, some prefer large lathe mounted abrasive wheels for speed and cost effectiveness.

The particle sizes of some the dust generated is less than 5μm which is inhalable the particle size varies depending upon the cutting geometry of the bur and the particle size of the filler.

Burs used include ruby abrasives, tungsten carbide and steel burs. The shapes used vary depending upon the form of the tray and the parts that need trimming. Holes are drilled for retention using rose-head shaped burs.

There is no published data on wear rates in relation to machining this material however many technicians report excessive wear whilst trimming this material as although the matrix of the material will be comparatively soft, the filler content however can include quartz (Moh's hardness 7 (15)).

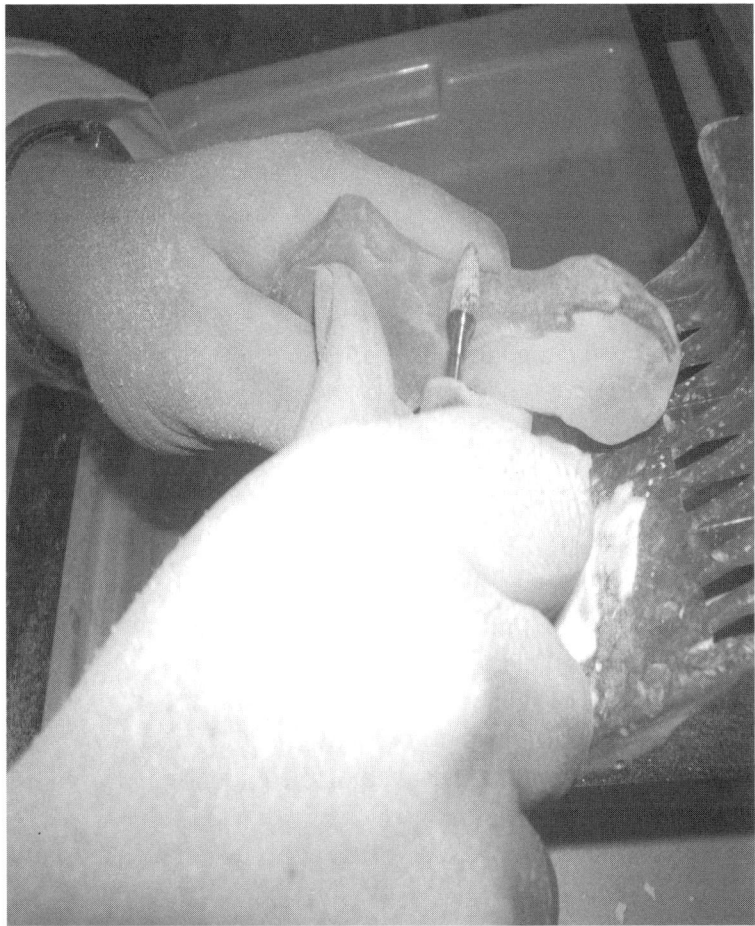

Fig. 8.8. Grinding of light curing polymer, note method of holding handpiece and dust generated

8.6.3 Materials for Dentures

Removable dentures are made on the cast of the oral structures. Teeth are mass-produced in acrylic or porcelain. A trial denture is constructed positioning the teeth into wax that is carved to reproduce the oral tissues. The trial denture is inserted into the mouth to assess the upper and lower jaw relationship, the appearance and tissue support. The wax allows for ease of alteration. When satisfactory the wax trial is returned to tee dental laboratory where a mould is made of it embedding it into plaster in a two-part flask, normally metal. The wax is eliminated in boiling water and dental polymers can be inserted either by compression moulding or injection moulding. The polymers are predominantly heat cured or chemically cured polymethyl methacrylate, or alternatively Nylon or other polymers. After curing, the denture is divested from the embedding plaster and trimmed using a dental drill (handpiece) and a range of dental burs. Dentures are polished using pumice slurry with various sizes and shapes of lathe brush or calico mops and buffed to a high lustre using bar compounds of tripoli followed by bar or liquid polishes containing whiting on a wool mop or worn calico mop at high speed.

The grinding of acrylic varies depending upon the type of polymer used or brand. The selection of bur to trim dentures and to prepare them for polishing and buffing is based up the experience of the individual dental technician. The type of bur used, the speed at which it is used and pressure with which it is applied is affected by many factors:

- The skills of the individual, the amount of trimming needed;
- The education and training of the technician;
- The price of the device, this influences the time available, and grade of technician undertaking the work;
- The expertise and experience of the dental technician making it; and
- Inexperienced technicians often apply excessive pressure and speed when grinding any device. An acrylic denture can be warped or burnt.

The hardness of acrylic resins, Knoop Hardness, 14–17.6 (15), Vickers hardiness 9–23 (16,17) means that there are a wide variety of burs and stones that can be used. However, excessive speed and pressure must be avoided. Working pressure for laboratory abrasives and tungsten vanadium alloyed tool steel is 1–5 N (18) Tungsten carbide has a recommended pressure of 3.3–7.5 N (18). The recommended speeds vary depending upon bur material and size of bur.

The burs used for trimming (summarised in Table 8.5) may include:

- Course, medium of fine cut tungsten carbide burs;
- Silicon carbide;
- Ruby abrasives are selected by many for fine grinding of acrylic resin dentures;

Table 8.5. Summary of materials and possible abrasives

Material	Grinding	Polishing	Buffing
Acrylic	• Cold worked steel • High speed steel • Tungsten carbide • Quartz • Normal grade corundum, white • Silicon carbide • Ruby Sapphire	• Fine sandpaper • Silicone polishers • Goat hair • Natural bristles • Synthetic bristles • Brass wire brushes • German silver wire brushes • Pumice slurry	• Felt • Flannel • Muslin • Felt cloth • Yarn • Whiting • Tripoli
Base metal alloys NiCr, CoCr	• Diamond, coarse or medium (fine, ultrafine, extra fine, very course for fine detail) • Tungsten carbide • Silicon carbide, Pink, brown • Normal grade corundum, pink, white	• Rubber wheels, cylinders, cones • black • green	• Synthetic bristles • Felt • Leather • Flannel • Muslin • Felt cloth • Yarn • Chrome oxide
Gold	• Tungsten carbide • Silicon carbide • Normal grade corundum, pink, white • Ruby Sapphire • Cubic boron nitride • Electrocorundum red • High speed steel • Cold worked steel	• Rubber wheels, cylinders, cones • black • green • white • silicone	• Goat hair • Natural bristles • Synthetic bristles • Leather • Felt • Flannel • Muslin • Felt cloth • Yarn • Tripoli • Rouge
Ceramic	• Diamond, medium fine, (ultrafine, extra fine, coarse, very course) • Silicon carbide green stones	• Diamond paste • Diamond impregnated polishers	

- White abrasives; and
- Fine abrading is done using sandpaper of various grit sizes, or abrasive impregnated silicone polishing burs (Figure 8.9).

Dentures should be trimmed, polished, and buffed to a high lustre because surface roughness enhances microbial adhesion and reduces the ability to be cleaned (19–25). Burs with a threshold surface roughness for bacterial retention of (RA = 0.2 μm) should be used. A number of studies have examined surface finishing techniques (26–30). The surface lustre produced in the dental laboratory provides the smoothest surface (27–29). Tungsten carbide burs produce a smoother non-grooved surface than the steel bur on acrylic resins (30).

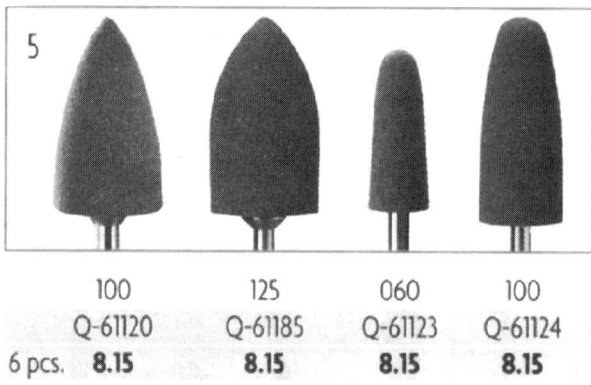

100	125	060	100
Q-61120	Q-61185	Q-61123	Q-61124
6 pcs. **8.15**	**8.15**	**8.15**	**8.15**

Fig. 8.9. Silicone polishers

8.6.4 Metal Components: Crowns, Bridges and Metal Partial Dentures

Metallic crowns, bridges, substructures and partial denture frameworks are made using the lost wax process. Wax forms of either the whole tooth or substructures for the later addition of aesthetic materials are created on casts and dies of high strength dental stone. These wax forms are attached to a sprue, removed from the die and attached to a casting cone former, they are invested in a refractory investment, and placed in a furnace to eliminate the wax and moisture form the investment. Dental alloys are melted and cast into the mould, after cooling they are divested and sand or bead blasted. They will then be cut from the sprue, trimmed, polished and buffed with a range of disks, burs, stones, rubber tools, and buffing mops. Metal partial dentures are made in a similar way but are a more complex shape and using a refractory model of the mouth and use multiple sprues.

The alloys used for metal components include high gold alloys (those containing more than 75% gold), low gold alloys, palladium silver alloys, and base metal alloys. The gold alloys have a Vickers hardness from 90 to 230 depending upon alloy type and brand. Low gold alloys and palladium silver alloys will have a Vickers hardness around 140 (15). Base metal alloys that may be cobalt chromium, cobalt chromium nickel, or nickel chromium have a Vickers hardness number of between 320 and 430. Titanium alloys are increasingly being used due to their lightness, they require specialised casting equipment, their Vickers hardness is about 140 and care is needed when finishing not to overheat the alloy.

The selection of the materials for machining and the shape of the tool the crown will depend upon the alloy used, the stage of finishing, the location of the grinding to the size and type of tooth. For example, for the inside of a lower incisor tooth very fine tools will be needed if any adjustment is needed. For optimum performance the device must be as smooth as possible.

8.6.5 Materials for Partial Denture Frameworks

Cobalt chromium alloys are most commonly used although these may also be made in gold alloys or titanium. The surface finish required is a high lustre to reduce microbial adhesion, surfaces if rough can also rapidly increase the wear of opposing or abutment teeth. There have been few papers evaluating finishing procedures of cobalt chromium denture frameworks. Aydin (31) determined that the best surface finish was obtained using a systematic approach after sandblasting, hard stone, medium silicone carbide disk, second sandblasting, electropolishing, hard rubber point, felt disk and soft brush with polishing paste although in the study cutting load was not standardised. Xenodimitropoulou and Radford (32) recommended a 6.5mm aluminium oxide pink stone as the most efficient and consistent due to low costs and cutting efficiency. The tungsten carbide bur was second however it lost 70% of its cutting efficiency over the duration of the machining. The diamond was the lest efficient, the diamond bur showed evidence of plucked out grit, being won flat and with the matrix contaminated by the machined sample. Ponnanna (33) determined the roughness generated by sandblasted and identified a systematic approach to finishing similar to Aydin (31). The amount of dust generated in the finishing of a cobalt chromium framework for a partially dentate patient using silicon carbide abrasives and rubber wheels can be up to 2 grams, 1 gram of cobalt chromium alloy and 1 gram of silicon carbide dust (34).

8.6.6 Titanium Alloys

Cast titanium alloys have excellent biocompatibility and are light, which is particularly useful for upper dentures (35, 36). The mechanical properties are slightly different to Co-Cr alloys and have to be accounted for during design and in manufacture (37,38). In finishing titanium castings the hard brittle reaction layers (α-case) on the cast surface must be removed as these layers are reported to reduce the ductility and fatigue resistance of the framework and clasps (39). Ohkubo et al. (39) identified that in comparing Ti-6Al-4V with Co-Cr and type iv gold alloys there was little difference in cutting effectiveness using silicon carbide burs finding no correlation between the hardness of the alloy and volume loss. However the opposite occurred when suing steel fissure burs. Steel burs will blunt, however SiC stones will present new abrasives particles as they wear off thus maintaining cutting efficiency although the diameter of the stone will reduce and thus reduce the cutting velocity unless speed of the motor is increased to compensate. Kikuchi et al. (40) compared different alloys compositions of Ti-Cu alloys assessed by their grindability using SiC burs. Hirarata (41) comparing the polishing of Ti and Ag-Pd-Cu-Au alloy with five dental abrasives, carborundum points, and silicone polishers found the Ti more difficult to polish and suggested the development of new abrasives for polishing of Ti. In the CAD/ CAM milling of titanium devices Hotta et al. (42) identified that tungsten carbide burs displayed chipping at the bur blade and gradual dulling of the tool and an increase in the average surface roughness on the crown. However, they concluded that the tungsten carbide burs could be used to fabricate up to 50 titanium crowns.

8.6.7 Materials for Metal Inlays, Crowns and Bridges

Crowns and bridges are made in a variety of alloys, including noble alloys varying from high gold soft alloys type I (VHN 60–90) (15)–type IV Extra hard (VHN 220), palladium silver alloys, porcelain fused to metal techniques (base metal or noble metal alloys), low gold alloys Ni-Cr, and Ti. The selection of bur type and technique will vary depending upon the alloy and local custom and practice, diamond burs, SiC and tungsten carbide the main options. Siegal (43) determined that cross cut diamond burs should be used for base metal alloys and medium grit diamond burs for high noble alloys. Miyawaki (44) identified the tungsten carbide bur generally superior in cutting effectiveness than diamond when using a dental air turbine to grind Ag-Pd-Cu-Au, Ag-Zn-In-Sn, NiCr or Ti alloys.

After grinding base metal crowns are finished using the same tools as for base metal denture frameworks however the tools sizes may include many with finer shapes to characterise teeth or trim inside the fitting surface. Noble alloy crowns being softer can use a wider range of polishers. The hard rubber wheels can be too abrasive for soft gold alloys or silver. Silicone polishers with finer abrasives or bristle brushes with polishing pastes are selected for polishing. Buffing is undertaken using wool, chamois, or cloth wheels with rouge or proprietary polishing compounds. A high lustre is developed for oral hygiene and on the occlusal surface to reduce wear on the opposing teeth (45–47).

8.6.8 Ceramics

Porcelain has been used for denture teeth since 1790 (15), it has excellent appearance, durability, and biocompatibility. Denture teeth are mainly feldspar with about 15% quartz and 4% kaolin (15). Dental crowns may be all ceramic or ceramic fused to metal for additional strength. The material is supplied as ground glass frits and the technician gradually builds up the crown on a removable die in the cast of the patients' mouth. A core of aluminous porcelain is applied on a platinum foil or metal substructure and fired in a vacuum furnace (Figure 8.10). Incrementally, small amounts of porcelain are then app-lied using small spatulas or paintbrushes to form the shape of the tooth, this is over built to allow for shrinkage during firing.

The crown is then trimmed and additional porcelain applied as required. The crown is shaped and characterised using rotary tools. The bur of choice for most ceramists appear to be diamond (although

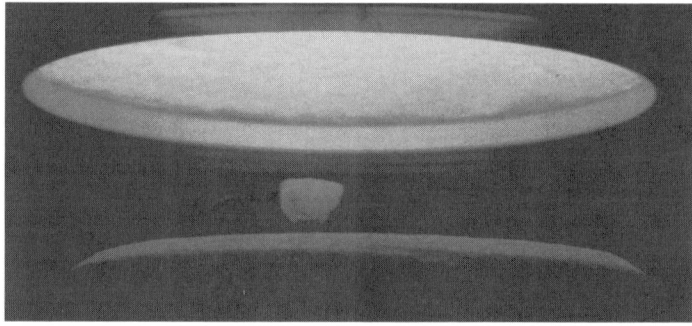

Fig. 8.10. Ceramic crown in furnace

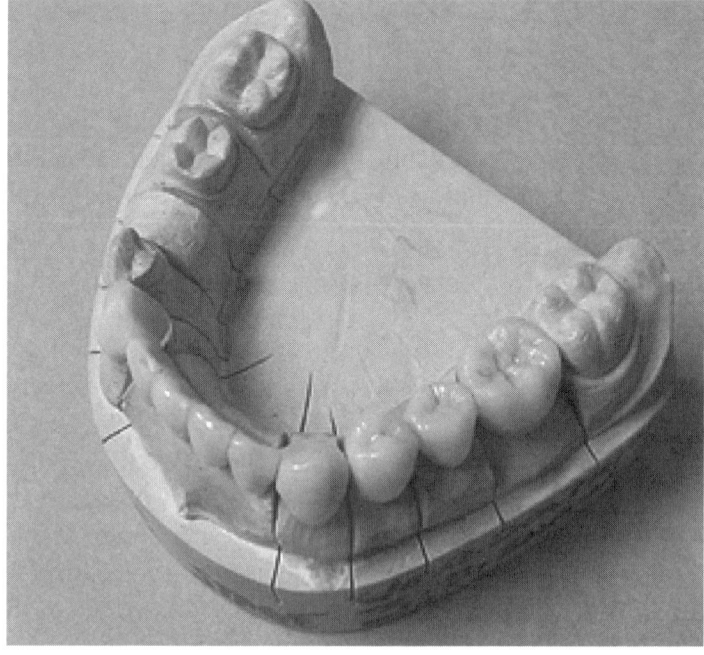

Fig. 8.11. Machining of crown with diamond points

there is no published data on this) alternatively silicon carbide abrasives are used. The selection of the shape of the diamond or silicon carbide points used, their grit size, and the speed used varies with local custom and practice (Figure 8.11). Finally, the crown glazed this can be a natural or a paint on glaze. Dental ceramics have a Vickers hardness of between about 600 and 700 (15).

The surface finish of the crown is very important for the dental health of the patient. Rough surfaces increase plaque retention and occlusally can cause considerable damage to the opposing teeth (48–62).

8.6.9 Machinable Ceramic Restorations

In the last 20 years CAD/CAM technology has been introduced in dentistry in some areas replacing lost wax techniques. These systems have been used for the manufacture of inlays, crowns, bridges, substructures and veneers. The tooth, or a cast of the tooth, is scanned with a laser or a probe, the device is milled from tooth coloured ceramics, zirconia, or alumina, using computer controlled milling machines. The machines use diamond tools to machine the ceramic. The Cerec system was introduced in the late 1980s (63,64) a cavity was is scanned stereo-photogrammetrically and a precision-fitting restoration milled from a standard ceramic block using a miniature three-axis milling device; driven by a water turbine unit. The CELAY system (65), a copy milling unit, was introduced to the market in 1991. Other systems have been introduced (66–71). The continual development of more powerful scanners, computing power and software, and improvements in milling technology software milling have increased accuracy (72–84). Evaluation of enamel wear of these materials has given conflicting results al-Hiyasat (85) found machinable ceramics were significantly less abrasive yet Ramp (86) and Imai (87) found greater wear. However, the surface roughness of the finished device is an important factor in wear (58).

Yin (88) classified the materials available as machinable dental bioceramics and difficult to machine dental bioceramics. The machinable dental bioceramics had a machinability similar to existing hand made dental ceramics, they could be easily adjusted in the clinic by the use of existing diamond tools. Abrasive damage and edge chipping during finishing procedures can be apparent which can be influenced by grit sizes and abrasive pressure of the diamond tools and is possibly influenced by the microstructure of the materials.

Difficult to machine dental bioceramics include glass-infiltrated alumina, and surgical grade zirconia. Finishing of these materials has been associated with short tool life, grit microfracture, wear flat, grit pullout and matrix abrasion have all been identified in conventional diamond burs (88,89). The small size of dental tools required for internal grinding of devices is the most challenging aspect of machining crowns and bridges (90) as in traditional techniques. The number and shape of diamond particles influences cutting rate and resulting surface roughness of the device (90,91). The clinical finish of these materials after adjustment is achieved using diamond pastes (92).

8.7 Dental Cutting Tools

8.7.1 Cutting Efficiency

Evaluation of cutting efficiency of burs has largely related to clinical use (93–98) and the recent developments in CAD/CAM with the few studies applied to dental technology. Diamond burs are manufactured in multiple layers by electrodeposition, sintering, or microbrazing to provide continuous cutting surface as they wear (95). Cutting efficiency is affected by the substrate being ground and can be affected by clogging (adhesive wear) of the bur as well as wear. Diamond burs and abrasives such as silicon carbide retain cutting efficiency as they wear (44) as the wear exposes new surfaces or particles. Particle size and distribution affect cutting performance (90,91). Tungsten carbide was found the most effective with Ag-Pd-Cu-Au alloy, Ag-Zn-In-Sn alloy Ni-Cr alloy and Ti, however, the cutting capability of the carbide bur declined whilst the diamond remained quasi-constant (44). Xeno-dimitropoulou (32) identified the silicon carbide bur as the most effective for grinding Co-Cr alloy in the laboratory but tungsten carbide in the surgery. Siegel and von Fraunhofer (43) identified that carbide burs sectioned the base metal alloy significantly faster than the diamond burs but the opposite was observed with noble alloys and medium grit diamond burs should be used. Rimondini et al. (99) identified that small particle diamond burs clogged and observed damage to tungsten burs when grinding titanium and that the most effective were 30 and 15 micron diamonds followed by finer grinding with tungsten carbide burs. Watanabe et al. (100) that when machining cast CP Ti and its alloys, carbide fissure burs possessed a greater machining efficiency than the diamond points. The selection of cutting tool material and the size of the cutting particles or edges is a balance between the speed of removal and the level of finish required. Large particles can remove material quickly but can lead to edge chipping of the substrate, greater surface roughness and poorer mechanical properties. Fine particles are slower, can clog more easily but give a smoother surface. The greater pressure and speed used the greater the heat and chance of clogging however as these tools are all hand controlled individual custom and practice affects selection and performance.

8.7.2 CVD Dental Burs

Cobalt cemented tungsten carbide burs used in dental technology are subject to edge chipping and wear reducing their performance. The sintered material is ground to create the cutting edges (Figures 8.12 and 8.13). Examination of the surfaces (Figures 8.14 and 8.15) shows edge chipping and the dulling of the surface reducing cutting efficiency.

Conventional diamond burs are manufactured embedding the diamonds into place using various techniques. The cutting efficiency of the diamond particles is related to the effectiveness of the method used to bind them to the bur, grit pullout and matrix abrasion have been identified as factors in the wear of dental diamond burs (95). Chemical vapour deposition of diamond films offer advantages in their uniformity of coating over complex surfaces and the nature of their bonding to the substrate.

In 1996 Trava-Airoldi et al. (102) published their work on the production of polycrystalline chemical vapour deposition (CVD) diamond from hot filament-assisted technique onto stainless steel dental burs. Further work using molybdenum as a substrate identified better wear characteristics than for a conventional diamond bur (103). The first report in the dental literature in Borges et al. 1999 (104) confirmed greater longevity and identified more efficient cutting ability and the

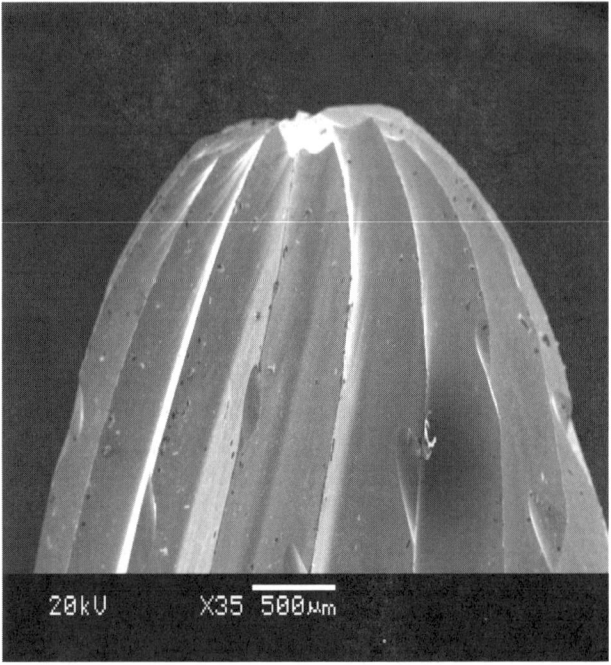

Fig. 8.12. Tungsten carbide bur showing toothing

Fig. 8.13. Barrel-shaped tungsten carbide bur

Fig. 8.14. Chip on edge of new tungsten carbide bur

Fig. 8.15. Chip on edge of tungsten carbide bur showing grains and dulling of surface

benefit of excluding the risk of metal contamination from the metallic binder used on conventional diamond burs. The unusual combination of housing the Centre for Dental Technology within the Department of Chemistry and Materials Science at Manchester Metropolitan University (MMU) stimulated work on the use of CVD on dental tungsten carbide work. CVD of diamond coatings onto the cemented carbide substrate was poor due to binder materials such as cobalt that can suppress diamond growth (105). The use of a pre-acid etched substrate surface and a modified HFCVD gave good adhesion (105), further work at MMU developed the CVD techniques for dental burs (106–109).

Studies on the performance of WC-Co CVD diamond coated burs (110–113) have supported earlier work (104). Ali et al. (110) identified that the coated WC-Co dental tools remained completely intact after drilling 500 holes into human teeth, the conventional bur had lost the majority of its embedded diamond particles. Sein et al (111) used flank wear to evaluate the wear rates of conventional and CVD diamond coated burs drilling into a range of substrates. Results show a 300% improvement with coated burs over conventional burs. Examples of adhesive wear were noted on the bur when grinding acrylic and borosilcate glass. Jackson et al. (112) supported these findings identifying evidence of adhesive and abrasive wear associated with increased rates of abrasion. Polini et al. (113) presented the first quantitative data on the cutting behaviour of uncoated and CVD coated TC burs. They used Co-Cr-Mo dental alloy as the workpiece material grinding at a speed of 20,000 rpm. Some of the uncoated burs failed catastrophically during the tests, the diamond-coated burs exhibited much longer life.

8.7.3 Shanks

The shanks of dental burs may be of the same material as the cutting head or the cutting head will be connected to the head by brazing. The shaft material can be made of metal e.g., steel or carbide. The type of materials and the treatment given to it is at the discretion of the manufacturer (BS EN ISO 1797–1:1995 Dental rotary instruments. Shanks. Shanks made of metals). The hardness shall be a minimum of 250 HV 5. The shanks are normally 2.35-mm diameter however some high-speed laboratory handpieces may require 3.0mm. If positioned incorrectly in the handpiece or running at excessive speed the shaft may be liable to deformation, figures 14, 15 there have been occurrences at MMU of the head separating from the shank. The force in these circumstances lead to the bur head embedding the wall across the laboratory. The implications of this are that ideally grinding occurs in closed chambers, this however is rare practice, then use of bench mounted guards and safety spectacles are normally required.

8.8 Health and Safety

The machining of dental devices exposes dental technicians to several hazards. Impact injuries to the eyes, vibration syndrome, and respiratory disease from the dust generated. Eye injuries can be caused by shank failure, abrasive disk failure, and debris from the grinding operations. Shields and safety spectacle are required in dental laboratories.

8.8.1 Vibration

The holding of dental handpiece for extended periods with the vibration from the motor itself and the effect of the vibration of the dental bur on the device being ground whilst has been reported to induce vibration syndromes in dental technicians (116–119). Hjortsberg et al. (117) identified damage to both myelinated and unmyelinated fibers in the fingers of subjects exposed to high-frequency vibration. Yoshida et al. (118) suggested that the usage of high- and low-speed machines may be a cause of vibration syndrome among dental technicians. Nakladalova et al. (119) in a study of 120 dental technicians reported paresthesiae in the hand fingers (47.4%) and pain in the joints of upper extremities (elbow 26.6%, shoulder 10.8%, wrist 6.6% and small joints of hand 6.6%). The EU Physical Agents (Vibration) Directive (2002) came into force 6 July 2005 (120) this requires employers to minimise risk and to address it. Mansfield (121) suggested selection of low vibration handpieces, training in correct usage and recognition of symptoms, health surveillance, and periodic reviews. The effect of bur toothing, vibration and its effect has not been investigated however increased cutting effectiveness and reduced blunting of tools would benefit. Initial investigations into the performance of CVD coated burs (113) would suggest their benefit in relation to this legislation.

8.8.2 Dust

Since 1967 there have been at least 70 published reports and studies of respiratory conditions associated with dust generated in the making of dental devices (122–133). The condition dental technicians pneumoconiosis was first specifically described in 1986 by De Vuyst et al. (128) who described it as complex pneumoconiosis distinct from silicosis, asbestosis, or hard metal disease and that Cr-Co-Mo alloys play a role in its pathogenesis. Since then a number of workers including the Centers for Disease Control and Prevention in the USA (133) have described a number of respiratory diseases associated with dental laboratory practices including:

- Silicosis (associated with dental ceramics and silica filled polymers);
- Occupational asthma;
- Auto-immune disorders associated with silica;
- Mineral-associated hepatic injury; and
- Sarcoidosis.

8.8.3 Particle Size

The manufacture of dental devices involve numerous techniques and materials and includes significant use of small hand held drills The drills operate at speeds of up to 40,000 rpm. The handpieces hold a range of burs that are used to grind, smooth and polish dental devices. The burs may be made of a range of abrasives including tungsten carbide, silicon carbide (SiC), carborundum (Al_2O_3), rubber, diamond, ruby abrasives or even sandpaper. The materials being ground include precious metal alloys, nickel chromium alloys, cobalt chromium alloys, ceramics, silica/polymer composites, acrylic resins and dental gypsum products. The elements in the dental alloys include, Co, Cr, Cu, Ag, Ni, Sn, Mo.

The dust generated by the processes should be collected by a local dust extractor system. This may be an individual machine or part of a centralised system as in the new development. In use the hand-piece and the device being ground are positioned close to the extractor port. The grinding of dental materials generates varying amounts of dust from the materials and the grinding stones. The particle size of these varies depending upon the materials being ground and the grinding tools, Brune and Beltesbrekke (134) identified ranges from 0.6 μm to 50 μm with a significant proportion below 5 μm that is respirable. This work has been confirmed in a number of undergraduate research projects at MMU. The weight of dust generated in the grinding and polishing of one metal denture base is approximately 1 gram. Brune and Beltesbrekke recommended a minimum extraction of 30 l/s (3 m/s) to remove the majority of this dust. Collard et al. (135) reported that a diamond bur created more respirable particles than the carbide bur for each composite tested although this will probably relate to abrasive particle size.

References

1. Ash, Claudius & Sons, A Centenary Memoir 1820–1921, Clownes London 1921.
2. Craig RG, Powers JM. Restorative Dental Materials 11th edition, Mosby, 2002.
3. Nuffield Report
4. Christensen, GJ. Dental laboratory technology in crisis, The challenges facing the industry. J Am Dent Assoc. 2005, Vol. 136, No 5, 653-655.
5. Glenner RA. Development of the dental drill. J Am Dent Assoc. 1974 88:712-727.
6. Vinksi I. Two hundred and fifty years of rotary instruments in dentistry. Br Dent J. Apr 1979, 3; 146(7):217-23.
7. Ring ME, Hurley N James Beall Morrison: the visionary who revolutionized the practice of dentistry. J Am Dent Assoc. Aug 2000; 131(8):1161-1167.
8. Siegel SC, von Fraunhofer JA. Dental cutting: the historical development of diamond burs. J Am Dent Assoc. Jun 1998; 129(6):740-5.
9. Crawford PR. The birth of the bur (and how a Canadian changed it all!). J Can Dent Assoc. Feb 1990; 56(2): 123-6.
10. Oral communication, G Needham, (Metrodent ltd, Huddersfield, England).

11. Oral communication, P Gough, Senior Lecturer Dental Technology, Manchester Metropolitan University.

12. Oral communication, J Lewis, Senior Lecturer Dental Technology, University of Wales Institute, Cardiff.

13. Siegel SC, von Fraunhofer JA. Dental burs–what bur for which application? A survey of dental schools. J Prosthodont. Dec 1999; 8(4):258-63.

14. Devlin H, Cash AJ, Watts DC. Mechanical behaviour and structure of light-cured special tray materials. J Dent. 1995 Aug; 23(4):255-9.

15. O'Brien WJ. Dental Materials and their Selection, 2003, Quintessence.

16. Neppelenbroek KH, Pavarina AC, Vergani CE, Giampaolo ET. Hardness of heat-polymerized acrylic resins after disinfection and long-term water immersion. J Prosthet Dent. 2005 Feb; 93(2):171-6.

17. Ilbay SG, Guvener S, Alkumru HN. Processing dentures using a microwave technique. J Oral Rehabil. 1994 Jan; 21(1):103-9.

18. Dental Laboratory catalogue, Metrodent Ltd, Huddersfield.

19. Verran J, Maryan CJ. Retention of Candida albicans on acrylic resin and silicone of different surface topography. J Prosthet Dent. 1997 May; 77(5):535-9.

20. Taylor R, Maryan C, Verran. J. Retention of oral microorganisms on cobalt-chromium alloy and dental acrylic resin with different surface finishes. J Prosthet Dent. 1998 Nov; 80(5):592-7.

21. Bulad K, Taylor RL, Verran J, McCord JF. Colonization and penetration of denture soft lining materials by Candida albicans. Dent Mater. 2004 Feb; 20(2):167-75.

22. Radford DR, Sweet SP, Challacombe SJ, Walter JD. Adherence of Candida albicans to denture-base materials with different surface finishes. J Dent. 1998 Sep; 26(7):577-83.

23. Morgan TD, Wilson M. The effects of surface roughness and type of denture acrylic on biofilm formation by Streptococcus oralis in a constant depth film fermentor. J Appl Microbiol. 2001 Jul; 91(1):47-53.

24. Bollen CM, Lambrechts P, Quirynen M. Comparison of surface roughness of oral hard materials to the threshold surface roughness for bacterial plaque retention: a review of the literature. Dent Mater. 1997 Jul; 13(4):258-69.

25. Zissis AJ, Polyzois GL, Yannikakis SA, Harrison A Roughness of denture materials: a comparative study. Int J Prosthodont. 2000 Mar-Apr; 13(2):136-40.

26. Ulusoy M, Ulusoy N, Aydin AK. An evaluation of polishing techniques on surface roughness of acrylic resins. J Prosthet Dent. 1986 Jul; 56(1):107-12.

27. Radford DR, Watson TF, Walter JD, Challacombe SJ. The effects of surface machining on heat cured acrylic resin and two soft denture base materials: a scanning electron microscope and confocal microscope evaluation. J Prosthet Dent. 1997 Aug; 78(2):200-8.

28. Sofou A, Emmanouil J, Peutzfeldt A, Owall B. The effect of different polishing techniques on the surface roughness of acrylic resin materials. Eur J Prosthodont Restor Dent. 2001 Sep-Dec; 9(3-4):117-22.

29. Rahal JS, Mesquita MF, Henriques GE, Nobilo MA Surface roughness of acrylic resins submitted to mechanical and chemical polishing. J Oral Rehabil. 2004 Nov; 31(11):1075-9.

30. Kuhar M, Funduk N. Effects of polishing techniques on the surface roughness of acrylic denture base resins. J Prosthet Dent. 2005 Jan; 93(1):76-85.

31. Aydin AK. Evaluation of finishing and polishing techniques on surface roughness of chromium-cobalt castings, J Prosthet Dent. Jun 1991; 65(6):763-7.

32. Xenodimitropoulou G, Radford DR. The machining of cobalt-chromium alloy in partial denture construction. Int J Prosthodont. 1998 Nov-Dec; 11(6):565-73.

33. Ponnanna AA, Joshi SM, Bhat S, Shetty P. Evaluation of the polished surface characteristic of cobalt-chrome castings subsequent to various finishing and polishing techniques. Indian J Dent Res. 2001 Oct-Dec; 12(4):222-8.

34. Pererra V, Maryan C. Evaluation of the dust generated finishing and polishing cobalt-chrome castings Manchester Metropolitan University BSc(Honours) dissertation 2002.

35. Kononen M, Rintanen J, Waltimo A, Kempainen P. Titanium framework removable partial denture used for patient allergic to other metals: a clinical report and literature review. J Prosthet Dent. 1995 Jan; 73(1):4-7.

36. Au AR, Lechner SK, Thomas CJ, Mori T, Chung P. Titanium for removable partial dentures (III): 2-year clinical follow-up in an undergraduate programme. J Oral Rehabil. 2000 Nov; 27(11):979-85.

37. Rodrigues RC, Ribeiro RF, de Mattos Mda G, Bezzon OL. Comparative study of circumferential clasp retention force for titanium and cobalt-chromium removable partial dentures. J Prosthet Dent. 2002 Sep; 88(3):290-6.

38. Srimaneepong V, Yoneyama T, Wakabayashi N, Kobayashi E, Hanawa T, Doi H. Deformation properties of Ti-6A1-7Nb alloy castings for removable partial denture frameworks Dent Mater J. 2004 Dec; 23(4):497-503.

39. Ohkubo C, Watanabe I, Ford JP, Nakajima H, Hosoi T, Okabe The machinability of cast titanium and Ti-6Al-4V. Biomaterials. 2000. Feb; 21(4):421-8.

40. Kikuchi M, Takada Y, Kiyosue S, Yoda M, Woldu M, Cai Z, Okuno O, Okabe T, Grindability of cast Ti-Cu alloys Dent Mater. 2003 Jul; 19(5):375-81.

41. Hirata T, Nakamura T, Takashima F, Maruyama T, Taira M, Takahashi J. Studies on polishing of Ti and Ag-Pd-Cu-Au alloy with five dental abrasives. J Oral Rehabil. Aug; 28(8):773-7.

42. Hotta Y, Miyazaki T, Fujiwara T, Tomita S, Shinya A, Sugai Y, Ogura H. Durability of tungsten carbide burs for the fabrication of titanium crowns using dental CAD/CAM Dent Mater J. Jun; 23(2):190-6.

43. Siegel SC, von Fraunhofer JA. Comparison of sectioning rates among carbide and diamond burs using three casting alloys. J Prosthodont. 1999. Dec; 8(4):240-4.

44. Miyawaki H, Taira M, Wakasa K, Yamaki M. Dental high-speed cutting of four cast alloys. J Oral Rehabil. 1993 Nov; 20(6):653-61.

45. Clayton J, Green E. Roughness of pontic materials and dental plaque. J Prosthet Dent 1970; 23:407-11.

46. Monasky GE, Taylor DF. Studies on the wear of porcelain, enamel and gold. J Prosthet Dent 1971; 25:299-306.

47. Hacker CH, Wagner WC, Razzoog ME. An in-vitro investigation of the wear of enamel on porcelain and gold in saliva. J Prosthet Dent. 1996; 75:14-17.

48. Newitter DA, Schlissel E, Wolff MS. An evaluation of adjustment and post-adjustment finishing techniques on the surface of porcelain-bonded-to-metal crowns. J Prosthet Dent. 1982; 43:388-95.

49. Schlissel ER, Newitter DA, Renner RR, Gwinnett AJ. An evaluation of postadjustment polishing techniques for porcelain denture teeth. J Prosthet Dent. 1980; 43:258-65.

50. Smith GA, Wilson NHF. The surface finish of trimmed porcelain. Br Dent J 1981; 151:222-4.

51. Sulik WD, Plekavich EJ. Surface finishing of dental porcelain. J Prosthet Dent. 1981; 46:217-21.

52. Zalkind M, Lauer S, Stern N. Porcelain surface texture after reduction and natural glazing. J Prosthet Dent. 1986; 55:30-3.

53. Wiley MG. Effects of porcelain on occluding surface of restored teeth. J Prosthet Dent. 1989; 61:133-7.

54. Raimondo RL, Richardson JT, Wiedner B. Polished versus autoglazed dental porcelain. J Prosthet Dent. 1990; 64:553-7.

55. Patterson CJW, McLundie AC, Stirrups DR, Taylor WG. Polishing of porcelain by using a refinishing kit. J Prosthet Dent. 1991; 65:383-8.

56. Scurria MS, Powers JM. Surface roughness of two polished ceramic materials. J Prosthet Dent. 1994; 71:174-7.

57. Jagger DC, Harrison A. An in vitro investigation into the wear effects of unglazed, glazed, and polished porcelain on human enamel. J Prosthet Dent. 1994 Sep; 72(3):320-3.

58. Jagger DC, Harrison A. An in vitro investigation into the wear effects of selected restorative materials on enamel. J Oral Rehabil. 1995 Apr; 22(4): 275-81.

59. Ramp MH, Suzuki S, Cox CF, Lacefield WR, Koth DL. Evaluation of wear: enamel opposing three ceramic materials and a gold alloy. J Prosthet Dent. 1997 May; 77(5):523-30.

60. al-Wahadni A, Martin DM. Glazing and finishing dental porcelain: a literature review. J Can Dent Assoc. 1998 Sep; 64(8):580-3.

61. Magne P, Oh WS, Pintado MR, DeLong R. Wear of enamel and veneering ceramics after laboratory and chairside finishing procedures. J Prosthet Dent. 1999 Dec; 82(6):669-79.

62. Clelland NL, Agarwala V, Knobloch LA, Seghi RR. Relative wear of enamel opposing low-fusing dental porcelain. J Prosthodont. 2003 Sep; 12(3):168-75.

63. Sirona Dental Systems, CEREC®,

64. Leinfelder KF, Isenberg BP, Essig ME. A new method for generating ceramic restorations: a CAD-CAM system. J Am Dent Assoc. 1989 Jun; 118(6):703-7.

65. Mikrona. Celay®, http://www.mikrona.com/mikrona_e.html March 2006.

66. Girrbach Dental GmbH. digiDENT®, www.girrbach.com , March 2006.

67. Nobel Biocare. Procera, http://www.nobelbiocare.com/global/en/Products/ Procera/default.htm March 2006.

68. DCS-Dental AG, Precident®, http://www.dcs-dental.com/eng/cadcam.htm March 2006.

69. Renishaw plc. http://www.renishaw.com/client/product/UKEnglish/PGP-1306. shtml, March 2006.

70. Concurrent Analysis Corporation, Cicero Dental Systems http://www.caefem. com/dental_crown.htm March 2006.

71. van der Zel JM, Vlaar S, de Ruiter WJ, Davidson C. The CICERO system for CAD/CAM fabrication of full-ceramic crowns. J Prosthet Dent. 2001 Mar; 85(3):261-7.

72. Molin M, Karlsson S. The fit of gold inlays and three ceramic inlay systems. A clinical and in vitro study. Acta Odontol Scand. 1993 Aug; 51(4):201-6.

73. Siervo S, Bandettini B, Siervo P, Falleni A, Siervo R. The CELAY system: a comparison of the fit of direct and indirect fabrication techniques. Int J Prosthodont. 1994 Sep-Oct; 7(5):434-39.

74. Siervo S, Pampalone A, Siervo P, Siervo R. Where is the gap? Machinable ceramic systems and conventional laboratory restorations at a glance. Quintessence Int. 1994 Nov; 5 (11):773-9.

75. Rinke S, Huls A, Jahn L Marginal accuracy and fracture strength of conventional and copy-milled all-ceramic crowns. Int J Prosthodont. 1995 Jul-Aug; 8(4):303-10.

76. Mormann WH, Schug J. Grinding precision and accuracy of fit of CEREC 2 CAD-CIM inlays. J Am Dent Assoc. 1997 Jan; 128(1):47-53.

77. Sulaiman F, Chai J, Jameson LM, Wozniak WT. A comparison of the marginal fit of In-Ceram, IPS Empress, and Procera crowns. Int J Prosthodont. 1997 Sep-Oct; 10(5):478-84.

78. Sturdevant JR, Bayne SC, Heymann HO. Margin gap size of ceramic inlays using second-generation CAD/CAM equipment. J Esthet Dent. 1999; 11(4): 206-14.

79. Boening KW, Wolf BH, Schmidt AE, Kastner K, Walter MH. Clinical fit of Procera All Ceram crowns. J Prosthet Dent. 2000 Oct; 84(4):419-24.

80. Molin MK, Karlsson SL. A randomized 5-year clinical evaluation of 3 ceramic inlay systems. Int J Prosthodont. 2000 May-Jun; 13(3):194-200.

81. Addi S, Hedayati-Khams A, Poya A, Sjogren G. Interface gap size of manually and CAD/CAM-manufactured ceramic inlays/onlays in vitro J Dent. 2002 Jan; 30 (1):53-8.

82. Yeo IS, Yang JH, Lee JB. In vitro marginal fit of three all-ceramic crown systems. J Prosthet Dent. 2003 Nov; 90(5):459-64.

83. Tomita S, Shin-Ya A, Gomi H, Matsuda T, Katagiri S, Shin-Ya A, Suzuki H, Yara A, Ogura H, Hotta Y, Miyazaki T, Sakamoto Y. Machining accuracy of CAD/CAM ceramic crowns fabricated with repeated machining using the same diamond bur. Dent Mater J. 2005 Mar; 24(1):123-33.

84. Reich S, Wichmann M, Nkenke E, Proeschel P. Clinical fit of all-ceramic three-unit fixed partial dentures, generated with three different CAD/CAM systems. Eur J Oral Sci. 2005 Apr; 113(2):174-9.

85. al-Hiyasat AS, Saunders WP, Sharkey SW, Smith GM, Gilmour WH. Investigation of human enamel wear against four dental ceramics and gold. J Dent. 1998 Jul-Aug; 26(5-6):487-95.

86. Ramp MH, Ramp LC, Suzuki S. Vertical height loss: an investigation of four restorative materials opposing enamel. J Prosthodont. 1999 Dec; 8(4):252-7.

87. Imai Y, Suzuki S, Fukushima S. Enamel wear of modified porcelains. Am J Dent. 2000 Dec; 13(6):315-23.

88. Yin L, Song XF, Song YL, Huang Jli. An overview of in vitro abrasive finishing & CAD/CAM of bioceramics in restorative dentistry. Int J Machine Tools & Manufacture, in press.

89. Yin L, Jahanmir S and Ives LK. Abrasive machining of porcelain and zirconia with a dental handpiece Wear, Volume 255, Issues 7-12, August-September, Pages 975-989.

90. Luthardt RG, Holzhuter MS, Rudolph H, Herold V, Walter MH. CAD/CAM-machining effects on Y-TZP zirconia Dent Mater. Sep; 20(7):655-62.

91. Yara A, Ogura H, Shinya A, Tomita S, Miyazaki T, Sugai Y, Sakamoto Y. Durability of diamond burs for the fabrication of ceramic crowns using dental CAD/CAM Dent Mater J. Mar; 24(1):134-9.

92. Finger WJ, Noack MD. Postadjustment polishing of CAD-CAM ceramic with luminescence diamond gel. Am J Dent. Feb; 13(1):8-12.

93. Wilwerding T, Aiello A. Comparative efficiency testing 330 carbide dental burs utilizing Macor substrate Pediatr Dent. May-Jun; 12(3):170-1.

94. Ayad MF, Rosenstiel SF, Hassan MM. Surface roughness of dentin after tooth preparation with different rotary instrumentation. J Prosthet Dent. Feb; 75(2):122-8.

95. Siegel SC, von Fraunhofer JA. Assessing the cutting efficiency of dental diamond burs. J Am Dent Assoc. 1996 Jun; 127(6):763-72.

96. Nishimura K, Ikeda M, Yoshikawa T, Otsuki M, Tagami J. Effect of various grit burs on marginal integrity of resin composite restorations. J Med Dent Sci. Mar; 52(1):9-15.

97. Siegel SC, von Fraunhofer JA. Cutting efficiency of three diamond bur grit sizes. J Am Dent Assoc. Dec; 131(12):1706-10.

98. Siegel SC, von Fraunhofer JA. Dental cutting with diamond burs: heavy-handed or light-touch? Prosthodont. Mar; 8(1):3-9.

99. Rimondini L, Cicognani Simoncini F, Carrassi A. Micro-morphometric assessment of titanium plasma-sprayed coating removal using burs for the treatment of peri-implant disease. Clin Oral Implants Res. 2000 Apr; 11(2):129-38.

100. Watanabe I, Ohkubo C, Ford JP, Atsuta M, Okabe T. Cutting efficiency of air-turbine burs on cast titanium and dental casting alloys. Dent Mater. 2000 Nov; 16(6):420-5.

101. de Jager N, Feilzer AJ, Davidson CL. The influence of surface roughness on porcelain strength. Dent Mater. 2000 Nov; 16(6):381-8.

102. Haselton DR, Lloyd PM, Johnson WT. A comparison of the effects of two burs on endodontic access in all-ceramic high lucite crowns. Oral Surg Oral Med Oral Pathol Oral Radiol Endod. 2000 Apr; 89(4):486-92.

103. Trava-Airoldi, VJ. Corat, EJ. Leite, NF. do Carmo Nono M. Ferreira NG and Baranaus V CVD diamond burrs – Development and applications Diamond and Related Materials, Volume 5, Issues 6-8, 1996 May, Pages 857-860.

104. Borges CF, Magne P, Pfender E, Heberlein J Dental diamond burs made with a new technology. J Prosthet Dent. 1999 Jul; 82(1):73-9.

105. Sein H, Ahmed W and Rego C. Application of diamond coatings onto small dental tools Diamond and Related Materials, 2002 Volume 11, Issues 3-6, March-June, Pages 731-735.

106. Sein H, Ahmed W, Jackson M, Ali N and Gracio J. Stress distribution in diamond films grown on cemented WC–Co dental burs using modified hot-filament CVD Surface and Coatings Technology, 2003. Volumes 163-164, 30 January, Pages 196-202.

107. Ahmed W, Sein H, Ali N, Gracio J and Woodwards R. Diamond films grown on cemented WC–Co dental burs using an improved CVD method Diamond and Related Materials, 2003. Volume 12, Issue 8, August, Pages 1300-1306.

108. Sein H, Ahmed W, Jackson M, Polinic R, Hassan I, Amara M and Rego C. Enhancing nucleation density and adhesion of polycrystalline diamond films deposited by HFCVD using surface treatments on Co cemented tungsten carbide. Diamond and Related Materials, 2004. Volume 13, Issues 4-8, April-August, Pages 610-615.

109. Ahmed W, Sein H, Jackson M and Polini R. Chemical vapour deposition of diamond films onto tungsten carbide dental burs. Tribology International Volume 37, 2004. Issues 11-12, November-December, Pages 957-964 Novel Carbons in Tribolgy.

110. Ali N, Cabral G, Neto VF, Sein H, Ahmed W and Gracio J. Surface engineering of WC– Co used in dental tools technology materials science and technology 2003. (19) 1273-1278.

111. Sein H, Ahmed W, Jackson M, Woodwards R and Polini R. Performance and characterisation of CVD diamond coated, sintered diamond and WC–Co cutting tools for dental and micromachining applications Thin Solid Films, 2004, Volumes 447-448, 30 January, Pages 455-461.

112. Jackson MJ, Sein H, Ahmed W. Diamond coated dental bur machining of natural and synthetic dental materials. J Mater Sci Mater Med. 2004 Dec; 15(12):1323-31.

113. Polini R, Allegri A, Guarino S, Quadrini F, Sein H and Ahmed W. Cutting force and wear evaluation in peripheral milling by CVD diamond dental tools. Thin Solid Films, 2004. Volumes 469-470, 22 December Pages 161-166.

114. Kataeva VA, Lakshin AM, Bol'shakov GV, Mamaev EN. Hygienic assessment of noise and vibration in the work of orthodontists and dental technicians. Russian Stomatologiia (Mosk). 1983 Jan-Feb; 62(1):56-8.

115. Hjortsberg U, Rosen I, Orbaek P, Lundborg G, Balogh I. Finger receptor dysfunction in dental technicians exposed to high-frequency vibration Scand. J Work Environ Health 1989 Oct; 15(5):339-44.

116. Yoshida H, Nagata C, Mirbod SM, Iwata H, Inaba R. Analysis of subjective symptoms of upper extremities in dental technicians. Sangyo Igaku 1991 Jan; 33(1):17-22.

117. Nakladalova M, Fialova J, Korycanova H, Nakladal Z. State of health in dental technicians with regard to vibration exposure and overload of upper extremities. Cent Eur J Public Health; 1995 3 Suppl:129-31.
118. DIRECTIVE 2002/44/EC OF THE EUROPEAN PARLIAMENT AND OF THE COUNCIL of 25 June 2002 on the minimum health and safety requirements regarding the exposure of workers to the risks arising from physical agents (vibration) (sixteenth individual Directive within the meaning of Article 16(1) of Directive 89/391/EEC). Official Journal of the European Communities 6.7.2002 L 177/13.
119. Mansfield NJ. The European vibration directive–how will it affect the dental profession? Br Dent J. 2005 Nov 12; 199(9):575-7.
120. Jedrzejewski T, Ulejska I. Air contamination in a dental laboratory as a result of mechanical procedures in the preparation of dental prosthesis. Protet Stomatol 1967; 11:65-9.
121. Hugonnaud C, Lob M. Risks incurred by dental technicians working on metallic prostheses. Soz Praventivmed 1976 Jul-Aug; 21(4):139.
122. Leclerc P, Fiessinger JN, Capron F, Ameille J, Rochemaure J. [Erasmus syndrome in a dental technician. Importance of the prevention of occupational hazards]. Ann Med Interne (Paris) 1983; 134(7):653-5.
123. Rom WN, Lockey JE, Lee JS, Kimball AC, Bang KM, Leaman H, Johns RE, Jr., Perrota D, Gibbons HL. Pneumoconiosis and exposures of dental laboratory technicians. Am J Public Health 1984 Nov; 74(11):1252-7.
124. Ichikawa Y, Kusaka Y, Goto S. Biological monitoring of cobalt exposure, based on cobalt concentrations in blood and urine Int Arch Occup Environ Health 1985; 55(4):269-76.
125. Morgenroth K, Kronenberger H, Michalke G, Schnabel R. Morphology and pathogenesis of pneumoconiosis in dental technicians. Pathol Res Pract 1985 Mar; 179(4-5):528-36.
126. De Vuyst P, Vande Weyer R, De Coster A, Marchandise FX, Dumortier P, Ketelbant P, Jedwab J, Yernault JC. Dental technician's pneumoconiosis. A report of two cases. Am Rev Respir Dis 1986 Feb; 133(2):316-20.
127. Sheikh ME, Guest R. Respiratory ill-health in dental laboratory technicians: a comparative study of GP consultation rates. J Soc Occup Med 1990 Summer; 40(2):68-70.
128. Choudat D, Triem S, Weill B, Vicrey C, Ameille J, Brochard P, Letourneux M, Rossignol C. Respiratory symptoms, lung function, and pneumoconiosis among self employed dental technicians. Br J Ind Med 1993 May; 50(5):443-9
129. Nayebzadeh A, Dufresne A, Harvie S, Begin R. Mineralogy of lung tissue in dental laboratory technicians' pneumoconiosis. Am Ind Hyg Assoc J 1999 May-Jun; 60(3):349-53.
130. Choudat D. Occupational lung diseases among dental technicians Tuber Lung Dis 1994 Apr; 75(2):99-104.
131. Centers for Disease Control and Prevention (CDC). Silicosis in dental laboratory technicians–five states, 1994-2000 MMWR Morb Mortal Wkly Rep. 2004 Mar 12; 53(9):195-7.

132. Brune D, Beltesbrekke H, Strand G. Dust in dental laboratories. Part II: Measurement of particle size distributions. J Prosthet Dent. 1980; 44(1):82-7.
133. Collard SM, McDaniel RK, Johnston DA. Particle size and composition of composite dusts. Am J Dent. 1989 Oct; 2(5):247-53.

9. Nanocrystalline Diamond: Deposition Routes and Clinical Applications

9.1 Introduction

Diamond is one of the most technologically advanced and potentially the most useful engineering material in existence today. The properties of synthetic diamond are very similar to that of single crystal diamond. Table 9.1 shows some of the key properties of synthetic diamond and single crystal diamond. It is well established that diamond has a unique combination of excellent physical, optical, chemical and biomedical properties [1–3]. Typically, each application area for diamond requires the optimum properties of the material. The optimisation of diamond properties can only be achieved by operating on the microstructure, since it is almost impossible to alter diamond's molecular structure or its chemical composition. It is interesting to note that majority of diamond's properties arise from the fact that carbon atoms, which give diamond its macromolecular structure, are relatively small and light. In addition, when the carbon atoms bond together to form the diamond structure, they form very strong covalent bonds. The C-C bonds and the giant covalent structure give diamond its immense strength and hardness. As a result, diamond is the hardest known material and possesses high wear resistant. Although, diamond has many outstanding properties, in actual fact, it cannot be engineered into many physical configurations required to fully exploit its unique combination of properties.

The development of the Chemical Vapour Deposition (CVD) technique has lead to the ability to deposit diamond in thin film form, i.e., as a coating. This enables the exploitation of more combinations of the extreme properties of diamond for specific applications. Generally, the two well-known and established technologies, in use today, for depositing diamond-based materials are (i) CVD and (ii) Physical Vapour Deposition (PVD). A typical CVD process involves many

Table 9.1. Properties of synthetic diamond and single crystal diamond

Property	Synthetic diamond	Single crystal diamond
Density (g cm^{-3})	2.8–3.51	3.515
Thermal capacity at 27°C (J mol^{-1} K^{-1})	6.9	6.195
Thermal conductivity at 25°C (W m^{-1} K^{-1})	2100	2200
Thermal expansion coefficient at 25–200°C ($\times 10^6$ °C^{-1})	~2.0	0.8–1.2
Band gap (eV)	5.45	5.45
Carrier mobility (cm^2 V^{-1}s^{-1})		
Electron (n)	1350–1500	2200
Positive hole (p)	480	1600
Electrical resistivity (Ω cm)	10^9–10^{16}	10^{16}
Dielectric constant at 45 MHz to 20 GHz	5.6	5.7
Dielectric strength (V cm^{-1})	10^6	10^6
Saturated electron velocity ($\times 10^7$ cm s^{-1})	2.7	2.7
Young's modulus[a] (GPa)	820–900 at (0–800°C)	910–950
Vickers hardness (GPa)	50–100	57–104
Index of refraction at 10 μm	2.34–2.42	2.40

[a]Young's modulus = 895 {1-1.04 $\times 10^{-4}$ (T-20)}, (GPa), where T in °C.

gaseous phase chemical reactions occurring above a solid surface, which cause film deposition onto that surface, whereas, PVD processes involve the transport of material from the target or source to the substrate surface where the material being transported is influenced by a physical driving mechanism.

However, to date, the most effective and the most successful methodology for depositing diamond-based coatings onto a range of materials is CVD. Although, there are different types of CVD processes, every process requires external energy sources to activate the chemical precursors. Activation can be accomplished in a number of different ways, i.e., thermal, resistive filaments, infrared, ultra-violet, radio frequency, laser and plasma powers [4–8]. Each individual method of activation has its own particular advantages and disadvantages. A method of activation is selected based on the requirements of a particular application. Subsequent to gas activation, a series of chemical reactions, such as adsorption, desorption, decomposition, reduction, oxidation, hydrolysis and transport occur, which ultimately lead to the growth of a solid film and reaction by-products. It is desirable to have the reaction by-products as stable and volatile. The volatile by-products usually desorb from the substrate and leave the reaction chamber through the pumping system and produce a film free from impurities.

Generally, microcrystalline diamond (MCD) films deposited using conventional CVD processes tend to exhibit high surface roughness mainly due to the columnar growth mode of the non-orientated poly-crystalline diamond films. One of the major limitations of the wide scale use of diamond coatings has been the high roughness of the as-grown diamond films. This has prevented its widespread use, especially in microelectronics, optical and biomedical applications. Generally, it is accepted that the mechanical, tribological and biological properties of CVD diamond films depend on the coatings micro-structure. For example, it was found that the friction coefficients of the diamond films are proportional to their grain size [9].

In addition, key properties; such as wear resistance and hardness, displayed by coatings are highly dependent on the crystallite size and smoothness of the coating surface [10]. Therefore, it is necessary to control film microstructure and roughness. Although, the existing methods currently used to control film microstructure have proven to be useful to some extent, however, the need for newer methods of pro-ducing smoother and nanocrystalline diamond (NCD) films is both urgent and timely. This chapter reviews and discusses the deposition routes for NCD formation. Further, it outlines the materials unique properties and potentials for use in biomedical applications and micro-electro-mechanical systems (MEMS).

9.2 Nanocrystalline Diamond

The term *"nano"* is used to refer to particles with dimensions ≤ 100 nanometers (nm). Properties, such as electrical, optical, chemical, mechanical, magnetic, biological etc., of nano-sized particles can be selectively controlled by engineering the size, morphology, and composition of the particles. After developing materials in this near-atomic size range, it becomes possible to combine and exploit the properties of the nano-structured surfaces to create new substances with improved or entirely different properties from their parent materials. For example, nano-particles can (i) render greater strength and hardness to metals; (ii) give ceramics better ductility and form-ability; (iii) make it feasible for insulating materials to conduct heat and/or electricity, and (iv) make thin film protective coatings trans-parent. Nano-sized particles enable the re-engineering of many existing products and designing novel new products/processes that function at unprecedented levels. Furthermore, nano-sized particles offer radical breakthroughs in areas such as materials and manufacturing; electro-nics; medicine and healthcare; environment and energy; chemical and pharmaceutical; biotechnology and agriculture; computation and information technology; and national security. In fact, many govern-ments in the western world have increased significantly their annual budgets for innovative nanotechnology research.

Recently, research interest has diverted considerably from conven-tional MCD films, consisting of an average grain size in the range of several hundred nm to several tens of µm [11–15], to NCD films, having a grain size in the range of 2 nm to a few hundred nm [11–13,16]. NCD films are known to display some superior properties to conventional MCD coatings. For example, NCD films exhibit smoo-ther surfaces; improved electron emission; less highly orientated grains; better wear resistance; and enhanced optical transmission. Therefore, nano-sized diamond coatings display greater potentials, and increased versatility for use in broader range of applications. As the size of the diamond grains changes from micrometer to nanometer, a factor of a million in volume, new properties start to emerge thus enabling the exploration of newer application areas [17], for example, in X-ray optics, X-ray physics, particle physics, etc. NCD has shown to be cost effective and presents superior performances in the areas of electron field emission [18–24], optical transparency and protective coatings [25–36], tribology [37–40,16,41–44], SAW devices [45] and MEMS [46–47].

9.2.1 Deposition Routes

Many attempts have been made to deposit NCD using a range of dif-ferent methods and techniques [11,48–49,40,16,50–53,32,41,17]. The most common and widely adopted approach in depositing NCD films is by performing deposition at moderately high methane (CH_4) partial pressures and/or lower hydrogen concentrations [54]. Deposition conducted at relatively high CH_4 concentrations favours the deposition of nano-sized diamond particles by inducing high nucleation rates and suppressing the growth of individual crystals. Generally, diamond deposition at CH_4-rich environments deteriorates the crystalline morphology of the depositing film and thus produces a much more disordered film. Typically, the resultant films produced at depositions in hydrogen deficient gas environments vary from the phase: pure NCD to NCD embedded in tetrahedral carbon (*ta*-C) or amorphous carbon (*a*-C) matrix. Michler *et al.* [55] demonstrated that if in a CH_4/H_2 mixture, the CH_4 content is continuously increased whilst the substrate temperature is kept constant, the morphology of the growing film changes from faceted micro-crystals to ball-shaped clusters of nano-crystals to graphitic feather-like morphologies with nano-sized diamond particle inclusions. A number of workers have employed CH_4 concentrations in the range 5–10% vol., in hydrogen, to produce NCD [27,56,41]. Catledge and Vohra [40,16,57] reported the synthesis of nanostructured diamond films onto Ti-6AL-4V materials.

They employed pressures of 95 Torr in a microwave CVD chamber and 5–15% CH_4 concentration in balanced hydrogen and nitrogen to obtain diamond nanocrystals of the average size 13 nm in a matrix of *ta*-C. Wu *et al.* [26] prepared NCD of 20–100 nm grain size and surface roughness of approx. 15 nm by using low pressure (5 Torr) and low microwave power (450W). Zarrabian *et al.* [28] utilised electron cyclotron resonance plasma to deposit NCD films of 4–30 nm grain size, which were embedded in a DLC matrix. Also, deposition of smooth composite diamond films using hot-filament CVD has been reported [58]. These films consisted of MCD and NCD film layers, and they had dielectric properties similar to conventional MCD films with smooth surfaces, thus making them potentially suitable for MEMS devices.

Konov *et al.* [59] performed NCD growth using $CH_4/H_2/Ar$ mixtures with a CH_4-content being varied in the range 10–100% using a dc arc plasma deposition system. The as-deposited NCD films on silicon substrates, seeded with 5 nm diamond particles, consisted of 30–50 nm sized diamond crystallites. Nistor *et al.* [60] deposited NCD films from $CH_4/H_2/Ar$ mixtures using d.c. plasma CVD system. Interestingly, Lin *et al.* [61] studied $CH_4/H_2/Ar$ mixtures in a hot-filament

CVD system and reported a change in microstructure from MCD to NCD, with a grain size smaller than 50 nm, at 95.5% Ar addition. Similarly, it was reported [62] that the surface morphology of the NCD films changed with increasing Ar concentration. The formation of NCD was observed when > 90% Ar was used in the plasma mixture. Amaratunga et al. [63–64] produced mixed phase films containing diamond crystallites, 10–200 nm in size, which were found embedded in a non-diamond carbon matrix whilst using CH_4/He plasmas. Bi et al. [45] also used $CH_4/H_2/Ar$ mixtures in the ratio 1:4:100 to deposit NCD films specifically for SAW devices.

In addition to using CH_4 as the carbon-containing precursor, fullerene can also be used to prepare NCD coatings [65–68]. Researchers at the Argonne National Laboratory, USA [65,46,11,69,37,70–73,9] successfully employed fullerene molecules (C_{60}) and argon-rich plasmas in a microwave CVD reactor to deposit NCD films. This process takes place in a specially designed microwave discharge chamber filled with a gaseous mixture of 1% C_{60}, 98% Ar and 1% hydrogen. The microwave energy converts the argon (Ar) gas into plasma. The Ar in the plasma collides with the C_{60} molecules and knocks an electron to create a C_{60}^+ ion. Electrons in the plasma react with the C_{60}^+ ions to initiate fragmentation, which produces a serious of carbon dimers, C_2. It is these dimers that are believed to be the critical plasma species that initiate the diamond growth process. The NCD films deposited on seeded substrates by this method are composed of 3-15 nm diamond crystallites with up to 1–10% sp^2 carbon residing at the grain boundaries [11].

Gas dopants, such as nitrogen [74,21–22], and oxygen [29–31,33] have also be used to deposit NCD. Such gas dopants are used to primarily dilute the CH_4 gas source during NCD deposition. The dilution approach alters significantly the nucleation processes occurring during diamond CVD and favours predominantly NCD film growth. Lee et al. [48–49] proposed a low temperature process, at low microwave powers, to deposit NCD at growth rates of up to 2.5 μm/hr. This process employs temperatures in the range of 350–500°C and uses CO/hydrogen mixtures to obtain NCD consisting of 30–40 nm grain size. Recently, Teii et al. [53] formed NCD films consisting of diamond crystallites of 20 nm size at 80 mTorr and 700°C by inductively coupled plasma employing $CO/CH_4/H_2$ and $O_2/CH_4/H_2$

gas mixtures. A positive bias of 20 V was imposed to the substrate in order to reduce the influence of ion bombardment. The as-deposited films consisted of ball-type grains (100 nm), where each ball-shaped grain was composed of approx. 20 nm NCD.

NCD films can also be deposited using a number of different techniques, including direct-ion beam deposition [75], two-stage growth method [76], dielectrophoresis/spraying coating [18,20,77–78], microwave CVD [79–84], radio-frequency plasma CVD [85], biased enhanced growth [86–88,17] and repetitive bias-enhanced nucleation (BEN) [89]. The bias-enhanced growth (BEG) process was developed by Sharda *et al.* [86] and it was designed to achieve higher diamond nucleation densities, similar to the BEN process. In the BEG process, the BEN stage is extended for the length of the deposition process in a microwave CVD system. The as-deposited films were NCD and these were prepared in 5% vol. CH_4 in balanced hydrogen whilst the bias current density was controlled in a special arrangement of the microwave CVD system. As a result of the BEG process, a number of other workers also adopted similar biasing techniques to produce NCD films [90,51–52,91–96]. Prawer *et al.* [97] produced a layer of NCD in fused quartz by the ion-implantation technique followed by annealing. This is a unique process, since it does not require any nucleation or any external high pressure, as is the case in the traditional high pressure high temperature (HPHT) technique, to produce diamond. Similarly, Wang *et al.* [98] produced NCD consisting of grains in the range approx. 2–70 nm by irradiation of graphite.

Recently, Yusa [99] has reported the growth of nano-sized diamond particles from a direct transformation from carbon nanotubes under high pressure. Multiwalled carbon nanotubes were heated in a diamond anvil cell by a CO_2 laser above 17 GPa and at 2500 K. The recovered product consisted of nano-sized octahedral diamond crystallites. Hirari *et al.* [35] produced NCD by transforming C_{60} fullerene by shock compression and rapid quenching. The resultant transparent NCD platelets consisted of a few nm-sized diamond crystallites. More recently, Gogotski *et al.* [100] introduced the synthesis of NCD in amalgamation with other types of carbon-forms, e.g. carbon nanotubes, nano-onion rings, amorphous carbon, graphite, etc., onto silicon carbide surfaces. The silicon carbide material is transformed into a number of structures of carbon at ambient pressures after the chlorination processes taking place below 1000°C. These carbide derived carbon (CDC) structures are currently been tested for tribological properties as well as their potentials for use in hydrogen gas storage applications.

In industry and in academia, generally, the principal method used for reducing the surface roughness of thin film coatings is by employing mechanical polishing procedures [101]. However, some difficulties arise during polishing the films using standard polishing procedures. As a result, techniques, such as chemical polishing, inert and oxygen ion-beam polishing have been developed to establish smooth coating surfaces [102–105]. Some researchers have employed pulsed bias procedures during diamond film growth, in order to produce highly orientated films, using different pulse bias duty cycles [106–107]. However, despite their efforts, greater detailed work is required before the full potentials of pulsed biasing can be realised. An *in-situ* method, which consists of sequential *in-situ* diamond deposition and planarisation in an electron cyclotron resonance plasma system, has been developed to produce smooth diamond films [108]. This method is believed to have the advantage of reducing processing time and costs, as well as maintaining a cleaner process environment. Silva *et al.* [109] attempted to grow smooth diamond films at lower temperatures by employing a two-step growth process. They proposed to promote non-diamond phase nucleation onto (111) faces. However, no significant progress concerning the smoothness of the film was obtained. Difficulties were encountered in promoting secondary nucleation on a particular facet at low deposition temperatures (approx. 550°C). Secondary nucleation occurs more favourably on (111) and (100) diamond facets [110-111]. Instead, they proposed to employ a gold interlayer in between two diamond layers in order to control surface roughness. Chen *et al.* [19] and Kumar *et al.* [113] employed a similar 2-step growth process, used by Silva *et al.* [109], to produce diamond-like-carbon films. These workers employed such processes to control the stress and improve the coating adhesion of the deposited films.

9.2.2 Time Modulated CVD

Although, during the relatively short bias-voltage pulses in bias-enhanced nucleation [114] where the CH_4 concentration is increased, only slightly, in almost all the methods described, the flow of CH_4 during film growth is kept constant. Diamond growth in a CVD vacuum reactor is conventionally performed under constant CH_4 flow, while the excess flow of hydrogen is kept constant throughout the growth process. In developing the new TMCVD process it was considered that diamond deposition using CVD consists of two stages: (i) the diamond nucleation stage and (ii) crystal growth stage. Diamond grains nucleate more efficiently at higher CH_4 concentrations. However, prolonged film growth performed under higher CH_4 concentration leads to the incorporation of non-diamond carbon phases, such as graphitic and amorphous. The TMCVD process combines the attributes of both growth stages. This technique has the potential to replicate the benefits obtained by using pulsed power supplies, which are relatively more expensive to employ. The key feature of the new process that differentiates it from other conventional CVD processes is that it pulses CH_4, at different concentrations, throughout the growth process, whereas, in conventional CVD, the CH_4 concentration is kept constant, for the full growth process. In TMCVD, it is expected that secondary nucleation processes occur during the stages of higher CH_4 concentration pulses. This can effectively result in the formation of a diamond film involving nucleation stage, diamond growth, secondary nucleation and the cycle is repeated. The secondary nucleation phase can inhibit further growth of diamond crystallites. The nuclei grow to a critical level and then are inhibited when secondary nuclei form on top of the growing crystals and thus fill up any surface irregularities. This type of film growth can potentially result in the formation of a multilayer type film coating. In such coating systems, the quality and the surface roughness of the film coatings is dependant not on the overall thickness of the film but instead on the thickness of the individual layer of the film coating. In demonstrating the CH_4 flow regimes, typically employed in conventional diamond CVD and TMCVD processes, Figure 9.1 shows, as an example, the variations in CH_4 flow rates during film deposition. In the CH_4 pulse cycle employed in Figure 9.1, the CH_4

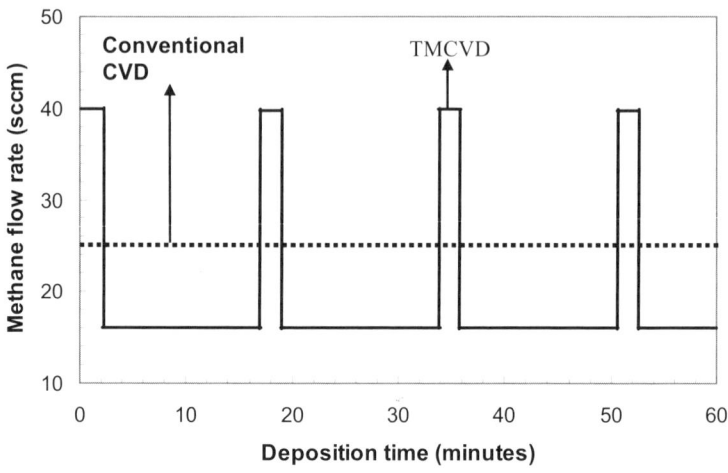

Fig. 9.1. Variations in CH_4 flow rates during film deposition in a typical time-modulated pulse cycle

flow rate remained constant throughout the conventional CVD process at 25 sccm. It is important to note that the hydrogen flow rate remains constant under both growth modes. CH_4 modulations at 9 and 40 sccm for 15 and 2 minutes, respectively, were performed during the TMCVD process using the microwave CVD system. Since higher CH_4 contents in the vacuum chamber results in the incorporation of non-diamond carbon phases in the film, such as graphitic and amorphous, and degrades the global quality of the deposited film, the higher CH_4 pulse duration was kept relatively short. The final stage of the time-modulated process ends with a lower CH_4 pulse. This implies that hydrogen ions will be present in relatively larger amount in the plasma and these will be responsible for etching the non-diamond phases to produce a good quality film.

Figure 9.2(a) & (b) displays the close-up cross sectional SEM images of diamond films grown using TMCVD and conventional CVD processes. The conventional MCD film displays a columnar growth structure. However, the time-modulated film displayed a somewhat different growth mode. Instead, the cross section consisted of many coarse diamond grains that were closely packed together. A pictorial model of the mechanism for the TMCVD process is depicted in Figure 9.2(c) as compared with the conventional CVD process (Figure 9.2d).

Primarily, diamond nucleation occurs first in both the TMCVD and conventional CVD processes. However, in TMCVD, diamond nucleates more rapidly as a result of the high CH_4 pulse at the beginning. The high CH_4 pulse effectively ensures the diamond grains to nucleate quicker to form the first diamond layer. The second stage, where CH_4 content is reduced to a lower concentration, the diamond crystallites are allowed to grow for a relatively longer period. This step enables the crystals to grow with columnar growth characteristics. The surface profile of the depositing film becomes rough, as expected.

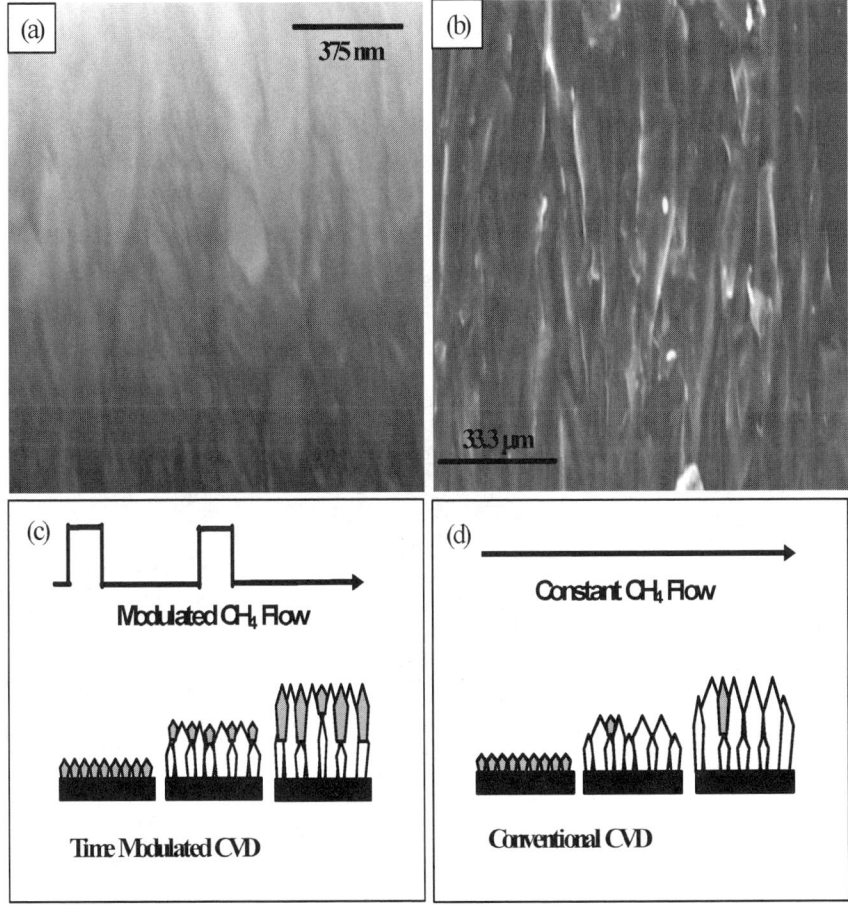

Fig. 9.2. Close-up cross sectional SEM images of diamond films grown using (a) TMCVD and (b) conventional CVD processes. In addition, the pictorial mechanisms for film growth using TMCVD (c) and conventional CVD (d) processes have also been shown

The third stage involves increasing the CH_4 flow back to the higher pulse. This enables further secondary nucleation of NCD to occur in between the existing diamond crystals, where the surface energy is lower. As a comparison, much less secondary nucleation occurs when the CH_4 flow is kept constant throughout the growth process. The distinctive feature of the TMCVD process is that it promotes secondary diamond-particle nucleation to occur on top of the existing grains in order to fill up any surface irregularities.

Figure 9.3 displays the SEM micrograph showing secondary nucleation occurring after a high CH_4 pulse. This result justifies the proposition of the mechanism for the TMCVD process, as shown in Figure 9.2. It can be expected that at high CH_4 bursts, carbon-containing radicals are present in the CVD reactor in greater amount, which favour the growth process by initiating diamond nucleation. The average secondary nucleation crystallite size was in the nano-meter range (≤ 100 nm).

Fig. 9.3. SEM micrograph showing secondary nucleation occurring after a high CH_4 pulse

It is evident that the generation of secondary nano-sized diamond crystallites has lead to the successful filling of the surface irregularities found on the film profile, in between the mainly (111) crystals. Figure 9.4 shows some randomly selected SEM images of as-deposited diamond films deposited using the TMCVD process at different CH_4 pulse duty cycles [115].

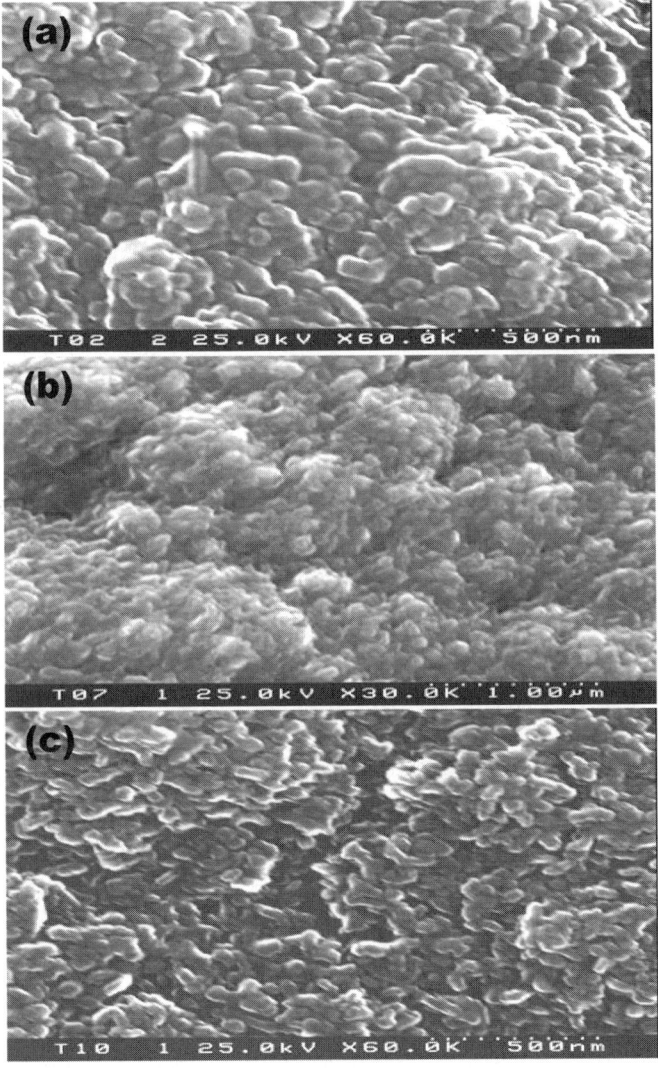

Fig. 9.4. SEM micrographs showing the morphologies of as-deposited diamond films deposited using TMCVD at different CH_4 modulation duty cycles [115]

Figure 9.5 shows the graph displaying the growth rates of conventional and time modulated films grown using hot-filament CVD (HFCVD) and microwave plasma CVD (MPCVD) systems. As expected, the MPCVD system gave much higher growth rates under both growth modes, conventional and time-modulated, compared to films produced using HFCVD. A growth rate of 0.9 μm/hr was obtained using the HFCVD system under constant CH_4 flow. The time-modulated films deposited using HFCVD were grown at a rate of 0.7 μm/hr. Whereas, with the MPCVD system, films grown using constant CH_4 flow were deposited at a rate of 2.4 μm/hr and using modulated CH_4 flow the films were grown at a rate of 3.3 μm/hr. Although, it is known that growth rates increase with CH_4 concentration, in the present case using the HFCVD system, the TMCVD process employs greater CH_4 flow than conventional CVD. Our results show that the growth rate of films deposited using constant CH_4 flow is slightly higher than similar films grown using timed CH_4 modulations.

However, films grown under both modes, conventional and time modulated, using the MPCVD system produced results that were contrary to those obtained using the HFCVD system. Using the MPCVD system, the trend observed was that the films were deposited at a higher growth rate using the TMCVD process than conventional CVD. The substrate temperature is a key parameter, which governs the growth rate in diamond CVD. Since the TMCVD process pulsed CH_4 during film growth, it was necessary to monitor the change in the substrate temperature during the pulse cycles. Figure 9.6 shows the graph relating substrate temperature to CH_4 concentration for both HFCVD and MPCVD systems.

For the HFCVD system, it was observed that the substrate temperature decreased with CH_4 concentration. In explaining the observed trend, it needs to be considered that the dissociation of CH_4 by the hot filament absorbs energy (heat) from the filament and is considered as a cooling process. In our case the filament power was kept constant, therefore, less heat can be expected to radiate to the substrate. In addition, only a small percentage of the thermally dissociated CH species reach the substrate and transfer kinetic energy to the substrate. It is known that the deposition of diamond films increases with substrate temperature [116].

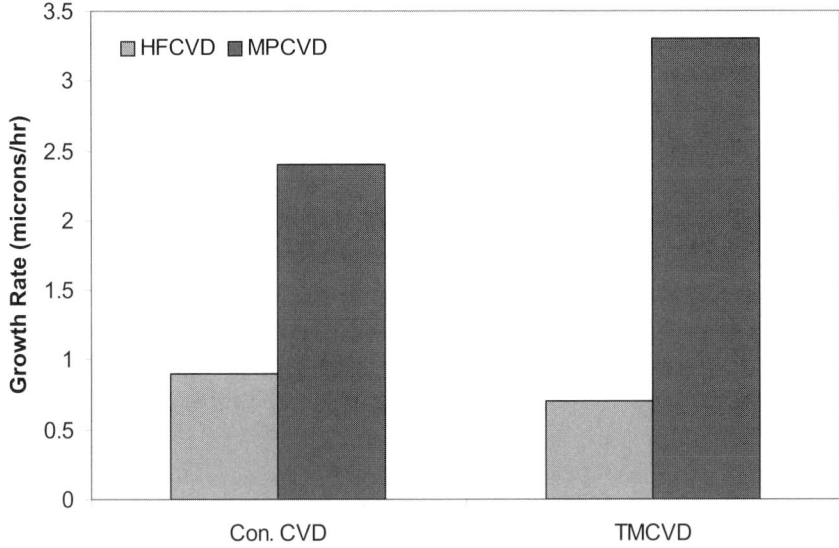

Fig. 9.5. Graph displaying the growth rates of conventional and time-modulated films grown using HFCVD and MPCVD systems

During the high CH$_4$ pulse in TMCVD, the lower substrate temperature may be sufficient to lower the growth rate significantly. Generally, in a MPCVD reactor, the substrate temperature increases with CH$_4$ concentration, as shown in Figure 9.6 [117]. As a comparison, H$_2$ is dissociated more extensively in a MPCVD reactor than in a HFCVD reactor to produce atomic hydrogen. Furthermore, in a MPCVD reactor, the plasma power is much greater, 3400W, than the plasma power in the HFCVD reactor. In fact, the plasma power used in MPCVD for growing diamond films was approximately 10 times greater than the power used in the HFCVD reactor. Therefore, the dissociation of CH$_4$ only absorbs lower percentage of the energy/heat from the plasma in a MPCVD compared to the HFCVD reactor. It is also understood that the reaction between atomic hydrogen and CH species at close vicinity to the substrate releases heat. Since in a MPCVD reactor there is a greater intensity of atomic H and CH species, there will be greater number of reactions between atomic H and CH species. This means that more heat will be released in a MPCVD reactor than in a HFCVD reactor due to such reactions.

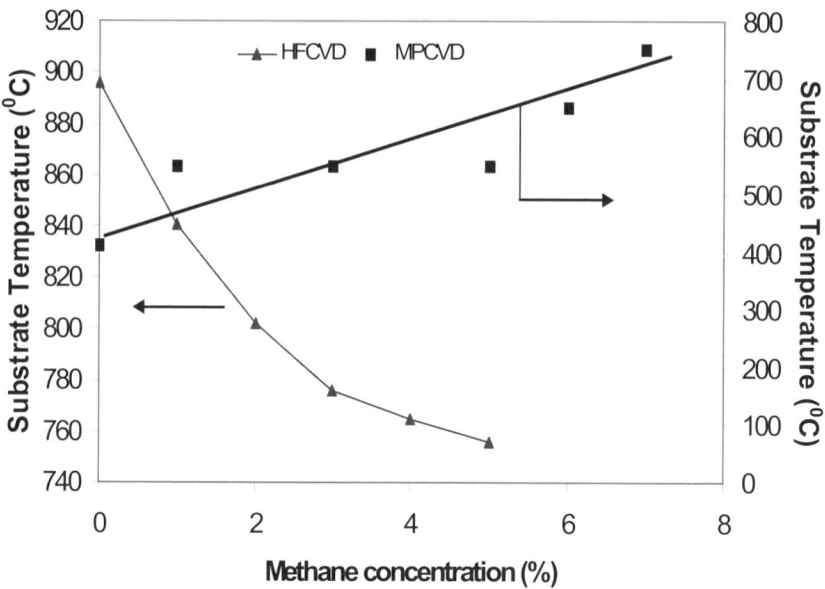

Fig. 9.6. Graph relating substrate temperature to CH_4 concentration for HFCVD and MPCVD systems

This effect contributes to the heating of the substrate. In a HFCVD reactor, the hot filament displays much lower ability to dissociate H_2. Since atomic hydrogen is a critical species that plays an important role in producing a good quality diamond film during CVD, the quality of the films grown using HFCVD is generally lower than similar films grown using MPCVD. In addition, the hydrogen atoms are required for the effective deposition of diamond onto the substrates. As mentioned earlier, since atomic H radicals are present in greater concentration in a MPCVD reactor than in a HFCVD reactor, the MPCVD process gives higher growth rates than the HFCVD process. In a separate study, reported elsewhere [118], we found that by controlling the temperature during the high/low CH_4 pulse cycles, a greater number of secondary diamond grains were generated and the resultant films displayed (i) smoother surfaces and (ii) higher growth rates.

9.3 Clinical Applications

Diamond coatings are used in applications such as optical lenses (ophthalmic lenses, aerospace screens), microelectronics (integrated circuits), engineering (piston rings, cylinder liners) and thermal management systems [119–91]. Although, amorphous diamond-like-carbon (DLC) coatings have been employed for use in hip-joint technology, unfortunately, there is one area where the great benefits of NCD have not been relished – the surface treatment of biomedical implants, such as artificial heart valves and hip prostheses, and dental tools such as burs, hip-joint reamers, orthodontic pliers and tweezers, all of which can benefit in terms of quality, safety and cost from the application of a CVD coating. Since diamond is a biocompatible material, both with human tissues and blood, it can find use in a wider range of biomedical applications. We now focus on four specific applications, namely, mechanical heart valves, dental burs, hip prostheses and MEMS and review the developments made in these key application areas.

9.3.1 Heart Valves

Heart disease is one of the most common causes of death in the world today, particularly in the western countries. There are various causes of heart disease, related most commonly to diet and exercise. The failure of heart valves accounts for about 25–30% of heart problems that occur today. Faulty heart valves need to be replaced by artificial ones using sophisticated and sometimes risky surgery. However, once a heart valve has been replaced with an artificial one there should be no need to replace it again and it should last at least as long as the life of the patient. Therefore, any technique that can increase the operating life of heart valves is highly desirable and valuable. Currently, pyrolytic-carbon (PyC) is used for the manufacture of mechanical heart valves. Figure 9.7 shows a typical PyC leaflet heart valve. Although, PyC is widely used for heart valve purposes, it is not the ideal material. In its processed form, PyC is a ceramic-like material and like ceramics, it is subject to brittleness. Therefore, if a crack appears, the material, like glass, has very little resistance to the growth/ propagation of the crack and may fail under loads.

In addition, its blood compatibility is not ideal for prolonged clinical use. As a result, thrombosis often occurs in patients who must continue to take anti-coagulation drugs on a regular basis [92]. The anti-coagulation therapy can give rise to some serious side effects,

Fig. 9.7. A pyrolitic carbon leaflet heart valve

such as birth defects. It is therefore extremely urgent that new materials, which have better surface characteristics, blood compatibility, improved wear properties, better availability and higher resistance towards breaking are developed. In artificial heart valve applications, a principal requirement is that the surface should essentially display a smooth surface, since surface roughness causes turbulence in the blood, which leads to the integrity of the red cells being damaged causing bacteria to adhere, and blood coagulation and clots. A possible method to increase the degree of PyC thrombo-resistance is by alloying the material with silicon [93].

Although, research on the surface engineering of mechanical heart valves has been limited and restricted, a number of researchers have attempted to develop biocompatible coatings, which could potentially be used for artificial heart valve purposes. For example, carbon nitride (CN) thin films have been investigated for biocompatibility and their properties strongly suggest their potentials for use in various surgical implants [94]. Generally, both the bio- and haemo- compatibilities of

DLC coatings have been extensively investigated and widely reported in the open literature [95]. Jones *et al.* [96] deposited DLC coatings, consisting of multilayers of TiC and TiN, onto titanium substrates and characterised the coatings for haemocompatibility, thrombogenicity and interactions with rabbit blood platelets. It was found that DLC produced no haemolytic effect, platelet activation or tendency towards thrombus formation. Furthermore, the platelet spreading correlated with the surface energy of the coatings. Thomson *et al.* [97] and Dion *et al.* [98] have also investigated DLC coatings and characterised their biological properties.

It is worth considering that titanium and its alloys have been used in biomedical implants for many years now and therefore it is sensible to look at the surface treatment of titanium alloys for producing superior surface characteristics, which could be ideal for heart valves. It should be noted that titanium alloys are not brittle like PyC. A large number of research scientists have deposited Ti-based coatings, using energised vapour-assisted deposition methods, and studied their potentials for use in biomedical areas, such as heart valves or stents. Yang *et al.* [99] deposited Ti-O thin films using plasma immersion ion implantation technique and characterised the anti-coagulant property employing in-vivo methods. They found that the Ti-O film coatings exhibited better thrombo-resistant properties than low-temperature isotropic carbon (LTIC), in long-term implantation. Chen *et al.* [100] deposited TiO coatings doped with Ta, using magnetron sputtering and thermal oxidation procedures, and studied the antithrombogenic and haemocompatibility of $Ti(Ta^{+5})O_2$ thin films. The blood compatibility was measured in-vitro using blood clotting and platelet adhesion measurements. The films were found to exhibit attractive blood compatibility exceeding that of LTIC. Leng *et al.* [101] investigated the biomedical properties of tantalum nitride (TaN) thin films. They demonstrated that the blood compatibility of TaN films was superior to other common biocompatible coatings, such as TiN, Ta and LTIC. Potential heart valve duplex coatings, consisting of layers of Ti-O and Ti-N, have been deposited onto biomedical Ti-alloy by Leng *et al.* [102] and their blood compatibility and mechanical properties have been characterised. The TiO layer was designed to improve the blood compatibility, whereas, TiN was deposited to improve the mechanical properties of the

TiO/TiN duplex coatings. They found that the duplex coatings displayed (i) better blood compatibility than LTIC; (ii) greater microhardness; and (iii) improved wear resistant than Ti6Al4V alloys. It has been reported that the TiO coatings display superior blood compatibility to LTIC [103].

9.3.2 Dental Burs

Dental burs are commonly used on patients as well as in the dental laboratories for removing dental material such as enamel etc. Conventional dental burs are manufactured by binding hard diamond particles onto the substrate surface using a binder matrix material. Figure 9.8 shows a typical SEM micrograph of a conventional diamond dental bur. Generally, there are certain limitations to dental tools and burs in particular. For example, the particles on some dental tools wear off quite quickly making the tools ineffective after only a short lifetime in operation. In addition, with conventional diamond dental burs, there is the heterogeneity of grain shapes and sizes, and the cutting

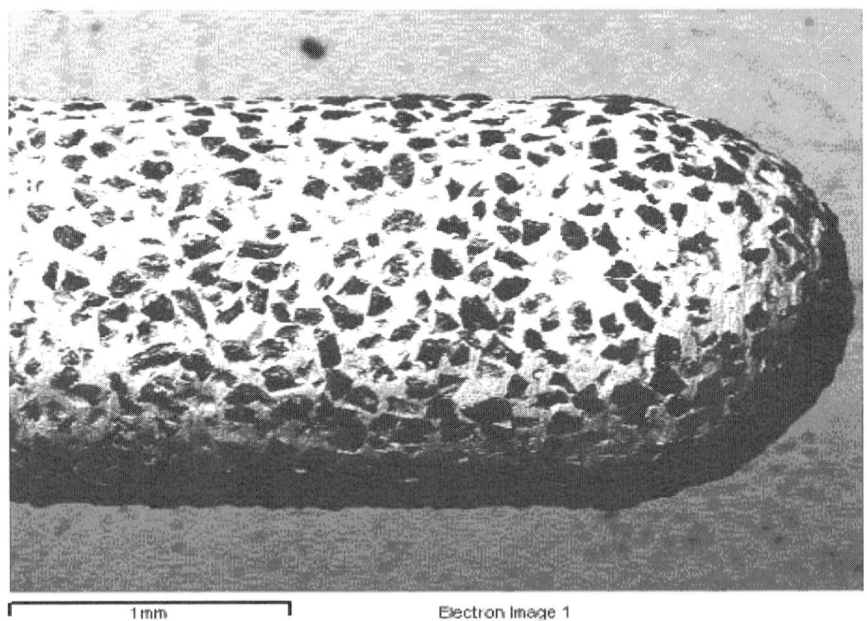

1mm   Electron Image 1

Fig. 9.8. A typical SEM micrograph of a conventional diamond dental bur

and trimming effectiveness decreases due to repeated sterilisation. Furthermore, there is also the health hazard problem associated with the imbedded diamond particles dislodging from the bur into the patient's mouth. This can result in potential release of Ni^{2+} ions from the metallic binder of the dental burs into the body fluids, which could potentially be toxic to the patient. This aspect not only poses a risk to the respiratory system of the patient, the dentist and the nurse, but also causes contamination of the ceramic during the laboratory manufacturing of dental restorations. There is a growing demand for better quality, long-lasting and more economical dental tools.

Cemented tungsten carbide (WC-Co) is also widely used for manufacturing dental burs. A number of researchers have attempted to grow diamond films onto cemented WC-Co substrates [104–112]. There are limited numbers of researchers who have investigated diamond deposition onto dental burs in any great detail. Borges *et al.* [113] employed a DC Arc Jet reactor to deposit diamond coatings onto dental burs. Airoldi *et al.* [114] deposited diamond films onto dental burs using a hot-filament CVD system, where the system configuration was such that two filaments were employed and the dental burs were placed on a rotary substrate holder in between the two filaments. The burs were rotated during film growth in order to obtain uniform film coatings. We have recently modified a conventional hot-filament CVD system to deposit thick diamond coatings onto small tools such as dental burs and micro-drills. The modification was such that the filament was placed vertically in the deposition chamber and the burs are inserted concentrically within the coils of the filament. The readers are encouraged to read some of our papers published on the surface engineering of cemented WC-Co tools [115–118]. However, further work is required before CVD diamond-coated dental burs can replace the conventional burs used in dental surgeries.

9.3.3 Hip Prostheses

The increasing life expectancy of the aging population and the need to surgically treat arthritis in growing numbers of people is placing greater demands on the durability and the expected clinical lifetime of the artificial hip joints. A painful hip can severely hamper a person's ability to live a full active life. The implantation of a hip prosthesis in a patient can eliminate the pain of the damaged hip joint, this being the major benefit of the surgery. It can also reduce disability and renders the patient greater mobility. Over the last twenty-five years, major advancements in hip replacement have improved the outcome of the hip implantation surgery greatly. Generally, modernised hip prostheses implanted in patients have a clinical success rate of up to approximately 15 years [119]. However, after this period the hip prosthesis begins to fail and thus a revision surgery is necessary. Although the modern hip prosthesis is adequately durable, tiny sized wear particles, in the micrometer and sub-micrometer range, are generated at the articulating surfaces and are released into the surrounding tissues, where they cause inflammation, joint loosening, severe pain and clinical failure [120].

The hip prosthesis is generally constructed of three main components: (i) cup, (ii) spherical head, and (iii) a stem. Suitable materials for hip-prosthesis are selected as a compromise, whilst considering tribological issues, corrosion environment, biocompatibility and difficulties encountered in manufacturing. The cups of conventional hip prostheses are typically constructed from ultra-high molecular weight polyethylene material [121]. Unfortunately, this material generates detrimental wear particles. Titanium alloys, such as TiAlV, are commonly used as stems because of their high tensile and fatigue strength, low modulus of elasticity, high corrosion resistance and good biocompatibility [122]. However, blackening of adjacent tissue commonly is observed with titanium alloy implants because of wear particles from bearing surfaces and mechanical instability of stems [123]. Cobalt alloys, such as CoCrMo, are used as stems and spherical heads in hip-prostheses. Although, they are less prone to wear and corrosion than titanium alloys, substantial amounts of wear particles are generated in vivo, especially during the initial running in period after surgery [124]. Researchers have attempted to surface engineer conventional hip-prostheses in order to overcome the major causes of concern. Methods such as ultrapassivation of titanium [125], nitriding [126], nitrogen ion-implantation of titanium alloys [127], titanium nitride [128], PyC [129], silicon nitride [130] and amorphous DLC coatings [131] have been employed. However, most of these methods failed to produce impressive results and acquired only limited market share.

NCD is a natural and trivial choice for use in hip-joint implants. Although, DLC coatings have similar properties to diamond, it is in fact inferior to diamond. Diamond has many superior properties, such as extreme hardness, corrosion-resistance, superior wear resistance and good biocompatibility, which makes it ideal for use in artificial hip joints [132]. Furthermore, it is non-irritating material and completely immune to human body fluids. It is highly desirable to transfer these unique properties to the surface of the hip prosthesis. However, before NCD films can succeed in overcoming the constraints facing artificial hip joints, a number of challenges need to be met. For example, it is essential that (i) an ultra-smooth diamond surface is produced; (ii) the strength at the hip/coating interface is sufficient and significant for the application; (iii) and the sp^3/sp^2 ratio in the NCD films is controlled in order to optimise the coating for the application.

9.3.4 Microfluidic Devices

With increasing demand for products associated with the medical, pharmaceutical, and analytical science industries over the past few years, much attention has been paid to the design and manufacture of microfluidic devices. Intensive research has been made on silicon-based microfluidic devices, in particular. Extremely delicate and complex structures such as microflow restrictors, microdroplet spraying nozzles, and micropumps can be manufactured in many industries and research institutes. However, application related issues on device performance such as long-term stability and reliability, bio-compatibility, low production costs, and high reproducibility must be considered.

Diamond-like carbon and nanocrystalline diamond coatings possess unique properties such as chemical inertness, bio-compatibility, and multifunctionality such as hydrophilicity or hydrophobicity. Deposition of diamond to silicon has been developed for use in such devices. Plasma enhanced CVD processes have been developed to deposit a typical layer thickness of approximately 500 nm with a surface roughness of 1 nm, which can be made hydrophilic or hydrophobic depending on the processing parameters. A layer is deposited on silicon and is immersed in KOH solution for 25 hours in order to check for signs of etching. If etching has not taken place then the coating is free from defects. Good conformity of the coating is observed and a cross section of a V-type microchannel is shown in Figure 9.9. Figure 9.10 shows a microflow restrictor made from silicon with a CVD diamond film deposited to it.

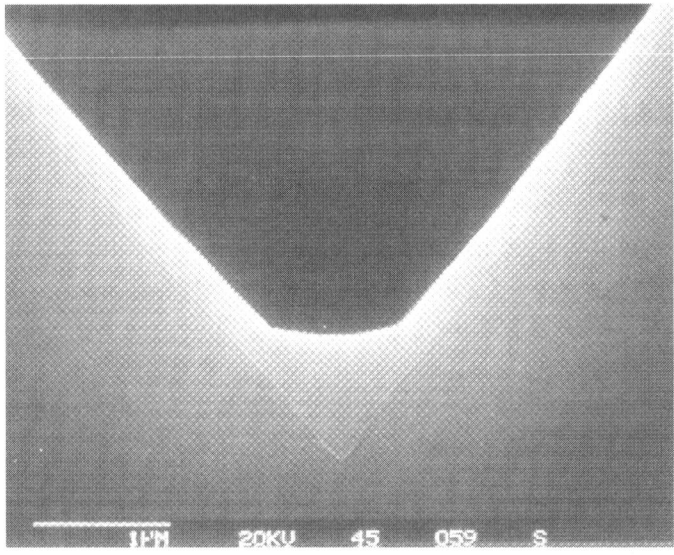

Fig. 9.9. Scanning electron micrograph of a cross-section of a V-type microfluidic channel showing diamond coating on top of a silicon substrate. Courtesy of CSEM Switzerland

9.4 Summary

The significance of NCD in relation to superior film properties, compared to MCD films, for various applications has been put forward. In addition, the different methods employed to deposit NCD films have been reviewed. The current and potential applications of diamond-based films have been briefly described.

In particular, the developments and key concerns relating to three specific biomedical applications, namely, artificial heart valves, dental burs and hip prostheses, have been discussed. A new TMCVD process for depositing improved, smoother, MCD and NCD films has been presented. The growth characteristics of films grown using the time-modulated process have been discussed. The growth rate

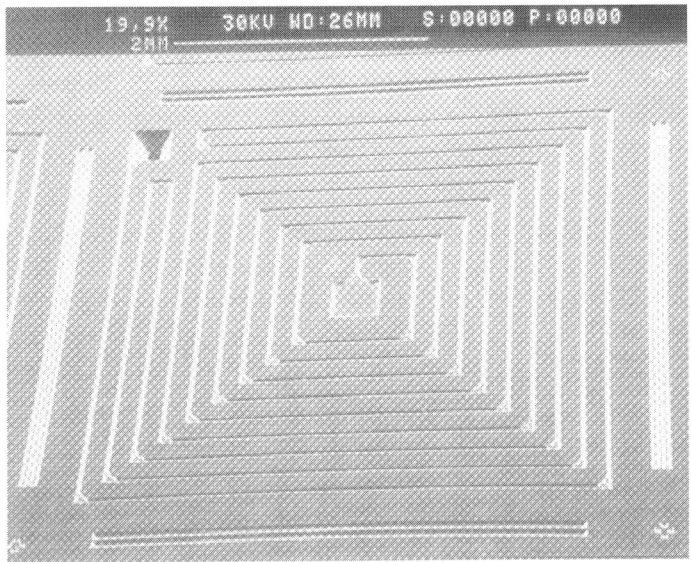

Fig. 9.10. Scanning electron micrograph of a microflow restrictor showing an array of diamond coated V-type channels. Courtesy of CSEM, Switzerland

trends observed using the hot-filament and microwave CVD processes have been discussed. As a concluding remark, it must be emphasised that the need for continued research effort in developing nano-sized diamond coatings, to expand the potentials of such coatings for use in newer biomedical applications, remains paramount.

References

1. P. W. May, *Phil. Trans. R. Soc. Lond.*, A (2000) 358, 473.
2. M. N. Ashfold, P. W. May, C. A. Rego and N. M. Everitt, *Chemical Society Reviews*, (1994) 23.
3. N. Ali, W. Ahmed, I. U. Hassan and C. A. Rego, *Surface Engineering*, 14 (4), (1998) 292.
4. N. A. G. Ahmed, *J. Physics E: Scientific Instruments*, 13, (1980) 295.
5. D. M. Mattox, *Proc 4th Int Colloquium on Plasma & Sputtering*, Nice, France, (Paris Societe Fracais du Vide) (1982), p.187.
6. D. M. Mattox, *Film Deposition Using Accelerated Ions*, Sandia Corporation, *Report*, SC-DR-28163 (1963).
7. W. Ahmed, D. B. Meakin, *J. Crystal Growth*, 79 (1986) 394.
8. P. J. Kelly, R. D. Arnell, W. Ahmed, *Materials World*, (1993) 161.

9. Q. H. Fan, E. Pereira, P. Davim, J. Gracio and C. J. Tavares, *Surface and Coatings Technology*, 96 (2000) 111.
10. W. Ahmed, C. M. J. Ackroyd, E. Ahmed and M. Sarwar, *Islamabad J. Sci.,* Vol. 11-9, No. 1-2, (1991-98) 29-34.
11. D. M. Gruen, *Annu. Rev. Mater. Sci.*, 29, 211 (1999).
12. D. Zhou, D. M. Gruen, L. C. Qin, T. G. McCauley, and A. R. Krauss, *J. Appl. Phys.*, 84, 1981 (1998).
13. T. Sharda, T. Soga, T. Jimbo, and M. Umeno, *Diamond Relat. Mater.*, 10, 561 (2001).
14. S. Saito, N. Fujimori, O. Fukunaga, M. Kamo, K. Kobashi, and M. Yoshikawa, "Advances in New Diamond Science and Technology." MYU, Tokyo, 1994.
15. K. Hirabayashi and S. Matsumoto, *J. Appl. Phys.*, 75, 1151 (1994).
16. S. A. Catledge and Y. K. Vohra, *J. Appl. Phys.*, 86, 698 (1999).
17. T. Sharda, M. Umeno, T. Soga, and T. Jimbo, *Appl. Phys. Lett.*, 80, 2880 (2002).
18. W. Zhu, G. P. Kochanski, and S. Jin, *Science*, 282, 1471 (1998).
19. J. Liu, V. V. Zhirnov, A. F. Mayers, G. J. Wojak, W. B. Choi, J. J. Hren, S. D. Wolter, M. T. McClure, B. R. Stoner, and J. T. Glass, *J. Vac. Sci. Technol.*, B 13, 422 (1995).
20. A Gohl, A. N. Alimova, T. Habermann, A. L. Mescheryakova, D. Nau, and G. Müller, *J. Vac. Sci. Technol.*, B 17, 670 (1999).
21. K. Wu, E. G. Wang, J. Chen, and N. S. Xu, *J. Vac. Sci. Technol.* B 17, 1059 (1999).
22. K. Wu, E. G. Wang, Z. X. Cao, Z. L. Wang, and X. Jiang, *J. Appl. Phys.*, 88, 2967 (2000).
23. C. Gu, X. Jiang, Z. Jin, and W. Wang, *J. Vac. Sci. Technol.*, B 19, 962 (2001).
24. O. Groning, L.-O. Nilsson, P. Groning, and L Schlapbach, *Solid State Electron,* 45, 929 (2001).
25. T. P. Ong and R. P. H. Chang, *Appl. Phys. Lett.*, 55, 2063 (1989).
26. R. L. C. Wu, A. K. Rai, A. Garscadden, P. Lee, H. D. Desai, and K. Miyoshi, *J. Appl. Phys.*, 72, 110 (1992).
27. R. Erz, W. Dotter, D. Jung, and H. Ehrhardt, *Diamond Relat. Mater.*, 2, 449 (1993).
28. M. Zarrabian, N. Fourches-Coulon, G. Turban, C. Marhic, and M. Lancin, *Appl. Phys. Lett.*, 70, 2535 (1997).
29. D. M. Bhusari, J. R. Yang, T. Y. Wang, S. T. Lin, K. H. Chen, and L. C. Chen, *Solid State Commun.*, 107, 301 (1998).
30. K. H. Chen, D. M. Bhusari, J. R. Yang, S. T. Lin, T. Y. Yang, and L. C. Chen, *Thin Solid Films*, 332, 34 (1998).
31. L. C. Chen, T. Y. Wang, J. R. Yang, K. H. Chen, D. M. Bhusari, Y. K. Chang, H. H. Hsieh, and W. F. Pong, *Diamond Relat. Mater.*, 9, 877 (2000).
32. L. C. Chen, P. D. Kichambare, K. H. Chen, J.-J. Wu, J. R. Yang, and S. T. Lin, *J. Appl. Phys.*, 89,753 (2001).
33. W. B. Yang, F. X. Lu, and Z. X. Cao, *J. Appl. Phys.*, 91, 10068 (2002).
34. T. Sharda, T. Soga, and T. Jimbo, *J. Appl. Phys.*, 93, 101 (2003) 368.

35. H. Hirari, K. Kondo, N. Yoshizawa, and M. Shiraishi, *Appl. Phys. Lett.*, 64, 1797 (1994); H. Hirai, K. Kondo, M. Kim, H. Koinuma, K. Kurashima, and Y. Bando, *Appl. Phys. Lett.*, 71, 3016 (1997).

36. R. Davanloo, T. J. Lee, H. Park, J. H. You, and C. B. Collins, *J. Mater. Res.*, 8, 3090 (1993).

37. A. Erdemir, G. R. Fenske, A. R. Krauss, D. M. Gruen, 1: McCauley, and R. 1: Csencsits, *Surf. Coat. Technol.*, 90-91, 565 (1999).

38. S. Hogmark, O. Hollman, A. Alahelisten, and O. Hedenqvist, *Wear*, 200, 225 (1996).

39. P. Hollman, O. Wanstrand, and S. Hogmark, *Diamond Relat. Mater.*, 7, 1471 (1998).

40. S. A. Catledge and Y. K. Vohra, *J. Appl. Phys.*, 84, 6469 (1998).

41. H. Yoshikawa, C. MoreI, and Y. Koga, *Diamond Relat. Mater.*, 10, 1588 (2001).

42. R. Davanloo, T. J. Lee, D. R. Jander, H. Park, J. H. You, and C. B. Collins, *J. Appl. Phys.*, 71, 1446 (1992).

43. R. DavanIoo, C. B. Collins, and K. J. Koivusaari, *J. Mater. Res.* 14, 3474 (1999).

44. N. Toprani, S. A. Catledge, Y. K. Vohra, and R. Thompson, *J. Mater. Res.*, 15, 1052 (2000).

45. B. Bi, W.S. Huang, J. Asmussen, and B. Golding, *Diamond Relat. Mater.*, 11, 677 (2002).

46. A. R. Krauss, O. Auciello, D. M. Gruen, A. Jayatissa, A. Sumant, J. Tucek, D. C. Mancini, N. Moldovan, A. Erdemir, D. Ersoy, M. N. Gardos, H. G. Busmann, E. M. Meyer, and M. Q. Ding, *Diamond Relat. Mater.*, 10, 1952 (2001).

47. J. E. Butler, D. S. Y. Hsu, B. H. Houston, X. Liu, J. Ignola, T. Feygelson, J. Wang, and C. T.-C. Nyguen, Paper 6.2, presented at the "*8th International Conference New Diamond Science and Technology 2002*," The University of Melbourne, Australia. The complete presentation is available online at http://www.conferences.unimelb.edu.au/icndst-8/presentations.htm; L Sekaric, J. M. Parpia, H. G. Craighead, 1: Feygelson, B. H. Houston, and J. E. Butler, *Appl. Phys. Lett.*, 81, 4455 (2002).

48. J. Lee, B. Hong, R. Messier, and R. W. Collins, *Appl. Phys. Lett.*, 69, 1716 (1996).

49. J. Lee, R. W. Collins, R. Messier, and Y. E. Strausser, *Appl. Phys. Lett.*, 70, 1527 (1997).

50. T. Sharda, M. Umeno, T. Soga, and T. Jimbo, *Appl. Phys. Lett.*, 77, 4304 (2000).

51. C. Z. Gu and X. Jiang, J. Appl. Phys., 88, 1788 (2000).

52. X. Jiang and C. L. Jia, *Appl. Phys. Lett.*, 80, 2269 (2002).

53. K. Teu, H. Ito, M. Hori, T. Takeo, and T. Goto, *J. Appl. Phys.*, 87, 4572 (2000).

54. D. M. Bhusari, J. R. Yang, T. Y. Wang, K. H. Chen, S. T. Lin and L. C. Chen, *J. Mater. Res.*, Vol. 13, No. 7, (1998) 1769-1773.

55. J. Michler, S. Laufer, H. Seehofer, E. Blank, R. Haubner, B. Lux [*Proc. 10th Int. Conf., On Diamond and Diamond-like Materials*, Prague, Czech Republic, 9-17 Sept. 1999, paper 5.231]

56. A. Heiman, I. Gouzman, S. H. Christiansen, H. P. Strunk, G. Comtet, L. Hellner, G. Dujardin, R. Edrei, and A Hoffman, *J. Appl. Phys.*, 89, 2622 (2001).

57. N. Jiang, S. Kujime, I. Ota, 1: Inaoka, Y. Shintani, H. Makita, A. Hatta, and A. Hiraki, *J. Cryst. Growth*, 218, 265 (2000).

58. H. W. Xin, Z. M. Zhang, X. Ling, Z. L. Xi, H. S. Shen, Y. B. Dai, and Y. Z. Wan, *Diamond Relat. Mater.*, 11, 228 (2002).

59. V. L. Konov, A. A. Smolin, V. G. Ralchenko, S. M. Pimenov, E. D. Obraztsova, E. N. Loubnin, S. M. Metev, and G. Sepold, *Diamond Relat. Mater.*, 4, 1073 (1995).

60. L. C. Nistor, J. V. Landuyt, V. G. Ralchenko, E. D. Obraztsova, and A. A. Smolin, *Diamond Relat. Mater.*, 6, 159 (1997).

61. T. Lin, Y. Yu, T. S. Wee, Z. X. Shen, and K. P. Loh, *Appl. Phys. Lett.*, 77, 2692 (2000).

62. T.-S. Yang, J.-Y. Lai, C.-L. Cheng, and M.-S. Wong, *Diamond Relat. Mater.*, 10, 2161 (2001).

63. G. Amaratunga, A. Putnis, K. Clay, and W. Milne, *Appl. Phys. Lett.*, 55, 634 (1989).

64. G. A. J. Amaratunga, S. R. P. Silva, and D. A. McKenzie, *J. Appl. Phys.*, 70, 5374 (1991).

65. D. M. Gruen, L. Shengzhong, A. R. Krauss, J. Luo and X. Pan, *Applied Physics Letters*, Vol. 64, 9, (1994) 1502.

66. D. Zhou, T. G. McCauley, L. C. Qin, A. R. Krauss and D. M. Gruen, *J. Appl. Phys.*, 83 (1), (1998) 540.

67. D. M. Gruen, *Annu. Rev. Mater. Sci.*, 29, (1999) 211.

68. T. M. McCauley, D. M. Gruen and A. R. Krauss, *Applied Physics Letters*, Vol. 73, No. 9, (1998) 1646.

69. D. M. Gruen, P. C. Redfem, D. A. Homer, P. Zapol, and L. A. Curtiss, *J. Phys. Chem.*, 103, 5459 (1999).

70. D. M. Gruen, X. Pan, A. R. Krauss, S. Liu, J. Luo, and C. M. Foster, *J. Vac. Sci. Technol.*, A 9, 1491 (1994).

71. D. Zhou, A. R. Krauss, L. C. Qin, T. G. McCauley, D. M. Gruen, T. D. Corrigan, and R. P. H. Chang, *J. Appl. Phys.*, 82, 4546 (1997).

72. D. Zhou, 1: G. McCauley, L. C. Qin, A. R. Krauss, and D. M. Gruen, *J. Appl. Phys.*, 83, 540 (1998).

73. S. Bhattacharyya, O. Auciello, J. Birrel, J. A. Carlisle, L. A. Curtiss, A. N. Goyette, D. M. Gruen, A. R. Krauss, J. Schlueter, A. Sumant, and P. Zapol, *Appl. Phys. Lett.*, 79, 1441 (2001).

74. D. Zhou, A. R. Krauss, L. C. Qin, T. G. McCauley, D. M. Gruen, T. D. Corrigan, R. P. H. Chang and H. Gnaser, *J. Appl. Phys.*, 82 (9), (1997) 4546.

75. X. S. Sun, N. Wang, W. J. Zhang, H. K. Woo, X. D. Han, I. Bello, C. S. Lee and S. T. Lee, *J. Mater. Res.*, Vol. 14, No. 8, (1999) 3204.

76. D. M. Bhusari, J. R. Yang, T. Y. Wang, K. H. Chen, S. T. Lin and L. C. Chen, *Mater. Lett.*, 36, (1998) 279.

77. N. S. Xu, J. Chen, Y. T. Feng, and S. Z. Deng, *J. Vac. Sci. Technol.*, B 18, 1048 (2000).
78. E. Maillard-Schaller, O. M. Kuettel, L. Diederich, L. Schlapbach, V. V. Zhirnov and P. I. Belobrov, *Diamond Relat. Mater.*, 8, 805 (1999).
79. H. Yagi, T. Ide, H. Toyota and Y. Mori, *J. Mater. Res.*, Vol. 13, No. 6, (1998) 1724.
80. J. Lee, B. Hong, R. Messier and R. W. Collins, *Appl. Phys. Letts.*, 69 (9), (1996) 1716.
81. T. Xu, S. Yang, J. Lu, Q. Xue, J. Li, W. Guo and Y. Sun, *Diamond and Related Materials*, 10, (2001) 1441.
82. S. P. McGinnis, M. A. Kelly, S. B. Hagstrom and R. L. Alvis, *J. Appl. Physics*, Vol. 79 (1), (1996) 170.
83. H. Yoshikawa, C. Morel and Y. Koga, *Diamond and Related Materials*, 10, (2001) 1588.
84. L. C. Chen, P. D. Kichambare, K. H. Chen, J.-J. Wu, J. R. Yang and S. T. Lin, *J. Appl. Physics*, Vol. 89, No. 1, (2001) 753.
85. S. Mitura, A. Mitura, P. Niedzielski and P. Couvrat, *Chaos. Solitons & Fractals*, Vol. 10 (9), (1999) 2165.
86. T. Sharda, M. Umeno, T. Soga and T. Jimbo, *Appl. Physics Letts.*, Vol. 77, No. 26, (2000) 4304.
87. T. Sharda, T. Soga, T. Jimbo, and M. Umeno, *Diamond Relat. Mater.*, 9, 1331 (2000).
88. T. Sharda, T. Soga, T. Jimbo, and M. Umeno, *Diamond Relat. Mater.*, 10, 1592 (2001).
89. B. D. Beake, I. U. Hassan, C. A. Rego and W. Ahmed, *Diamond and Related Materials*, 9 (2000) 1421.
90. S. N. Kundu, M. Basu, A. B. Maity, S. Chaudhuri, and A. K. Pal, *Mater. Lett.*, 31, 303 (1997).
91. X. T. Zhou, Q. Li, F. Y. Meng, L. Bello, C. S. Lee, S. T. Lee, and Y. Lifshitz, *Appl. Phys. Lett.*, 80, 3307 (2002).
92. O. Groning, O. M. Kuttel, P. Groning, and L. Schlapbach, *J. Vac. Sci. Technol.*, B 17, 1970 (1999).
93. T. S. Yang, J. Y. Lai, M. S. Wong, and C. L. Cheng, *J. Appl. Phys.*, 92; 2133 (2002).
94. T. S. Yang, J. Y. Lai, M. S. Wong, and C. L. Cheng, *J Appl. Phys.*, 92, 499 (2002).
95. X. T. Zhou, Q. Li, R. Y. Meng, I. Bello, C. S. Lee, S. T. Lee, and Y. Lifshitz, Paper P1.01.11, presented at the "Eighth International Conference New Diamond Science and Technology 2002," The University of Melbourne, Australia.
96. N. Jiang, K. Sugimoto, K. Nishimura, Y. Sbintani, and A. Hiraki, *J. Cryst. Growth*, 242, 362 (2002).
97. S. Prawer, J. L. Peng, J. O. Orwa, J. C. McCallum, D. N. Jamieson, and L. A. Bursill, *Phys. Rev. B*, 62, R16360 (2000).
98. Z. Wang, G. Yu, L. Yu, R. Zhu, D. Zhu, and H. Xu, *J Appl. Phys.*, 91, 3480 (2002).

99. H. Yusa, *Diamond Relat. Mater.*, 11. 87 (2002).

100. Y. Gogotsi, S. Welz, D. A. Ersoy, and M. J. McNallan, *Nature*, 411, 283 (2001).

101. P. Malshe, B. S. Park, W. D. Brown, H. A. Naseem, *Diamond Rel. Mater.*, 8 (1999) 1198.

102. C. Tokura, F. Yang and Yoshikawa, *Thin Solid Films*, 29, (1992) 49.

103. T. Zhao, D. F. Grogan, B. G. Bovard and H. A. Macleod, *Appl. Opt.*, 31, (1992) 1483.

104. A. Hirata, H. Tokura and M. Yoshikawa, *Thin Solid Films*, 29, (1992) 43.

105. D. G. Lee and R. K. Singh, in *Beam-Solid Interactions for Materials Synthesis and Characterization*, edited by D. E. Luzzi, T. F. Heinz, M. Iwaki and D. C. Jacobson (*Mater. Res. Soc. Symp. Proc.*, 354, Pittsburgh, PA, 1995) 699.

106. S. D. Wolter, F. Okuzumi, J. T. Prater and Z. Siter, *Phys. Stat. Sol.*, (a) 186. No. 2, 331 (2001).

107. I. U. Hassan, N. Brewer, C. A. Rego, W. Ahmed, B. D. Beake, N. Ali & J. Gracio, Proc. of *New Developments on Tribology: Theoretical Analysis and Application to Industrial Processes*, Editors: J. Gracio, P. Davim, Q. H. Fan and N. Ali, ISBN- 972-789-059-8, University of Aveiro, Portugal, May 2002, page 153.

108. D. R. Gilbert, D.-G. Lee and R. K. Singh, *J. Mater. Res.*, Vol. 13, No. 7, (1998) 1735.

109. F. Silva, A. Gicquel, A. Chiron and J. Achard, *Diamond and Related Materials*, 9 (2000) 1965.

110. A. Gicquel, K. Hassouni, F. Silva, *J. Electrochem. Soc.*, 14716 (2000) 2218.

111. W. Zhu, A. R. Badzian, R. Messier, in *Diamond Opt.* 111, San Diego, California, 1990 (SPIE. The *Int. Soc. For Opt. Eng.*), 187.

112. C. F. Chen and T. M. Hong, *Surf. Coat. Technol.*, 5 (1993) 143.

113. S. Kumar, P. N. Dixit, D. Sarangi and R. Bhattacharyya, *J. Appl. Phys.*, 85, (1999) 3866.

114. X. Li, Y. Hayashi and S. Nishino, *Jap. J. Phys.*, (1997) 36:5197.

115. N. Ali, V. F. Neto, Sen Mei, D. S. Misra, G. Cabral, A. A. Ogwu, Y. Kousar, E. Titus and J. Gracio, *Thin Solid Films*. Volumes 469-470, 22 (2004) 154.

116. Y. Hayashi, W. Drawl and R. Messier, *Jpn. J. Appl. Phys.*, Vol. 31, (1992) L194.

117. N. Ali, V. F. Neto and J. Gracio, *Journal of Materials Research*, Vol. 18, No. 2, (2003) 296-304.

118. N. Ali, Y. Kousar, Q. H. Fan, V. F. Neto and J. Gracio, *Journal of Materials Science Letters*, 22, 2003, 1039-1042.

119. J.E. Field (ed.), *Properties of Natural and Synthetic Diamond* (Academic Press, San Diego, CA, 1992, p.667).

120. J. C. Angus and C. C. Hayman, *Science* 241, 913 (1988).

121. W. Ahmed, N. Ali, I. U. Hassan and R. Penlington, *Finishing*, (1), (1998) 22.

122. K. Barton, A. Campbell, J. A. Chinn, C. D. Griffin, D. H. Anderson, K. Klein, M. A. Moore and C. Zapanta, *Biomedical Engineering Society (BMES) Bulletin*, Vol. 25, No. 1, (2001) 3.

123. S. L. Goodman, K. S. Tweden and R. M. Albrecht, *J. Biomedical Materials Research*, Vol. 32, (1996) 249-258.

124. F. Z. Cui and D. J. Li, *Surface and Coatings Technology*, 131 (2000) 481-487.

125. J. McLaughlin, B. Meenan, P. Maguire, N. Jamieson, *Diamond and Related Materials*, 8 (1996) 486-491.

126. M. I. Jones, I. R. McColl, D. M. Grant, K. G. Parker and T. L. Parker, *Diamond and Related Materials*, 8 (1999) 457-462.

127. A. Thomson, F. G. Law, N. Rushton, J. Franks, *Biomaterials*, 9 (1), (1991) 37.

128. I. Dion, C. H. Roquey, E. Baudet, B. Basse, N. More, *Biomed. Mater. Eng.*, 3, (1993) 51.

129. P. Yang, N. Huang, Y. X. Leng, J. Y. Chen, H. Sun, J. Wang, F. Chen and P. K. Chu, *Surface and Coatings Technology*, 156 (2002) 284-288.

130. J. Y. Chen, Y. X. Leng, X. B. Tian, L. P. Wang, N. Huang, P. K. Chu and P. Yang, *Biomaterials*, 23 (2002) 2545-2552.

131. Y. X. Leng, H. Sun, P. Yang, J. Y. Chen, J. Wang, G. J. Wan, N. Huang, X. B. Tian, L. P. Wang and P. K. Chu, *Thin Solid Films*, 398-399, (2001) 471-475.

132. Y. X. Leng, P. Yang, J. Y. Chen, H. Sun, J. Wang, G. J. Wang, N. Huang, X. B. Tian and P. K. Chu, *Surface and Coatings Technology*, 138 (2001) 296-300.

10. Environmental Engineering Controls and Monitoring in Medical Device Manufacturing

10.1 Introduction

The trend toward an aging population in the highly developed countries of the world has the demand for innovative biomedical devices and tools at record levels. The products desired in this market are typically smaller and more portable than their predecessors, and require more sophisticated components and allied manufacturing technologies and automation techniques. In essence, similar to traditional consumer products, biomedical devices such as patient monitors, drug deliver systems, therapeutic devices, and life assisting devices have all shrunk in size yet still have market expectations of enhanced performance characteristics and features [1–35]. An example of a modern day portable medical device, an implantable pacemaker, is shown in comparison to its first generation predecessor in Figure 10.1.

As a consequence of these market realities, many biomedical device companies have begun modeling their manufacturing environments in a similar fashion to the more traditional industries. For example, a typical biomedical device manufacturing facility starting up today might include the capabilities for fine pitch component placement as a part of a high volume automation line, which is additionally equipped with the necessary advanced testing instrumentation to ensure product quality and assurance. Unfortunately, as similarities increase in manufacturing design, it should also be expected that some of the negative consequences inherent to traditional manufacturing environments should become real issues as well in the biomedical device industry. One major area of concern must be the proper control and monitoring of environmental and worker exposures to potentially harmful chemical, biological, and physical stressors found in increasing concentrations in biomedical device manufacturing.

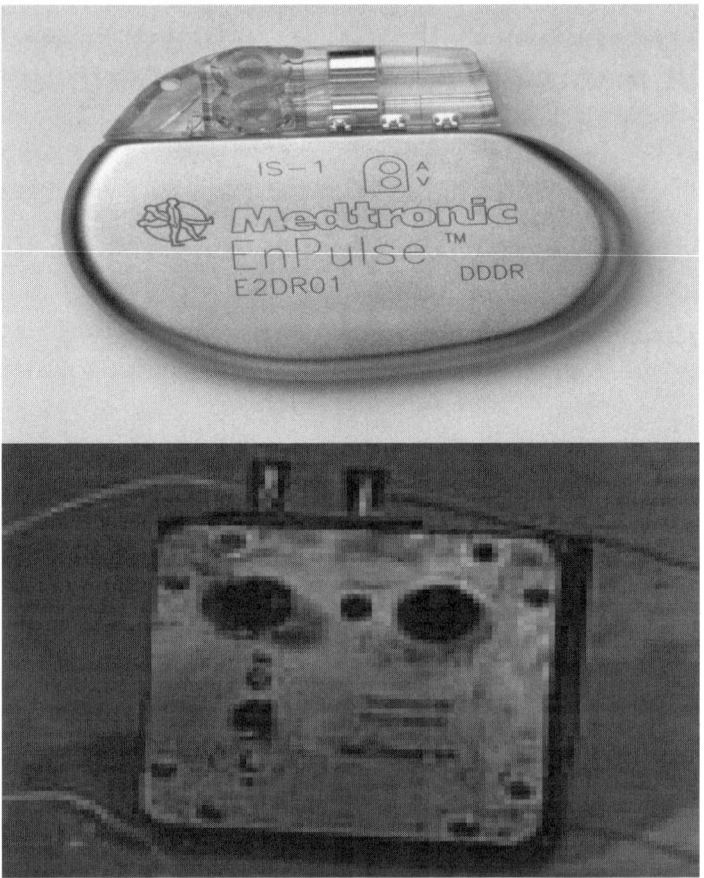

Fig. 10.1. Modern implantable pacemaker (left) versus original portable external device (Permission: Medtronics Canada and Minnesota Technology)

The purpose of this chapter is to present some of the strategies that could be employed to effectively control and monitor for workplace hazards associated chemical, biological, and physical agents in the biomedical device industry. The chapter begins with a presentation of a comprehensive list of the stressors found in the industry, with an overview of the properties, toxicity/exposure limits, and other pertinent characteristics of each, respectively. The chapter continues with an introduction to the typical environmental and engineering control methods and personal protective equipment that should be implemented to help alleviate (or eliminate) the concern for any overexposures to any of these stressors [1–35].

10.2 Stressor Source, Properties, and Characteristics

It is expected that the biomedical device market will significantly grow globally over the next couple of years, with 34.5% growth rates estimated in the nanoscale market alone through 2007 [1–35]. With the increased production rates brought on by these pressures, it can only be predicted that the use of additional chemical, biological, and physical agents associated with manufacturing these products will also rise substantially. Thus, it can easily be argued that an overall understanding by manufacturing personnel of the capabilities and limitations as well as the potential benefits and detriments of their usage should be imperative. This section attempts to delimit the list to pertinent stressors, with a detailed coverage for each provided on normal source and usage, chemical and physical properties, and toxicity characteristics. A comprehensive list of the stressors provided in this section and deemed the most commonly found in the biomedical device industry, along with source, properties, and toxicity characteristics for each, has been summarized in Table 10.1. The major manufacturing processes found in the biomedical industry that are related to this discussion can be generalized to the following categories:

1. Sterilization
2. Cleaning, etching, and surface preparation
3. Adhesive application
4. Coating application
5. Drilling, grinding, cutting, and other light production machining
6. Welding and soldering
7. General maintenance activities
8. Laboratory research and testing

10.3 Sterilization

Sterilization activities in the biomedical device industry are required by the European Union (EU) and the United States Federal Department of Agriculture (USFDA) and validation and routine control

Table 10.1. Common Stressors Found in Biomedical Device Manufacturing

Stressors	Source	Properties and Toxicity Characteristics
Ethylene oxide (EtO)	Sterilization	EtO is a colorless, flammable gas at room temperature and pressure with an ether-like odor, which has been linked to leukemia and peritoneal cancer. Acute exposures to >800 parts per million (ppm) can results in severe mucous membrane irritation and tion and edema.
Ionizing Radiation (IR)	Sterilization, lab instruments	Ionization radiation exposures can be from primarily gamma ray, x-ray, beta particle, alpha particle, and electron beam exposures. Gamma and x-rays are the most penetrating, with beta particles being intermediate and alpha particles depositing energy over only a short traverse.
Non-Ionizing Radiation (NIR)	Sterilization, instruments, surface prep, cutting, etc.	The most common non-ionizing radiation exposures will be from UV and lasers. UV can cause damage to the skin and eyes while laser energy primarily targets the eyes.
Ozone	Welding, sterilization	Ozone is a colorless to blue colored gas at room temperature and pressure, with a very pungent odor. It is non-flammable but a powerful oxidizer that severely irritates the eyes, mucous membranes, and respiratory tract at levels greater than 5ppm.
Hydrogen peroxide	Sterilization	Hydrogen peroxide is a noncombustible, colorless liquid with a slightly sharp odor. It is totally miscible in water and a powerful oxidizer with the potential to cause severe damage to the respiratory tract at concentrations greater than 75 ppm.
Isopropanol (IPA)	Cleaning, disinfecting	IPA is a colorless liquid with the odor of rubbing alcohol. It is flammable and miscible in water with a vapor pressure of 33mmHg. At concentrations greater than 4000 ppm, severe dizziness and drowsiness can occur in those exposed.
Methanol	Cleaning	Methanol is a colorless, flammable liquid at room temperature and pressure with a characteristic pungent odor. It has a vapor pressure of 96 mmHg and is miscible in

		water with sever dizziness, drowsiness, and blindness occurring at levels greater than 6000 ppm.
Ethanol	Cleaning	Ethanol is a colorless, flammable liquid. It has a vapor pressure of 44 mmHg and totally soluble in water. It can cause severe respiratory and CNS effects at concentrations greater than 1%.
Trichloroethylene (TCE)	Degreasing	TCE is a colorless, combustible liquid with a chloroform-like odor. It has a vapor pressure of 58 mmHg, a specific gravity of 1.46 (sinker), and 0.1% solubility. At levels >1000 ppm, nausea, convulsions, and death can occur. Skin contact can lead to dermatitis.
1,1,1-Trichloroethane	Degreasing	1,1,1 – TCA is a colorless, combustible liquid with a mild, chloroform-like odor. It has a vapor pressure of 100 mmHg, a specific gravity of 1.34 (sinker), and 0.4% solubility. At levels >700 ppm, severe respiratory tract irritation, poor equilibrium, and liver damage can occur. Skin exposures can lead to dermatitis.
Acetone	Degreasing, cleaning	Acetone is a colorless, flammable liquid with a mint-like odor. It has a vapor pressure of 180 mmHg, is miscible, and has a specific gravity of 0.79 (floater). High airborne concentrations >5000 ppm can cause CNS depression. Skin exposures can cause dermatitis.
Perchloroethylene (Perk)	Degreasing	Perk is a colorless, noncombustible liquid with a chloroform-like odor. It is heavier than water, with a low solubility in water (0.02%) and a low vapor pressure (14 mmHg). Exposures above 150 ppm can cause respiratory problems, dizziness, and liver damage. Exposures have resulted in liver tumors in animals.
Hydrofluorocarbons	Degreasing	Hydrofluorocarbons are non-flammable solvents of very low toxicity. They are recyclable and have no effect on the ozone layer. However, emissions do contribute significantly to global warming.
Hydrofluoroethers	Degreasing	Same as for hydrofluorocarbons.
Perfluorocarbons	Degreasing	Same as for hydroflourocarbons and hydrofluoroethers.

(Continued)

Table 10.1 (Continued)

Stressors	Source	Properties and Toxicity Characteristics
Sulfuric acid	Etching, anodizing	Sulfuric acid is a strong corrosive liquid. It is colorless to dark brown in color, with little odor and an oil-like appearance. The target organs for this miscible, non-combustible liquid include the eyes, skin, and respiratory tract and the IDLH is 15 mg/m^3.
Nitric acid	Etching, anodizing	Nitric acid is a noncombustible and color-less, yellow, or red fuming liquid with an acrid, suffocating odor. It is miscible, with a vapor pressure of 48 mmHg. At concentrations >25 ppm, irritation to the eyes, skin, and respiratory tract can occur.
Phosphoric acid	Etching, anodizing	Phosphoric acid is a noncombustible, colorless solid with no odor. It is miscible and causes skin burns and dermatitis on contact.
Chromic acid	Etching, anodizing	Chromic acid is an odorless, noncombustible red solid, normally used in the flake or powder form. It is a known human carcinogen (septum/lung). Acute reactions include irritation of the respiratory tract and continuous exposures can cause sensitization dermatitis.
Potassium hydroxide	Etching, anodizing	Potassium hydroxide is a strong corrosive solid (pH > 13.0) with mainly skin and eye contact concerns. It is found as white or yellow lumps, flakes, rods, sticks, or pellets and in aqueous solutions. Inhalation exposures can be a concern if it gets airborne.
Sodium Napthenate	Etching, anodizing	Sodium napthenate is a corrosive solid with mainly skin and eye contact concerns. It has a low vapor pressure and solubility in water and can be inhaled as an aerosol.
Particulate matter	Surface prep, maintenance, welding, general activities, etc.	Particle matter has many sizes, shapes, and origins. It seems that those particles lesser than 2.5 micron have the most detrimental impact on human health. Inhalation of respirable particles can cause severe fibrosis and chronic manifestations such as silicosis.

Polymer adhesives	Adhesive application	Exposures to airborne vapors from adhesives can lead to dizziness and headaches in those exposed. Many of the glues dry on contact and can negatively impact the skin and eyes.
Heavy metal fumes and oxides	Welding and soldering	Chronic exposures to the various heavy metals can cause severe central nervous system malfunctions. Metal fumes have been linked to metal fume fever, with some metals such as cadmium and nickel classified as probable or known human carcinogens.
Fluoropolymers	Surface coatings	Recent studies involving fluropolymers support a linkage to cancer.
Fluorides	Welding and soldering	Significant exposures to fluorides can lead to fluorosis, a severe condition that results in bone and enamel embrittlement.
Acetylene	Welding	Acetylene is a very explosive gas that can also be a simple asphyxiant in high enough concentrations.
Various aliphatic hydrocarbons	Coatings, adhesives	Dermal exposures to aliphatic hydrocarbons can lead to dermatitis. Hexane is the biggest airborne concern, however, most of those in this chemical group have relatively low toxicity. These are typically very flammable liquids.
Various aromatic hydrocarbons	Coatings, adhesives	Dermal exposures to aromatic hydrocarbons such as benzene, xylene, and toluene can dry out the skin and cause dermatitis. Benzene is a know carcinogen (blood cancer) and high concentrations can negatively impact the CNS and respiratory tract.
Heat stress	Maintenance, some production	Heat stress can be an issue when personal protective equipment is being worn or when various hot working operations are being conducted.
Noise	Maintenance, some production areas	Unhealthy levels of noise exposures are typically considered to be in excess of an average of 85 decibels for 8 hours or more in duration. Hearing protection is mandated at 85 or 90 decibels.

procedures are outlined in such documents as Association for the Advancement for Medical Instrumentation (AAMI)/International Organization for Standardization (ISO) 11135 for ethylene oxide sterilization and AAMI/ISO 11137 for radiation sterilization. There are literally dozens of ways to sterilize biomedical devices and new techniques are currently being developed and tested in research labs throughout the world. However, the most common methods of sterilization at this time are with ethylene oxide, gamma rays, and electron beam radiation. The following paragraphs provide an overview of each of these as well as a few of the lesser used techniques (i.e., ozone, vapor-phase hydrogen peroxide, plasma, microwave, and steam) and the associated health effects that could be realized if overexposed.

10.3.1 Ethylene Oxide Sterilization

Ethylene oxide (EtO) is widely used as a sterilizing agent in the biomedical device industry, due primarily to its potency in destroying pathogens and material compatibility characteristics. It is estimated that nearly one-half of all medical devices produced are currently sterilized by EtO. EtO kills microbes by alkylation. Alkylation is the process by which EtO takes the place of the hydrogen atoms on molecules needed to sustain life. With enough time and concentration, this proves lethal to all of the microbial life that is present on the device.

Conventional wisdom would lead one to believe that if EtO is toxic to microbes, it would also likely be a human toxin. This is certainly the case. Ethylene oxide is a colorless gas at room temperature, with a flashpoint of below 0 degrees Fahrenheit and a flammability range of 3 to 100 volume percent in air. It has an ether-like odor and is considered a regulatory concern primarily due to its flammability and/or explosivity as well as its acute and chronic human toxicity characteristics. It is classified as a probable human carcinogen (A2), with a United States Occupational Safety and Health Administration (USOSHA) permissible exposure limit (PEL) of 1 part per million averaged over an 8-hour shift and 5 parts per million as a 15-minute excursion.

Fig. 10.2. Illustration of ethylene oxide (EtO) leaking from a sterilization chamber (Permission: Japanese Advanced Information Center for Safety and Health)

The EtO sterilization process typically includes five steps: conditioning, sterilization, evacuation, air wash, and aeration. While human exposures to EtO during any of these stages is normally unlikely, there is always a chance of a process system leak or an operator making a deviation from the normal protocol or standard operating procedures. In addition, during set-up and changeover periods, there is always a possibility of unsafe airborne exposures to personnel of this highly toxic gas. An illustration showing the precarious safety concerns involving the use of the EtO sterilization process is shown in Figure 10.2.

10.3.2 Gamma Ray Sterilization

Gamma rays, typically emitted from a source of cobalt-60 (Co-60), are also a common means for sterilizing biomedical devices. In fact, it is estimated that nearly 50% of all single-use medical supplies (e.g., syringes, catheters, IV sets) have been sterilized by this technique.

The gamma radiation emitted by the Co-60 source destroys any residual microbe by attacking the DNA of the molecules.

The main advantages to using ionizing radiation to sterilize include optimal device penetration, process repeatability, and no product residues. And, from a health and safety standpoint, the typical, properly shielded cobalt-60 source has just enough energy to kill the microorganism of concern, but yet does not have enough energy to impart any harmful radioactivity to the surrounding workers or the environment. While the threat to overexposure to gamma radiation may be minimal to the biomedical device production worker, careful attention must be still taken to minimize the impact of any exposures and to always follow proper ALARA (as low as reasonably achievable) guidelines.

Cobalt-60 is solid substance that has a radioactive half-life of 5.27 years and decays by a beta/gamma scheme. Since it is gamma emitter, external exposures to large sources of this radionuclide can cause severe skin burns, acute radiation syndrome, and death. While careful safeguards have been put into place to prevent any worker exposures to ionizing radiation in the biomedical device industry, accidental emergency releases are always possible. Unlike EtO, the main regulatory responsibility for gamma radiation in the United States is not OSHA or the Environmental Protection Agency (USEPA) but the Nuclear Regulatory Commission (NRC). The acceptable annual dose limit for a non-nuclear energy worker in the U.S is 1 milliseivert (100 mrem) dose equivalent while nuclear energy workers are allowed 50 milliseivert (5 rem) per year, with 100 milliseivert (10 rem) allowed accumulated exposure over a five-year period. The USOSHA PEL is currently set at 0.1 mg/m^3 for the non-radioactive component.

10.3.3 Electron Beam Radiation Sterilization

Another growing means for sterilization in the biomedical device manufacturing industry is by an electron beam radiation technique. Like gamma ray sterilization, this technique employs a beam of ionizing radiation that alters the DNA of the microorganism it attacks. Commercial electron beam accelerators range in energy from about 3 MeV to 12 MeV (million electron volts) and usually operate at only one energy level. The main advantages to this technique include shorter product exposure times, higher production rates, and less material oxidation.

One key aspect of both gamma ray and electron beam sterilization is the concept of dosimetric release. Dosimetric release is a procedure accepted by the USFDA and detailed in ANSI/AAMI/ISO 11137-1994. Dosimetric release is based upon readings from dosimeters placed on devices during processing. Verification of the minimum and maximum doses applied provides the mechanism for release and shipment. As will be discussed later in the monitoring section of the chapter, radiation dosimeters also provide useful information and control for worker health and safety biomedical device industry as well.

The same dose equivalent standards are used for electron beam sterilization as those in effect for the gamma ray techniques. USOSHA standards for ionizing radiation used in general industry can be found in 29 CFR 1910.1096. The European Union (EU) regulations for ionizing radiation were set forth in the Council Directive 96/29/EURATOM of 13th of May 1996.

10.3.4 Other Sterilization Techniques

While less common than those already addressed, device sterilization methods using ozone gas, steam, plasma, vapor-phase hydrogen peroxide, and microwave radiation have also been piloted in the laboratory and field settings. While proper safety controls have been normally put into place during circumstances employing one or more of these techniques, there is always a chance of an accidental release or a deviation from normal protocol that could result in an overexposure to either a single individual employee or a group of workers.

Ozone is a toxin that is a significant acute respiratory stressor. The immediately dangerous to life and health (IDLH) guideline for ozone is set a 5 ppm and USOSHA and ACGIH both set the occupational exposure limit for an 8-hour shift at 0.1 ppm. In addition, USEPA regulates ozone emission to the environment and considers it to be one of the six National Ambient Air Quality Standards (or criteria pollutants), and the EU sets the ambient air standard for ozone at 0.12 mg/m^3. Hydrogen peroxide vapors are also considered toxic and most be controlled. The target organs for vapor-phase hydrogen peroxide would be the eyes, skin, and respiratory tract, with a USOSHA PEL of 1 ppm established. Guidelines for non-ionizing radiation, including microwaves, are provided based upon various frequency ranges and are delimited in such sources as ACGIH's TLVs for Chemical Substances and BEIs and the ICNIRP's General Approach to Protection Against Non-Ionizing Radiation. Regulations and guidelines for plasma processes are in the currently research and findings stage and are not well established for worker health and safety.

10.4 Cleaning, Etching, and Surface Preparation

The effectiveness of the surface cleaning and preparation processes followed both during the manufacturing of the device as well as with the finished product will significantly impact the ultimate reliability and overall quality of the device in the field. For example, it is imperative that electronic medical devices have a clean surface in order to ensure good bonding and coating. In addition, compromised surface preparation can lead to the existence of chemical contaminants that can cause corrosion, and the non-removal of particulate matter may result in an undesirable electrical conductance path and short circuits.

Some of the most common cleaning processes used in biomedical device manufacturing include methanol, ethanol, isopropanol, chlorinated hydrocarbon solvents, fluorinated hydrocarbon solvents, acetone, and deionized water. Common etching or anodizing agents include sulfuric acid, phosphoric acid, chromic acid, sodium napthenate, and potassium hydroxide. Mechanical surface preparations that many times cause unwanted particle residues and potential airborne contamination include surface and scuff sanding as well as grit blasting. Table 10.2 provides the target organs for each of these potential stressors and the approximate vapor hazard ratio, based upon the worldwide average occupational exposure limits (OELs), for some of the more commonly used solvents. The vapor hazard ratio (VHR), or vapor hazard index (VHI) as it is sometimes called, is found as follows:

$$\text{VHR (or VHI)} = C_{vp}/\text{OEL} \qquad (10.1)$$

Where

C_{vp} = concentration at the saturation vapor pressure in ppm

and

OEL = occupational exposure limit in ppm

The VHR provides a convenient means for comparing the potential exposure impact to various solvents. Essentially, the VHR describes by how many times a saturated vapor volume must be diluted by this same volume of air so that the OEL is not exceeded.

Table 10.2. Target Organs and Vapor Hazard Ratio (VHR) for Selected Chemical Stressors

Chemical Stressor	Target Organs (NIOSH 2005)	Vapor Hazard Ratio
Isopropanol	Eyes, skin, and respiratory system	100
Methanol	Eyes, skin, respiratory system, and central nervous system (CNS)	700
Ethanol	Eyes, skin, respiratory system, CNS, liver, blood, and reproductive system	75
Trichloroethylene	Eyes, skin, respiratory system, heart, CNS, liver, and kidney	1750
1,1,1 Trichloroethane	Eyes, skin, CNS, cardiovascular system, and liver	800
Perchloroethylene	Eyes, skin, respiratory system, liver, kidneys, and CNS	550
Acetone	Eyes, skin, respiratory system, and CNS	450
Potassium hydroxide	Eyes, skin, and respiratory system	–
Particulates (not otherwise regulation)	Eyes, skin, and respiratory system	–
Sulfuric acid	Eyes, skin, respiratory system, and teeth	–
Phosphoric acid	Eyes, skin, and respiratory system	–
Chromic acid	Blood, respiratory system, liver, kidneys, eyes, and skin	–

10.4.1 Alcohols

There are two types of surface contamination that are produced during the production of biomedical devices: polar and nonpolar. The majority of polar contaminants found on biomedical devices during manufacturing include various inorganic compounds, with the source being primarily from flux activators or finger salts. Since alcohol is a polar compound and by taking into consideration the rule that "likes dissolve likes", the alcohols are many times used to remove polar contamination from the surface of biomedical devices. It has been a widely accepted premise that alcohol is the most effective and economical solvent available for removing ionic residues from biomedical devices, and thus, its use has grown concurrently with the increases in market demand over the years.

Employee exposure to airborne concentrations of alcohol can be irritating to the eyes, nose, and respiratory tract, with significant doses having been linked to manifestations of the central nervous system, liver, blood, and reproductive system. Obviously, efforts should be made to limit the exposure of alcohols to the employee's skin, due to its solvent nature. Methanol is considered to be more toxic than both ethanol and isopropanol and its use should be limited under most circumstances.

However, as might be expected, methanol is deemed superior to the other two alcohols in the removal of significant ionic surface residues. Typical OELs have been set at 1000 ppm for ethanol, 200 ppm for methanol, and 400 ppm for isopropanol. The alcohols are not currently considered to be either a known or probable human carcinogen by any regulatory authority.

10.4.2 Chlorinated and Fluorinated Hydrocarbons

Chlorinated and fluorinated hydrocarbons are used as nonpolar solvents in the industry. Nonpolar solvents such as methyl chloroform (1,1,1-TCA), trichloroethylene (TCE), and tetrachloroethylene (Perk) have been traditionally used because of their outstanding capabilities of removing oils, greases, rosin flux, etc. during the surface preparation process. However, requirements set forth by occupational and environmental regulatory authorities regarding these highly toxic and flammable compounds had increased the popularity of fluorinated solvents, chloro-fluorohydrocarbons (CFCs), and various blends throughout the 1970s and 1980s. Still, with even more recent regulations, biomedical device producers have been dissuaded from using such fluorinated solvents such as trichlorotrifluoroethane. For example, Freon[TM], once a widely used industrial fluorochlorohydrocarbon, was identified as a major precursor contaminant that contributed significantly to stratospheric ozone depletion in the atmosphere, and thus, its use has been all but completely eliminated in most modern countries. As a matter of fact, the further production of this CFC in the U.S. was banned completely in 1996.

Today, the majority of biomedical device manufacturers use one or more of the following classes of nonpolar solvents in the production process to remove primarily oils and grease from the devices:

1. Hydrofluorocarbons (HFCs)
2. Hydrofluoroethers (HFEs)
3. Perfluorocarbons (PFCs) or perfluoropolyethers (PFEs)
4. Chlorinated hydrocarbons (e.g., TCE, Perk, 1,1,1-TCA)
5. trans-1,2-dichloroethylene
6. Brominated solvents
7. Hydrocarbons and oxygenated solvents

The current trend for cleaning nonpolar compounds from the surface of medical devices appears to be moving away from the traditional chlorinated hydrocarbons and more toward HFCs and HFEs. These compounds are considered relatively benign in toxicity to animals and humans, however, are considered to be significant stressors to the environment, potentially enhancing global warming and greenhouse gas effects with a warming potential as much as five orders of magnitude greater than carbon dioxide gas.

As was mentioned previously, the chlorinated hydrocarbons are very closely scrutinized by both the environmental and occupational regulatory bodies due to both toxicity and flammability detriments. As would be expected, this family of chemicals affects the skin by removing all natural oils and potentially leading to severe dermatitis conditions. They are heavier than water and vapor pressures that range from 14–100 mmHg. The relative toxicity of the main three chemicals in the class (TCE, Perk, and 1,1,1-TCA) varies based on the concentration and the extent of the exposure (i.e., acute or chronic).

TCE is a colorless liquid with a chloroform-like odor, with chemical incompatibilities and reactivities to strong caustics and chemically active metals. The USOSHA PEL established for TCE is 100 ppm and has been linked to causing both liver and kidney tumors in animals. Along with the liver and kidneys, the main target organs include the respiratory tract and the central nervous system. The flash point for TCE is 160 degrees F, which classifies it as a combustible liquid.

Perk is a colorless liquid with a mild, chloroform-like odor and considered to be a strong oxidizer and incompatible with chemically active metals. It has an USOSHA PEL equal to 100 ppm and targets the liver, kidneys, respiratory tract, and central nervous system. Perk has been classified as animal carcinogen by causing liver tumors in test subject. While it is not classified as either a flammable or combustible liquid, it will decompose in the event of a fire to significant concentrations of hydrochloric acid (a corrosive vapor) and phosgene (a highly toxic gas).

Like both TCE and Perk, methyl chloroform (or 1,1,1-TCA) is a colorless liquid with a mild, chloroform-like odor, with incompatibilities to chemically active metals. It will also react with strong oxidizers, caustics, and water. 1,1,1-TCA targets the respiratory tract, skin, central nervous system, and liver but has not been linked to causing cancer in humans or animals. The EU OEL has been established for the stressor at 100 ppm and it is classified as a combustible liquid.

The majority of the HFCs, HFEs, PFCs, and PFEs do not have any occupational exposure established at this time. This is both due to the fact that these chemicals have been determined to be primarily nontoxic as well as their increased usage is a rather recent phenomenon, and thus, have not warranted up to this point much concern to the health and safety community. However, their collective effect on the environment, and particularly, on global warming impacts will undoubtedly be an issue to contend with as emissions are expected to only increase in the future. Figure 10.3 provides an example of a cell washer for in-line process degreasing.

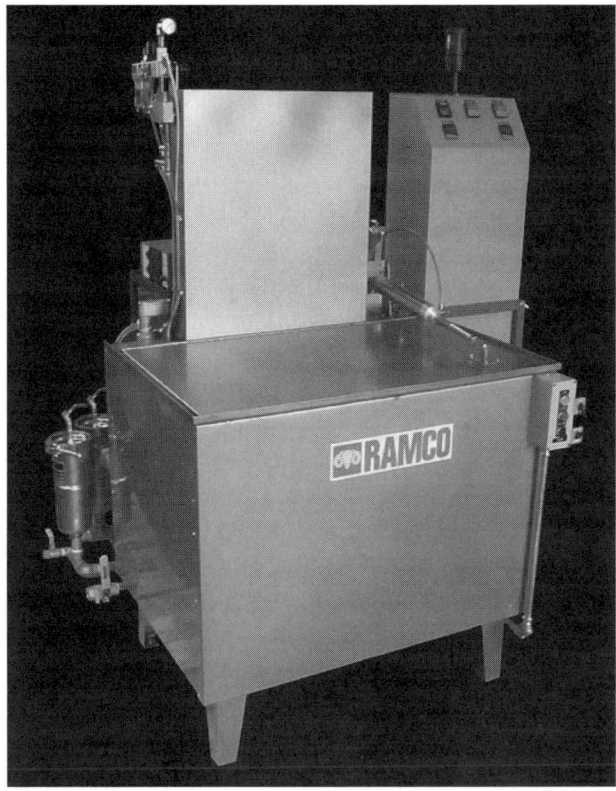

Fig. 10.3. Vapor degreaser used in biomedical device manufacturing (Permission: Ramco, Inc.)

10.4.3 Acids and Alkalis

The primary acids used to surface etch biomedical devices include chromic, phosphoric, nitric, and sulfuric acid. Polymer surfaces are typically etched with a strong oxidizing agent such as chromic acid. The chromic acid provides the means to oxidize the substrate surface, resulting in an optimal surface for further treatments such as adding coatings or adhesives. Chromic, along with sulfuric and phosphoric acid, are many times employed as an etching or anodizing agent for metals such as titanium, stainless steel, and nickel. The surface preparation for fluoropolymers (or materials to be coated with fluoropolymers) is a difficult task. Treatments that have been effectively used in the industry include sodium naphthenate and potassium hydroxide etches as well as flame treatments.

Due to its highly acute and chronic toxicity, airborne and dermal chromic acid exposures to workers in the biomedical device industry must be of an utmost concern. Chromic acid has been linked to both human septum and lung cancer and is currently classified as a known human carcinogen. Since it is a strong acid, its corrosive nature can severely affect the respiratory tract as well as the skin of any exposed worker(s). Chromic acid is normally found in a red-colored, aqueous solution in industry and is reactive with most readily oxidizable materials. The USOSHA ceiling exposure limit has been established for this chemical at 0.1 mg/m^3.

While the other acids typically used as etching and anodizing agents in biomedical device manufacturing are not considered known human carcinogens, they do exhibit comparable corrosivity. Thus, each of these agents target the respiratory tract and the skin, with the possibility of severe burns being a reality upon even minimal exposure. The occupational limits established are 1 mg/m^3 for sulfuric, 5 mg/m^3 for nitric, and 1 mg/m^3 for phosphoric. Substance incompatibilities and reactivities include caustics (phosphoric), organic materials, metals, and even water (sulfuric and nitric).

Sodium naphthenate and potassium hydroxide are strong caustic etching agents used effectively on some polymer surfaces. Like their acid counterparts, they are very corrosive materials and target primarily the skin, eyes, and respiratory tract. They are not currently classified as probable human carcinogens and are typically found in industry in an aqueous solution. There is currently no USOSHA PEL or EU OEL established for either of these agents. However, potassium hydroxide has a recommended TLV ceiling recommended by the ACGIH of 2 mg/m^3. These substances are reactive with acids, water, and metals.

10.4.4 Acetone

Acetone is ubiquitous ketone solution used as a general purpose cleaning solvent in many industries. Its use varies in the biomedical device industry, with toxicity and exposure concerns typically minimal. The established EU OEL for this agent is 500 ppm, with incompatibilities existing between it and acids and oxidizers.

10.4.5 Particulate Matter

Surface preparations such as sanding and blasting can cause the generation of particles and aerosols of various sizes, constituents, and morphology. Not only can their existence potentially compromise the quality and integrity of the completed device, but it can also be troublesome to those workers exposed to potential unhealthy airborne concentrations in their respective breathing zones. Depending primarily on the particle type and shape, there exists a myriad of factors that can be used to help determine the potential for negative impacts on employee health and well-being, both acute and chronic in nature.

Particulate matter is typically classified by aerodynamic diameters in the micron or submicron size ranges. Traditionally, any particles or aerosols of sizes less than 10 microns are considered to be respirable. Particulate matter greater than 10 micron in diameter is considered essentially benign because of the assumption that it will be eventually removed by other body defenses such as nose hair, cilia, and mucus before it reaches the inner respiratory tract. Due primarily to sufficient research on the exponential growth in adverse effects on human health from exposures to particulate matter in the 1 micron to 2.5 micron range, the USEPA has recently introduced a new tighter standard for ambient air exposures of this stressor.

Long-term exposures to some types of particulate matter found in the workplace have resulted in various forms of fibroses and pneumoconiosis ("dusty lung") in otherwise healthy workers. Free silica from chronic exposures to sand has resulted in a chronic condition known as silicosis and exposures to airborne beryllium dust has initiated a manifestation known as berylliosis in those exposed. Recently, concerns have been expressed about the increasing exposures of workers in the biomedical device industry that may come into contact with airborne particles in the nanometer range. The human health effects from exposures to particles in the submicron ranges are not well understood at this time. Currently, USEPA considers particulate matter as one of the six criteria pollutants and occupational exposure limits in Spain have been set at 3 mg/m^3 for the respirable fraction. More stringent restrictions have been set for some particulate matter types that have been linked to chronic conditions such as fibrosis and cancer.

10.4.6 Spent Solvents

Whether it is a benign substance like deionized water or a toxic substance such as chromic acid, once the agent has been used in the production process the resulting waste must be dealt with by following local and/or federal guidelines. The USEPA mandates a normal protocol to follow per RCRA and CERCLA provisions for hazardous waste generation and transport based upon the quantity generated. In the U.S., State-run EPA programs handle the management of wastes generated from "cradle-to-the-grave". Depending on the jurisdiction, other countries, territories, provinces, and local governing groups may have different protocols to follow in order to effectively deal with their respective spent surface preparation agents.

10.4.7 The Future of Surface Preparation Techniques

As have been previously discussed, most of the conventional cleaning methods and surface preparations have employed wet chemical

Fig. 10.4. Plasma cleaning apparatus (Permission: UCP Processing Ltd)

techniques. However, the current trend seems to be moving away from chemical treatments and toward such modern techniques as cold plasma, corona discharge, and laser cleaning.

Corona discharge is a process by which high voltage electricity is discharged into an airstream, producing large concentrations of ozone to the oxidize the device surface while plasma cleaning employs a ionized, equally-charged oxygen gas stream to chip apart the surface contaminants. Laser cleaning is still yet another recent technology used in the industry. Lasers are, of course, a concentrated form of light energy and are considered nonionizing in nature. The employee exposure concerns of various lasers will be covered in greater detail later on in this section.

Essentially, the health and safety concerns for workers in this environment have been switched from chemical stressors, for the most part, to physical stressors such as electricity, electrical/magnetic fields, and

Fig. 10.5. Laser cleaning operation (Permission: Adapt Laser Systems)

non-ionizing radiation. Figure 10.4 provides an example of the plasma-cleaning device that might be implemented in the biomedical device industry while Figure 10.5 shows a laser cleaning method in use.

10.5 Adhesive Applications

Many of the previously discussed surface preparation techniques were completed in order to effectively and efficiently apply adhesives to various device substrates. While various mechanical fasteners as well as welding, brazing, and soldering techniques can be used to join many biomedical device materials together, there are still other materials, such as thermoplastic and thermosetting polymers, that are considered incompatible to these types of joining processes. Adhesive bonding conditions include:

1. Bonding of dissimilar materials.
2. Joining to promote optimal stress distributions or impact resistance.
3. Joining of very thin materials.
4. Joining of outsourced subassemblies.
5. Bonding for mechanical joint augmentation.

The most common adhesives used in biomedical device manufacturing include urethanes, cyanoacrylates, acrylics, epoxies, and silicones. Collectively, the adhesives primarily attack the skin, eyes, and respiratory tract. While sometimes quite odiferous, their vapor pressures are usually very low and most do not contain ingredients that are carcinogenic. Some of adhesives have components that are recognized as chemical sensitizers, and exothermic polymerization is always a concern if they should ever come into contact with incompatible materials. Occupational exposure limits have not been established specifically for any of these polymer groups but yet exposure standards have been determined for any hazardous ingredients that might be used as a product component.

The commercially available urethanes have a variety of ingredients; however, common to most products will be less than 5% isocyanates and 5% naptha. They are not carcinogenic but have been classified as chemical sensitizers, and overexposures can cause severe respiratory tract ailments including pulmonary edema and bronchitis. In case of a fire, urethane-based decomposition products of concern include carbon monoxide, nitrogen oxides, hydrochloric acid, and trace amount of hydrogen cyanide. Cyanoacrylates (or the "superglues") are eye and mucous membrane irritants and tissue bonders. Their vapors are lachrymatory and, if decomposed during a fire, produce a dense, choking smoke. They are incompatible with water, alcohols, and amines, sometimes producing a significant exothermic polymerization event.

The majority of the acrylic adhesive formulations used in biomedical device manufacturing can cause skin dermatitis as well as allergic reactions for those sensitive individuals. At high processing temperatures, it is possible for some employees exposed to acrylic adhesives to exhibit flu-like conditions known as "polymer flu". As for the epoxies, the main health and safety concerns include skin, eye, and respiratory tract irritation and chemical sensitization while silicone formulations being irritants which many times have a major toxic aliphatic or aromatic hydrocarbon component such as n-heptane or xylenes.

10.6 Coating Applications

The most common coatings applied to biomedical devices are urethane-based, fluoropolymers, or polyimide laminates. The techniques with a prove track record for device surface coating varies in sophistication and applicability. Depending on the type of material, as well as product size and configuration, one technique may prove to be more effective than another in ensuring coating quality and repeatability. The required coating thickness is also a major factor in the decision process. In any case, the coating techniques employed have the potential of producing aerosols of varying sizes and shapes of which significant employee exposures could be realized.

Since the majority of the coatings are polymer-based, many of the health and safety concerns with coatings are shared with the common adhesives that were discussed in the previous paragraphs. However, what differs significantly is the potential for toxic aerosols to build up to a significant concentration in the employee-breathing zone during the specific coating application.

Like adhesives, there are certainly concerns for employee skin and mucous membrane exposures to the common coatings applied. Additionally, they can be chemical sensitizers and have other toxic ingredients in their formulation. However, what differs substantially between the two involves the potential for exposures to harmful levels of particulate matter and toxic metal pigments that can be inhaled by the associated production worker. Heavy metals linked to human cancer, such as chromium and cadmium, are used to provide color to many industrial coatings. Another major health concern related to biomedical device coating applications involves fluoropolymers such as Teflon®, which is a very common device coating due to its biocompatibility. Recent studies have linked fluoropolymers to increased incidences of cancer and teratogenesis for those exposed to a particular raw material, perfluorooctanoic acid (PFOA), used in its production, yet claims are still very controversial.

10.7 Drilling, Grinding, Cutting, and Machining

Drilling, grinding, and other light machining operations produce fine and course particulate matter, as well as significant concentrations of aerosols from the use of cutting oils and fluids. These particles range from just a few nanometers to well over 10 micron in sizes. Exposures to particles from the near micron range to around 10 micron have resulted in various lung ailments such as bronchitis, emphysema, anthracosis, and silicosis. The kinetics of these particles is pretty well understood and related health effects data is rather complete. However, the effects on human health of particles in the ultrafine particle size region of 100 nanometers, or less, is not well understood. As a matter of fact, the particles generated in this region behave more like gases than they do particles with regard to motion and kinetics.

While recent efforts have been directed toward removing from manufacturing many oils and cutting fluids with toxic ingredients and replacing them with human and environmentally friendly alternatives, there are still a vast amount of these necessary lubricants available for use in the biomedical industry. While there are an increasing number of aqueous based fluids becoming commercially available, many of these lubricants are still petroleum-based, with the associated ill effects related to over exposures to aliphatic and aromatic hydrocarbons still a daunting reality.

Traditional cutting techniques also produce particulate matter and aerosols in the micron and submicron ranges. However, advanced techniques, such as laser cutting, have been piloted in the field and are gaining popularity.

10.7.1 Laser Cutting

Lasers are used for cutting in many manufacturing industries and the biomedical industry is no exception. Additionally, lasers can be found in welding, sealing, and coating operations as well as in medical micromachinery, lab instrumentation, and in the final device itself. Thus, a discussion on the environmental, health and safety issues regarding its proper use is imperative.

Lasers and laser equipment may be potentially dangerous to eyes and skin of the employee. The relative degree of risk depends on the type of beam, the power frequency, beam divergence, beam intensity, and duration of exposure. The eye is the most susceptible to damage, with retina burns resulting in the possibility of total blindness. Given certain levels and wavelengths of laser radiation, coupled with adequate duration, skin reddening, swelling, blistering, and even charring can occur.

Exposure guidelines are based on the characteristics of the type of laser and are expressed as maximum permissible exposure or MPE. The guidelines most often used involving the safe use of lasers has been published by ANSI, ACGIH, ICNIRP, and IEC (Hitchcock et al. 2003). Traditionally, the USFDA has used a laser classification scheme using four roman numbers I, II, IIIa, IIIb, and IV. The Class I laser was considered the most benign and eye safe whereas the Class IV was the most dangerous for eye and skin exposures. However, the USFDA and ANSI and other industrialized countries are currently in the process of adopting, if they have not already done so, the International Electrotechnical Commission standard, IEC 60825–1. Table 10.3 summarizes the laser classes, with power, duration, and relative hazards provided. It should be pointed out that the

"M" designation after the class is for "magnification" while the "R" is for "reduced requirements".

Table 10.3. Laser Classification Scheme and Characteristics

Laser Class (IEC)	Laser Class (old USFDA)	Allowable Power (watts)	Emission Duration (sec)	Hazard Description
1	I	0.39E-60	>10,000	Not a known eye or skin hazard
1M	I	0.39E-60	>10,000	Eye safe with no optical aids
2	II	<1.0E-3	>0.25	Potential eye hazard for chronic viewing
2M	IIIa (low irradiance)	<5.0E-3	>3.8E-4	Potential eye hazard for chronic viewing and may be so with optical aids
3R	IIIa (high irradiance)	<5.0E-3	>3.8E-4	Marginal hazard for intra-beam viewing
3B	IIIb	<5.0E-1	>0.25	Known intrabeam viewing hazard
4	IV	>0.5	NA	Known eye and skin hazard

10.8 Welding and Soldering

Welding and soldering activities can sometimes pose an exposure risk for those not wearing the proper personal protective equipment and using adequate ventilation control. The main hazards associated with soldering include skin burns and airborne contaminant exposures, primarily from the solder, soldering flux, and any surface pre-or post-cleaning solutions. The primary welding hazards include exposures of air contaminants from sources such as the base material, welding rod, welding flux, and inerting gases. Potential physical hazards encountered during the process include non-ionizing radiation, heat stress, and electricity.

Traditionally, the biggest concern for occupational solderers was the likely exposures to significant concentrations of lead in the air. Lead is an extremely acute and chronic toxin linked to a myriad of potential manifestations. Fortunately, most of the lead-based solder has been removed from manufacturing in the developed countries. However, exposures to the fumes generated from solders even without lead should be kept to a minimum. Many formulations of solder flux provide a substantial potential for unhealthy doses of fluorides. Additionally, normally low vapor pressure cleaners used during the process can become heated and emit higher than normal levels of gases and vapors into the breathing zone of the biomedical manufacturing worker.

Depending on the type of welding and the level of engineering controls implemented, a wide range of contaminant exposures can be realized. Airborne levels of ozone, nitrogen oxides and fluorides are normally troublesome and heavy metal fumes of varying types and concentrations may also be of paramount concern. Typical metal fumes and oxides include iron, zinc, copper, cadmium, aluminum, magnesium, nickel, chromium, and manganese. The majority of these heavy metals are considered to be chronic toxins, targeting the central nervous system and lungs, and several of these toxins are either known or probable human carcinogens. Exposures to heavy metal fumes have resulted in a condition known as "metal fume fever".

Workers must be shielded from the non-ionizing radiation exposures possible from some welding processes. Harmful, high frequency ultraviolet radiation has caused a manifestation known as "welder's flash" in some welders and vicinity workers. The symptoms of this condition include visual impairment, the feeling of sand or grit in the eye, and a severe headache with malaise. Additionally, welding should never be conducted near chlorinated cleaning solvents due to the potential of sparks initiating dangerous phosgene gas accumulations. Of course, burns and electrical shock are additional possible physical stressors of which the biomedical device production welder may become exposed.

10.9 General Maintenance Activities

There is a plethora of general maintenance activities possible at each respective biomedical device manufacturing facility, with many of these activities such as welding, drilling, cleaning, etc. having already been covered in the preceding paragraphs. However, the severity and the distinct nature of some of these tasks have merited a separate discussion.

Maintenance activities differ from those in normal production in potentially several ways. For one, the maintenance activities are many times not planned some time in advance before the actual event occurs. This means that exposures to non-expecting workers may be intensified. Secondly, maintenance activities that create potential environmental stressors are normally completed in a shorter duration, with higher activity levels and robustness. Thus, the potential for exposures is once again enhanced. Finally, maintenance and set-up activities many times lead to process control changes that may, after completion, result in unusual and significant short-term exposures to physical and chemical agents by area production workers.

Machining activities accomplished by maintenance workers sometimes produce noise levels that exceed the occupational limits and heat stress can also be of concern. Increased maintenance activities can produce airborne contaminants such as particulate matter/fibers and noxious gases, vapors, and fumes. Obviously, many the activities of facility maintenance personnel can produce unsafe slip, trips, and falls as well as electrical wiring and pneumatic/hydraulic plumbing concerns for all associated workers.

10.10 Laboratory Research and Testing

The research and quality control laboratories in the biomedical device manufacturing industry provide a rather unique environment for employee exposures. Lab technicians are typically working with a wide range of instruments with a variety of operational characteristics and potential to exposures from various biological, chemical, and physical stressors. Instruments found in the laboratory environment include, among others, sources such as x-rays, gamma ray, lasers, and some wet chemistry components that could be harmful if not properly controlled.

While concerns involving biological stressors are possible in any of the before mentioned areas or during certain previously discussed activities, the research and quality control laboratories are the places where pilot runs are conducted and final devices are tested for contamination. Experimental procedures might include such activities as biocompatibility testing, with uses of bloods and other body fluids common. Depending on the device, ISO clean room status may be desired at a certain level in the labs as well as on the production floor; thus, various testing for bacteria and fungal contamination may be required. While there are currently no occupational mandates for personal exposure to most biological agents, governmental agencies on the environment and food and drugs have set some standards for microbial contamination.

10.11 Environmental and Engineering Controls

Administrative actions, engineering controls, and personal protective equipment are considered as the three main approaches to controlling environmental emissions and employee exposures. The intent of this section is to elucidate the common environmental and engineering controls that could be implemented in the biomedical device manufacturing climate to help protect the employee and the environment. Any of the following implemented individually or in combination are viable engineering control techniques that could work in the biomedical device industry:

1. Substitution
2. Process controls (continuous or automation)
3. Enclosure/Isolation
4. Process elimination
5. Process change
6. Ventilation controls (local exhaust or dilution)

While prevention is not included as one of the above environmental and engineering controls, it should always be the first consideration taken when there is the potential for employee exposure to chemical, biological, and/or physical stressors. In essence, the control assessment should always begin with an evaluation of whether or not the situation that apparently requires control can just be totally prevented instead by some means.

Once the condition that has the potential to adversely impact human health or the environment has been recognized, then a proper controls implementation scheme should be followed. An effective protocol should be systematic and involve a series of steps taken to identify and characterize the hazard, exposure source, worker involvement, and air movements, as well as identifying all alternatives, with the ultimate goal of implementing and testing/maintaining the best of these alternatives.

10.12 Substitution

Substitution is the process by which a more environmentally-friendly substitute is made for a known hazardous substance, process, or piece of equipment. For example, it has been argued that the plasma cleaning process is superior to organic solvent cleaning when it comes to the potential for harmful exposures and negative human health impacts. Another example of substitution in the biomedical device industry might involve the use of lab instruments that employ sensors which work on the principal of non-ionizing radiation instead of their predecessors, either gamma or x-ray radiation, which have more damaging characteristics on the cellular or tissue levels.

One must be careful that the substitution made does not result in such an inferior replacement to the original product, process, or equipment that it might compromise the quality and integrity of the final device. An example of this undesirable event might include a process/material change by a company from a non-polar solvent to the use of a surfactant and water to clean a particular medical device, resulting in ineffective removal of contamination. Economics can also limit the benefits of substitution. For instance, while many biomedical companies have realized successes by substituting automated processes for those that were once somewhat labor intensive, others have failed to accomplish this goal. Quite frequently, this is due to the significant up-front costs associated with automation and the inability of the company to reach any economies of scale because of their size.

10.13 Process Controls

There are times when the current process controls need to be evaluated for their merit. As a general rule, intermittent or batch processes are typically considered to be more hazardous than those that are more continuous in nature. In essence, the automated line takes some of the human component out of the process, and thus, typically also reduces the potential for human exposures. However, as was discussed in the previous subsection, automated processes can have their own set of downfalls and shortcomings.

10.14 Enclosure/Isolation

The use of enclosures and isolation techniques will be found in almost all biomedical device manufacturing facilities. The sophistication of these control alternatives varies widely, with the intent to separate the potentially exposed employee from the hazardous event. A very good example of a type of enclosure being used in the industry is the glovebox set-up. These units are found in a myriad of operations to protect the worker from the workpiece and process during applications of various physical, biological, and chemical stressors. For example, a glovebox apparatus might be used in the laboratory of a facility to test a particular medical device's biocompatibility or potential for rapid oxidation when coming in contact with a substance. Another common enclosure used in the industry is the particle and fume

Table 10.4. Enclosure/Isolation Techniques Used in Biomedical Device Manufacturing

Type of Medical Device or Process	Enclosure/Isolation Technique
Co-60 use in teletherapy	Operator kept at safe distance from source and lead shielding from gamma radiation
Irradiator for instrument sterilization	Lead shielding with operator in a secured room
Ethylene oxide use for device sterilization	Specialized containment and time without employee in close proximity
Use of biological agents in the laboratory for testing	Glovebox with proper hazard classification characteristics
Radioactive iodine syringe preparation and assembly	Syringe shields and Plexiglass^{TM} shielding for beta radiation during preparation
Acid etching of the substrate on devices and tools	Special acid enclosure and fume hood use
Controlled storage of chemicals	A regulatory-approved chemical storage cabinet
Automated production line	Clear acrylic safety shields to protect worker from moving parts and potential stressors

hood, typically required to keep microbial and particle concentrations at levels to meet or exceed ISO clean room standards.

The principle of isolation can be by either space or time. The use of walls or complete rooms to separate employees from a potential hazard, of course, is an example of the former. In contrast, conducting a special cleaning operation to remediate facility mold contamination on either the weekend or overnight would be an excellent example of the latter. Table 10.4 provides some examples of potential enclosure or isolation techniques that could be used effectively in biomedical device manufacturing.

10.15 Process Change or Elimination

Sometimes it is possible to change a process to make it less hazardous to the employee. For instance, a surface coating application could be applied through a dipping process rather than one that requires spraying. This change could eliminate the majority of the coating aerosols from ever entering the breathing zone of the worker. Additionally, the dipping process should be considerably more manageable from the aspect of environmental control and regulatory affairs.

It is even possible that a process of concern could be completely eliminated. Given our coating application example, it might be possible to eliminate the coating operation or simply apply a much thinner layer of coating using isolation and an advanced nanotechnology technique. Fortunately, many studies in general industry show that process changes (or eliminations) made in order to increase hazard control have actually resulted in enhanced productivity, as well as an improvement in overall product quality.

10.16 Ventilation Controls

A discussion on environmental and engineering controls would not be complete without a significant effort being put on covering the proper design, development, and implementation of the site ventilation system(s). Without adequate ventilation, there would be, in many cases, no other alternative but to put workers on respirators to eliminate their exposure potential. This is an issue that most companies should try to avoid; mandating respirators for protection complicates production and regulatory matters and adds a significant cost to the company. Fortunately, the use of proper ventilation will eliminate the necessity for respirator use by most biomedical device production workers during their normal work-related activities.

There are two types of ventilation: dilution ventilation and local exhaust ventilation (LEV). Most facilities have both of types of ventilation, with additions and changes to these networks occurring at least periodically, if not frequently. The following paragraphs attempt to elucidate the benefits that can be realized in controlling the production environment with an optimally designed ventilation network.

10.16.1 Dilution Ventilation

Dilution ventilation, also known as general exhaust ventilation, controls the level of airborne stressors by removing the potentially contaminated air and replacing it with fresh dilution air before concentrations reach unhealthy levels. Under certain assumptions and constraints, the resulting equilibrium concentration of any airborne contamination can be estimated as follows: $C = E/Q$, where C is the concentration, E is the emission rate, and Q is the ventilation rate.

Traditionally, ventilation experts have used the units of air changes per hour to express dilution ventilation exchange rates, and the notion of an "acceptable concentration" has been used to indicate a safe or comfortable level of exposure. Since a heavy reliance on adjacent sources of outdoor air for dilution exists, it is a must that this fresh air source has less contamination than what is realized on the production floor. Therefore, careful attention must be given to where air intakes are located to minimize the effects from outdoor sources of such ambient air contaminants as ozone, sulfur dioxide, particulate matter, and nitrogen dioxide. Also, during summer and wet months of the year, the dilution air could have significant levels of mold spores that could be brought into the biomedical device manufacturing environment.

Obviously, relying solely upon dilution ventilation to control airborne contaminants in the biomedical device manufacturing can be problematic. It is important to evaluate the potential for various stressor exposures and the relative toxicity of each of these. Inevitably, there will be some operations conducted during the development, testing, and manufacturing of a device that will not allow for engineering control only through dilution ventilation efforts. The use of dilution ventilation independently as a means for environmental and engineering control should be avoided if the following conditions exist:

1. The contaminants realized are highly toxic chemical, biological, or physical stressors.
2. The concentration levels are higher than established action levels or guidelines.
3. The emission rates are variable.
4. There exist only a few, high concentration discharge points for any contaminants.
5. The outdoor air might be suspect for various reasons.
6. The existing HVAC system is not adequate to provide "controllable" dilution air.
7. The worker's breathing zone is in close proximity to the emission point(s).

10.16.2 Local Exhaust Ventilation (LEV)

Local exhaust ventilation (LEV) is many times coupled with effective dilution ventilation to keep airborne contaminant levels down to acceptable concentrations. The LEV commonly employs the use of a properly dimensioned hood, plumbed with the necessary ductwork to a series of mechanical components, with its endpoint being an emissions stack. At a minimum, a properly designed LEV will have an air cleaning device capable of removing the stressor(s) of concern as well as a fan designed to drive the air of the given volume and desired flow rate. Figure 10.6 shows the components of a typical LEV system.

The LEV is the primary means for removing gases, vapors, fumes, and particles from the immediate breathing zone of workers in the biomedical device manufacturing industry. It is called "local" exhaust ventilation because the physical location of the system is always in the immediate proximity of the point source emission. In essence, the face of the hood is placed close enough to the workpiece/process

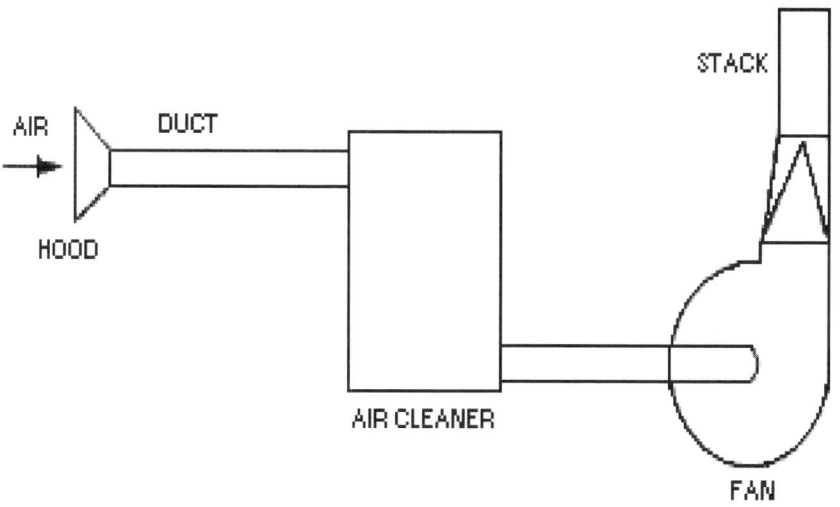

Fig. 10.6. Components of a local exhaust ventilation (LEV) system

to allow for the suction created from the mechanical fan to capture the contaminant and remove it up and out of the associated ductwork. Consequently, it has been written that capturing and removing an airborne stressor, such as an acid etchant vapor or a hydrocarbon solvent vapor, at its source is the principle objective of LEV systems. Typical operations that merit the use of LEVs in the industry include solvent cleaning, acid etching, sterilization, welding, soldering, coating applications, adhesive applications, clean room/laboratory activities, and any other industry-specific operation that produces point source air emissions.

A variation on the traditional LEV, commonly found in the biomedical device industry, is the kind designed specifically to meet international clean room standards for particulate and microbial contamination. In reality, this type of environmental and engineering control is really a combination of both local exhaust ventilation and a form of isolation/enclosure. Unlike the traditional units, the clean room system not only protects the worker from any stressor that may be inside its hood, but also serves the dual purpose of protecting the device/process from any external stressors, which if allowed to be present, could contaminate or corrupt. Many of these types of control devices are of the ductless variety, with installed high efficiency particulate air (HEPA) or ultra-low penetration air (ULPA) filters to effectively collect and remove any airborne particles.

The local exhaust hood, in a broad sense, is any suction opening that is intended to draw the contaminant into the control system. Generalized, the three basic types of hoods are capture, enclosing, and canopy. Two important parameters unique to the local exhaust hood are the face velocity and the capture velocity. The face velocity, quite simply, is the air velocity at the hood opening. In contrast, the capture velocity is the air speed at any point in front of the hood (or at the hood opening) required to overcome any opposing air currents and capture the contaminated air, causing it to flow into the hood. The point in space at which this occurs is called the capture point. Other velocity parameters important to proper hood and LEV design include slot velocity, plenum velocity, duct velocity, and minimum design duct velocity. Table 10.5 provides approximate capture velocities required to properly remove common contaminants found in the biomedical device manufacturing.

Table 10.5. Approximate Hood Capture Velocities (adapted from NSC and ACGIH Vent Manual)

Conditions of Contaminant Release	Approximate Capture Velocity in m/sec (fpm)	Examples of Processes or Operations in Biomedical Device Manufacturing
Release with no significant velocity into quiet air	0.25-0.5 (50-100)	Degreasing, cleaning, etching, anodizing, adhesive application
Released with low velocity into moderately quiet air	0.5-1.0 (100-200)	Coating applications, welding, and soldering
Released with considerable velocity or into area of rapid air movement	1.0-2.5 (200-500)	Light surface preparations, some spray coatings, and machining operations
Released with high velocity or into zone of rapid air movement	2.5-10 (500-2,000)	Some maintenance operations, grinding, and other heavy machining operations

Normal capture velocities vary widely in the industry and are based mainly on the spatial characteristics of the LEV system and process interface as well as on the physical and chemical properties of the contaminant(s). Normal face velocities found on hoods in the biomedical device manufacturing industry usually are in the range of 80 feet per minute (fpm) to 100 fpm.

After the contaminated air has been captured by the local exhaust hood, the ductwork serves as the carrying conduit on to the other mechanical components of the LEV, and finally, on up the stack and out. The ducts are typically made from sheet metal, with rectangular or circular cross-sections of varying dimensions. The mass flow of contaminated air in a duct system is based on the duct velocity and the cross-sectional and can be calculated as follows: $Q = v/A$ where Q is the mass flow, v is the duct velocity, and A is the cross-sectional area. Typically, the economically optimal duct velocity found for systems in the biomedical device industry ranges from 1000 fpm for most cleaning, etching, and adhesive application processes up to around 2000 fpm for some welding or soldering operations. The maximum duct velocity that could possibly be encountered in the industry would be for heavy maintenance and machining operations, where velocities may reach as high as 4000 fpm.

Three critical parameters that require an understanding when calculating and controlling duct velocities include the static pressure, the velocity pressure, and the duct friction losses. The static pressure is the energy source of the system and is created by the fan while the velocity pressure is the pressure creating by the air in flux. It can be said that velocity pressure is what is realized by converting static pressure into air movement within the duct. These important parameters are typically measured by performing what is known as a pitot traverse. Friction losses (and other losses) must be considered for optimal LEV system design, and values for these losses are based on such characteristics as length, diameter, and configuration. Values for LEV duct friction losses can be found tabulated in various sources including manufacturer specification sheets.

Air cleaners are a necessary component of a properly designed LEV system. Some of the more commonly found types of cleaners commercially available include:

1. Electrostatic precipitators
2. Simple setting chambers
3. Wet and dry centrifugal collectors
4. Venturi scrubbers
5. Washers
6. Fabric filters (e.g., HEPA or ULPA)
7. Packed tower or scrubber

8. Carbon adsorption
9. Catalytic units

The type of cleaner chosen for the LEV system must be able to effectively remove the contaminant of concern. Thus, the type of cleaner capable of removing particulate matter at a known efficiency will be more than likely considerably different from one which is effective at removing a vapor or fume. Along with the nature of the contaminant, the other major factors considered when choosing one type over another include airborne concentration, outflow cleanliness, and cost. Air cleaners for removing gases and vapors normally either work on the principle of absorption, adsorption, condensation, or catalytic conversion. For metals fumes, a cloth filter, high efficiency wet collector, or electrostatic precipitator provides the best removal efficiencies. Particles (or dust) of sizes greater than 1 micron can usually be cleaned effectively by such control technologies as cyclones, precipitators, venture scrubbers, settling chambers, settling chambers, and cloth filters. Particles smaller than 1 micron are typically controlled with HEPA or ULPA filter set-ups. A considerable control challenge is presented by significant concentrations of particles, smoke, of fumes of less than one micron. These stressors show active Brownian movement because of their small size and do not tend to settle out like those particulate contaminants that are greater than 1 micron or so.

The required removal efficiency is an important parameter that needs to be covered on any discussion on air cleaners. Many operations in the device industry require robust collection efficiencies in order to keep the atmosphere at "clean" levels. For example, HEPA and ULPA filters are rated on their ability to remove a certain percentage of particles at a particular diameter. The efficiency and particle sizes specified vary from manufacturer to manufacturer, however, a typical claim for a HEPA filter may be something like '99.99% removal efficiency of particles 0.3 micron and larger'. Likewise, for a ULPA filter it may claim a '99.997% removal efficiency of particles greater than 0.1 micron'.

The fan is the mechanical driver of the LEV system and typically is of a centrifugal or axial configuration. The fan is rated based on the required volumetric air-flow and static pressure. In addition, specifications for such parameters as voltage, current, revolutions per minute (rpm), outlet velocity, and brake horsepower (BHP) are typically made by the designer, and capacity tables are available from the fan manufacturer to assist in the decision making process. Important criteria that should be taken into consideration in the fan selection process include:

1. The characteristics of the airstream such as contaminant identity and physical state.
2. An evaluation of the physical constraints within its environment.
3. An analysis of the proper drive arrangement and potential configuration.
4. A prediction of the additional noise to be generated by the fan.
5. The dealing with any potential safety concerns posed by its use.
6. An assessment of the requirements for any supplemental accessories.

The exhaust stack is the final component of the complete LEV system, and is simply just an extension of the ventilation system's ductwork above the building rooftop. A properly designed stack will serve two important purposes. Firstly, it should aid in the adequate dispersion of the gas stream stressors well above and beyond the roof-line of the facility. Secondly, the mere existence of the stack in the LEV system causes a reduction in the velocity pressure at the outlet, and therefore, and an overall increase in fan performance. Some rule-of-thumbs that should be considered for optimal stack design include:

1. The stack should be configured as a straight cylinder to avoid any mechanical losses.
2. The use of rain caps or screens is not recommended.
3. If possible, the location of the stack should be on the highest rooftop.
4. The stack should be kept as far away as possible from any plant air intakes.
5. Stack height increases in lieu of good emission controls should be avoided.
6. Stack height requirements should be increased rather than gas stream exit velocities in order to realize the necessary control.

For those interested in more detailed information on how to design an effective and efficient LEV system for their specific process, a myriad of excellent resources are available for guidance through the design process.

10.17 Personal Protective Equipment and Clothing

A final control option for consideration in biomedical device manufacturing is the use of personal protective equipment (PPE) or personal protective clothing (PPC). However, the paradox lies in the reality that the use of PPE or PPC is many times the easiest contaminant control solution. For example, the device manufacturer might use a dermal barrier device (e.g., neoprene gloves, barrier cream, etc.) long before ever considering a change of process or materials. Other common examples of PPE and PPC usage in biomedical device manufacturing include ear plugs and muffs for noise protection, cooling vests for heat stress protection, and safety glasses for general eye protection.

The PPE and PPC usage in the industry varies widely and is based upon the type of device being manufactured. A detailed discussion on the various PPE and PPC usage in the various device manufacturing environments is beyond the scope of this chapter and will not be covered in any further detail.

10.18 Control Strategies in Device Manufacturing

The choice of the optimal control strategy to follow depends on a multitude of different factors. Obviously, the type of device being manufactured and the potential stressors associated with its production top this list. However, such additional factors as economics, regulatory requirements, workforce characteristics, and facility design must also be given due attention.

Typically, the best strategy involves the combination of two or more of the discussed control strategies. For example, a barrier or isolation control technique may be used in conjunction with an adequately designed local exhaust ventilation system in order to prevent environmental releases of worker exposures to ethylene oxide during the sterilization process. Another example might be the reliance on dilution ventilation to remove the majority of facility contaminants, with LEVs installed at questionable operations in enclosed areas. Table 6 provides the typical strategy (or strategies) followed to control some of the more common stressors found in the biomedical device manufacturing environment.

Control of the clean room environment in biomedical device manufacturing has been realized primarily through the standardization of equipment, facilities, and operational methods. These methods include procedural limits, operational limits, and testing procedures aimed at achieving internationally the desired environmental attributes to minimize microscale contamination. Clean rooms can have localized and enclosed forms of ventilation and contaminant removal or robust area HVAC systems capable of minimizing particle and microbial contamination. A comparison of the current international standards for classifications of clean rooms is given in Table 10.7.

Table 10.6. Control Strategies in Device Manufacturing

Category of Stressor(s)	Control Strategies
Ionizing radiation (Co-60 or various gamma and beta sources)	Time, distance, shielding, PPE, LEV
Organic degreasing solvents (TCE, 1,1,1 – TCA, perchloroethylene)	LEV, dilution ventilation, substitution
Acid etching agents (e.g., sulfuric acid, phosphoric acid, chromic acid)	LEV, PPE
Particles or aerosols from various operations	HEPA or ULPA filtration
Microbial stressors from indoor air quality issues	Dilution ventilation with control over HVAC; HEPA or ULPA
Adhesives and coatings	Process control (i.e., continuous operations over batch), robotic application
Nonionizing radiation (e.g., lasers, microwaves)	PPE, enclosure/isolation
Welding and soldering contaminants (e.g., metal fumes, flux, ozone)	LEV, substitution
Toxic gases and vapors (e.g., EtO, methanol)	Isolation/enclosure, LEV, substitution

Table 10.7. Comparison of Clean Room International Standards

International Standard Organization	Germany 2083	VDI USA 209D	Britain BS 5295	Australia AS 1386	France AFNOR NFX 44-101
3	1	1	C	0.035	–
4	2	10	D	0.35	–
5	3	100	E or F	3.5	4000
6	4	1000	G or H	35	–
7	5	10000	J	350	400000
8	6	100000	K	3500	4000000

10.19 Monitoring

The use of instruments and techniques to monitor for various hazards common to the device industry is an essential and complementary part of the overall exposure control process. While environmental monitoring will be conducted for various specific reasons, the real impetus behind this activity is to determine the extent of the facility contamination that exists in relationship to the worker and environment. The way that it is performed will normally depend on the actual (or perceived) stressors that are present as well as the existence of any outside pressures, such as regulatory compliance. The overall goal of an effective monitoring program, of course, is to keep the biomedical device employee and the environment free from the potential adverse health impacts from exposures to associated stressors.

While several authors have attempted to segregate or classify the various monitoring techniques and instrumentation in different ways, the simplest strategy for this discussion might be to just break these down into categories by environmental media (i.e., air, water, soil, artifact) and exposure target (i.e., personal or area). For example, if one wants know the extent of the contamination of a groundwater source from a spill of a tanker filled with chlorinated hydrocarbons, several monitoring wells could be installed with the necessary sensors for measuring these contaminants in real-time. This would be classified as an area (target) monitoring event for groundwater (media) contaminants. Since the scope of this chapter is on the engineering control and monitoring of stressors in the biomedical device manufacturing industry, the majority of the techniques and instruments covered will be for personal exposures to primarily airborne stressors.

Airborne chemical, physical, and biological stressors can be classified as primarily either particles, fumes, vapors, gases, or electromagnetic radiation, with the techniques or instrumentation used dependent upon the particular category. Additionally, monitors for sound pressure energy and heat stress merit adequate coverage due to their potential importance in the industry. The following sections provide an overview of the techniques and instrumentation commonly used in the device industry to monitor for environmental stressors.

10.20 Particle, Fumes, and Aerosol Monitoring

The generation of significant concentrations of particles and dust will occur anywhere there is human activity; obviously, the biomedical device industry is not immune. With such particle-producing activities required in medical device production as surface preparations and coatings, light machining, and welding/soldering, the appropriate particle control and monitoring efforts must be implemented and followed in order to protect employees from the potential harmful effects associated with airborne exposures.

Some of the more commonly used techniques for area particle monitoring involve the use of either laser optics or condensation nuclei counting. In contrast, the current best practice in measuring employee exposures to airborne particulate concentrations is to use a personal sampling device to collect a representative volume of potentially contaminated air, typically conducted over an eight-hour workshift. The monitoring is conducted in compliance to approved analytical methods, with subsequent shipment of the completed samples on to an accredited laboratory for analysis. The metric most often used to determine a relative exposure to microscale particles is the time-weighted mass concentration of each particular aerosol. Table 10.8 provides a summary of particle measurement techniques that are either currently in the developmental stages or have already been implemented in the workplace. The table includes the method, the metric measured, the sensitivity, and the major capabilities and limitations of each.

Table 10.8. Summary of Particle Measurement Techniques

Method or Instrument	Measurement Metric	Sensitivity (10^{-9} meters)	Major Capabilities and Limitations
Personal Sampler with accessories (e.g., cyclone, impactor, etc.)	Mass	0.02 mg/m^3	Acceptable for exposure compliance; no size fraction cut-off in nm size
Laser Particle Counter and other optics counters	Number Concentration	300	Portable and easy to operate; mainly for microscale use
Condensation Particle Counter (CPC)	Number Concentration	10	Portable and easy to operate; not size selective
Scanning mobility particle sizer (SMPS)	Number Concentration	3	Excellent sensitivity; not portable or user friendly and cost
Nanometer Aerosol Size Analyzer	Number Concentration	3	Excellent sensitivity; not portable and in development stage
MiPac Particulate Classifier	Number Concentration	10	Ease of use; no detection under 10 nanometers
Electrical low pressure impactor (ELPI)	Number Concentration	7	Successful use studies; cost and not portable
Epiphaniometer	Surface Area	NA (surface area)	Successful use studies; bulky, complex, and costly
Gas adsorption	Surface Area	NA (surface area)	Well understood technology; large samples sizes needed for validity
Scanning electron microscopy (SEM)	Number, size, and morphology	5	Excellent sensitivity and resolution; sophisticated instrumentation
Transmission electron microscopy (TEM)	Number, size, and morphology	1	Excellent sensitivity and resolution; complicated sampling routine
Laser Induced Plasma System	Composition	3	Outstanding for composition studies; composition information only

The first method discussed is a personal sampling device that is size selective. Currently, most analytical methods for particulate matter are based on the collection on a pre-weighed filter of any additional mass sampled at a known air flow rate. This is typically weighed on a laboratory balance and the full production shift (i.e, eight hours) detection limit is approximately 0.02 mg/m^3. Obviously, the use of this technique would present a problem in analysis if the air sample comprised mainly of just particles in the nanoscale range, weighing normally only a fraction of this amount. However, with all of this said, it has still been suggested that a size selective personal sampler could be developed with, for instance, a 100 nm cut-off point. This could provide some meaningful accuracy for measuring any coating- or surface preparation-originating aerosols above approximately 50 nanometers in diameter. Figure 10.7 provides an example of the type of personal air sampler that would be used to monitor for respirable dust. Note that a cyclone is attached in order to collect the respirable fraction of the mass.

The second through seventh methods provided in Table 10.8 are based on the number of particles counted. These methods include laser optical particle counter, condensation nuclei counters, scanning mobility particle sizers, and electrical low-pressure impactors. These are primarily real-time counters and range in relative portability, and subsequent applicability, to workplace exposure assessments. Also, several of these methods are still in the developmental stages.

Due to its portability, versatility, and lower detection size limit, laser particle counters have been traditionally used to measure particles down in the low microscale range. However, particles that are less than 300 nanometers will not be detected by this method. This

Fig. 10.7. Personal sampler used to monitor for respirable dust (Permission: SKC, Inc.)

limits the applicability in the biomedical device manufacturing industry, in particular, the clean room environment, where particles of concern are quite frequently found at an order magnitude smaller. There
are more sophisticated optical samplers but these are not currently
portable devices, and therefore, would not typically be used in this
industry to measure workplace exposures.

The most commonly used instrument capable of measuring ultrafine particles employs condensation particle counting technology. The
condensation particle counter (CPC) condenses vapor onto the sampled particles in order to "grow" them to a detectable size range.
This type of instrument is usually very portable and easy to operate.
The main disadvantage to using this type of instrument is that it is
not size selective and only provides the total particle counts above
the detection limit, which ranges from 3-100 nanometers on commercially available units. Figure 10.8 shows an example of a typical
CPC used to characterize ultrafine particles.

The measurement methods that are currently available, which provide both size-selective information as well as number concentration,
are inherently more complicated to use as well as not being very

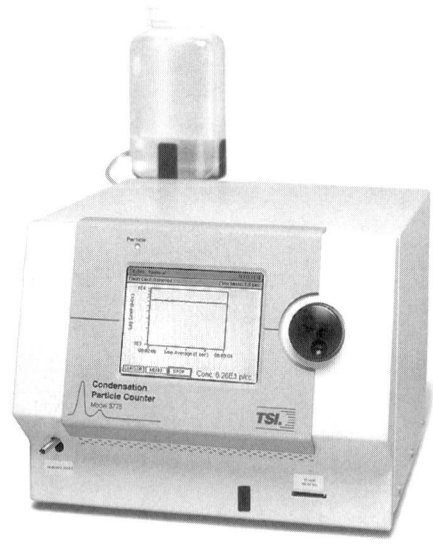

Fig. 10.8. Condensation particle counter (Permission: TSI, Inc.)

Fig. 10.9. Electrical Low Pressure Impactor (Permission: Dekati, Ltd.)

portable or versatile for field exposure assessments. In addition, their higher costs typically eliminate applicability altogether in the workplace. The best instrument examples of these methods are the scanning mobility particle sizer (SMPS) and the electrical low pressure impactor (ELPI). Both of these instruments can provide size-selective concentration data of particles all the way down to less than 10 nanometers in diameter. Examples of both an ELPI and a SMPS are provided in Figure 10.9 and Figure 10.10.

While microscale particles typically do not agglomerate significantly, the majority of nanoscale-sized particles generated do agglomerate to some extent. Therefore, it can be argued that the best way to characterize nanoscale particles is by the measurement of its surface area. The only instrument that has been currently employed to measure surface area is called an epiphaniometer. This instrument uses radioactive tagging to determine the particle's surface area. Again, this instrument is very complicated and lacks versatility for field use. Gas adsorption techniques that require rather large sample sizes have also been used infrequently as a bulk method of ascertaining particle surface areas.

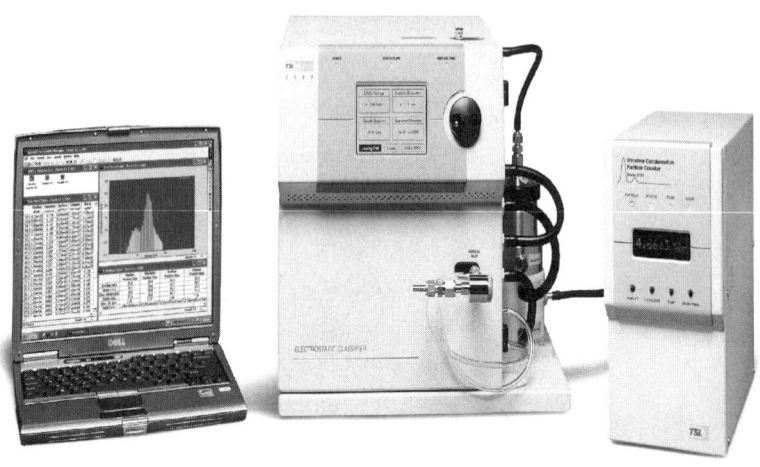

Fig. 10.10. Scanning Mobility Particle Sizer (Permission: TSI, Inc.)

Scanning Electron Microscopy (SEM) and Transmission Electron Microscopy (TEM) also provide the means for determining particle and dust characteristics. While these instruments provide the morphology of the particles and excellent resolutions (e.g., TEM = 1 nanometer; SEM = 5 nanometers), they are very expensive and usually require an expert technician or specialized training to use effectively. However, recent studies point to the merit of this technique to characterize exposures in the workplace.

Micro- and nanoparticle composition measurements are normally essential components for detailed particle research studies. Not unlike many of the number, size-selective, and surface area techniques previously discussed, most composition techniques are currently in the developmental stages. The laser-induced plasma system and the high temperature nanoparticle measurement systems can detect the composition of nanoscale particles as small as 3 nanometers.

Each of the portable methods for area monitoring that have been discussed has their own set of merits and limitations. In order to alleviate the negative impact posed by some of these specific limitations, the use of two or more of these techniques in combination may be considered. While the more sophisticated instruments have excellent resolution and many times both concentration and size-selectability, they are primarily limited to research settings due to their complexity, size, and costs. And, for the majority of end users, personal exposure sampling devices, like the one shown in Figure 10.7, coupled with one clean room portable particle measuring device, will be more than adequate.

10.21 Vapors and Gases

Techniques for monitoring the air for gases and vapors are essentially the same, thus, including them together in a section is appropriate. Both active and passive sampling methods exist to measure concentrations of many of the typical stressors found in device manufacturing such as ethylene oxide, trichloroethylene, phosphoric acid, and isopropanol (IPA). In order to conduct personal exposure monitoring for gases and vapors, an active sampling train such as the one shown in Figure 10.11 should be used. Once this sampling train is calibrated to a known volumetric flow rate, it can be attached to the worker in order to monitor breathing zone air or placed at a site of concern to conduct an area monitoring event. In each case, the active set-up will include a sampling pump, a calibrator, flexible hose, connectors, and some form of sampling media. The sampling media typically will capture the gas or vapor through sorbent action (e.g., adsorption, absorption). Depending on the compliance standard and protocol, a worker will normally be evaluated for his/her exposure for the whole workshift. Like particle exposure monitoring, once the sampling event is completed to accepted protocol, the filter is sent off to an accredited lab for quantification. Normal media material for monitoring airborne contaminants includes activated charcoal, silica gel, or a series of organic polymers.

There are now also passive methods for monitoring workplace and environmental exposures to some gases and vapors. For example, ethylene oxide, the common sterilizer, has a fully validated passive method of monitoring. The passive monitors are typically worn as badges or dosimeters. After the monitoring event duration is complete and in similar fashion to both loaded filters and absorbent media samples, the dosimeter or badge is packaged up and sent off to an accredited lab for subsequent analysis. While lab quantification is

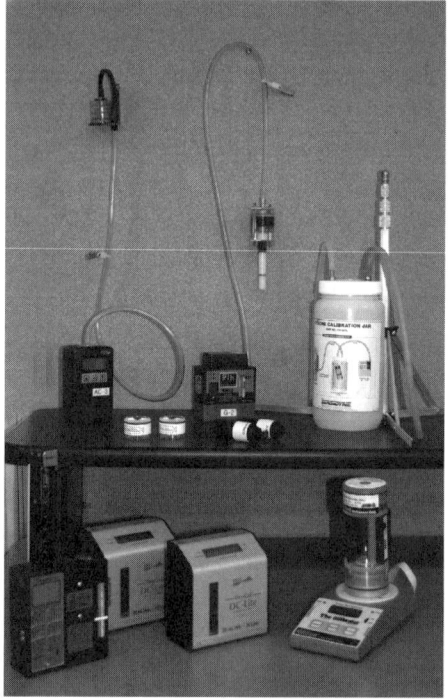

Fig. 10.11. Personal exposure air sampling equipment

still the usual means of analysis for compliance testing, there is a current impetus to test and approve direct-reading techniques for monitoring workplace exposures. Before a company considers conducting a comprehensive personal exposure assessment, it is recommended that an evaluation using direct-reading instruments and/or colorimetric techniques be performed to screen the potentially contaminated areas and develop a concentration profile of the facility.

There are several techniques for conducting this "snapshot" monitoring scheme, with normally a significant payoff realized by the company due to the useful information obtained. These instruments work on such principles as x-ray fluorescence, ionization potential, and chemiluminescence. These techniques vary on their relative accuracies and resolution and either provide qualitative, quantitative, or both qualitative and quantitative data to the end user. Due to their significance and usefulness, the following paragraphs elucidate the principles and operational characteristics of these instruments and techniques with respect to the common gases and vapors found in biomedical device manufacturing.

10.21.1 Detector (Colorimetric) Tubes

The use of detector tubes to provide a concentration profile in the workplace is common due primarily to economics. These tubes are available for a wide variety of organic and inorganic vapors as well as for common gases found in industry and in the environmental field. A monitoring event can be conducted with one or even several colorimetric tubes for a fraction of the cost of some of the available survey instruments calibrated to measure the same stressor. Typically, ten tubes are included in a package with easy instructions on how to use them effectively in the field. The principle of operation involves a colorimetric reaction between the sorbent material in the tubes and the gas or vapor being monitored. While this technique is an excellent means of determining whether or not a contaminant exists in appreciable concentrations, it cannot be used for compliance purposes for employee exposures. The tube has other colorimetric chemical interferences and a normal concentration accuracy of only about 60–70%. However, this technique provides an inexpensive means for detecting several common airborne contaminants in the biomedical industry. Colorimetric tubes are considered to be both a qualitative and pseudo-quantitative in nature. The list of contaminants identified and partially quantified includes methanol, isopropanol, ethanol, trichloroethylene, methyl chloroform, perchloroethylene, sulfuric acid, and phosphoric acid, to name a few.

For those just getting into the characterization phase of the worker exposure assessment program, the use of colorimetric tubes make the most sense initially. Additional instrumentation can be acquired subsequently, if needed, as the level of survey and assessment procedure becomes more sophisticated. Figure 10.12 provides an example of a detector tube used to measure methanol concentrations. Figure 10.13 (top, right) shows the latest generation of colorimetric detection device.

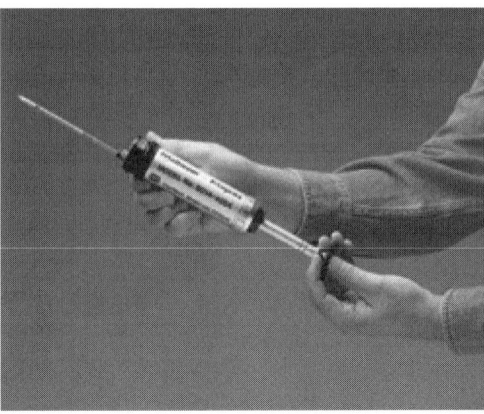

Fig. 10.12. Colorimetric tubes for field surveys (Permission: Terra Universal)

10.21.2 Photoionization Detectors (PIDs)

The photoionization detector, or PID, is another common means for detecting organic vapors and gases in the workplace. The PID works on the principle that vapors and gases will ionize if a sufficient source of energy (e.g., UV radiation from a lamp) is allowed to come into intimate contact with them in a chamber. The PIDs have a range of ionization energy potentials, with 9.5 eV, 10.2 eV, and 11.7 eV being the most common. The different energies are a means for differentiating between two or more contaminants coexisting in a particular airspace. However, the instrument does not have the ability to determine the different contaminants by any direct means. Thus, it is considered to be quantitative but not qualitative in nature. The ionization potential of common device industry stressors include the values of 10.56 eV for EtO, 10.10 eV for IPA, and 9.45 eV for TCE (NIOSH 2005). An example of a common PID is given in Figure 10.13 (top, left).

The biomedical device manufacturing professional might use this instrument if a known chemical hazard is possibly present at unsafe levels. As long as there are no other significant concentrations of different gases or vapors, which ionize at or under the ionization energy output of the device, a calibrated device should provide an accurate (e.g., within +/– 2 ppm or 10% of the reading) representation of the concentration existing at any point in time. Of course, this assumes that the known gas being evaluated ionizes at or under the lamps ionization energy. The normal resolution of a commercially available PID is 0.1 ppm. Common industry contaminants that can be characterized by a PID include the various sterilizers, cleaning solvents, coatings, and adhesives.

10.21.3 Flame Ionization Detectors (FIDs)

The flame ionization detector, or FID, is another means for detecting primarily organic vapors and also works on the principle of ionization potential. However, in contrast to the PID, the FID energy source is a hydrogen gas-initiated flame. Because of the hot flame, this instrument is capable of a wider range of ionization potentials. For this reason, it is used many times in conjunction with a PID out in the

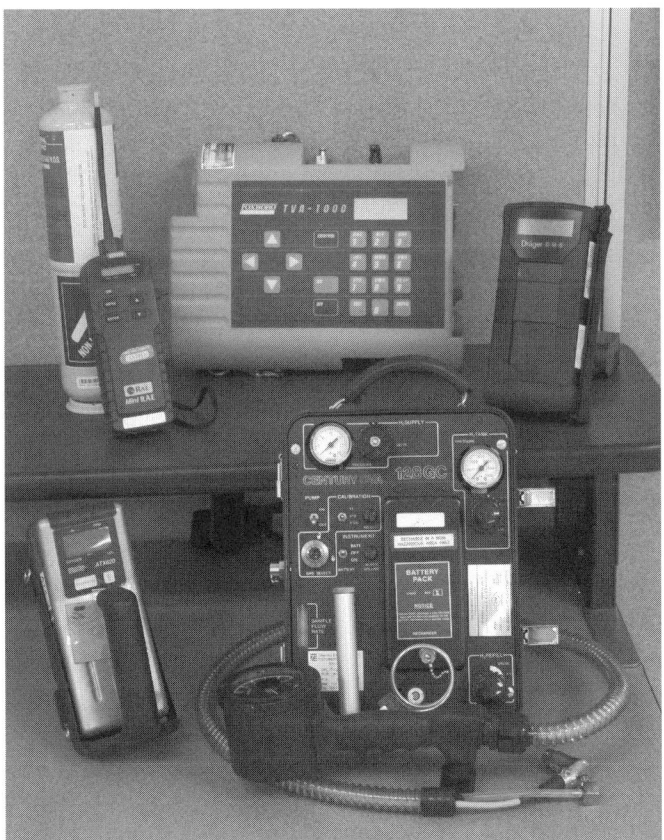

Fig. 10.13. Examples of field portable air monitoring instrumentation

environmental field where substantial concentrations of methane gas, with an ionization potential of 12.98 eV, may exist. Essentially, the two instruments are used in conjunction to identify both the methane concentration of the air as well as the concentration of organic vapors that exist. In essence, the trained user can differentiate the concentrations by subtracting the PID concentration (i.e., parts per million of organic vapors other than methane only) concentration from the FID output (i.e., parts per million of both organic vapors and methane) to get the total methane. Due to its use of an explosive gas as the energy source and complicated operational characteristics, this instrument is used mainly in the environmental field and would unlikely ever be considered for use in the biomedical device industry. Figure 10.13 (bottom, right) shows an example of a flame ionization detector coupled with a gas chromatograph column. Figure 10.13 (top, middle) also shows a combination PID/FID instrument for field surveys.

10.21.4 Electrochemical Sensor Monitors

Electrochemical sensor monitors are available from the original one gas monitor all the way up to the present day, five-sensor model. Typically, these monitors measure percent oxygen, percent lower explosive limit, hydrogen sulfide concentration, carbon monoxide concentration, and an end user gas concentration of choice. However, any of the sensors can be interchanged and the programming functions allow for customizing to the application. A common electrochemical multigas monitor to be used in the biomedical device industry might include sensors for ethylene oxide, % oxygen, hydrogen peroxide, ozone, and carbon monoxide. This device is considered to be both qualitative and quantitative, with accuracies of approximately +/– 5% and resolutions of 0.1 ppm. Figure 10.13 (bottom, left) provides an example of a multigas monitor.

10.21.5 Infrared Spectrophotometers

Portable infrared spectrophotometers are also available to measure airborne organic and inorganic contaminants. These are sensitive instruments with a series of mirrors that direct significant wavelengths within the unit. These are bulkier instruments that cost several times the amount of a PID, and thus, are currently in limited use in the biomedical device industry. However, this may change in the near future due to decreases in size and pricing, coupled with the instrument's inherent capability of providing relatively accurate qualitative and quantitative concentration measurements for many of the common airborne contaminants found in the industry.

10.21.6 Gas Chromatographs (GCs)

While gas chromatographs (GCs) have been used for years as lab bench top instruments to identify and quantify many organic compounds, the technology has only been portable for the last couple of decades. Like the infrared spectrophotometers, the inherent cost and complexity of operating these units have limited their usage in the field. Still, with only a few known contaminants and an experienced operator, this instrument provides a viable option to monitor many of the airborne contaminants found in the device industry. Figure 10.13 (bottom, right) provides an example of a commercially available GC unit, couple with a FID.

10.21.7 X-ray Fluorescence (XRFs)

Portable x-ray fluorescence instruments provide an alternative for measuring heavy metals in the environment. The units are very portable and use a radioactive gamma-emitting source to produce the characteristic x-rays for each heavy metal of concern. While being relatively costly, they have the added benefit of being approved for usage on an airborne lead compliance method, with additional methods targeting the measurement of other heavy metals currently in the validation stage. This unit would be used to monitor concentrations of heavy metals during such operations as welding or machining. An example of a typical portable XRF is provided in Figure 10.14.

10.22 Ionizing Radiation

The main types of ionizing radiation found in the biomedical device industry include gamma ray, x-ray, beta particle, alpha particle,

d

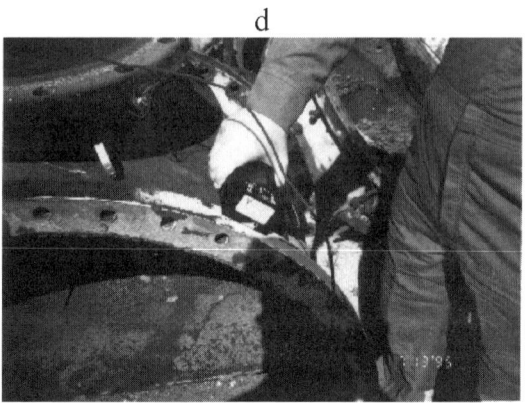

Fig. 10.14. Portable XRF being used in the field to characterize heavy metals (Permission: Niton)

and electron beam. The worker's occupational exposure to the various forms of ionizing radiation is typically monitored by the use of personal dosimeters and portable survey meters. Some of the most common operations for sterilizing medical devices employ the use of radioactive sources. In addition, radioactive tracers are many times used during the research phase of the device development, with sources also included as integral components in some the lab instrumentation.

Worker exposure could be monitored during the workshift by personal dosimeters, such as a thermoluminescent dosimeter (TLD). This is simply worn for the time period and then removed and analyzed by a qualified technician. Additionally, there are other active and passive dosimeters designed specifically to measure the dose of certain types of ionizing radiation to which an individual may be exposed. Another means for determining worker exposure is by performing a bioassay after the event. A detailed discussion on these various dosimeters and their appropriate uses is beyond the scope of this chapter.

As for the survey-type instruments, the type of detector will be chosen based on whether or not the source of radiation is gamma/ x-ray, beta, alpha, or electron beam. Normally, a Geiger-Mueller survey meter will be the instrument of choice, with various added capabilities such as data logging and programmability. Both a common

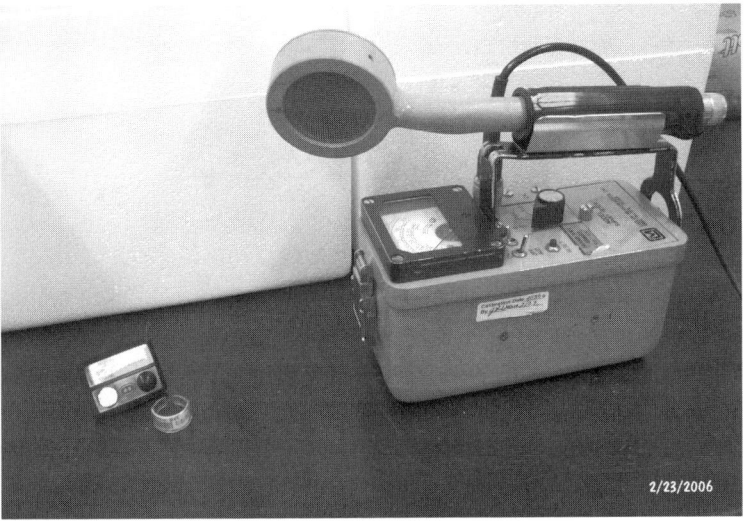

Fig. 10.15. Radiation dosimeter (left) and Geiger-Mueller counter for radioactivity surveys

radiation dosimeter and a Geiger-Mueller counter, with associated detector, are shown in Figure 10.15.

10.23 Non-Ionizing Radiation

The major forms of non-ionizing radiation that a worker may be exposed to in this industry include, but are not limited to, laser, ultra violet, microwave, and infrared. The majority of non-ionizing radiation monitors provide an output on the magnitude of both the electric and magnetic fields associated with their operations. Other important parameters that must be ascertained include the frequency, wavelength, duration of both signal and exposure, and power of the source. These are determined by various instruments and techniques and are not outputs from just a single monitor. Figure 10.16 shows a monitor capable of measuring the associated fields produced by a particular non-ionizing source.

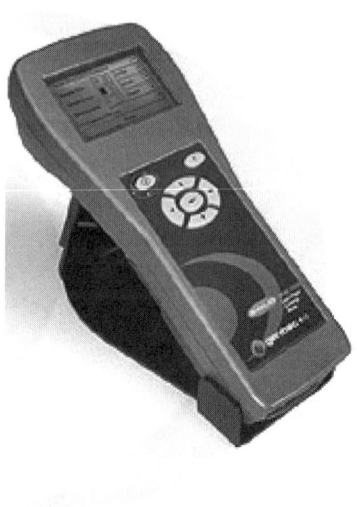

Fig. 10.16. Monitor for measuring electromagnetic radiation (Permission: Gentec-EO)

10.24 Noise and Heat Stress

Personal exposures to noise are monitored by noise dosimeters. They are typically attached to the worker and allowed to collect the data for the entire workshift. The value is then integrated over the duration and then compared to the acceptable time-weighted average for the time frame. To determine area noise levels or environmental noise, an instrument known as a sound level meter (SLM) is typically used to log the data. This instrument provides a means for determining a noise profile for a whole facility. Once the problem areas are identified, personal exposure monitoring is conducted for those potentially overexposed. Figure 10.17 provides examples of both typical noise dosimeters (middle) as well as a common type of SLM (right).

Heat stress monitoring is normally conducted using a wet bulb globe temperature apparatus. This instrument provides a combination measure of the effects of dry air temperature, radiant heat transfer, and humidity. While there is no universal standard for heat stress, there are guidelines that normally involve work-rest regimens and different work rates. While the use of this monitor in the biomedical device manufacturing environment would be atypical, there still could be conditions necessitating its use (e.g., laborious maintenance activities). A portable heat stress monitor is also shown in Figure 10.17 (left).

10.25 Microbial Environmental Monitoring

Significant concentrations of contaminants originating from biological origin are often found in the device industry. In particular, there is a major concern to limit the amount of microbial activity in the facility air (and on surfaces) due to the nature of the product. Biomedical device manufacturers use microbial environmental monitoring programs to evaluate the effectiveness of facility-wide cleaning and disinfection procedures as well as to assess the overall microbial cleanliness of their manufacturing environment. The ultimate goal must be to minimize the bioburden on the biomedical device being manufactured. If allowed to exist, undesirable bioburden spikes on the finished product can cause a reduction in the sterility assurance level for the device.

In order to manage and control the indoor environment with respect to biological stressors, usually it is necessary to get a baseline of contamination through an approved microbial environmental monitoring procedure. This procedure normally includes a sampling train made up of a calibrator, vacuum pump, Anderson impactor (or comparable), and necessary tubing and accessories. The sample media is usually an agar plate prepared with the proper substrate of microbial subsistence. Figure 10.18 gives an example of a typical biological stressor monitoring apparatus.

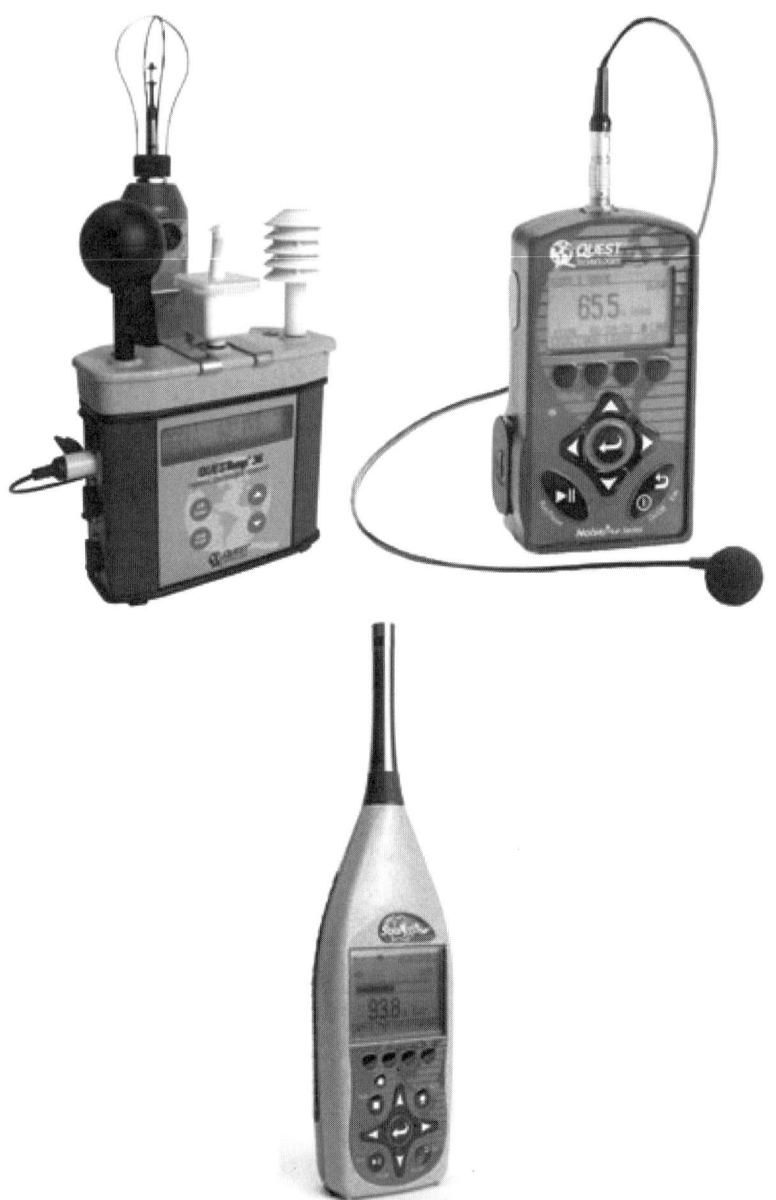

Fig. 10.17. Monitors for heat stress (left), noise dosimetry (center), and sound level measurement (Permission: Quest Technologies, Inc.)

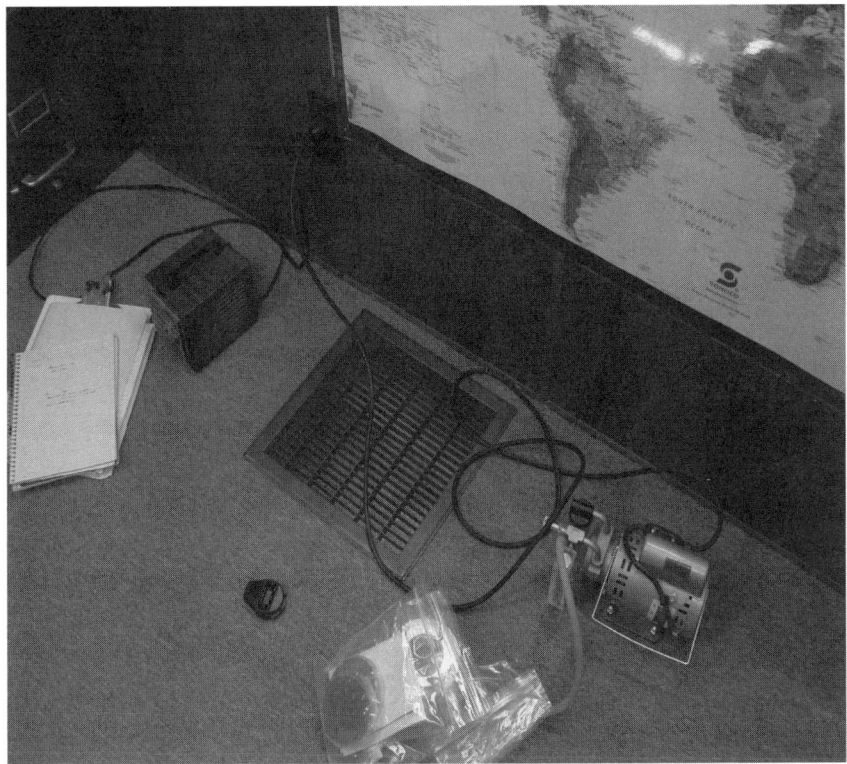

Fig. 10.18. Viable contaminant monitoring apparatus with agar plates

The available monitors, like the one shown in Figure 10.18, typically work on the principle of collecting a known volume of air at a certain flow rate and then, in much the same fashion as for chemical samples, the sampling media is shipped off to an accredited lab for subsequent analysis. The monitoring duration varies based on the collection procedure and objective for monitoring, but typically is less than twenty minutes per location. Since personal exposure standards are essentially nonexistent at this time from most biological contaminants, area monitoring is the only type currently conducted in the majority of cases.

10.26 Clean Room Monitoring Requirements

In the majority of biomedical device manufacturing settings, there will exist at least one condition or operation that requires a mandated clean room environment. Therefore, it is appropriate to discuss the monitoring requirements in order to meet compliance standards as a separate topic altogether in this section. Current international regulations require the monitoring of particulate matter and biological agents as a part of good manufacturing practice. The following paragraphs provide an overview of details surrounding proper clean room monitoring.

Table 10.9 provides the airborne classification used by the European Union as the guide to good manufacturing practice (EU 2003). Grade A and B correspond to a class 100 (USFDA) or ISO 5 clean room while Grade C corresponds to a class 10000 or ISO 7 clean room and a Grade D corresponds to a class 100000 or ISO 8 clean room. The particle monitoring requirements can either be completed manually or automatically, and typically measurements of the critical sizes (i.e., 0.5 micron and 5.0 micron) are taken with a laser-based detector with an isokinetic sampling probe. Normal flow rates and sampling duration at each monitoring location are 1.0 cubic feet per minute and twenty minutes, respectively; however, this rate and duration will vary on case-to-case basis. For clean rooms of ISO 5 and lower, the mandatory testing time interval to demonstrate compliance to particle limit standards is every six months, with those greater than ISO 5 required to be tested every 12 months. Air pressure differentials and air flow velocity is also required at 12 month intervals. Portable devices which measure both air velocity and pressure are commercially available and normally used to quantify these parameters. The number of monitoring locations per site varies based on class level and regulatory authority, respectively.

There are also limits to the microbial contamination that exists in a clean room environment. Thus, the monitoring of the clean room air for total microbes is essential to attain and maintain compliance. Figure 10.18 provides an excellent example of the monitoring set-up typically used to collect the colony forming units (cpu) for each air sample. Additionally, there are cpu sampling requirements for settle plates, contact plates, and gloves. The EU GMP for recommended limits at each grade level for microbial contamination for all four of these is shown in Table 10.10.

Table 10.9. Airborne Particle Classification in the EU Guide to Good Manufacturing Practice (Maximum Number of Particles Permitted Per Cubic Meter)

Grade	>=0.5 micron (at rest)	>=5.0 micron (at rest)	>=0.5 micron (in operation)	>=5.0 micron
A	3500	0	3500	0
B	3500	0	350000	2000
C	350000	2000	3500000	20000
D	3500000	20000	not defined	not defined

Table 10.10. Airborne Microbial Classification in the EU Guide to Good Manufacturing Practice

Grade	Air Sample cfu/m^3	Settle Plates 90mm diameter cfu/m^3	Contact Plates 55mm diameter cfu/m^3	Glove Print (5 fingers) cfu/glove
A	<1	<1	<1	<1
B	10	5	5	5
C	100	50	25	–
D	200	100	50	–

10.27 Monitor Selection in Device Manufacturing

The choice of the best monitor for the application is based on such factors as the type and physical state of the stressor, environmental media contaminated, concentration of the contaminant, goal of the monitoring effort, and overall economics. If the monitoring event is just to provide an area characterization of the extent of contamination, many of the direct-reading instruments and techniques may suffice. However, if the goal is to show environmental or occupational regulatory compliance, then personal monitoring techniques, with subsequent accredited lab testing and analysis, is the correct choice. Table 11.11 pairs up some of the more common industry stressors with the instrument or technique normally used to identify and/or quantify each. The second column of Table 11.11 also identifies whether or not the instrument or technique is typically used for area, snapshot/screening, or personal monitoring events.

Table 10.11. Common Stressor and Instruments/Techniques Normally Used to Characterize

Chemical, Physical, or Biological Stressor	Instrument or Technique
Cleaning and disinfecting alcohols (e.g., methanol, isopropanol, ethanol, etc.)	PID, detector tubes (area or snapshot/screening) Sampling pump with silica gel media (personal or area)
Chlorinated hydrocarbons for cleaning and degreasing (e.g, TCE, Perk, etc.)	PID, detector tubes (area or snapshot/screening) Sampling pump with activated charcoal media (personal or area)
Ethylene oxide from sterilization	Electrochemical sensors (area or snapshot/screening) Sampling pump with activated charcoal media or passive badge (personal or area)
Ozone from sterilization, welding activities, and air cleaners	Electrochemical sensors, detectors tubes (area or snapshot/screening) Sampling pump with coated glass filter media (personal or area)
Hydrogen peroxide from disinfecting and sterilization	Electrochemical sensors, detector tubes (area or snapshot/screening) Sampling pump with impinger (personal or area)
Heavy metals from welding and machining operations	X-ray fluorescence (area, snapshot/screening, or personal) Sampling pump with 0.8 micron mixed cellulose filter media (personal)
Particulate matter and dust from a myriad of sources	Laser counters, condensation nuclei counters (area or snapshot/screening) Sampling pump with cyclone and filter media (personal)
Acids and alkalis used for etching and anodizing	Detector tubes, IR spec (area or snapshot/screening) Sampling pump with activated charcoal media (personal)
Gamma radiation from sterilization and instrument-specific sources	Geiger-Mueller survey meter with gamma detector (area and survey) TLD, film badge, or other dosimeter (personal)
Fungal and bacterial species from indoor air quality problems	Impactor with high volume sampling pump and agar plate (area and survey)
Noise from plant and maintenance activities	Sound level meter (area and survey) Dosimeter (personal)

10.28 Summary

The effective monitoring and control of chemical, biological, and physical stressors in the biomedical device manufacturing environment is imperative. This chapter began with an overview of the sources, properties, and characteristics of some of the more common contaminants found in the industry. Once these stressors were identified and characterized, examples of some of the more typical techniques implemented for controlling their presence in the workplace and the potential risk of exposures associated with each were discussed. The chapter concluded with a synopsis of the potential monitoring techniques and instruments that may be used as a means to qualify and quantify the extent of contamination and potential employee exposures to the hazardous stressors inherent to biomedical device manufacturing.

References

1. American Conference of Governmental Industrial Hygienists (2005) Threshold limit values for chemical substances and biological exposure indices. ACGIH, Cincinnati
2. American Conference of Governmental Industrial Hygienists (2001) Air sampling instruments 9th ed. ACGIH, Cincinnati, OH
3. American Conference of Governmental Industrial Hygienists (1995) Industrial ventilation: a manual of recommended practice 22nd ed. ACGIH, Cincinnati
4. Burgess WA et al. (1989) Ventilation for control of the work environment. John Wiley & Sons, New York
5. Burton DJ (2003) General methods for the control of airborne hazards. In The Occupational Environment: Its Evaluation, Control, and Management 2nd Ed. (S.R. DiNardi). AIHA, Fairfax, VA
6. Conviser S (2000) The future of ethylene oxide sterilization. Infection Control Today – 06/2000, http://infectioncontroltoday.com/articles/061feat.html
7. Environmental Science and Technology Online (2005) It's in the microwave popcorn, not the Teflon® pan. Science News report November 16th
8. European Agency for Safety and Health at Work (2006) Indicative Occupational Exposure Limit Values http://europa.eu.int/comm/employment_social/health_safety/docs/oels_en.pdf
9. European Environmental Agency, http://www.eea.eu.int
10. European Union (1996) Council directive 96/29/EURATOM. Official Journal L159 – 29/06/1996, pp 0001-0114

11. European Commission Guide to Good Manufacturing Practice Revision to Annex 1(2003) Manufacture of Sterile Medicinal Products, http://pharmacos. eudra.org
12. Gaggeler HW et al. (1989) The epiphaniometer, a new device for continuous aerosol monitoring. J Aerosol Science 20(6):557-564
13. George DK et al. (2003) An introduction to the design of local exhaust ventilation systems. In The Occupational Environment: Its Evaluation, Control, and Management 2nd Ed. (S.R. DiNardi). AIHA, Fairfax, VA
14. Institute of Occupational Medicine for Health & Safety Executive 2004 Research Report 274, http://www.hse.gov.uk/research/rrpdf/rr274.pdf
15. International Commission on Non-Ionizing Radiation Protection (2002) General approach to protection against non-ionizing radiation. Health Physics 82(4): 540-548
16. International Organization for Standardization, http://www.iso.org
17. International Organization for Standardization (2000) Cleanrooms and associated controlled environments-part 2: specifications for testing and monitoring to prove continued compliance with ISO 14644-1, ISO 14644-2:2000. ISO, Geneva
18. International Organization for Standardization (1999) Cleanrooms and associated controlled environments-part 1: classification of air cleanliness, ISO 14644-1:1999. ISO, Geneva
19. Heinsohn RJ (1991) Industrial ventilation: engineering principles. Wiley-Interscience, New York
20. Hitchock et al. (2003) Nonionizing radiation. In The Occupational Environment: Its Evaluation, Control, and Management 2nd Ed. (S.R. DiNardi). AIHA, Fairfax, VA
21. Kittleson DB (1998) Engines and nanoparticles: a review. J Aerosol Science 29(6):575-578
22. Maslansky CJ, Maslansky SP (1993) Air monitoring instrumentation. Van Nostrand Reinhold, New York
23. Maynard AD (2004) Examining elemental surface enrichment in ultrafine aerosol particles using analytical scanning transmission electron microscopy. Aerosol Science & Technology 38(4):365-381
24. National Institute of Safety and Hygiene at Work, http://www.mtas.eu/insht/en/principal/insht_en.htm
25. National Institute of Occupational Safety and Health (2005) Pocket guide to chemical hazards. US Dept. of Health and Human Resources, Washington, DC
26. Nickerson R, Sheu M (2000) Plasma cleaning of medical devices. Critical Cleaning in Precision Manufacturing
27. Reitz V (2004) Sterilization for beginners. Medical Design Magazine – 06/2004, http:www.medicaldesign.com/articles/ID/12003
28. Ritter GW (2000) Using adhesives effectively in medical devices. Medical Device and Diagnostic Industry Magazine, http://devicelink.com
29. Salditt P (2004) Trends in medical device design and manufacturing. J SMT 17(3):19-24

30. SKC, Inc. (2004) Comprehensive catalog and sampling guide. SKC Inc., Eighty Four, PA
31. Soule RD (1991) Industrial hygiene engineering controls. In Patty's Industrial Hygiene and Toxicology 4th ed. (Clayton and Clayton). John Wiley & Sons, New York
32. The InfoShop (2003) Biomedical applications of nanoscale devices. The InfoShop report September 12th
33. US Environmental Protection Agency, http://www.epa.gov
34. US Occupational Safety and Health Administration, http://www.osha.gov
35. Washington University-St. Louis, Missouri (December 2005) Dr. Da Ren Chen's faculty webpage, http://mesun4.wnstl.edu/me/faculty/dc/dc_reseach.pdf

11. Biomaterial-Cell-Tissue Interactions In Surface Engineered Carbon-Based Biomedical Implants and Devices

11.1 Introduction to Surface Engineered Carbon-Based Materials

Implantable prosthesis and medical devices are subjected to several interacting forces whenever they come in contact with the physiologic systems (blood, immune, musculoskeletal, nervous, digestive, respiratory, reproductive, and urinary) and organs of the human body. These interactions include the effects of core body temperature (and/or variable temperatures in the oral cavity), the body physiologic fluids containing several ions and bio-molecules, proteins and cells of various progeny and functions. Pathological diseases of the arteries and the heart that cause life threatening blood flow restrictions for example can be treated either by open heart surgery intervention, by implantation of intra coronary stents and/ or by use of artificial devices like artificial heart valves in heart valve pathologies. However, in spite of considerable advances in improving the mechanical properties of stents, advances in implants techniques, and advances in anti-thrombosis therapy, the use of stents and heart valves are still complicated by substantial cases of thrombotic occlusions/stenosis and restenosis [1–4] due to platelet activation resulting from the release of metallic particles/ions (in metallic stents), shear forces and blood contacting of the metallic surface [5–8]. Likewise, thromboembolism (valve thrombosis and systemic embolism) remains the major draw back in the management of implanted mechanical heart valve prostheses [9–10]. Patients with these implanted prostheses are faced with life threatening bleeding problems because they are kept under life-long anticoagulant therapies in order to reduce the risk of thromboembolism. Platelet aggregation in these prostheses is the key factor in thrombus formation and dissemination as emboli which can be life threatening, if not promptly managed. In order to reduce the risk of platelet aggregation/thromboembolism and complications following the life-long course(s) of anticoagulants, the biomaterials need to be improved in order to achieve better biocompatibility/ hemocompatibility [11–13].

Apart from thrombosis, other problems associated with the failure of medical implants and devices that need to be overcome are problems of mechanical failure, wear, tear and or fatigue; the problem of chemical degradation, corrosion and oxidative degeneration; the problem of calcification and the problem of excessive immune response and/ or infection as triggered by these biomaterial implants. Metallic implants may have good fatigue life and may be cheap (stainless steel for example) but they can release metallic ions and wear debris into the surrounding tissues leading to osteolysis, loosening and/ or failure of the implants. All these problems encountered with implanted prostheses and medical devices could be solved with carbon thin film surface coating modifications using appropriate, durable and biocompatible biomaterial.

The passive nature of carbon in tissues has been known since ancient times. Charcoal and lampblack were used for ornamental and official tattoos by many. Other forms of carbon have been studied for possible use for biomedical applications stimulated primarily by Gott's original studies [14]: artificial graphite, vitreous or glassy carbons, carbon fibres, pyrolytic carbons, composites, and vacuum vapour deposited carbon coatings [15]. The fundamental nature of these carbon materials and their interactions with the living tissues needs to be explored therefore. Likewise, pyrolytic carbon coated heart-valve leaflets have been successfully applied as artificial clinical heart valves [16]. Pyrolytic carbon has the major advantage of being resistant to thrombus formation, which was the biggest limitation to the earlier generation of stainless steel artificial heart-valves. There was always a need to use anticoagulant drugs by patients who had the earlier stainless steel heart valves to prevent clot formation on the stainless steel, but this had the potential to suppress the beneficial effect of the natural blood clotting mechanism in patients. Pyrolytic carbon is an artificial material made of carbon microcrystals with a high-density turbostatic structure, originally engineered for use in nuclear reactors and may not be readily available for large scale use. When pyrolytic carbon is alloyed with silicon, it shows excellent thrombo-resistance [16]. Pyrolytic carbon has also been used as a coating on different types of implant prostheses, such as dental implants, per-cutaneous devices, and tendon tracheal replacements. This chapter tends to reports the state of the art in the potential of a more readily available synthetic carbon based thin film coatings (DLC and its doped hybrids) for thrombo-resistant applications and various biomedical applications as stated below.

Surface coating modification is essential because it is now known that the outermost layer of a biomaterial (few nanometre scale range) is most crucial in its interfacial interaction in-vivo. Baier *et al* (1972) have shown that exposure of an organic-free surface to fresh flowing blood for as little as 5 seconds leads to its complete coating by a very uniform, tenaciously-adherent proteinacious thin film [17]. The biomaterials outer surface dictates the configuration of this attached protein film that in turn plays an important role in determining the fate of the biomaterial in the host via a series of cascade- interactions. Thus intensive research has been focused on DLC over the last few decades due to the ability of forming ultra-thin films of DLC and due to the promising characteristics of DLC like attractive tribological, electrical, chemical, and optical properties. The issue of biocompatibility and hemocompatibility of DLC when used as implants and medical devices will no doubt be expected to stem from possibly favourable tissue-biomaterial surface and interactions. Generally, two main pathways could be feasible, in an effort to create a biocompatible material: creating a material with surfaces that are bioactive (like the host tissue) that can actively support the body's control mechanisms; and or creating materials that are 'inert or passive' to the body's control mechanism in order to avoid triggering an adverse reaction (though there being nothing exactly like absolute inertness in a hostile physiological environment [18]). Moreover, it is impossible to exaggerate though, that among a large number of events occurring during the process of blood coagulation (including perhaps thrombus formation and possible thromboembolism) and de-coagulation, the physicochemical adsorption is virtually the main reaction that can be readily regulated unless a bioactive material is used [19]. Thus the present chapter reports on researches into the use of DLC thin films (and its doped hybrids) as an approach to modify the physico-chemical properties of biomaterials and to create a material that may be 'passive' and/ or 'bioactive' in the tissue.

The endothelial lining has been reported to be the best non-thrombogenic surface [20–21]. According to Herring *et al* [22] and [23] various methods including improvement of physico-chemical properties, pre-treatment with proteins and incorporation of negative charges have been proposed in order to reduce the surface thrombogenicity of vascular prostheses. Pesakova *et al* [24] and [25] have also stated that the biocompatibility of materials can be influenced by factors such as surface charge, hydrophobicity, and topography. It has been reported by Ahluwalia *et al* [26], and [27] based on surface potential measurements using the vibrating Kelvin probe method, that positively charged surfaces enhance cell adhesion in comparison to neutral or negatively charged surfaces. The hydrophyllic or hydrophobic nature of a surface has also been associated with extent of cell interactions with the surface [26, 28]. Altankov and Groth (1997) [28] have reported that wettable (hydro-philic) surfaces tend to be more conducive for cell adhesion. Grinnell (1978) [29] also reported observing cell adhesion to occur preferentially to water wettable surfaces. Van Wachem *et al* [30–31] carried out investigation of in-vitro interaction of human endothelial cells with polymers of different wettabilities in culture, and reported observing optimal adhesion of cells with moderately wettable polymers. The biocompatibility and hemocompatibility of Diamond-Like Carbon (DLC) films has been investigated in the literature [32–40]. Jones *et al* (1999) [33] deposited DLC coatings prepared by PECVD on titanium substrates and tested them for hemocompatibility, thrombogenicity and interactions with rabbit blood platelets, and reported that the DLC coatings produced no haemolytic effect, platelet activation or tendency towards thrombus formation and that platelet spreading correlated with the surface energy of the coatings. Cytotoxicity tests have also been carried out on DLC coatings by [41–43] amongst others, and they all reported observing no negative effects by DLC coatings on the viability of cells which showed normal metabolic activities like cell adhesion and spreading. Mouse fibroblasts grown on DLC coatings for 7days showed no significant release of lactate dehydrogenase [44], an enzyme that catalyses lactate oxidation, often released into the blood when tissue is damaged, compared to control cells, indicating no loss of cell integrity. It has also been reported by Allen *et al* [45–46] that mouse macrophages, human fibroblasts and human osteoblast-like cells grown on DLC coatings on various subst-rates exhibited normal cellular growth and morphology, with no in vitro cytotoxicity.

Szent-Gyorgyi [47–48] suggested that proteins may have an electronic structure similar to that of semiconductors. Eley *et al* [49] reported semi-conductivity in certain proteins. This was later corrobo-rated by works from others like that of Postow and Rosenberg [50]. Bruck in 1973 suggested that intrinsic- semi-conduction and electronic conduction may be involved in blood compatibility of polymeric systems [51–54], instead of mere ionic interaction, after the compa-tibility of the blood with surfaces has been associated chiefly with ionic charges, based on the observation that endothelial wall, platelets, and plasma proteins carry net ionic charges in normal physiologic conditions. Bruck [51–52] observed clotting times six to nine times longer than those observed with non-conducting polymers and also observed little or no platelet aggregation in electro-conducting poly-mers, when compared to non-conducting control samples based on his study with pyrolytic polymers. Bruck concluded that 'it is possible that electro-conduction and semi-conduction is involved in the interaction of surfaces with plasma proteins in the activation of the Hageman factor and the platelets by an unknown mechanism' [53].

Since changes in surface energy [27], surface charge conditions [26] and electronic conduction [52] have all been suggested to have an effect on the biocompatibility and hemocompatibility of materials, the present investigation is directed at understanding the effect of the above identified factors in DLC biomaterials on especially the human micro-vascular endothelial cell compatibility, the platelet interaction (thrombus formation) and the cytotoxicity of various cell lines (e.g., retinal pericytes, V79, L132, etc). These factors were changed by modifying the DLC films by doping with e.g., Si, N, for example which is known to alter the surface energy, surface charge condition and electronic conduction in DLC films. This investigation has the final aim of assisting with furthering the understanding of the under-lying physics of material interactions in a biological environment, and the potential to develop Si-DLC as a more readily available alter-native to pyrolytic carbon and to discover and exploit the fundamental role of the electrical property and the surface energy of DLC bio-material, so as to use it as a key to turn the potential biocompatibility and hemocompatibility of DLC into any desirable (particular) bio-medical application. Surface Topography is another important factor in biomaterial-cell-tissue interaction. Surface engineered and patterned surfaces are now commonly created and used model for studying biomaterials-cell-tissue interaction. This philosophy is based on the spatial distribution/extension of cellular appendages (cell spreading) and cell-cell/cell-material interactions based on this communication channels as one of the cell signalling modalities. Cell-cell interactions and cell-material interactions are thought to be physicochemical and electrochemical.

The physicochemical surface properties of the outermost interface of bacteria for example and other particles as well as phagocytic cells, can essentially be of only two kinds: (a) interfacial tension and (b) electrical surface potential [55]. When a foreign surface, solid, liquid, or gas, is brought into contact with the body tissue fluid/protein solution, a certain amount of the dissolved protein will be adsorbed to the surface. This process is consistent with the Gibbs theory of surface energy and may be described by the adsorption isotherm of Freudlich or Langmuir. The amount of protein adsorbed and the characteristics of the protein monolayer depend mainly on the nature of the foreign surface and the structure and the concentration of the proteins in solution [56]. The protein needs to first approach a distance to the foreign surface that will allow interaction between the molecular forces associated with the foreign surface and the protein- this is governed by diffusion. Then the characteristics of the foreign surface determines the nature of bonds or the type of changes that takes place in the configuration of the protein and the biologic molecule present.

In this chapter the biocompatibility and hemocompatibility of carbon based thin film materials is reviewed and cathegorised based on certain models already existing as well as models being proposed by the author, under specific interaction with various proteins and cells due to possible differences in particular cell behaviour, and also taking into consideration whether the cells used for the test is human or animal cells and if the test is in-vitro, ex-vivo and or in-vivo. These tests can equally be categorised under direct cell-biomaterial interactions to estimate the number of cells adherent (cell adhesion and/ or cell aggregation for platelets), cell characteristics on the surfaces (cell spreading for example) and assays to determine the level of enzymes from the intracellular compartment. It is to be noted that some cells that actually exist in the blood when used for toxicity test or examined for cellular proliferation on the biomaterial are actually a test of biocompatibility rather than hemocompatibility, although almost all hemocompatibility test may equally be used to assess degree of biocompatibility, biocompatibility tests may not be used to assess hemocompatibility.

11.2 Potential Biomedical Applications of DLC

Very well adherent and appropriate DLC coating could be chemically inert and impermeable to liquids. They could therefore protect biological implants against corrosion as well as serve as a diffusion barriers. DLC films are considered for use as coatings of metallic as well as polymeric biocomponenets to improve their compatibility with body fluids [57–59]. The potential biomedical applications of DLC and modified DLC include protection of surgical prostheses of various kinds: intra-coronary stents [60–61]; prosthetic heart valves [62, 42, 9, 10, 63]. The new prosthetic heart valve designed by FII Company and Pr. Baudet is composed of a Ti6Al4V titanium alloy coated with DLC [42]. When artificial heart organ polymers used for making heart organs are compared to DLC-coated polymers, these polymers seem to show higher complement activation compared to DLC counterpart (Polycarbonate substrates coated with DLC, PC-DLC compared with Tecoflex, polyurethane) [64]. DLC and modified DLC can be used in blood contacting devices e.g., rotary blood pump [65]. DLC is now being investigated for anti-Prion protective coating on surgical instruments and as well as anti-MRSA bug in hospital utensils where appropriate due to its known hydrophobic and low surface energy properties.

DLC is equally being investigated as a template/model surface for DNA writing and immobilisation for biochemical sensor applications. Specific bio-molecules could be easily immobilised on the DLC. Surface immobilisation of DNA was reported to have several advantages: (1) Minimizing the amount of DNA needed to achieve a desired effect and enhancing effective concentration vector;

(2) Preventing DNA/vector aggregation; (3) reducing toxicity and degradation of delivered particles; and (4) Delivering DNA to specific cellular sites [66]. DLC-Ag-Pt nano-composites were reported to exhibit significant antimicrobial efficacy against staphylococcus bacteria, and to exhibit low corrosion rates at the open-circuit potentials in a PBS electrolyte [67].

In orthopedics DLC can be used for coating orthopedic pins [57] and coating of hip implants (e.g., femoral heads) [68–70]. DLC can reduce the wear of the polyethylene cup by a factor of 10–600 when used on metal implants to form a DLC-on-DLC sliding surface. The wear (and the amount of particles causing a foreign body reaction) is 10^5–10^6 lower compared to metal on metal pairs. The corrosion of a DLC-coated metal implant can be 100, 000 times lower than in an uncoated one. DLC can diminish the bone cement wear by a factor of 500, which can improve the bone cement to implant bonding [71, 69, 70].

In urological dialysis (hemodialysis), DLC coated micro-porous polycarbonate and DLC coated dialysis membranes show that DLC imparted an enhanced enzyme electrode performance [72–73]. DLC has also been reported to do well in both organ [74] and cell culture [75] when compared to the materials conventionally used for this purpose. DLC can be used as active barrier against attack by micro-organisms and against bio-deterioration of advanced technological devices operating in closed spaces of satellites, air-crafts, and sub-marines for examples [76] and as good protectors against environmental pollutants and atmospheric wastes [77]. In addition nanocrystallite copper modified DLC has been reported to have a fungicidal effect [78].

11.3 Definitions and General Aspects of Biocompatibility

Biocompatibility can be defined as 'the ability of a material to perform with an appropriate host-response in a specific application' [18]. Four components of biocompatibility have been identified:
1.   The initial events that take place at the interface, mainly including the adsorption of constituents of tissue fluids onto the material surface [79];

2. Changes in the material as a result of its presence in the tissues, usually described under the headings of corrosion or degradation [80];

3. The effects that the material has on the tissue, the local host response [81];

4. The sequalae of the interfacial reaction that are seen systematically or at some remote sites [82].

Possible tests useful for evaluation of biocompatibility are listed below:

LEVEL I:

- Initial Screening and Quality Control of Polymers
- Agar overlay response of materials
- Agar overlay response of materials extracts
- Inhibition of cell growth by water extracts of materials
- Intradermal irritation test for materials extract and leachable components

LEVEL II:

- Initial Evaluation of Novel Biomaterial
- Tissue culture test on materials
- Tissue culture test on materials extracts
- Cell growth in contact with test materials
- Hemolytic activity test
- Intramuscular implantation of material
- Test of osmotic fragility of erythrocytes
- In vitro mutagenicity test
- Test of material extracts by perfusion of isolated rabbit heart

11.3.1 Specie Differences

1. The dog model is relatively inexpensive and convenient, but may lack relevance to humans;

2. The baboon model is relatively expensive and unavailable, but is relevant to humans; and

3. Similarity of platelet function, the concentration and activity of clotting factors, and the hematocrit should be the primary determinants for deciding which species are most relevant to humans.

11.3.2 Cell Specificity

It is important to note that different cells perform different functions and thus their interactions with same biomaterial may differ. Specific biomaterial-cellular interaction can be compared with that of another cell-line if the cell type, origin and function are very similar.

11.4 Blood

The blood is a fluid connective tissue with a matrix called plasma. The plasma proteins are in solution unlike the other connective tissues that occur in insoluble forms like fibers, thus the proteins in solution in the plasma makes the plasma slightly denser than water. The blood is composed of plasma (46–63%) and formed elements (37–54%). The plasma is composed of water (92%), plasma proteins (7%), and other solutes (1%). The formed elements of blood is composed of red blood cells 99.9%, and the remaining 0.1% platelets and white blood cells. The water 'dissolves' and transports organic and inorganic molecules, formed elements and heat from one part of the body to the other. The plasma proteins are composed of albumins (60%), globulins (35%, transports ions, hormones, lipids; immune function), fibrinogen (4%) and regulatory proteins (<1%: enzymes, proenzymes, hormones). Other solutes of blood is composed of electrolytes (major ones are Na^+, K^+, Ca^{2+}, Mg^{2+}, Cl^-, HCO_3^-, HPO_4^{2-}, SO_4^{2-}), organic nutrients (lipids:fatty acids, cholesterol, glycerides; carbohydrates mainly glucose; and amino acids) and organic wastes (eg., urea, uric acid, creatinine, bilirubin, ammonium ions). Albumin is the major contributor to osmotic pressure of plasma and transport lipids and steroid hormones. Fibrinogen is an essential component of blood clotting system that can be converted to insoluble fibrin [84].

11.4.1 Definitions and General Aspects of Hemocompatibility

The European Society for Biomaterials Concensus Conference following the considerations of the Macromolecule Division of the International Union of Pure and Applied Chemistry (IUPAC) thought that definition of blood compatibility should take into account the following [81]:

- The activation of the blood coagulation system at the blood-material interface;
- The response of the immune system induced after the blood-material contact;
- Other tissue responses which appear as consequences of the blood-material contact.

At the conference they proposed to define four properties characteristic of the biomaterial's blood compatibility:

♦ Thrombogenicity
♦ Antithrombogenicity
♦ Complement activation ability
♦ Complement inhibition ability

The understanding of the process of coagulation occurring during injury to the blood vessel wall like a cut may give some background idea of the events that may apply to the blood-material interaction. The process of haemostasis, the cessation of bleeding and establishment of a framework for tissue repair consists of three phases: the vascular phase, the platelet phase and the coagulation phase, all of which occurring in a chain reaction. The vascular and platelet phases occur within a few seconds after the injury, but the coagulation phase does not start until 30 seconds or more after the vessel wall has been damaged. When the blood vessel wall is cut in an injury for example, contraction of the vessel's smooth muscle is triggered locally (vascular spasm) and this decreases the diameter of the vessel wall. This vascular spasm and constriction helps to stop the loss of blood and lasts for about 30mins- a period known as the vascular phase. During this phase also changes occur in the local vessel endothelium: the endothelial cells contract and expose the underlying basement membrane to the blood stream; the endothelial cells begin releasing chemical factors and local hormones (e.g., ADP, tissue factor, prostacyclin, endothelins); and then the endothelial cell membranes become 'sticky'. Platelets now begin to attach to the sticky endothelial surfaces, to the basement membrane, and to the exposed collagen fibres. This attachment marks the start of the platelet phase of the haemostasis- platelet adhesion, activation, platelet aggregation and the formation of a platelet plug. Platelet aggregation begins within 15 seconds after an injury occurs. As the platelets arrive at the injury site, they become activated, change shape, become spherical and develop cytoplasmic processes that extend towards adjacent platelets. The platelets begin releasing ADP (stimulates platelet aggregation and further secretion from the platelets), thromboxane A2 and serotonin which stimulate vascular spasm, clotting factors, PDGF (platelet derived growth factor, a peptide that promotes vessel repair) and calcium ions (required for platelet aggregation and by several steps in clotting process). The platelet phase proceeds rapidly, because the ADP, thromboxane and calcium ions that each arriving platelet releases stimulate further platelet aggregation. Finally the blood coagulation occurring during the coagulation phase involves a complex sequence of steps that leads to the conversion of circulating fibrinogen into the insoluble protein fibrin which forms a growing network that covers the surface of the platelet plug [84]. Listed below are various hypotheses proposed for blood-biomaterial interactions by several authors.

The interfacial blood-biomaterial interactions: some 'conventional wisdom' and some 'unresolved hypotheses' adapted partly from [85–87]:

11.4.2 General Hypothesis

The overall process of in vivo thrombogenesis, thromboembolization and subsequent endothelialisation on a foreign surface are dominated by surface properties rather than hemodynamics (or by hemodynamics rather than by surface properties).

11.4.3 Material

1. A material with a critical surface tension of about 25dyn/cm will have a low thrombogenic potential.
2. A small negative surface charge lowers material thrombogenicity.
3. High water content materials have a low thrombogenic potential due to the lowered free energy of the hydrated interface.
4. High water content materials may continually expose a fresh foreign interface, leading to a high thrombogenic potential;

however, they also tend to exhibit low thromboadherance due to their low interfacial free energy.

5. H-bonding group in a surface lead to strong interactions with biological species and therefore endow a surface with a high thrombogenic potential.

6. A surface with a high apolar/ polar ratio is desirable for low thrombogenic potential

7. Thrombus is nucleated in regions of the surface where a specific spatial distribution of specific chemical (electrostatic) groups is present.

8. Flexible (as opposed to stiff) polymer chain ends and loops in the material interface lower the thrombogenic potential of the foreign surface.

11.4.4 Material and Hemodynamics

1. Thrombi will always be generated at surface imperfections due to flow disturbances, surface compositional differences, and/or trapped gas bubbles.

2. Smooth surfaces in arterial flow conditions may release thromboemboli before they grow too large to be dangerous (corollary: high shear rates can detach thromboemboli before they have grown too large).

3. Certain rough and textured surfaces may form and retain fibrin, thrombus, leading to a 'passivated' surface.

11.4.5 Hemodynamics

1. Thrombi will always be generated in regions of low flow or flow separation.

2. Low shear rates can lead to regional accumulation of activated protein coagulation factors and subsequent thrombogenesis on a nearby surface.

3. High shear rates can be destructive to blood cells (e.g., shear rates can initiate platelet activation and lead to thrombogenesis).

4. In a tubular flow field, the platelets tend to accumulate preferentially near the wall and the red cells near the central core

(Corollary: the red cells enhance the rate of collision of platelets with the wall).

11.4.6 Erythrocytes and Leucocytes

1. The role of leucocytes in thrombogenesis may be related to their ability to recognize a particular biomaterial surface as 'foreign' after certain proteins and or platelets have adhered to that surface.
2. Red blood cells may play only a minor role in the thrombogenic process.

11.4.7 Blood Cells and Protein Surface Tensions

1. Surface tensions of cells and proteins can be measured by a variety of techniques.
2. Surface tension of cells and proteins are relatively high, they tend to be hydrophilic in their natural state [88]

11.4.8 Heparinised Surfaces and Drugs

1. The natural endothelium is non-thrombogenic because endothelial cells produce the powerful anti-platelet aggregation agent prostacyclin (PGI_2).
2. There may be synergistic interaction between specific drug therapies and specific biomaterials such that reduced drug regimens may be indicated in combination with the use of specific biomaterials in devices or implants.
3. Heparinised surfaces must leach heparin to be non-thrombogenic. (General corollary: 'immobilised' anti-thrombogenic drugs are ineffective unless they leach into the flowing blood.)
4. Heparinised surfaces that bind antithrombin III do not need to leach heparin to be non-thrombogenic.

11.4.9 Calcification

1. Calcification may be initiated at points of high mechanical strain in a foreign material (such as blood pump diaphragm).

2. Calcification in foreign materials is a biological process; γ-carboxy glutamic acid is a necessary amino acid in one key protein involved in this process.

11.4.10 Surface Charges

1. Under normal conditions, the blood vessel wall and blood cells are negatively charged (potentials across the blood vessel wall were measured using Ag-AgCl reference electrodes: under normal conditions the inner electrode was negative with respect to the outer electrode) [89]
2. Injury to the blood vessel wall reduces the magnitude of the negative charge density and very often even causes a reversal in the sign of the surface charge (injury is generally accompanied by thrombus formation) [90].
3. A decrease in pH reduces the negative charge density of the blood vessel wall and of blood cells. The isoelectric point occurs at pH ~ 4.7 (The pH of the electrolyte has a significant effect on the surface charge of the blood vessel walls. At pH ~ 4.7 the blood vessel wall has zero surface charge and below this pH the blood vessel wall is positively charged) [91].
4. Antithrombogenic drugs increases the magnitude of the negative charge density, whereas thrombogenic drugs have the opposite effect and in many cases even reverse its signs (electrophoresis measurements conducted on RBC and WBC in the presence and absence of anti-thrombogenic and coagulant drugs shows similar actions of these drugs on both blood cells and blood vessel wall) [92].
5. Positively charged prosthetic materials are thrombogenic whereas negatively charged surfaces tend to be non-thrombogenic-the higher and the more uniform the negative charge density, the better is the chance of the material being non-thrombogenic (tubes of various metals were implanted in the canine thoracic aorta or the canine thoracic inferior vena cava- metals which have negative standard electrode potential tended to function longer in dogs than those which registered positive potential; with insulator materials using streaming potential to determine surface charge characteristics and using untreated,

chemically treated and electrically treated Teflon tubes, the more negative the surface the more useful the material) [93].

11.5 Cell Culture/Seeding Peculiar to Each Cell

11.5.1 Human Microvascular Endothelial Cells (HMEC-1)

Human Microvascular Endothelial Cells (HMEC-1) were recovered from the molecular biology departmental bank of the University of Ulster. Cell cultures were maintained in MCDB-131 supplement with L-Glutamine (200mM), 10% fetal Calf Serum (FCS), epidermal growth factor (EGF) (10ng/ml) and Penicillin (20IU/ml), streptomycin (20μg/ml). Cells were grown as mono layers in tissue culture flasks at 37°C under 5% CO_2 / 95% air. Proteins are removed with two washings of phosphate buffered saline (PBS: 8.2g/l NaCl, 3.1g/l $Na_2HPO_4.12H_2O$, 0.2g/l $NaH_2PO_4.2H_2O$; pH 7.4). Harvesting of cells for sub-culturing or testing was performed with a trypsin solution (0.05% trypsin/0.02% EDTA), shortly afterwards, the trypsin was inactivated with the culture media and by centrifugation supernatants separated from the cells. Cells were used when cells were about confluent under exponential growth phase. The samples were sterilized with 70% ethanol before they are taken into the hood and given sufficient time to dry inside the hood for the experiment and afterwards rinsed with PBS or distilled water. Every normal culturing sterility precaution was taken through out the experiment. Approximately 4.0×10^5 cells/ml were seeded on top of the silicon wafers (placed inside the Petri dishes) of the a-C:H samples. About 5.5 $\times 10^5$ cells/ml were seeded into the rest of the samples, and about 1×10^3 cells /ml were seeded into the wells of the 96-well culture plates that were coated with a-C:H and Si-DLC. The un-coated samples were used as control. For the MTT-assay some control wells were also created and marked blank by the computer program. The MTT-assay was carried on for about 56-hours, some of the test using silicon wafer substrates were carried on for up to about 36-hours, and the rest for 6-hours (length of time indicated as the case may be).

11.5.2 Human Platelets

Whole blood was taken from normal healthy individuals into a standard tube with anticoagulant (3.8% sodium citrate). Centrifugation at 1200rpm for 5mins to get PRP (platelet rich plasma) and at 3000rpm for 10mins to get PPP (Platelet poor plasma) was done as soon as possible. The platelet number in the PRP was diluted with PPP to ~1×10^8cells/ml by mixing PRP with PPP. The samples were sterilized with 70% ethanol before they were taken into the hood and given sufficient time to dry inside the hood. Every normal culturing sterility precaution was taken through out the experiment. About 1×10^8 cells/ml were seeded on top of the a-C:H and Si-DLC samples placed inside the Petri-dish. Incubation was done at 37°C under 5%CO_2 / 95% air for the 15mins, 30mins and 75mins. Afterwards proteins were removed with two washings of phosphate buffered saline (PBS) before the fixation. The cells on the silicon wafer substrates were fixed with 2.5% glutaraldehyde in 0.1M phosphate solution, followed by 1% osmium tetraoxide in 0.1M phosphate solution. The samples were dried with increasing concentrations of ethanol successively and finally with hexamethyldisilaxane (HMDS).

11.5.3 Pericytes Cell Line

The cells are normal bovine retinal pericytes isolated from the eye at the University of Ulster biomedical science department (see appendix for procedure). The cells used for the test was of passage number 4 (not to exceed passage number 6 in this particular cell line). The media used for cell culturing was made up with DMEM (500mls), FCS (100mls), fungizone (5mls) and Pen/Strept (2mls). Cells were grown as mono layers in tissue culture flasks at 37°C under 5% CO_2 /95% air. Proteins are removed with two washings of phosphate buffered saline (PBS: 8.2g/l NaCl, 3.1g/l $Na_2HPO_4.12H_2O$, 0.2g/l $NaH_2PO_4.2H_2O$; pH 7.4). Harvesting of cells for sub-culturing or testing was performed with a trypsin solution (0.05% trypsin/0.02% EDTA), shortly afterwards, the trypsin was inactivated with the culture media and by centrifugation supernatants separated from the cells. Cells were used when cells were about confluent under exponential growth phase. About 1–2×10^3 cells/ml were seeded into the various 96 well plates (coated with a-C:H, or Si-DLC, or uncoated control TCPS) for MTT-assay. The samples were sterilized with 70% ethanol before they are taken into the hood and given sufficient time to dry inside the hood for the experiment. Every normal culturing sterility precaution was taken through out the experiment.

11.5.4 Human Embryonic Lung, L132 Cell Lines

The cell line originally purchased from ATCC (CCL-5), is epithelial and normal but with a Hela-characteristics. The passage number is important so the passage number (P33) of the cells used for the test were under the normal passage number. The culture media was composed of MEM (500mls, with L-glutamine), FCS (50mls), sodium pyruvate (5mls), Penicillin/streptomycin (5mls) and non-essential amino acids, NEAA (5mls). Cells were grown as mono layers in tissue culture flasks at $37^{\circ}C$ under 5% CO_2/95% air. Proteins are removed with two washings of phosphate buffered saline (PBS: 8.2g/l NaCl, 3.1g/l $Na_2HPO_4.12H_2O$, 0.2g/l $NaH_2PO_4.2H_2O$; pH 7.4). Harvesting of cells for sub-culturing or testing was performed with a trypsin solution (0.05% trypsin/0.02% EDTA), shortly afterwards, the trypsin was inactivated with the culture media and by centrifugation supernatants separated from the cells. Cells were used when cells were about confluent under exponential growth phase. The samples were sterilized with 70% ethanol before they are taken into the hood and given sufficient time to dry inside the hood for the experiment, and afterwards rinsed with PBS or distilled water. Every normal culturing sterility precaution was taken through out the experiment. About 12×10^3 cells/ml were seeded into the various 96 well plates (coated with a-C:H, or Si-DLC, or uncoated control TCPS) for MTT-assay.

11.5.5 V79 Cell Lines

The Chinese hamster fibroblast like normal cell-line (V79) was originally bought from ATCC (V79–4; CCL–93). The culture media was composed of DMEM (500mls), FCS (50mls), Penicillin/streptomycin (5mls), NEAA (5mls) and sodium pyruvate (5mls). Cells were grown as mono layers in tissue culture flasks at $37^{\circ}C$ under 5% CO_2/95% air. Proteins are removed with two washings of phosphate buffered saline (PBS: 8.2g/l NaCl, 3.1g/l $Na_2HPO_4.12H_2O$, 0.2g/l NaH_2PO_4. $2H_2O$; pH 7.4). Harvesting of cells for sub-culturing or testing was performed with a trypsin solution (0.05% trypsin/0.02% EDTA), shortly afterwards, the trypsin was inactivated with the culture media and by centrifugation supernatants separated from the cells. Cells were used when cells were about confluent under exponential growth phase. About $1-2 \times 10^3$ cells/ml were seeded into the various 96 well plates (coated with a-C:H, or Si-DLC, or uncoated control TCPS) for MTT-assay. The samples were sterilized with 70% ethanol before they are taken into the hood and given sufficient time to dry inside the hood for the experiment. Every normal culturing sterility precaution was taken through out the experiment.

11.6 Statistics and Counting of Cells

11.6.1 Cell Fixation and Drying

The cells on the silicon wafer samples were fixed with 2.5% Glutaraldehyde in 0.1M phosphate solution for 5 minutes, followed by 1% Osmium Tetraoxide in 0.1M phosphate solution for 5 minutes. The samples were successively dried with increasing concentrations of ethanol and finally with hexa-methyldisilaxane (HMDS).

11.6.2 Gold-Platinum Coating for Charging Compensation

The samples in silicon wafer substrates were coated with a conducting material (gold-platinium), after cell fixation and drying, to about 30nm thickness, using Polaron E5000 SEM coating unit, in order to reduce charging and obtain a better contrast during SEM imaging.

11.6.3 SEM Imaging of Cells

A low vacuum scanning electron microscope (SEM) Hitachi S-3200N was used for the observation of the cells interaction with the Si-DLC (a-C:H:Si) films on silicon substrates. The conditions for SEM-imaging were high secondary electron (HSE), aperture 3 (#3), $0°$-tilt, scan–4, 5.0kV, and $\times 200$ magnification.

11.7 Stereological Investigations

The cells number counting over an area of 600μm–400μm on the SEM image of various samples were performed using the UTHSCSA, ImageTool programme developed in the department of dental diagnostic science at the University of Texas Health Science Center, San Antonio, by C. Donald Wilcox, S. Brent Dove, W. Doss McDavid and David B. Greer [94].

11.7.1 Stereological Investigation and Statistical Analysis (Endothelial & Other Cells)

Human microvascular endothelial cells (HMEC-1) were obtained from the molecular biology department of the University of Ulster at Jordanstown, Northern Ireland. Cell cultures were maintained in MCDB-131 supplement with L-Glutamine (200mM), 10% foetal calf serum (FCS), epidermal growth factor (EGF) (10ng/ml) and Penicillin (20 I.U/ml), streptomycin (20μg/ml). Cells were grown as monolayers in tissue culture flasks at $37°C$ under $5\%CO2/95\%$ air. Proteins were removed with two washings of phosphate buffered saline (PBS). Harvesting of cells for sub-culturing or tests was performed with a trypsin solution. Shortly afterwards, the trypsin was inactivated with the culture media and by centrifugation supernatants were separated from the cells. Cells were used when they were about confluent and under exponential growth phase. The samples were sterilised with 70% ethanol before they were taken into the hood and given sufficient time to dry inside the hood (and afterwards rinsed with PBS or distilled water). Every normal culturing sterility precaution was taken throughout the experiment. Approximately, 4×10^5cells/ml were seeded on top of the DLC and Si-DLC samples placed inside the petri-dishes, and about 1×10^3cells/ml were seeded into the wells of the 96-well culture plates that were coated with DLC and Si-DLC.

The uncoated samples were used as control. For the MTT-assay some control wells were also created and marked blank. The cells for the MTT-assay were seeded for a total of about 56-hours. The cells on the silicon wafer substrates were fixed with 2.5% glutaraldehyde in 0.1M phosphate solution for 5mins, followed by 1% osmium tetraoxide in 0.1M phosphate solution for 5minutes. The samples were dried with increasing concentrations of ethanol successively and finally with hexamethyldisilaxane (HMDS).

11.7.2 Stereological Investigations and Statistical Analysis (Platelets)

The cell number counting over an area of 600μm×400μm on the SEM image of various samples were performed using the Image-Tool program [94]. In order to cover a statistically reasonable large area of the tested film area, several x200 (higher magnifications cover only very small sample area) images of the SEM were acquired (the SEM stage with the mounted samples were moved ≥2mm in both X and Y direction before SEM was acquired in order to avoid scanning same area twice) for counting of the number of platelet aggregates. This is because platelets are numerous (~300,000 cells/μl) and the average size of platelets is very small (~2μm). In estimating the number of adherent platelets, high magnification (x1500) is used to enable visibility of platelet full morphology and counting. However, at such high magnifications, the sample areas covered by the SEM scan become so small that the estimation of the number of adherent platelets may become subjective if several images were not taken at different spots for each sample and the analysis of these painstakingly recorded and averaged. The effect of silicon doping on the mean of the number of platelet aggregates counted on the films examined was tested using a two-tailed heteroscedastic t-test to compare group means for the un-doped and silicon doped films that had an unequal variance in the number of platelet aggregates counted. Statistical significance was defined as a p-value of < 0.05.

11.8 Photo-Fluorescent Imaging of Cells/Tissues

In photo-fluorescent imaging, fluorophores (fluorescent compounds: e.g., fluoroscein, rhodamine, luminol) are incorporated to the cell/tissue to be examined in order to label the sample. Fluorophores naturally absorb light at one wavelength and emit at a longer wavelength following some energy transitions. A fluorophore on absorbing a photon has its outer shell of electron excited from a ground to an excited energy level, this energy can be released as a thermal radiation (with the electron returning to the ground state) or part of this energy can be transferred to the molecular environment (e.g., cell/tissue) with the remainder photon energy emitted (photon with less energy and longer wavelength). This absorption and emission of photons creates contrasts which can be utilized in the imaging process to image structures at a molecular level in either live or fixed cells/tissues. The temperature and the micro-molecular environment do affect the fluorescent process, making different molecules of the same fluorophore to release different amounts of energy and creating a range of emission spectrum. These days, the molecular structure of the fluorescent material can be modified to target specific regions or molecules within a cell or tissue in order to study the entity of interest. Thus the chosen photo-fluorescent molecule can be used to label specific proteins and/ or transfect (genetically alter) cells in order to insert for example a fluorescent peptide called green fluorescent protein (GFP) in the native protein of the cell.

11.8.1 Typical Sample Preparation for Photo-fluorescent Microscopy

Place samples in 6 well plate in triplicate and seed with 1×10^5 cells (3 ml media) and incubated at 37°C and 5% CO_2 for 24 h. Fix cells using a solution of 4% paraformaldehyde and 2% sucrose in phosphate buffered saline (PBS) for 20 mins; bovine serum albumin (BSA) in PBS and then permeabilise using buffered 0.5% Triton X-100 (0.5% Triton supplemented with 20 mM Hepes buffer, 300 mM sucrose, 50 mM NaCl and 3 mM $MgCl_2$) by chilling to 0°C for 5 min. Wash samples again with 1% BSA/PBS. Add TRITC-conjugated phalloidin solution at a concentration of 10μg/ml, in 1% BSA/PBS, for 20 min at room temperature. Then remove the phalloidin, wash samples and mounted on microscope slides with glycerol. Observed samples with, for example, a Confocal laser scanning microscope (CLSM). Finally analyse cell shape factor and cell spreading area from the micrographs.

The software calculates the shape factor using the formula 4 π (area/perimeter)^2 which gives a value between zero and one, a value of one being a perfect circle.

11.9 Biocompatibility and Hemo-compatibility Models

Several models for assessing biocompatibility and haemo-compatibility of materials in-vitro exist (Figure 1). These models are used in an attempt to find a platform for explaining what actually goes on at the material-bio-interfaces. Interestingly all these are influenced by electron and ionic exchanges at these interfaces. The balance (ratio) of non-adhesive to adhesive proteins for example say albumin-fibrinogen ratio are thought to depict how these models tend to overlap and therefore all together may be important in interpreting material-bio-interfacial interactions.

11.9.1 Proteins-adhesive and Non-adhesive Proteins

This model is based on adhesive verses non-adhesive protein interactions on surfaces, and/or Vroman effect of protein adsorption, where proteins and the adhesion molecules compete for specific binding sites. It is now known that proteins either present in serum or secreted by the cells play a key role in the adhesion and spreading of the cells on the substrate biomaterial. The existing hypotheses are as follows [85–87]:

1. Protein adsorption comprises the initial interaction of a foreign material with blood (only the outermost ~1nm range of the surface is involved in the interfacial interaction. Exposure of an organic-free surface to fresh flowing blood for as little as 5 seconds, leads to its' complete coating by a very uniform, tenaciously-adherent proteinacious film. Thus the focus of attention has shifted from the substrate as the inducers of thrombogenicity, to the substrate as the dictators of a special configuration of adsorbed protein molecules that will favor or inhibit the subsequent events, including activation of the clotting mechanism and adhesion of platelets [95–96].

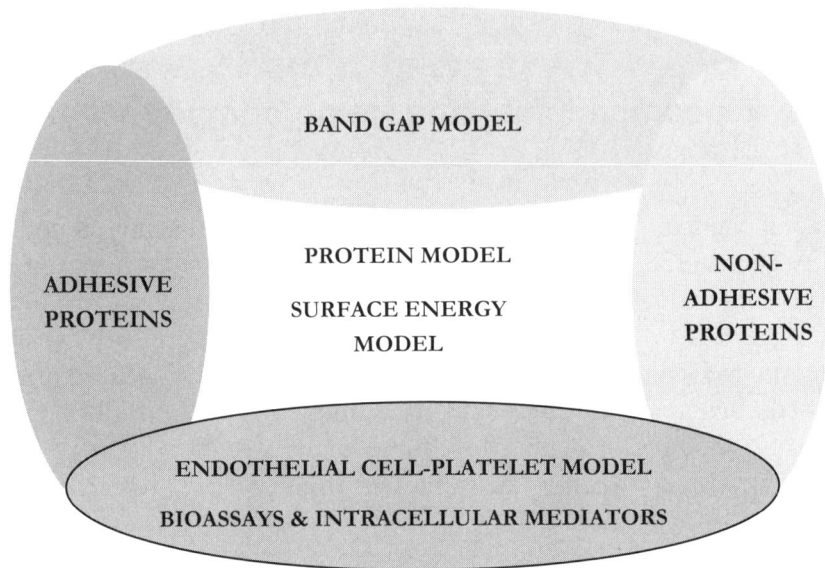

Fig. 11.1. Overlapping models for assessing material-bio-interactions

2. The composition and organization of the initial protein layer is determined by the surface properties of the material.

3. The composition and organization of the initial protein layer mediates subsequent platelet interactions in vivo, and may also determine long-term effects.

4. In vitro protein adsorption studies are relevant to in vivo behaviour in humans.

5. The heat evolved on adsorption of proteins can lead to their denaturation on the surface; the magnitude of the heat evolved may be determined by the surface composition

6. Hydrophobic surfaces will tend to adsorb protein more 'strongly' than hydrophilic surfaces, leading to greater denaturation of proteins on the hydrophobic surfaces.

7. Fibrinogen dominates the initial protein layer on most foreign materials and fibrinogen adsorption leads to high thrombogenic potential for that surface.

8. A layer of adsorbed albumin reduces in vitro platelet adhesion; materials that preferentially adsorb albumin will be antithrombogenic in vivo.

9. Certain other specific proteins may also adsorb and have a significant influence on subsequent events (e.g., CIG, or fibronectin, VWF, complement factors, high molecular weight kininogen, lipoproteins etc.)

10. The various carbohydrate components of adsorbed glycolproteins may play an important role in the recognition of the biomaterial as foreign and in the subsequent events leading to thrombus deposition.

11.9.2 Surface Energy Model

Materials surface energy (and/or contact angle) is known to influence protein and cellular interactions on these surfaces. The actual interaction superseding depends on the properties of the surface and/ or the bio-molecule(s) arriving on the surface. Some of the existing hypothesis is stated above (Section 4) for example: 'a material with a critical surface tension of about 25dyn/cm will have a low thrombogenic potential'. It is important to know that that the contact angle based of single liquid cannot give a good or accurate indication of the material's surface chemistry. There is need to use more than one liquid for measuring the contact angle, essentially a polar liquid on one hand and a dispersive liquid on the other. Surface energy of the material can be calculated from the contact angles values obtained using more than one liquid if the appropriate equations are applied. The calculated surface energy can equally be resolved into the surface energy components (by the use of appropriate equation/method e.g., DLVO (Derjaguin-Landau-Verwey -Overbeek), Owens-Wendt, Van Oss-Chaudhry-Good): the Lifshitz-van der Waals dispersive, the polar and the acid-base components. Equally the hydrophobic and the hydrophilic forces can be assigned. It may be possible to relate the biocompatibility of various materials to the appropriate surface energy components (and/ or the total surface energy), the hydrophobic or hydrophilic energies and to determine the major contributing component in the biocompatibility behavior.

11.9.3 Band-gap Model

Both materials' surfaces and the interacting bio-molecules have got some electronic properties and more so when the two different interacting surfaces come into close ranges/contact. The band gap of the biomaterial under investigation can be related to that of the interacting bio-molecule or protein. It is possible to predict the biomaterial-bio-molecule interactions based on the distribution of electrons (or density of states, DOS), or contact potential difference, work function and/or band gap. The complexity and dynamic nature of these interactions have to be taken into account. If some electrons move from their occupied valence band level in the bio-molecule to the free state of the biomaterials surface, it is expected that the bio-molecules morphology could change or denature. This is only possible where the energy gap at the biomaterial-bio-molecule interface allows a charge transfer. Chen *et al* [97] studied the hemocompatibility of $Ti(Ta^{+5})O_2$ and reported an improved biocompatibility based on the band gap of $Ti(Ta^{+5})O_2$ being 3.2eV compared to 1.8eV of fibrinogen arriving on the surface. Thus because the band gap of fibrinogen is within that of $Ti(Ta^{+5})O_2$, it is not possible to effect electron transfer form the protein to the materials surface, and less amount of fibrinogen become adherent on the surface, which subsequently led to less adherent platelets.

11.9.4 Surface Topography, Roughness and Patterning

Surface topography, roughness and patterning have been implication in altering protein adhesion and conformational changes. Studies creating various patterns on surfaces have indicated the implication of having various features of shape, size and depth dimensions on the surfaces to the degree of information gained on cellular and developmental biology [98]. It is almost impossible to change surface pattern without changing the chemical, physical and biological interactions. The author is of the opinion that this parameter is important in understanding the processes of developmental biology rather than directing dictation biocompatibility interactions. The experimental data presented in this chapter by the author is based on using samples with same and similar ultra smooth surfaces having a non-statistically different surface topography, roughness and patterns.

11.9.5 Endothelial-Platelet Model

The seeding of bovine thoracic endothelial cells on cellulose surfaces with increasing hydrophobicity resulted in increased endothelial cell adhesion and proliferation and decreased migration [99]. Investigation on endothelial-specific cell adhesion to peptide sequences on different extra cellular matrix (ECM) molecules grafted on to various surfaces reveal that the arg-glu-asp-val (REDV) sequence from fibronectin was selective for the adhesion of endothelial cells but not fibroblasts, smooth muscle cells, or activated platelets where other sequences like arg-gly-asp (RGD), tyr-ile-ser-gly-arg (YISGR), or pro-asp-ser-gly-arg (PDSGR) were implicated [100]. The material-micro-vascular endothelial cellular interaction could be related inversely to those of platelets (in-vitro and in-vivo) since increased platelets aggregation/adhesion on a material could be associated with increased potential of a material to induce clotting [36], whilst increased endothelial-material adhesion could be associated on the other hand with an increased potential of a material not to induce clotting [40] (Figure 2). Further detail of this model is implied in the discussions below [35–36, 39–40].

11.10 Carbon-based Materials Interaction with Selected Proteins and Cells

When a cell is coming in contact with a biomaterial the degree of interaction can generally be taken as adsorption, contact, attachment and or spreading [29]. For activated platelets five stages of spreading can be described according to the increasing degree of activation [101]:

1. Round or dicsoid with no pseudopodia;
2. Dendritic, early pseudopodial with no flattening;
3. Spread-dendritic, intermediate pseudopodial with one or more flattened pseudopodia, but with no spreading of the cell body;
4. Spreading, late pseudopodial with the cell body beginning to spread;
5. Fully spread morphology, the cell is well spread with no distinct pseudopodia.

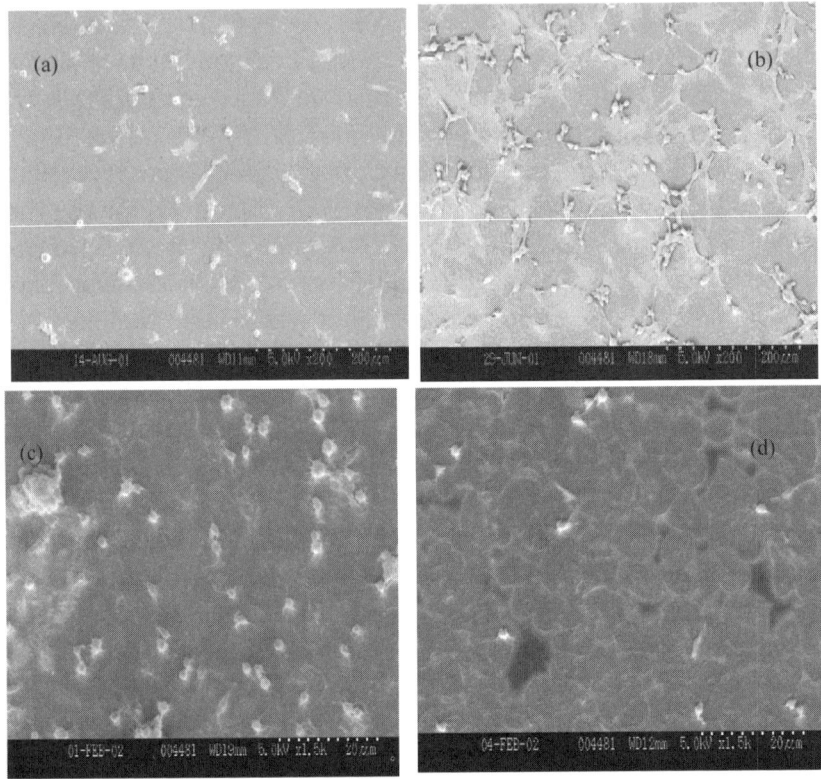

Fig. 11.2. Endothelial-Platelet model: (a) Endothelial Cells on DLC, (b) Endothelial cells on Si-DLC, (c) Platelets on DLC and (d) Platelets on Si-DLC thin films

11.11 DLC Interactions with Fibroblasts In-Vitro

The body's connective tissues can be classified as connective tissue proper (subdivided into loose and dense connective tissue proper), fluid connective tissues (subdivided into blood and lymph) and supporting connective tissues (subdivided into the cartilage and the bone). Fibroblasts are the most abundant permanent residents of the connective tissue proper and are the only cells always present in it. Fibroblasts secrete hyaluronans (a polysaccharide derivative) and proteins, both of which interact in the extracellular fluid to form proteoglycans that make ground substances viscous. They also secrete protein subunits that interact to form large extracellular fibers which could create loose/open framework or densely packed framework.

11.11.1 Human Fibroblasts

Dowling *et al* **(1997)** [102] carried out a cell adhesion test on a DLC (obtained by saddle beam deposition) partially coated 2.8cm diameter stainless steel disc using human fibroblast cell line and reported a very good cell adhesion and good spreading of the cells on the coated as well as the non-coated surfaces of the disc [102].

Allen *et al* **(1994)** [103] tested DLC coated polysterene (coating obtained by the low temperature dual ion beam technique using a saddle field source) and control uncoated polysterene tissue culture plates with primary cultured human synovial fibroblast (HSFs) and reported that there was no statiscally significant difference in the cell growth on both samples [103]. The LDH assay of the fibroblasts also indicate that DLC has not caused any significant level of cell toxicity when compared to the un-coated samples [103].

Mouse fibroblasts

Hauert *et al* **(1997)** [104] examined the interaction of mouse fibroblast (3T3 cell line) with a-C:H:Si (0.2 at% to 22.5 at% silicon addition) obtained with PACVD system and reported that the cells proliferated well on the coated culture dishes, that no influence like any toxic effect was observed from the Si-C bonds present on the surface to the growth and proliferation of the cells after two days of incubation [104]. Hauert *et al* [104] concluded that the toxic effect described by Allen *et al* [105] is caused by bulk Si-C not present in the a-C:H:Si thin film.

McColl *et al* **(1993)** [106] and **Parker** *et al* **(1993; 1994)** [107–108] studied the interaction of DLC with 3T3-1 mouse fibroblasts in vitro. The 3T3-1 cell viability in inserts Millicell-PCF membrane with DLC coating and without coating were detremined in their study by Trypan Blue dye exclusion and reported that the cells grew well in both control membrane inserts and DLC coated sample hich implies that DLC is not cytotoxic to the growing 3T3-1 cells.

Thomson *et al* **(1991)** [109] following exposure of mouse fibroblast (C3H10T1/2 cell line) to DLC coated 24 well culture plate obtained by saddle field source (using acetylene, butane or propane source gas) over a period of 7 days, reported that there was no significant difference in release of lactate dehyrogenase (LDH) in any of the coated samples and the uncoated control sample. Also their photmicrographic morphological examination confirmed that there was no cellular damage in the coated samples when compared to the uncoated control samples [109].

Murine fibroblast

Dowling *et al* (1997) [102] did a cytotoxicity study using murine fibroblast to examine DLC-coated alloy, titanium alloy and a plastic sample using SEM after 6-, 24- and 48-h incubation period. According to Dowling et al (1997) there was good cell morphology, adhesion, density and spreading observed on both the DLC coated alloy and on the plastic, but the tinanium alloy surface exhibited many cell death, thus the DLC coating acts as a barrier between the titanium alloy and the murine fibroblast cell line and demonstrates a low leve of cytotoxicity [102].

11.11.2 DLC Interaction with Osteoblasts In-Vitro

Osteoblast are the bone forming cells and have major role in mineralisation leading to osseointegration of a prosthesis.

Allen *et al* (2001) [110] investigated the effect of DLC coatings (obtained by fast-atom bombardment from a hexane precursor) deposited on polystyrene 24-well tissue culture plates on two osteoblast-like cell lines cultured on the uncoated and DLC-coated plates for periods of up to 72 h and by measuring the production of three osteoblast-specific marker proteins: alkaline phosphatase, osteocalcin, and type I collagen. According to Allen *et al* (2001) there was no evidence that the presence of the DLC coating had any adverse effect on any of these measured parameters which are indicative of metabolic processes in these osteoblast-like musculoskeletal system cells [110].

Schroeder *et al* (2000) [111] evaluated a new surface coating for bone-related implants by combining the hardness and inertness of a-C:H (DLC, obtained by a combination of radio-frequency plasma and DC magnetron sputtering deposition techniques) films with the biological acceptance of titanium. They incorporated different amounts of titanium (7–24 atm. %) into a-C:H films by a combined radio frequency (RF) and magnetron sputtering set-up. Their X-ray photoelectron spectroscopy (XPS) of air-exposed a-C:H/titanium (a-C:H/Ti) films revealed that the films were composed of TiO_2 and TiC embedded in and connected to an a-C : H matrix. They performed cell culture tests using primary adult rat bone marrow cell cultures (BMC) to determine effects on cell number and on osteoblast and osteoclast differentiation. According to Schroeder *et al* (2000) addition of titanium to the carbon matrix, leads to cellular reactions such as increased proliferation and reduced osteoclast-like cell activity, while these reactions were not seen on pure a-C:H films and on glass control samples, thus they concluded that a-C:H/Ti could be a valuable coating for bone implants, by supporting bone cell proliferation while reducing osteoclast-like cell activation [111].

Du *et al* **(1998)** [74] reported based on their study of interaction between osteoblasts (isolated from 4 day old Wistar rats) and DLC as well as CN (carbon nitride) thin films (obtained by IBAD technique), that the osteoblasts attach, spread and proliferate on both DLC and CN sample surfaces without apparent impairment on the cell physiology [74].

Allen *et al* **(1994)** [103] have also reported that DLC interact well with human 'osteoblast-like' cell line SaOS-2. When they compared the growth of human osteoblasts in both the DLC coated and non-coated polysterene plates, they found out that there was similar level of growth observed in both samples, and the osteoblasts adhered well to the DLC samples and produced extensive filopodia when viewed under the scanning electron microscope. The LDH assay of the osteoblast-like cells also indicated that DLC has not caused any significant level of cell lysis/toxicity when compared to the un-coated samples [103].

11.11.3 DLC Interaction Kidney Cells In-Vitro

Human embryonic kidney (HEK-293) cells

Lu *et al* **(1993)** [112] observed the interaction of DLC (obtained by ion beam assisted deposition) with HEK-293 cells using a haemocyto-meter for cell counting and trypan blue dye exclusion for assessing HEK-293 cell viability in DLC coated P-35 dishes. According to Lu *et al* (1993) HEK-293 cells grew well, there was no delay attachment to the DLC coated dishes compared to the control and that both the cells growing in the DLC coated and the control dishes had cell viability of 60% at the first day of incubation which increased to >90% at the second day of incubation [112].

Baby hamster kidney cells

Evans *et al* (1991) [113] examined the interaction of DLC obtained using saddle field source and baby hamster kidney cells and reported good cell adhesion on the coated surfaces indicating good cell compatibility.

11.11.4 Mutagenicity Evaluation of DLC

Dowling *et al* (1997) [102] perfomed a mutagenicity test (Ames test) on DLC coatings on stainless steel samples coated on both sides and uncoated samples using five strains of *Salmonella typhimurium* bacteria (TA-98, TA-100, TA-1535, TA-1537, TA-102) with and wothout metabolic activation in accordance with the method originally reported by Ames *et al* (1975) [114]. According to Dowling *et al* (1997) both the DLC and the stainless steel samples were not mutagenic as they induced no significant increase in the number of revertants of the five strains of Salmonella typhimurium tested [102].

11.11.5 DLC Interaction with Specific Cells (Hemocompatibility)

Bruck (1977) [115] has pointed to the importance of specie related haematological differences of experimental animals in the proper assessment of biomaterials for human use. He pointed out that 'the terms 'biocompatibility' and 'hemocompatibility' are often used inaccurately to denote the performance of biomaterials based on single or few in-vitro tests; these tests frequently ignoring considerations of hemorheological parameters, damages to the reticuloendothelial system, and haematological species-related differences' [115]. Diamond-like carbon, deposited on stainless steel and titanium alloys used for components of artificial heart valves has been found to be biologically and mechanically capable of improving their perfomance [116].

Devlin *et al* (1997) [117] has shown improvement of carbon/carbon composite prosthesis by DLC coating [117].

11.11.6 DLC Interaction with Endothelial Cells

Endothelium is nature's haemocompatible surface, and the performance of any biomaterial designed to be haemocompatible must be compared with that of the endothelium [118]. Endothelial hemocompatibility can be considered under three areas: the interaction between the endothelium and circulating cells (mainly platelets and leucocytes – close interactions between erythrocytes and endothelium are rare); the modulation of coagulation and fibrinolysis by endothelium; and other activities that affect the circulating blood or the vascular wall. Under normal circumstances, platelets do not interact with the endothelial cells – that is platelet adhesion to the vessel wall and the formation of platelet aggregates do not normally take place except when required for haemostasis. Hence, the surface of endothelial cells does not promote platelet attachment [118]. The formation of platelet aggregates in close proximity to the endothelium is also rendered difficult by prostacyclin (PGI2), a powerful inhibitor of platelet aggregation secreted by the endothelial cells. Prostacyclin can be secreted by endothelial cells in culture as well as by isolated vascular tissue [119]. The vascular endothelium is now known to be a dynamic regulator of haemostasis and thrombosis with the endothelial cells playing multiple and active (rather than passive) roles in haemostasis and thrombosis [120–121]. Many of the functions of the endothelial cells appear to be anti-thrombotic in nature. Several of the 'natural anticoagulant mechanisms,' including the heparin-antithrombin mechanism, the protein C-thrombomodulin mechanism, and the tissue plasminogin activator mechanism, are endothelial-associated. Among the proteins on the endothelial surface is antithrombin III [122] which catalyse the inactivation of thrombin by heparin. Endothelial cells also have heparan sulphate and dermatan sulphate (glycosaminoglycans) on their surfaces [123] which are known to have anticoagulant activity. On the other hand, the endothelial cells also appear capable of active prothrombotic behaviour in some extreme conditions of anticoagulation, because endothelial cells synthesise adhesive cofactors such as von Willibrand factor [124], fibronectin and thrombospondin [125]. Endothelial cells are now known to play crucial roles in a large number of physiological and pathological processes [126–135]. Most of these physio-pathologic events take place at the microvasculature (capillary beds) which constitutes the vast majority of the human vascular compartment. Thus it becomes vital to conduct hemocompatibility studies using microvascular endothelial cells. This is also vital because not all endothelial cells are alike. Endothelial cells derived from the microvascular structures of specific tissues differ significantly from large-vessel endothelial cells [136–141]. The study of human microvascular endothelial cells has been limited due to the fact that these cells are difficult to isolate in pure culture, are fastidious in their in vitro growth requirements, and have very short life span undergoing senescence at passages 8–10. Ades *et al* (1992) overcame these problems by the transfection and immortalization of human dermal microvascular endothelial cells (HMEC) [142]. These cells termed CDC/EU.HMEC-1 (HMEC-1) do retain the characteristics of ordinary endothelial cells (HMEC) and could be passage up to 95 times, grow to densities 3–7 times higher than ordinary HMEC and require much less stringent growth medium [142]. HMEC-1 is just like ordinary endothelial cells and exhibits typical cobblestone (or polyhedral) morphology when grown in a monolayer culture.

Van Wachem *et al*, 1985, reported that in their investigation of in-vitro interaction of human endothelial cells (HEC) and polymers with different wettabilities in culture, optimal adhesion of HEC generally occurred onto moderately wettable polymers. Within a series of cellulose type of polymers, the cell adhesion increased with increasing contact angle of the polymer surfaces [143]. Moderately wettable polymers may exhibit a serum and/or cellular protein adsorption pattern that is favourable for growth of HEC [143]. Van Wachem *et al* (1989) reported that moderately wettable tissue culture poly(ethylene terephthalate) (TCPETP), contact angle of 44° as measured by captive bubble technique, is a better surface for adhesion and proliferation of HEC than hydrophobic poly(ethylene terephthalate) (PETP), contact angle of 65° suggesting that vascular prostheses with a TCPETP-like surface will perform better in vivo than prostheses made of PETP [144].

11.11.7 Nitrogen-doped DLC Interaction with Endothelial Cells

This section reports the initial response of atomic nitrogen doped diamond like carbon (DLC) to endothelial cells in-vitro. The introduction of nitrogen atoms/molecules to the diamond like carbon structures leads to an atomic structural changes favorable to the thriving of human micro-vascular endothelial cells, thus the bio-response of ordinary diamond like carbon could be improved with atomic nitrogen doping. Whilst the semi-conductivity and stress relieving properties of nitrogen in DLC is thought to play a part, the increase in the non-bonded N atoms and N_2 molecules in the atomic doped species (with the exclusion of the charged species) seems to contribute to the improved bio-response [39–40]. The bio-response is associated with a lower work function and slightly higher water contact angle in the atomic doped films, where the heavy charged particles are excluded, as confirmed by SIMS analysis. The films used in the study were synthesized by RF PECVD technique followed by post deposition doping with nitrogen, and afterwards the films were characterized by XPS, Raman spectroscopy, SIMS and Kelvin probe. The water contact angles were measured, and the counts of the adherent cells on the samples were carried out. This study is relevant to improving bio-compatibility of surgical implants and prostheses.

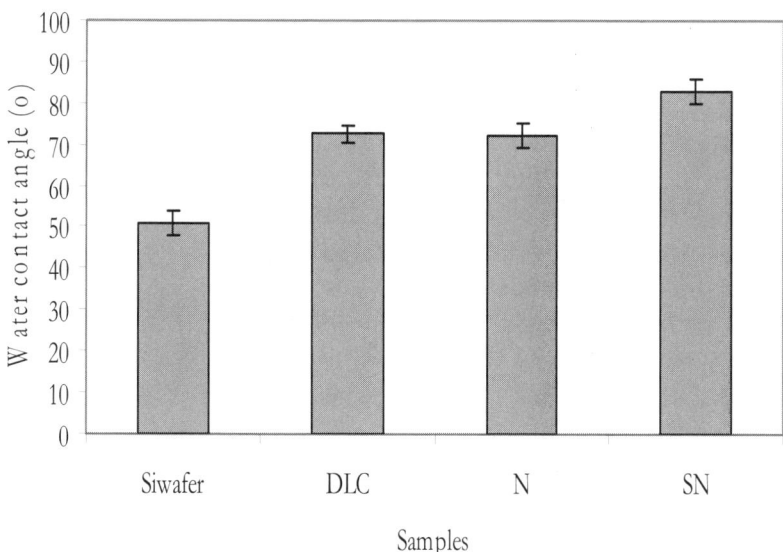

Fig. 11.3. Water contact angle of control silicon wafer, DLC and N doped DLC obtained by the water drop optical technique

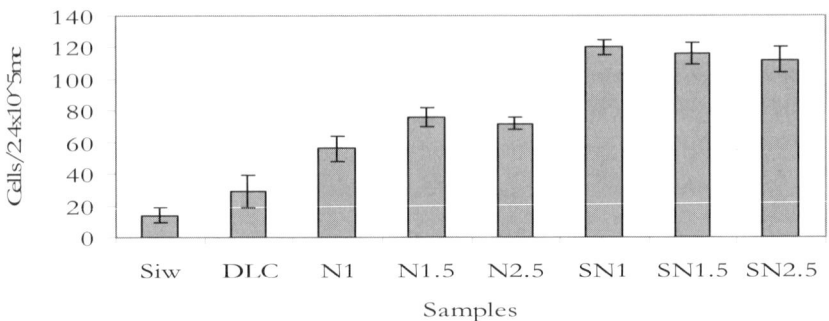

Fig. 11.4. HMEC attachment per 2.4x10⁵μ² of a-C:H:N (SN1-2.5 hours- with use of sweep plate to remove ions, and N-films doped with ions included), a-C:H (DLC) thin films and bare silicon wafer (Siw) control samples

The results in Figure 11.3 show the water contact angle of uncoated, DLC coated, N-DLC coated as well as 'SN'-DLC coated samples. The water contact angle increased with DLC coating when compared with uncoated sample. However, the contact angle value for the N-doped film is slightly lower than that of DLC whereas the values for the doping where the sweep plate (to remove ions) were employed is higher. The adherent endothelial cells are shown in Fig. 11.4. The number of adherent cells seems to be highest for the doped films (where the sweep plates were employed), followed by the doped films including the ionic species, DLC and finally, uncoated sample. This shows that the trend in the endothelial behaviour seems not to be directly related to the degree of hydrophobicity. These preliminary results seem to suggest therefore that hydrophobic films, with additional properties like decreased compressive stress, increased atomic networks and decreased graphitic clusters as well as semi-conductivity [36] may be favourable to human micro-vascular endothelial cellular attachment and proliferation. Other researchers have reported surface properties expressed in terms of hydrophobicity to be the key factor dictating the type and conformation of adsorbed proteins and therefore the cell adhesion. The seeding of bovine thoracic endothelial cells on cellulose surfaces with increasing hydrophobicity resulted in increased endothelial cell adhesion and proliferation and decreased migration [99]. Investigation on endothelial-specific cell adhesion to peptide sequences on different extra cellular matrix

(ECM) molecules grafted on to various surfaces reveal that the arg-glu-asp-val (REDV) sequence from fibronectin was selective for the adhesion of endothelial cells but not fibroblasts, smooth muscle cells, or activated platelets where other sequences like arg-gly-asp (RGD), tyr-ile-ser-gly-arg (YISGR), or pro-asp-ser-gly-arg (PDSGR) were implicated [100]. The material-micro-vascular endothelial cellular interaction could therefore be related inversely to those of platelets (in-vitro and in-vivo) since increased platelets aggregation/adhesion on a material could be associated with increased potential of a material to induce clotting [36], whilst increased endothelial-material adhesion could be associated on the other hand with an increased potential of a material not to induce clotting. Figure 11.4 and 11.5 compares directly the cell adhesion results obtained with both types of nitrogen species used for the doping over the duration of ~1–2.5 hours. The doping changes with time seem to be insignificant (1 hour compared with 2.5 hours). This is not surprising as only a small amount of impurities are usually required to effect a change in the micro-structure, and the doping effects seems peak after some time. It seems that the films obtained with the used of sweep plates (to remove ions) encouraged more endothelial growth and proliferation compared to its counterpart. This is thought to be due to some changes in the films microstructure and chemical bonding as revealed by XPS and SIMS techniques.

The results of the atomic chemical bonding inferred from the XPS peak assignment, suggests an increased atom percentage of the non-bonded N atoms (Peak at 399.6eV) and N_2 molecules (Peak at 401.1eV) in the films where the sweep plates were used to remove the ions. The non-bonded N atoms in these films are up to five times (x5) higher when compared to their counterparts [39].

The SIMS analysis of the films is shown in Figure 11.6, and is displayed as normalized intensity of mass to charge ratios of various ions detected. The negative ion scans for both the 'N' and 'SN' type films are very similar. These ions are of low m/z ratio (<25 amu), Fig. 11.6a & b. However, the relative/normalized intensity of the positive ions seems to be different, and the detected ions include heavier particles of higher m/z ratio (Fig. 11.6b & c). The relative intensity of the heavier particles of higher m/z ratio (>25<73 amu) seems to be higher in the 'N' type films. The heavier ions associated with higher plasma energies may be important during processing in increasing the films density, and establishing the integrity of the film's surface barrier to gas/moisture percolation, as the peak at 73 [$H_3O(H_2O)_3^+$] seems to be smaller in the 'N' type films (Both films were subjected to deionised water drops that were dried up afterwards, before SIMS analysis). The pattern of the SIMS depth profile with time is shown in Figure 11.7. The positive ions depth profile were performed to probe for m/z corresponding to 14 (N^+, CH_2^+), 28 (N_2^+, $CHNH^+$, CO^+), and 30 (CH_3NH^+); and the negative ions depth profile for m/z corresponding to 14 (N^-, CH_2^-), 26 (CN^-) and 38 (C_2N^-). The depth profile result shows that the non-bonded N_2 (28 amu) species are relatively more intense and steadily distributed in the 'SN' type films (Fig. 11.7b), compared to its counterpart.

The structural vibration information gained by the Raman spectroscopy shows a slight difference. The Raman D- and G-peak positions shifted slightly to a higher energy as a result of the inclusion of the ionic species [39]. This may be indicative of an increased sp^3/sp^2 fraction in the film, also suggested by the XPS results [39].

The relative work function of the films as measured by the Kelvin probe technique shows that the relative work functions changes from higher values ('N') films to lower values ('SN') films {Au-Au:250mV, Au-N2:200mV, Au-SN2.5:100mV, Au-SN1:75mV}, Fig. 8. Kaukonen *et al* [145] suggested based on their density function theoretical (DFT) calculation that a single N atom substitution at sp^3 or sp^2 site in a-C subsurface layers increases the total density of states (TDOS) below the energy gap resulting in Fermi energy (E_f) level moving down and the work function increasing. Whereas substitution on the sp^1 and sp^2 rings in the outer surface leads to TDOS increase near the conduction band edge with the Fermi level moving up and the work function decreasing. This decrease in the work function is thought to be dependent on the new states formed above the E_f following N substitution and a redistribution of the surface charges resulting in changes in the surface dipoles. Based on this interpretation it seems that N substitution in this study is dominant at the sp^1, sp^2 rings in the outer surface for 'SN' type N-doped films, that is, the 'SN' atomic species substitute preferentially at the sp^1 and sp^2 rings compared to the ionic N species. Nitrogen doping of a-C is known to increase the sp^2/sp^3 ratio and density function theoretical (DFT) calculations

[145] suggests that N atoms positioned at an sp^3 site decreases their coordination number with resulting sp^2 N (or N with a non-planar threefold coordination). It therefore seems logical that N atoms positioned at sp^1 or sp^2 sites could increase the co-ordination number with resulting sp^2 N or sp^3 N respectively. The Raman and XPS analysis, suggest that both sp^3 C and sp^2 C are higher in the 'N' films, with the I_D/I_G ratios being equally higher. Thus its seems that in the 'N' type films where higher energy of the impinging N species are concerned, the more substitutions occur at the sp^3 sites, and thus decreasing the co-ordination number with resulting sp^2 N. On the other hand it seems that relatively less substitution occur in the 'SN' films. It is therefore not surprising that nitrogen incorporation into DLC as in the 'N' type species reduces the contact angle (increases the surface energy), whereas the N-neutral specie ('SN') doping increased the contact angle (slightly) in this study (Fig. 11.3).

Typical SEM imaging of the adherent cells on the film surfaces are shown in Figure 11.5. Figure 11.5a & b shows the endothelial cell attachment on bare (uncoated) substrate and DLC coated substrate respectively. The worst bio-response occurs on the bare substrate (Fig. 11.5a), followed by DLC coatings (Fig. 11.5b) and the best response is seen on the nitrogen doped counterparts. However, it can be seen from the SEM images that there are better endothelial cell adhesion and proliferation on the 'SN' films (Fig. 11.5c & d) compared to 'N' films (Fig. 11. 5e & f). Statistical analysis at 95%

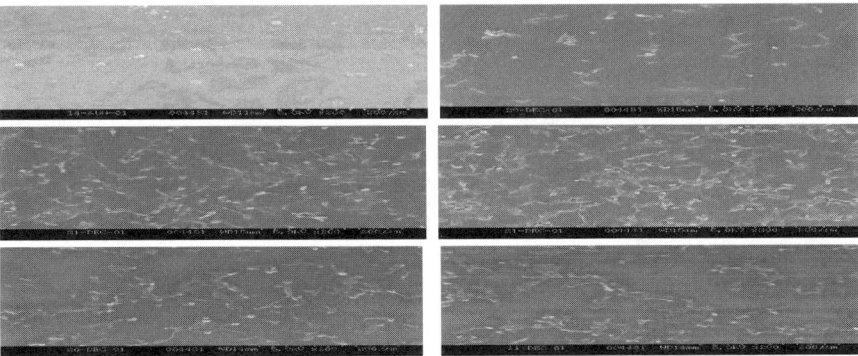

Fig. 11.5. SEM micrographs of endothelial cells attached to a-C:H:N thin films, x200; (a)Uncoated substrate, (b) DLC coated substrate, (c) sample 'SN1' (1 hr exposure to atomic nitrogen), (d) sample 'SN2.5', atomic species for 2.5hrs; (e) 'N1' (1 hr, atomic and charged species) and (f) 'N2.5' (2.5 hrs, atomic and charged species)

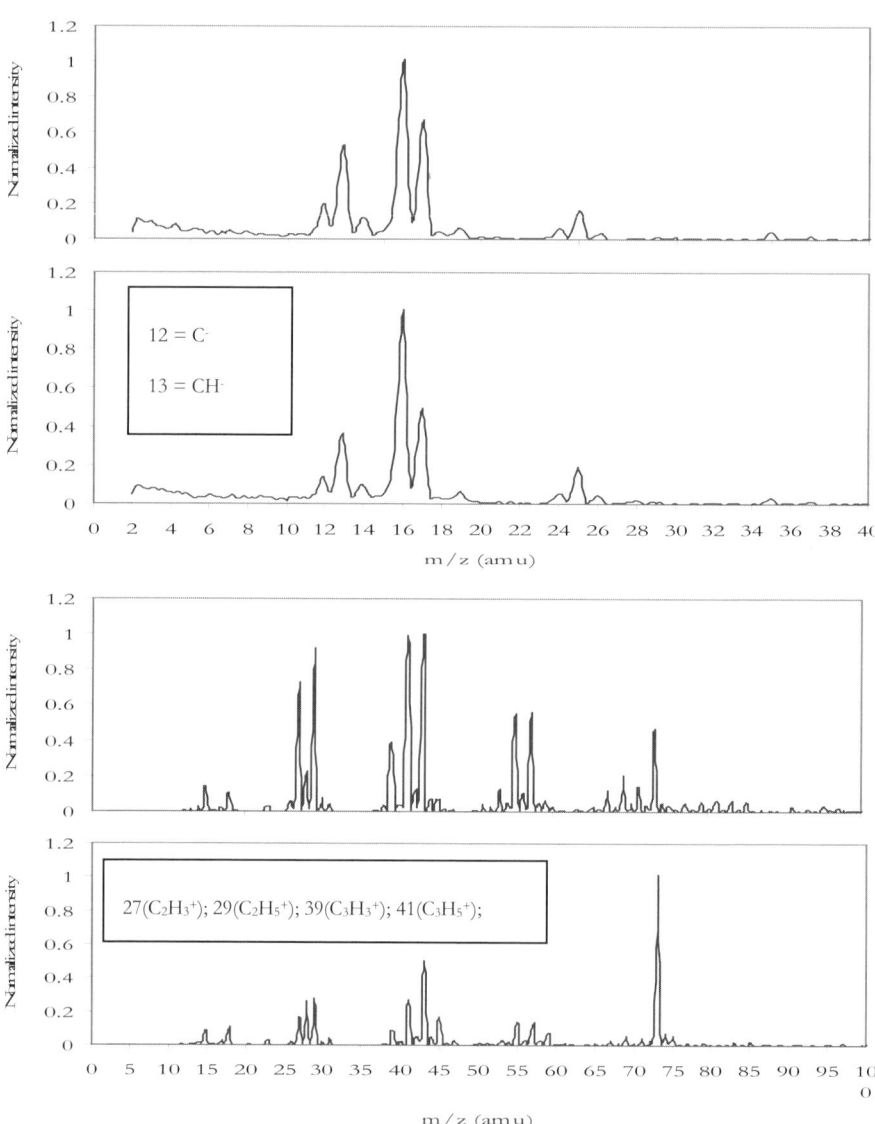

Fig. 11.6. SIMS analysis of nitrogen doped DLC thin films: negative ions (a) = 'N', (b) = 'SN'; positive ions (c) = 'N', (d) = 'SN'

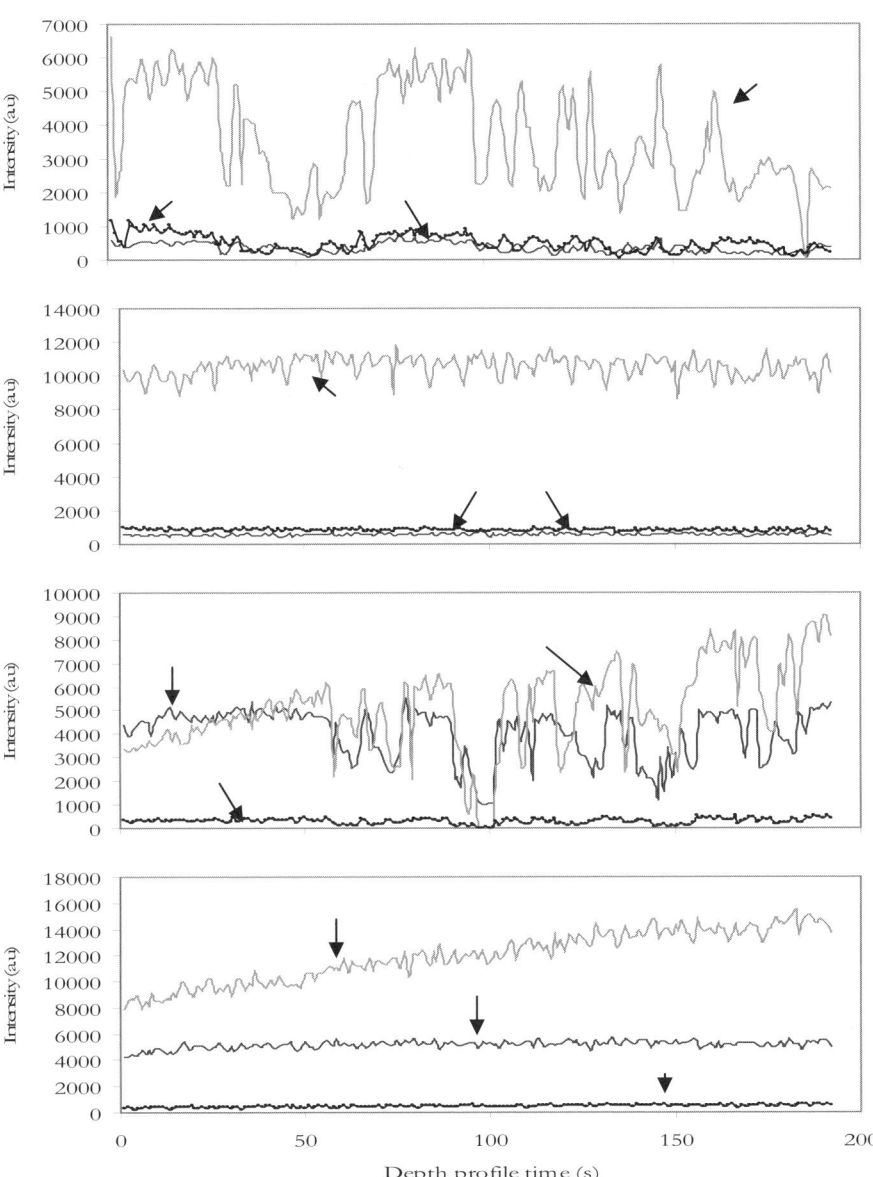

Fig. 11.7. SIMS depth profiling of some positive and negative ions: (a) + ions sample 'N', (b) + ions of sample 'SN'; (c) negative ions of sample 'N' and (d) sample 'SN'

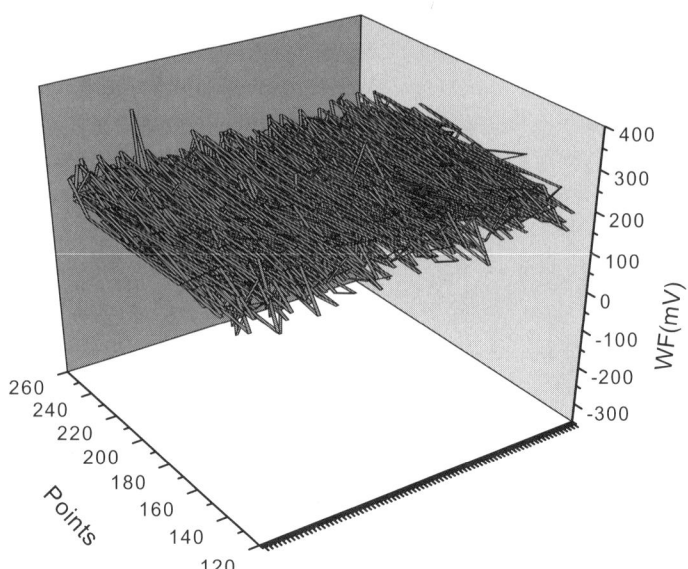

Fig. 11.8. The Work Function ('N2' film) over a scan area of $60 \times 30 \mu m$ to $80 \times 30 \mu m$ as determined by Scanning Kelvin Probe

confidence interval with 2-tailed paired sample analysis shows that there is a significant difference between uncoated samples and DLC coated samples, between DLC and either 'N' or 'SN' type films, as well as between 'N' type and 'SN' type films.

Nitrogen doped DLC thin films and their interaction with microvascular endothelial cells were examined. The nitrogen doped DLC

Table 11.1. Operating conditions for SVTA RF plasma source for atomic nitrogen post deposition doping of a-C:H thin films

Operating conditions	SVTA RF plasma source	PECVD
Power (W)	500	140
Reflected power (W)	25	5
Sweep plate voltage (V)	100–120	–
Bias voltage (V)	–	400
Deposition temperature (°C)	<100	25
Source temperature (°C)	300	–
Pressure (torr)	0.3	$\sim 7 \times 10^{-3}$
Atomic flux (atoms/sec)	0.8×10^{18}	–
Exposure time (hours)	1–2.5	~0.1

seems to have a better bio-response compared to the un-doped DLC thin films even though some of the N doped films (where the ions were not excluded) have lower contact angle (more hydrophilic). Amongst the nitrogen-doped films, the films synthesized with the exclusion of ions seems to improve the bio-response compared to its counterpart. The non-bonded N atoms in these films are higher and the contact angles obtained are also higher. Biocompatibility is therefore not simply related to hydrophobicity but may also relate to other materials micro and atomic structural changes induced by atomic nitrogen, like semi-conductivity and films stress relieve.

11.11.8 DLC Interaction with Platelets

Platelets are small, granulated bodies 2-4μm in diameter. They are round to spindle shaped cytoplasmic fragments containing enzymes and prenzymes but no nuclues. There are about 300,000/μl of platelets in the circulating blood and they normally have a half life of about 4 days. The membranes of platelets contain receptors for collagen, vessel wall von Willebrand factor (vWF), and fibrinogen. When a blood vessel wall is severed platelet adhere to the exposed collagen, laminin and vWF in the wall via intergrins. This process of platelet adhesion, unlike platelet aggregation does not require platelet metabolic activity [146]. However, binding of platelets to collagen initiates platelet activation (this can also be induced by ADP or thrombin). The activated platelets change shape, put out pseudopodia, discharge their granules which attract other platelets to stick to other platelets, a process known as platelet aggregation. The cytoplasm of platelets contains actin, myosin, glycogen, lysosomes and two kinds of granules (dense-granules and α-granules) which may be release after platelets are activated. The dense granules contain non-protein substances (includes serotonin, ADP and other adenine nucleotides) that are secreted in response to platelet activation and the α-granules contain secreted proteins other than hydrolases in lysosomes which includes clotting factors and platelet-derived growth factor (PDGF). Released platelet granules generate inflammatory response to injury, the white blood cells (leucocytes) are attracted by selectins and bind to integrins on endothelial cells leading to their extravasation through the blood vessel walls [146].

1. **Platelet adhesion hypothesis:**
   Platelets adhesion on a foreign surface is a necessary precursor to platelet aggregation on that surface
2. High platelet adhesion on a foreign surface is bad
3. In vitro platelet adhesion is related directly to in vivo thrombus formation and embolisation on foreign surfaces.
4. Platelets adhere with different strengths on different sites
5. Platelet adhesion and release reactions at foreign interfaces occur when specific platelet membrane receptor sites 'recognise' specific groups on the foreign surface (corollary: platelet adhesion is not a random process)
6. Some platelet release factors (e.g., serotonin and ADP from dense granules) enhance platelet aggregation on foreign surfaces, while the roles of others (platelet factor IV with its heparin neutralising activity, HNA, and β-thomboglobulin from α-granules) remain to be clarified.

Chen *et al* (2002) [147] varied acetylene to argon flow ratios in the DLC obtained with plasma immersion ion implantation deposition technique in their study. According to Chen *et al*, the blood compatibility of DLC depends on the sp^3 to sp^2 ratio rather than absolute sp^3 or sp^2 content and that the blood compatibility becomes worse with larger sp^3 to sp^2 ratio [147].

Krishnan *et al* (2002) [148] did a quantitational analysis of I-125 radio-labeled platelet to examine platelet adhesion (using radiscintigraphy techniques) on titanium and DLC coated titanium and have shown that DLC coated titanium exhibited a lower platelet adhesion compared to uncoated titanium [148].

Cui and Li (2000) [75] also studied platelet interaction with DLC coated PMMA intra-occular lens and uncoated PMMA intra-occular lens (IOLs). Their result shows that DLC coated PMMA-IOLs had a lower platelet adhesion compared to the uncoated sample [75].

Gutensohn *et al* (2000) [149] studied the interaction of stents coated on both inner and outer surfaces with DLC (obtained by plasma induced cold deposition technique) and human platelets. In their study they used DLC to achieve a uniform protective coating to reduce the release of metal ion, platelet activation and thrombogenicity. Flow cytometric analyses revealed significantly higher increase of mean channel fluorescence intensity for the platelet activation dependent antigens CD62p and CD63 in non-coated stents compared to DLC-coated stents (p<0.05). Patients undergoing coronary angioplasty and stenting procedures are known to be at higher risk for reocclusion and restenosis of vessel when plate express increased numbers of activation dependent antigens [2–3]. Of all the antigens they analysed P-selectin (GMP-140) seems to play a key role and appears to be most closely associated with an increase in thrombotic risk [2–3]. By using atomic adsorption spectrophotometry and inductively coupled plasma mass spectrometry analyses they have shown that there was a significant release of metallic ion in the non-coated stents compared to DLC coated stents. As a consequence of reduced metal ion release due to DLC coating platelet activation was significantly lower in DLC coated stents compared to the non-coated stents under otherwise identical experimental conditions [149].

Alanazi *et al* **(2000)** [64–65] evaluated the interaction of polycarbonate coated DLC (obtained with CVD under varied deposition conditions), segmented polyurethane (SPU, usually used for fabrication of medical devices including artificial heart) and an amphiphilic block copolymer composed of 2-hydroxyethylmethacrylate (HEMA) and styrene (St) (HEMA/St; an excellent non-thrombogenic polymer was used as a negative control) with platelets in whole human blood. They used the parallel plate flow chamber and epi-fluorescent video microscopy (EVM) using whole human blood containing Mepacrine labelled platelets perfuse at a wall shear rate of $100s^{-1}$ at 1min intervals for a period of 20mins. In their assessment of the optical penetration of their EVM system and the activation/adhesion of platelets, they concluded that the activation of platelets on PC-DLC compared with the other biomaterials was minimal, the surface roughness before and after the coating is applied to blood contacting devices is insignificant (16–23nm), the contact angle is improved after DLC coating, the contact angle and chemical composition is independent of film thickness, defects of DLC films can be caused by elevated substrate, and blood compatibility depends on deposition conditions [64–65].

Jones *et al* **(1999)** [33] studied interaction between rabbit platelets and components of a Ti-TiN-TiC-DLC multilayer system. They adopted an interlayer approach in order to achieve adequate adhesion between DLC coatings deposited by plasma-assisted CVD and titanium substrate. The substrate, inter-layers and DLC were assessed for haemo-compatibility and thrombogenicity using a dynamic blood method and interactions with rabbit blood platelets, respectively. The adhesion, activation and morphology of the platelets were determined by stereological techniques using SEM. The coatings produced no significant haemolytic effect compared to the medical grade polysterene control. In contrast to the DLC coating, all of the inter-layers showed a slight tendency towards thrombus formation during the later stages of the incubation [33].

Dion *et al* **(1993)** [42] evaluated the in vitro platelet retention of the new prosthetic heart valve that has been designed by FII Company and Pr. Baudet which is composed of Ti6Al4V titanium alloy coated with DLC (obtained by chemical vapour deposition technique). The retention/adhesion of platelets was evaluated by analyzing radioactivity on the exposed wall of test or control tubes through which a blood cell suspension containing ^{111}In-labelled platelets had circulated. Their results show that on DLC/Ti6Al4V platelets adhere twice the amount that they do on the reference material (a silicone medical grade elastomer the behaviour of which in contact with blood is the same as that observed with the NIH recommended polydimethyl siloxane) [42].

Okpalugo *et al* [36, 40] The human blood platelets interaction with a-C:H:Si films has revealed a relation between the microstructure of a-C:H:Si and its level of platelets aggregation. An increase in contact angle (or lowering of surface energy) of a-C:H and a moderate increase in the intrinsic electron conduction (semi-conduction)/decrease in work function may lead to decreased platelets aggregation implying an increase in clotting time (decreased rate of clotting), high surface energy (low contact angle), non-conduction (insulating films), and graphitisation seems to lead to an increase in platelet aggregation implying a decrease in clotting time and an increase in the rate of blood clotting. Thus the hemocompatibility and biocompatibility of a-C:H and a-C:H:Si seems to be dependent on its electrical properties as well as the surface energy and the microstructure of the thin film biomaterials.

Silicon doped films show a much lower level of platelet aggregation [36]. This seems to correlate with the lowering of surface energy, CPD/work function, resistivity and degree/rate of graphitisation due to silicon doping. Bruck [52–54] observed clotting times six to nine times longer than those observed with non-conducting polymers and also observed little or no platelet aggregation on electro-conducting polymers, when compared to non-conducting control samples. In this study, thermal annealing of a-C:H below 400°C led to excellent performance similar to that observed in a-C:H:Si. This is attributed to the increased electro-conduction without graphitisation at the lower annealing temperature of a-C:H (Fig. 4.2.1–4). Chen *et al* [97] varied acetylene to argon flow ratios during the deposition of a-C:H obtained by plasma immersion ion implantation. According to Chen *et al*, the blood compatibility of a-C:H depends on the sp^3 to sp^2 ratio rather than absolute sp^3 or sp^2 content and that the blood compatibility becomes worse with larger sp^3 to sp^2 ratio [97]. The results of the surface roughness as obtained with the AFM have shown that all the films analysed have smooth surfaces [40].

The results of the Raman spectroscopy show an increase in the I_D/I_G ratios [34–40] with annealing temperature for the a-C:H films in agreement with earlier reports in the literature [150–151] and also for the a-C:H:Si thin films. The increase in I_D/I_G ratio on the annealing of a-C:H has been associated with the growth of crystallites structure in the a-C:H thin film. In the a-C:H films the increase in the I_D/I_G ratio with thermal annealing seems to be linear whereas in a-C:H:Si the increase occurred only at relatively higher annealing temperatures above 300°C. The I_D/I_G ratio decreases with increasing amount of silicon in the films [34–40].

Physical and chemical changes in carbon materials resulting in graphitisation may occur during thermal annealing of carbon materials. Graphitisation occurred at higher annealing temperatures greater than 400°C [34–40] as revealed by the bimodal shoulders. Silicon doping seems to lower the degree of graphitisation [34–40]. Graphitisation is associated with increased sp^2 content, while silicon doping seems to increase the sp^3 sites. Shoulder peaks associated with annealing and graphitisation appeared on a-C:H and lightly doped a-C:H:Si (TMS=5sccm) at 400°C-600°C [34–40]. A much higher annealing temperature is required to give a shoulder on films with higher amounts of silicon. Silicon does not form π-bond and it therefore increases the amount of sp^3 bonds in the film.

Typical x-ray photoelectron spectroscopy (XPS) chemical analysis for the as deposited and thermally annealed a-C:H and a-C:H:Si films are as indicated [34–40]. The peak binding energies of the films are consistent with those reported in the literature by Dementjev [152–153], Grill [154], Miyake [155] and Baker and Hammer [156]. There was only a slight change in the values of the binding energies for the silicon-modified films even after annealing to 600°C. Also in agreement with Demichelis *et al* [157] there was an increase in sp^3/sp^2 ratios after peak de-convolution. Silicon does not form π-bonds, thus silicon doping of a-C:H films would be associated with increase in sp^3 bonds and delayed graphitisation observed in a-C:H:Si films annealed between 200°C and 600°C, which is associated with decreased platelet aggregation observed in these a-C:H:Si films. The contact angle measurement results obtained using the optical method and the surface energy measured (and calculated) by the Wilhemy plate technique of films deposited on silicon substrates reveal that silicon doping leads to an increase in the contact angle and a lowering of the surface energy which could be associated with decreased platelet aggregation [36, 40].

The histograms below shows the summary of the result of the platelet aggregation on the a-C:H and Si-a-C:H thin films (as deposited and thermally annealed films) seeded with platelets for 15 minutes (Figure 11.9) and 30 minutes (Figure 11.10) respectively. Figure 11.11 shows the SEM image of platelet seeding for 75 minutes on as deposited DLC and silicon doped DLC thin films.

The electrical properties of the a-C:H and a-C:H:Si samples as well as the contact potential differences/work function have shown a correlation with the microstructure of the films as revealed by the Raman spectroscopy and XPS investigation, and the platelet aggregation on the films (Fig. 11.12–13). Typical I-V curves of the metal-semiconductor-metal (MSM) sandwich show that the electrical conduction mechanism is not simple ohmic but semi-conducting. The resistivity curve of the as deposited shows that silicon addition to a-C:H lowers the resistivity.

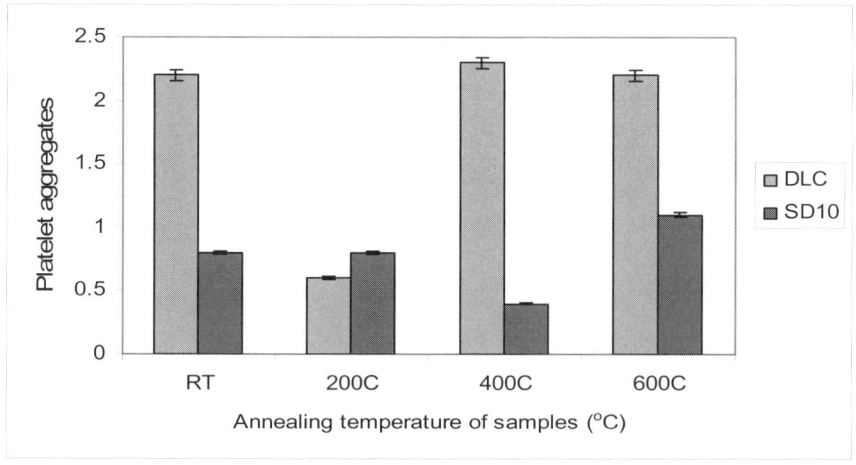

Fig. 11.9. Averaged human platelets aggregates/2.4x10^{5}μ^2 on a-C:H and a-C:H:Si (As obtained and thermally annealed) samples (seeded for 15mins)

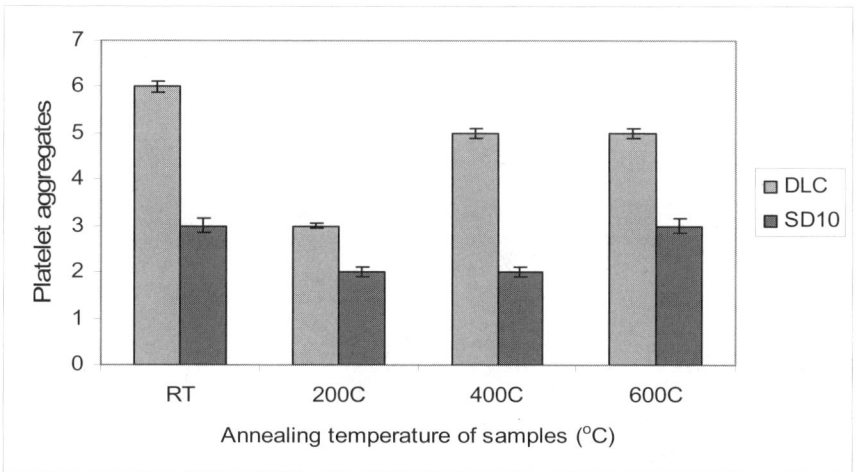

Fig. 11.10. Averaged human platelet aggregates/2.4 x 10^{5}μ^2 on a-C:H and a-C:H:Si (As obtained and thermally annealed) samples (seeded for 30mins). RT=Room temperature

Thermal annealing of both a-C:H and a-C:H:Si leads to a decrease in the resistivty (Fig. 11.12–13) which is likely to be associated with micro-structural changes monitored by Raman spectroscopy as indicated by the sp^3/sp^2 ratios. At 600°C of annealing temperature, the

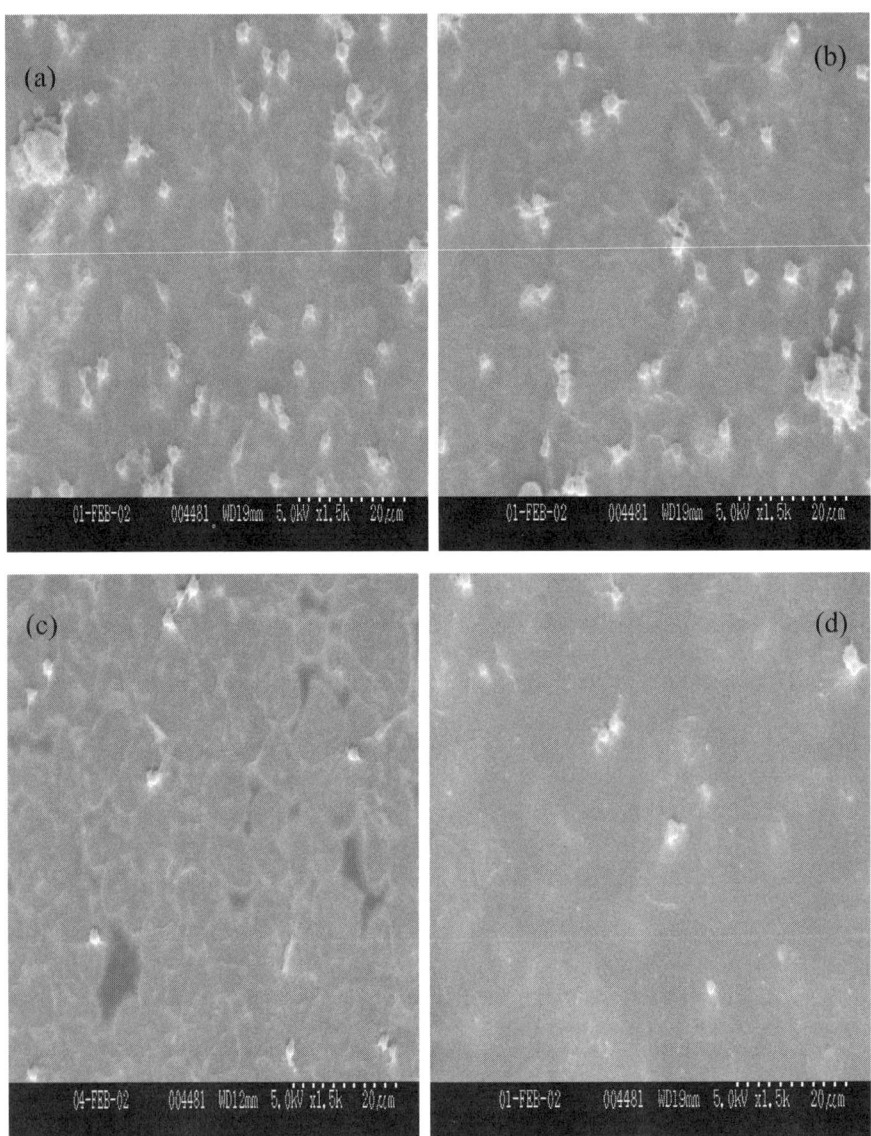

Fig. 11.11. SEM images of platelets seeded for 75mins on as obtained a-C:H (a & b), SD10 (c) and SD5 (d); x1500

conductivity in both a-C:H and a-C:H:Si becomes simple ohmic [38, 40]. This result is consistent with the Raman spectroscopy investigation, which revealed graphitisation at this annealing temperature. Though decreased resistivity and decreased work function resulted

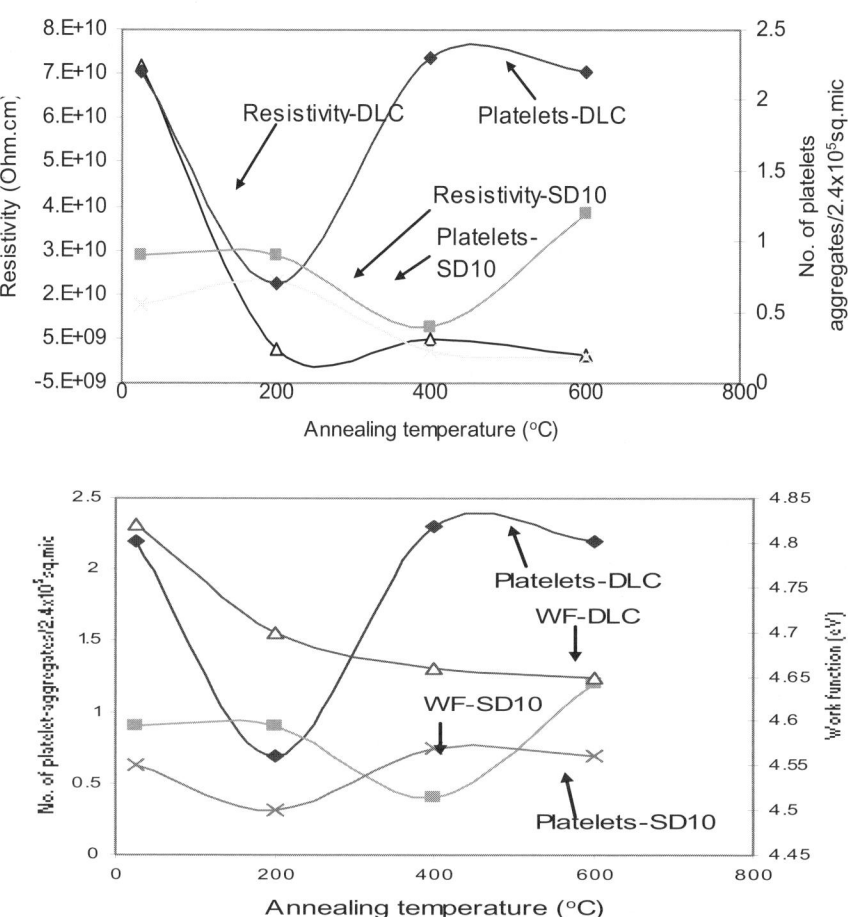

Fig. 11.12. Number of platelets aggregation (Seeded for 15mins) per $2.4 \times 10^5 \mu^2$ as a function of resistivity and work function of a-C:H (DLC) and a-C:H:Si (SD10) thin films

in less platelet aggregates graphitisation leads to increased number of platelet aggregates in the film [35–36, 38].

The increasing graphitic content of the films at an annealing temperature of $600^{\circ}C$ increases the proportion of de-localised π-bonded electrons and therefore increases the electrical conductivity of the films as well, which hence result in the ohmic-behaviour. The contact potential difference (CPD) and the work function (WF) of a-C:H thin films decreased with increasing amount of silicon doping and with increasing annealing temperatures as shown in Fig. 11.12–13. These electrical properties are related to the observed changes in the platelet aggregation on the a-C:H and a-C:H:Si thin films (Fig. 11.12–13).

11.11.9 DLC Interaction with Blood Cells Not Involved in the Clotting Process

Going strictly by the definition of hemocompatibility, apart from endothelial cells and platelets that are directly involved in the process of

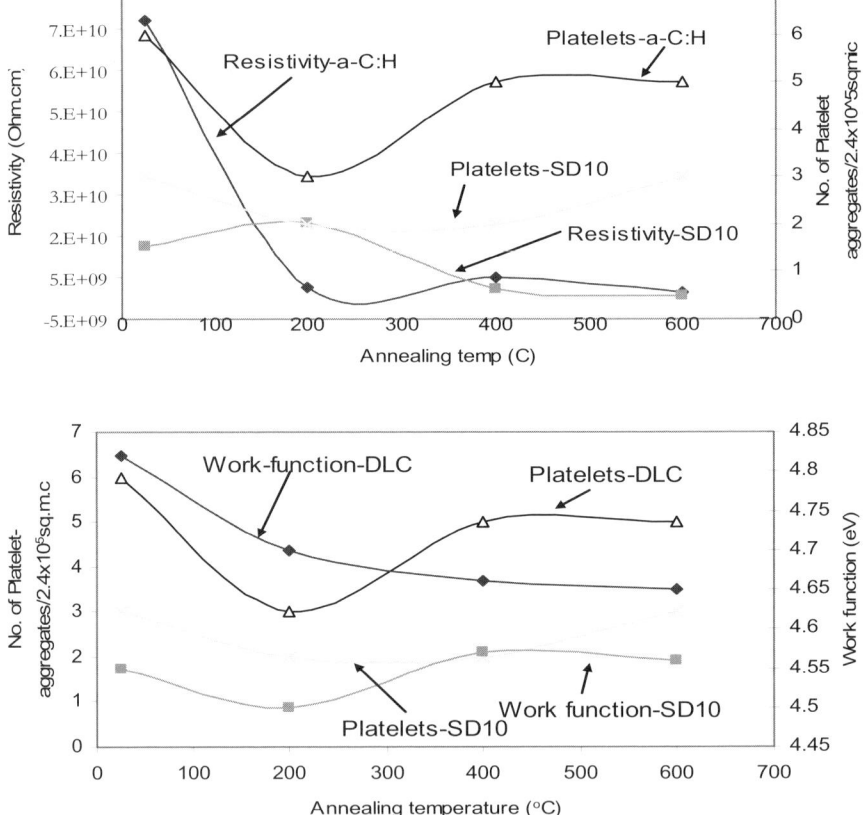

Fig. 11.13. Number of platelets aggregation (Seeded for 30mins) per $2.4 \times 10^5 \mu^2$ as a function of resistivity and work function of a-C:H (DLC) and a-C:H:Si (SD10) thin films

thrombosis and antithrombosis, the other blood cells though present in the blood can only be used to assess biocompatibility instead of hemocompatibility since they are not directly involved in thrombus formation and or prevention of thrombus fromation. Thus the interaction of DLC with the other blood cells like neutrophils, lymphocytes, monocytes, erythrocytes (RBC) etc can only give an indication of biocompatibility and not really 'hemocompatibility'.

11.11.10 DLC Interaction with Erythrocytes (Red blood cells, RBC)

Higson *et al* [72–73] examined the biocompatibility and substrate diffusion limiting properties for a range of DLC (obtained from a saddle field source) coated porous polycarbonate membrane (nominal pore size of 0.01µm) and hemodialysis membranes in whole blood. According to Higson *et al* [72–73] after 30minutes exposure to unstirred whole blood, the uncoated polycarbonate did show red blood cells (and possibly other blood cells) adherent to the membrane surface, while in the DLC coated membranes there was much reduced surface adherence of red blood cells and proteins, though some deposition of amorphous materials occurred. They also found that glucose sensors employing DLC coated outer covering membranes were found to have suffered smaller losses of response following exposure to whole blood [72–73].

Dion *et al* **(1993)** [42] also evaluated the in vitro RBC retention of the new prosthetic heart valve that has been designed by FII Company and Pr. Baudet which is composed of Ti6Al4V titanium alloy coated with DLC (obtained by chemical vapour deposition technique). The retention/adhesion of RBC was evaluated by analyzing radioactivity on the exposed wall of test or control tubes through which a blood cell suspension containing ^{99m}Tc-labelled red blood cells had circulated. According to Dion *et al*, the red cell retention, which may be due to either poor rinsing or morphological irregularities, leads to greater platelet retention; that mechanical entrapment accounted for only 0.07% of red cell retention for silicone and 0.08% for DLC/Ti6Al4V [42].

11.11.11 DLC Interaction with Human Haematopoietic Myeloblasts In-Vitro

The haematopoietic myeoblasts are young blast cells derived from the projenitor cells in the bone marrow responsible to generation of new white blood cells-granulocytes (neutophils, basophils and eosinophils).

ML-1 cells

ML-1 human haematopoietic myeloblasts was used in the study of **Lu** *et al* [112] for assassing the biocompatibility of DLC. By using the hemacytometer for cell counting and Trypan Blue dye exclusion for determining cell viability, Lu *et al* [112] found out that ML-1 cells cultured in DLC coated P-35 dishes proliferated very well when compared to ML-1 cells growing in the uncoated control dishes and there was no sign to indicate cellular differentiation occurring among the ML-1 cells growing on the DLC coating dishes [112].

11.11.12 DLC Interactions with Granulocytes (neutrophils or basophils or eosinophils) In-Vitro

Granulocytes are white blood cells containing granules (granules can either be acidic, basic or neutral thereby giving rise to different types of granulocytes, neutrophils, basophils and eosinophils, see table 2.8.1 above), that are also responsible for combating infections. Neutrophils and eosinophils are also called microphages.

Neutrophils (polymorphonuclear leucocytes or PMNs, polymorphs)

Neutrophils are the most abundant white blood cell numbering an average of 4150 cells/μl (50–70% of total number of white blood cells). They are round cells with a lobed nuclues (usually 2–5 lobes) that may resemble a string of beads and cytoplasm containing large pale inclusions, their granules being neutral and difficult to stain with either acidic or basic dyes. Their function is phagocytic, engulfing pathogens or debris in tissues and releasing cytotoxic enzymes and chemicals (lysosomal enzymes and bactericidal compounds). They measure about 12μm in diameter and are the first of the WBCs to arrive at an injury site. High neutrophil adhesion produce negative effect on tissue biocompatibility?

Li *et al* (1999) [158] reported based on their study of neutral granu-locytes/neutrophils ( ) interaction with DLC (obtained by ion beam assisted deposition, IBAD technique) coated polymethylmethacrylate (PMMA) intraoccular lens (IOLs), that DLC coated PMMA-IOLs exhibit a lower neutophils adhesion compared to the uncoated PMMA-IOLs [158].

Eosinophils (acidophils)

Eosinophils are another type of granulocytes with cytoplasms that contains large granules that generally stain bright red with a red dye eosin, and can also stain with other acid dyes, hence also known as acidophils. The size of eosinophils is similar to that of neutrophils (~12μm). Eosinophils have bilobed nucleus and make up ~2–4% of the white blood cell population. They are phagocytic and engulf anti-body coated or marked foreign substances. Their primary mode of attack is the exocytosis of toxic compounds, including nitric oxide, and cytotoxic enzymes, onto the surface of their targets. They are att-racted to site of injury and so the increase in their number may indi-cate inflammation, allergy etc.

Basophils

These are another type of granulocytes with numerous granules that stain darkly with basic dyes. They measure 8–10μm in diameter and make up only about 1% of the white blod cells. Basophils migrate to injury sites and cross the capillary endothelium to accumulate in the damage tissue where they discharge their granules which contains histamine (dilate blood vessels) and heparin (prevents clotting). They enhance local inflammation at the sites of injury and other chemicals that thy release attract eosinophils and other basophils in the area of injury.

11.11.13 DLC Interaction with Monocytes (macrophages) In-Vitro

The monocytes also falls into the white blood cell category primarily produced in the bone marrow. The monocytes are 'agranular-leuco-cytes' (agranulocytes), that is they lack abundant, deeply stained granules, though they also contain vesicles and lysosomes which are much smaller compared to those of the granulocytes [146]. They are very large cells (~15µm in diameter, nearly about 2x the diameter of red blood cells) with kidney bean-shaped nuclues and abundant pale cytoplasm. They constitute about 2–8% of the population of the circulating white blood cells. Monocytes move from the flowing blood to the tissues after 1–2days. When monocytes enter the tissue they become known as macrophages and are responsible for fighting foreign bodies or pathogen and debris by engulfing and inactivating and digesting them in a process known as phagocytosis. They are aggressive phagocytes, often attempting to engulf as large or larger than their size. When activated they release chemicals that attract and stimulate neutrophils, monocytes and other phagocytic cells, as well as fibroblasts to the region of injury. The fibroblasts then begin producing scar tissues which walls off the injured area.

Linder *et al* **(2002)** [159] studied the adhesion, cyto-architecture and activation of primary human monocytes and their differentiated derivatives, macrophages, on DLC (obtained by radiofrequency PACVD using methane/helium mixture: 1.5 vol%/98.5 vol%) coated glass coverslips using immunofluorescence technique. According to Linder *et al* [159] the adhesion of primary monocytes to a DLC-coated coverslip is slightly, but not significantly, enhanced in comparison to uncoated coverslips, while the actin and microtubule cytoskeletons of mature macrophages show a normal development. The activation status of macrophages, as judged by polarization of the cell body, was not affected by growth on a diamond-like carbon surface, thus diamond-like carbon shows good indications for biocompatibility to blood monocytes in vitro. It is therefore unlikely that contact with a diamond-like carbon coated surface in the human body could cause the cells to elicit inflammatory reaction [159].

Allen *et al* **(1994)** [103] reported that DLC coated polysterene (coating obtained by the low temperature dual ion beam technique using a saddle field source) culture plates produced good macrophage (murine macrophage cell line, IC-21) cell proliferation with a statistically significant faster growth rate (observed at the 48 and 72 h time-points) when compared to the uncoated polysterene, and with no evidence of cytoplasmic vacuolation, membrane damage or excessive macrophage cell death [103]. Their LDH assay also indicate that there was no significant increase in LDH release from cells grown on DLC-coated surfaces as compared with cells grown on control surfaces, thus DLC has not caused any significant level of cell toxicity when compared to the un-coated samples [103].

Thomson *et al* (1991) [109] by measuring the level of lysosomal enzymes N-acetyl-D-glucosaminidase released (enzyme is usually released as part of inflammatory reaction) in cell culture medium by macrophages (mouse peritoneal macrophage) after the cells interacted with DLC, reported that there was no significant different in the amount of enzyme detected in the DLC coated samples (using saddle field source and different source gases: acetylene, butane or propane) compared with the uncoated control tissue culture sample (24 well tissue culture plates) [109]. This implies that DLC is not cytotoxic and may not have elicited an inflammatory reaction. This was also corroborated by their lactate dehydrogenase, LDH assay which indicated that there was no statistically significant different level of LDH detected on the DLC coated samples when compared to the non-coated samples.

11.11.14 DLC Interaction with Lymphocytes

The lymphocytes are also 'agranular-leucocytes' (agranulocytes) lacking abundant and deeply stained granules. In a blood smear they are seen as a thin hallow of cytoplasm around a relatively large nucleus. In diameter they are slightly larger than the RBCs. Lymphocytes account 20–30% of the WBC population of blood. Some lymphocytes are in circulation (small percentage) while others are in various tissues, organs and lymphatic system. Three classes of lymphocytes exist in the circulating blood, the T-cells (responsible for cell-mediated immunity), the B-cells (responsible for humoral immunity) and the NK-cells (natural killer cells responsible for immune surveillance, are important in preventing cancer, sometimes known as large granular lymphocytes).

Plasma proteins and cell adhesion proteins/molecules (CAM)

Plasma, the fluid portion of the blood is a remarkable solution containing an immense number of ions, inorganic molecules, and organic molecules that are in transit to trasnport substances to various parts of the body. The plasma proteins consist of albumin, globulin, and fibrinogen fractions. Cells are attached to the basal lamina and to each other via the cell adhesion molecules (CAM) which are adhesion proteins. Many of these proteins pass through the cell membranes of the cells and are anchored to the cytoskeleton. Many CAMs have been characterised biochemically, and their functions are being investigated. Some of the CAMs bind to like molecules on other cells (homophilic binding) and the others to other molecules (heterophilic binding). Many CAMs bind to laminins, a family of large cross-shaped molecules with multiple receptor domains in the extracellular matrix. Nomenclature of CAM is still chaotic due partly of rapid growth in the field and also due to extensive use of acronyms in modern biology, however 4 broad-families of CAM are known. The CAM categories are (1) the 'integrins' which are heterodimers that bind to various receptors; (2) the adhesion molecules of the 'IgG superfamily' of immunoglobulins some of which bind to their molecules and some binding homophilically; (3) the 'cadherins', Ca^{2+}-dependent molecules that mediate cell-to-cell adhesion by homophilic reactions and (4) 'selectins', which have lectin-like domains that bind carbohydrates [146]. Goodman *et al*, 1991 [160] reported that since shape-change responses of in vitro column-purified platelets to several polymeric biomaterials are very similar to those of circulating non-anticoagulated platelets and because column separation significantly reduces the protein content [161] of in vitro platelets, it appears that differences in platelet spreading on various materials are not entirely dependent on the preferential adsorption of plasma proteins from blood. They also observed that in vitro shape-changes on Formvar and glass are the same in the presence and absence of protein-containing buffer. Also cell spreading is similar in the presence and absence of serum proteins [162–163, 160]. A 'shine-through' hypothesis has been proposed to explain this biomaterial-protein-cellular interaction [164]. Additionally, fibroblasts and platelets appear to remove or move (sweep aside) adsorbed proteins [165–168], it is therefore conceivable that platelets alter adsorbed albumin layer(s) and adhere directly or via particular adhesion proteins on platelets surface to the material surface, if this occurs it would be expected that platelets would exhibit little differences in spreading on uncoated and albumin-coated surfaces, regardless of the thickness of an albumin layer. It has also been observed that initially adsorbed proteins tend to be more denatured than subsequent layers [169], and that the initial layer also appears to be more important in determining thrombotic responses [170]. Park *et al*, 1985 have also shown that platelet deposition in ex vivo circulation is determined by the first layer of adsorbed proteins and not subsequent layers, suggesting that platelets remove secondarily adsorbed layers of proteins. Another hypothesis is that platelets may not make direct contact with the substrate but are nonetheless influenced by surface character through an intervening layer of proteins, which is itself influenced by the surface. Different orientations or conformations of adsorbed proteins, as determined by the substrate, would then influence the behavior of adherent platelets and other cells [160]. How these proteins may interact with DLC and DLC-like materials is yet to be explored.

Non-adhesive proteins: Albumin, Transferrin, -like proteins

Albumin and tranferin like proteins with non-adhesive functions tend to decrease subsequent thromboembolic events [160, 171].

Dion *et al* (1993) [42] have examined ^{131}I labelled albumin plasma protein adhesion on DLC coated Ti6Al4V and siliconee elastomer and reported that DLC can adhere more albumin than the medical grade elastomer.

Adheisve proteins: Fibrinogen, Fibronectin, VWF, and CAM

In general these plasma proteins with adhesive functions tend to increase thrombosis [160, 172]. Adhesive proteins and likely increased expression of cell adhesion molecules (CAM) e.g., ICAM-1, VCAM-1, ELAM-1, E-selectin, GMP-140 (P-selectins) and other molecules/ligands from the immunoglobulin and selectin superfamily has been shown to be important in cascade reactions like the platelet-leucocyte and leucocyte-endothelial cell adhesion and activation reactions [173–175]. When expressed on the cell surface the NH_2-terminal lectin-like domains of the selectins bind with their counter-receptors (specific carbohydrate ligands on white blood cells and platelets).

Dion *et al* (1993) [42] have also examined ^{125}I labelled fibrinogen plasma protein adhesion on DLC coated Ti6Al4V and silicone elastomer and reported that DLC can adhere slightly more fibrinogen than the silicone elastomer.

Non-adhesive/Adhesive protein ratios: Albumin/Fibrinogen ratios

It has been shown that platelet adhesion depends on the albumin/ fibrinogen ratio: the higher the albumin/fibrinogen ratio the lower the number of adhering platelets and hence less risk of platelet aggregation and less risk of thromboembolism. The albumin/fibrinogen ratio for DLC is 1.24 and 0.76 for silicone elastomer [42]. According to Dion *et al* [42] these two ratios allow us to consider that platelet adhesion would be weaker on DLC than on silicone elastomer but, infact the opposite occurred, which they thought could be explained by the large dispersion of results in percentage of platelet retained due to the device concept itself they added [42].

Cui and Li (2000)[75] also studied the adhesion of plasma proteins on DLC-coated, CN-coated PMMA, and uncoated PMMA using radioactive targed proteins. They reported the albumin/fibrinogen ratio of 1.008 for DLC, 0.49 for CN and 0.39 for PMMA [75].

11.12 Endothelial Pre-seeding on Biomaterials for Tissue Engineering

Preliminary study on the platelet interaction with a-C:H:Si and a-C:H thin film samples pre-seeded with endothelial cells in-vitro is reported in this section to corroborate the reported results in the other sections where a-C:H:Si and a-C:H were interacting with endothelial cells and platelets separately. These joint interactions were investigated in order to confirm the reports in the literature on the opposing but complementary interactions of endothelial cells and platelets in blood coagulation/thrombus formation regulation in the body. The role of micro-vascular endothelial cell in preventing platelet aggregation and acting as a simple model in haemo-compatibility assessment is hereby affirmed. This investigation also tries to demonstrate that since these interactions do not occur either separately or in sequence but, at about the same time, the micro-structural finish of the biomaterial before implantation becomes crucial therefore in determining the fate of the biomaterial in-vivo.

11.12.1 Endothelial Cell-Platelet Interactions on a-C:H and a-C:H:Si Thin Films

Human micro-vascular endothelial cells were seeded on a-C:H and a-C:H:Si thin films for about six hours (6 hours), washed with PBS twice, followed finally by seeding of human platelets for 30mins. This section of the study acts as an adjunct to the reports already presented in this chapter.

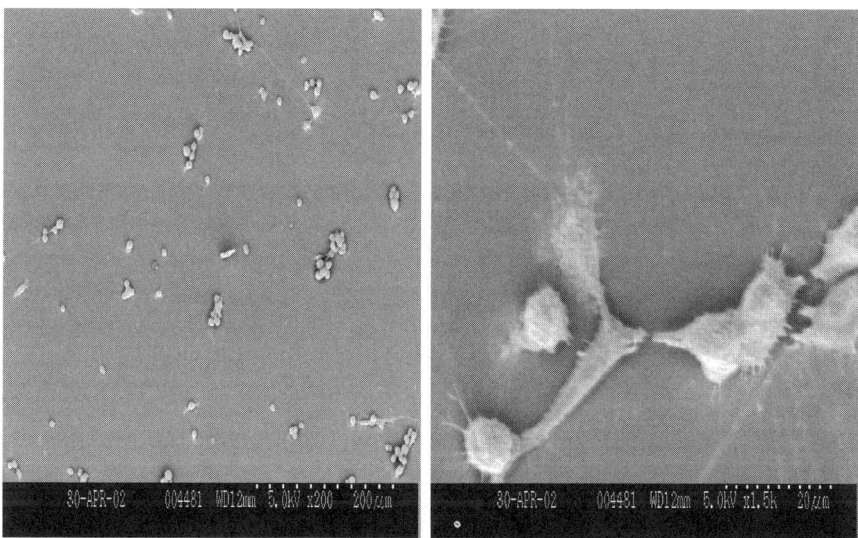

Fig. 11.14. Endothelial cell seeded on as obtained a-C:H for 6hrs and followed by platelets seeding for 30mins; x200 (left), x1500 (right)

In the above micrograph (Fig. 11.14), there seems to be no platelet aggregation present on the as obtained a-C:H. Instead there seems to be endothelial cell activation, and or aggregation in response to potential platelet attachment and or aggregation. Also there seems to be some changes in the morphology of the endothelial cells. However, the exact relation of this change in morphology in the presence of platelets is not known to the author. That is whether the endothelial cells engulf the platelets and or by its secretions wall off platelets where ever they seem to appear is not really understood by the author. The other possibility could be that the endothelial cell do not allow the platelets to adhere at all, such that they could all be washed off by the PBS washing (to remove proteins) before samples were prepared for imaging.

Under normal circumstances, platelets do not interact with the endothelial cells – that is platelet adhesion to the vessel wall and the formation of platelet aggregates do not normally take place except when required for haemostasis. Hence, the surface of endothelial cells does not promote platelet attachment [118]. The formation of platelet aggregates in close proximity to the endothelium is also rendered difficult by prostacyclin (PGI2), a powerful inhibitor of platelet aggregation secreted by the endothelial cells. Prostacyclin can be secreted by endothelial cells in culture as well as by isolated vascular tissue [119]. The vascular endothelium is now known to be a dynamic regulator of haemostasis and thrombosis with the endothelial cells playing multiple and active (rather than passive) roles in haemostasis and thrombosis [120–121].

Figures 11.15–17 show both the doped (Silicon, SD10) and non-doped DLC films, as deposited and thermally annealed (at low temperature, < 400C) pre-seeded with endothelial cells (6 hours) and then seeded with platelets. At both low (x200) and high magnifications (x1.5K) the features seen are neither that of platelet aggregation nor individual platelet adhesion (Fig. 11.15–17). Many of the functions of the endothelial cells appear to be anti-thrombotic in nature. Several

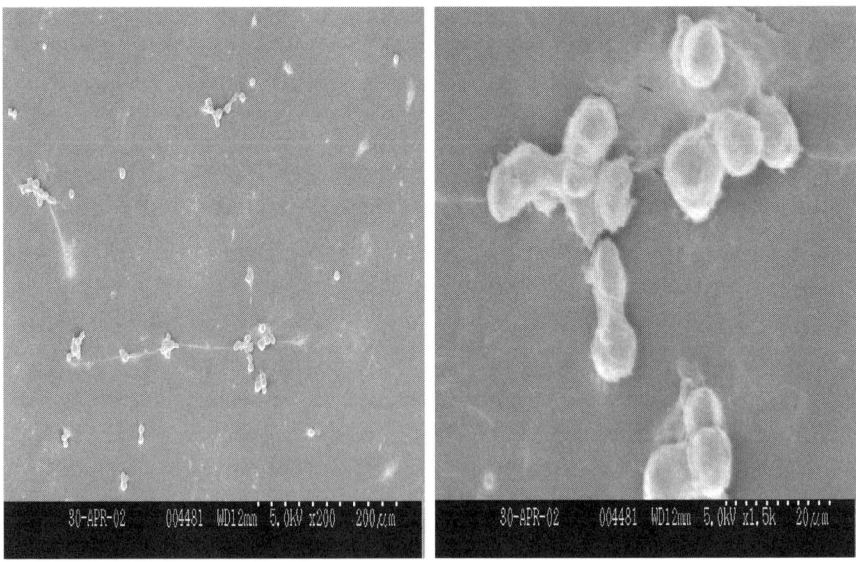

Fig. 11.15. Endothelial cell seeded on as obtained a-C:H:Si (SD10) for 6hrs and followed by platelets seeding for 30mins; x200 (left), x1500 (right)

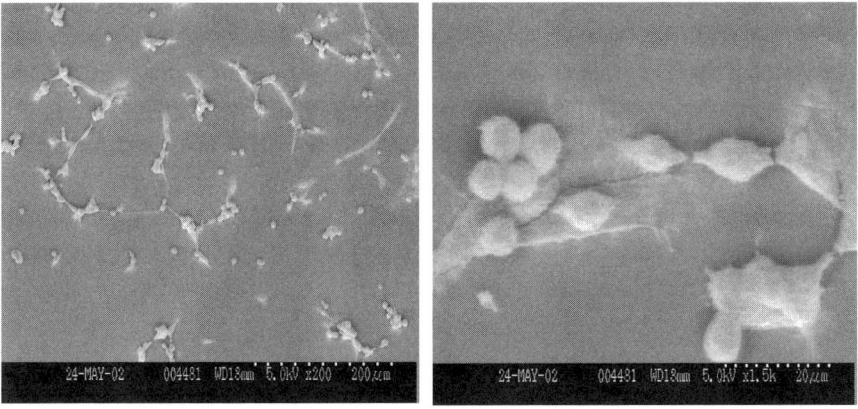

Fig. 11.16. Endothelial cell seeded on a-C:H (thermally annealed at 200°C) for 6hrs and followed by platelets seeding for 30mins; x200 (top), x1500 (bottom)

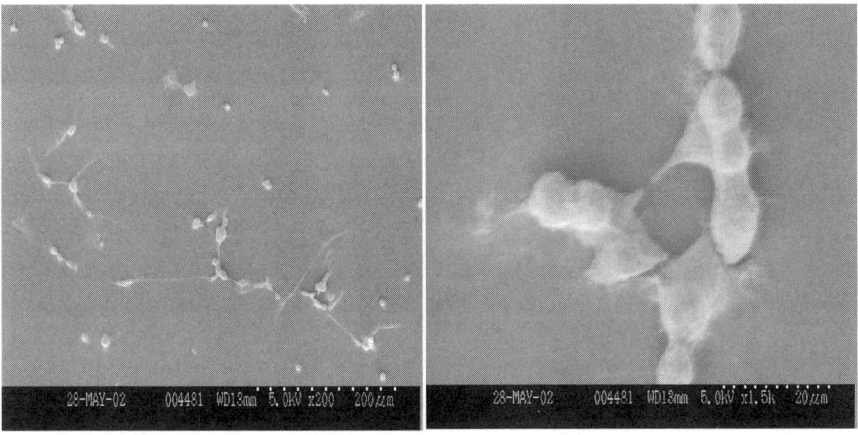

Fig. 11.17. Endothelial cell seeded on a-C:H:Si (SD10, thermally annealed at 200°C) for 6 hrs and followed by platelets seeding for 30mins; x200 (left), x1500 (right)

of the 'natural anticoagulant mechanisms,' including the heparin-antithrombin mechanism, the protein C-thrombomodulin mechanism, and the tissue plasminogin activator mechanism, are endothelial-associated. Among the proteins on the endothelial surface is antithrombin III [122] which catalyse the inactivation of thrombin by heparin. Endothelial cells also have heparan sulphate and dermatan sulphate (glycosaminoglycans) on their surfaces [123] which are known to have anticoagulant activity.

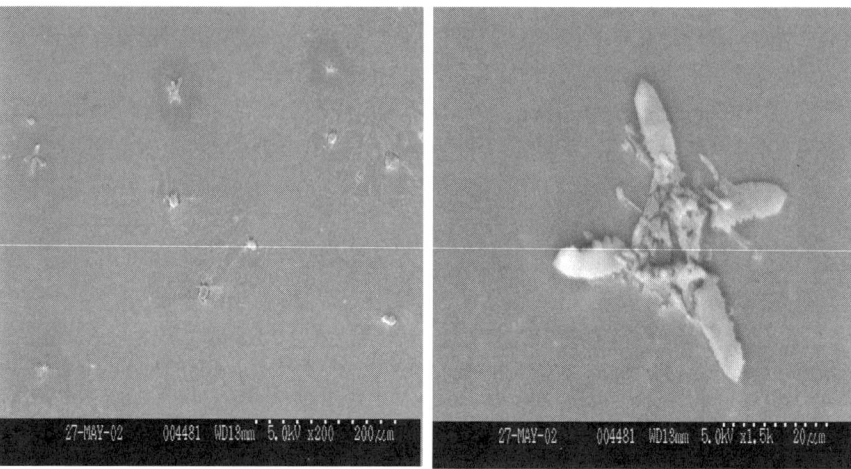

Fig. 11.18. Endothelial cell seeded on a-C:H (thermally annealed at 400°C) for 6hrs and followed by platelets seeding for 30mins; x200 (left), x1500 (right)

In this micrograph (Fig. 11.18) a bit strange star-shaped features that seemingly engulfing platelet aggregates are seen in this 400°C thermally annealed a-C:H. These could possibly be endothelial cell(s) in 'extreme activation'. Based on the results from previous chapters, fewer endothelial cells combating with more platelet aggregations is expected in this supposedly graphitized film. This could possibly lead to extreme endothelial cell activation and extreme flattening/spreading of the few endothelial cells expected to be interacting with platelet aggregations.

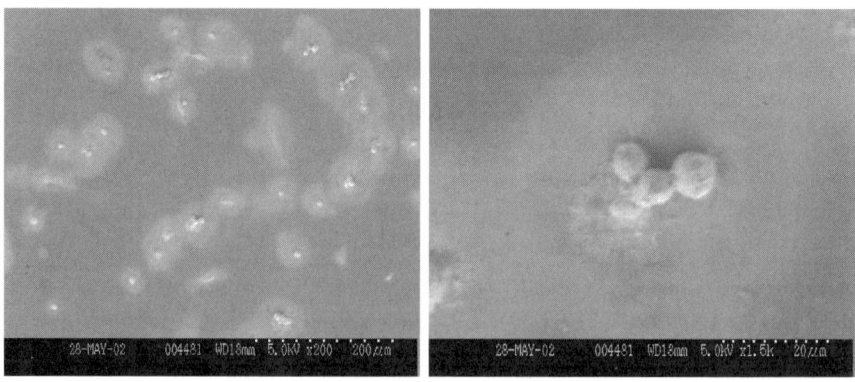

Fig. 11.19. Endothelial cell seeded on a-C:H:Si (SD10, thermally annealed at 400°C) for 6 hrs and followed by platelets seeding for 30mins; x200 (left), x1500 (right)

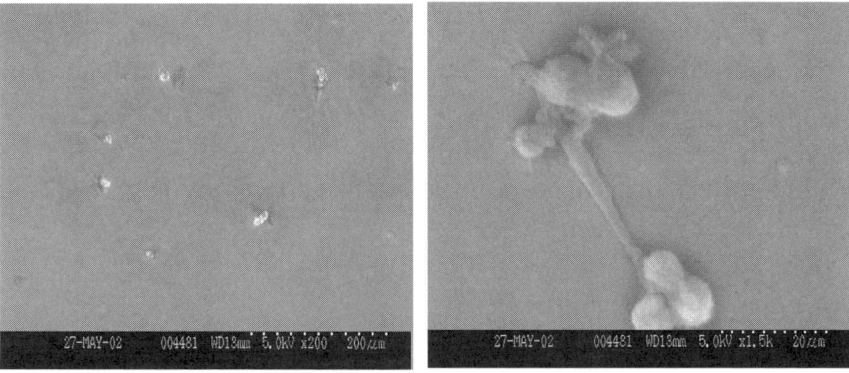

Fig. 11.20. Endothelial cell seeded on a-C:H (thermally annealed at 600°C) for 6hrs and followed by platelets seeding for 30mins; x200 (left), x1500 (right)

The clump seen above (Fig. 11.19) seems to be a platelet aggregate walled off by endothelial cell or endothelial secretions. No individual platelet or platelet aggregates are identifiable.

In this micrograph (Fig. 11.20) is seen also 'endothelial aggregates' and star-shaped endothelial cells in 'extreme activation'. This film is also expected to be graphitized.

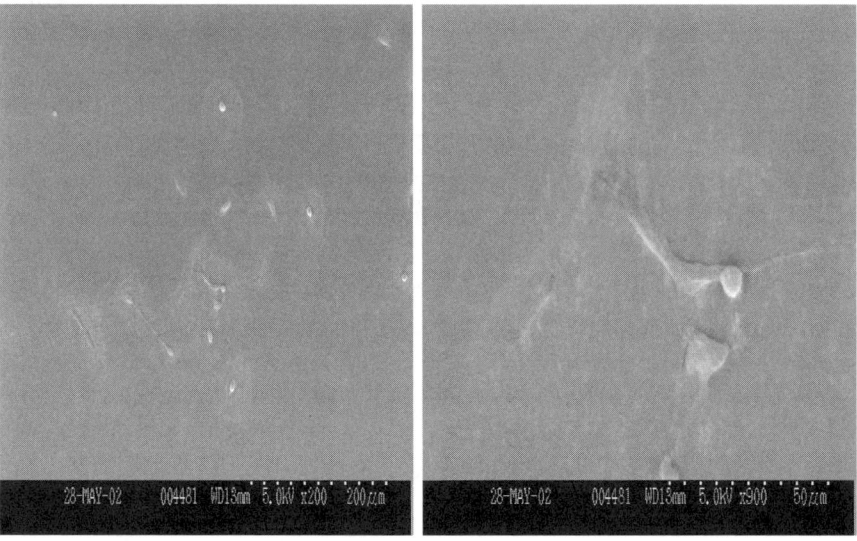

Fig. 11.21. Endothelial cell seeded on a-C:H:Si (SD10, thermally annealed at 600°C) for 6 hrs and followed by seeding with platelets for 30mins

The top left micrograph (Fig. 11.21) seems to show some platelet aggregates surrounded by a halo of what could be endothelial secretion. This preliminary study seems to show that in the presence of endothelial cells, platelet aggregation could be almost impossible. Therefore any biomaterial with good surface properties suitable for endothelial cell adhesion could be confirmed as haemo-compatible biomaterial until proven otherwise.

11.13 Bio-assays and Assessment of Intracellular Activities

Bioassays are generally adopted biochemical protocols for assessing intracellular activities such as free ions concentration, radicals or even membrane potentials. These bioassays are based on the assumption that biomaterial-cellular interactions occurring over a reasonable length of time at the cell-biomaterials interfaces are able to triggers some intracellular event/processes via the cell mediators. Cell viability, cell proliferation, cell morphology, and cytotoxicity, are commonly determined and some live-cell functions such as gene expression, apoptosis, level of protein phosphorylation, chemotaxis, endocytosis, cell secretions, cell transduction or production of cell signaling molecules such as nitric oxide or flux of Ca^{2+} and expression of cell adhesion molecules can be determined by some bioassays. The effect of the biomaterials surface features on first the overlying protein biofilms and the morphological changes induced on these proteins result in some cellular events. Various bioassays exist and involve in most cases the use of various biochemical agents and reagents to assess some of the expressed biochemical mediators and/ or enzymes (proteins). The results of the bioassays depends on the cell function being investigated and may be detected by microscopy, microplate (e.g., ELISA) readers, and flow cytometry. There is some challenge in using bioassay to detect early biomaterial-cellular interactions occurring at the cell-material interface. Assessment of cell proliferation can be achieved by various bioassays: (a) the detection of proliferation associated antigens by immunohistochemical techniques, e.g., Ki-67 antigen [176], proliferating cell nuclear antigen, PCNA; (b) quantification of DNA synthesis by measuring tritiated thymidine, ^{3}H-thymidine or bromodeoxyuridine, BrdU uptake); (c) measurement of changes in total DNA content with DNA specific dyes (e.g., Hoechst 33258); (d) Determination of intracellular metabolic activity or reduction state by Tetrazolium salts (MTT, XTT, MTS), or Alamar-Blue reduction; (e) Determination of the optical absorbance of Neutral Red stained cells. Cell viability can be assessed by various methods: (a) Trypan Blue exclusion and Propidium iodide exclusion as these are excluded from the viable cells; (b) CFDA staining, Crystal violet inclusion and Neutral red staining of viable cells; (c) Quantification of cell mediated cytotoxicity by measurement of the LDH (lactate dehydrogenase) activity, measurement of the release of ^{51}Cr or Europium Titriplex V from labeled cells; (d) AlamarBlue reduction.

11.13.1 MTT-Assay

MTT (3-(4,5-dimethylthiazol-2-yl)-2,5-diphenyl tetrazolium bromide; Sigma) was dissolved in PBS at 5mg/ml concentration and filter sterilised inside the hood to remove a small amount of insoluble residues usually present in some batches of MTT. About 5 hours before the end of the incubation period 20µl of MTT solution were added to each well including the blank wells (wells with added media, but no cells added). The plates were then transferred back to the incubator (37°C) for 5 hours. After the incubation period, the media were gently removed from all the wells with a syringe and 200µl of DMSO were added to each of the wells (DMSO was handled in the dark, because it is unstable in the light). The plates were returned to the incubator for 5 minutes in order to dissolve air bubbles. These samples were then transferred to the Titertek+plus MS2 Microelisa reader and the optical density were read using a test wavelength of 550nm and a reference wavelength of 660nm. The plates were read within one hour of adding DMSO.

11.13.1.1 The Interaction of a-C:H and a-C:H:Si Thin Films with Bovine Retinal Pericytes

Another carefully selected cell in this study is the pericytes which are cells supporting the blood vessel walls. The bovine pericytes used in this study is from the eye compartment (retinal) which is a special compartment of the body. The pericytes and the other cells (human embryonic lung cells, L132, and Chinese hamster-V79 cell lines) used in this study were essentially used to examine the possible cyto-toxicity/biocompatibility of the a-C:H:Si thin film biomaterial. The results (Fig. 11.22) show that when pericytes are seeded on 96-well plates coated with a-C:H and a-C:H:Si, the cells viability in these coated wells are comparable (and/ or generally slightly better) to those in the non-coated TCPS (control) wells [37]. This is in agreement with reports in the literature using some other cell lines [74–75]. a-C:H has also been reported to do well in both organ [74] and cell culture [75] when compared to the materials conventionally used for this purpose. It should be noted however that the MTT-assay of different cell line could be different due to the origin, nature, function and rate of proliferation/metabolic activation state of the cells in culture. Growing pericytes on the surfaces of these as deposited thin films coated on silicon wafers are shown below, a-C:H (Fig. 11.23) and SD10 (Fig. 11.24).

Pericytes are intimately associated with the vasculature and appear to be present in most tissues. They are generally considered to be res-tricted to the microvessels (arterioles, venules and capillaries where

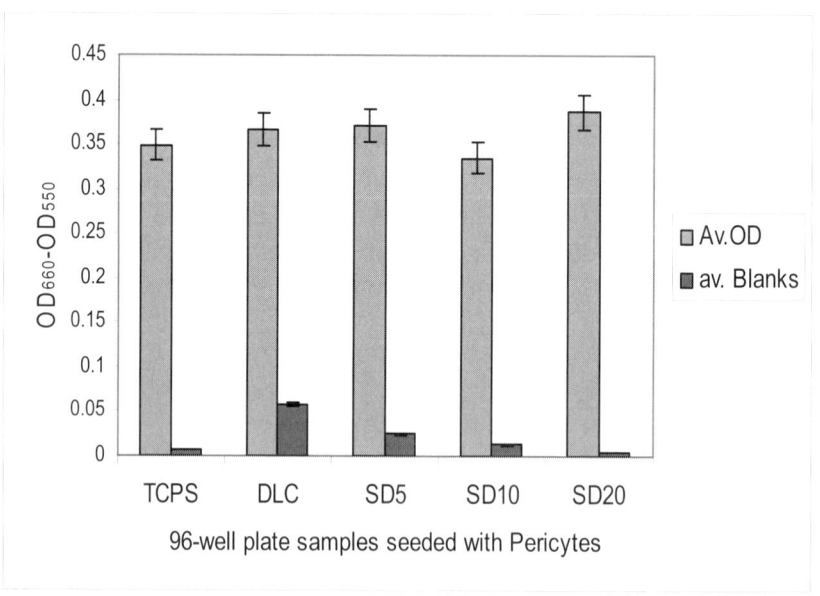

Fig. 11.22. MTT-assay of pericytes seeded on standard tissue culture 96-well plates coated with a-C:H and a-C:H:Si (SD5-20), TCPS=control (uncoated) [37]

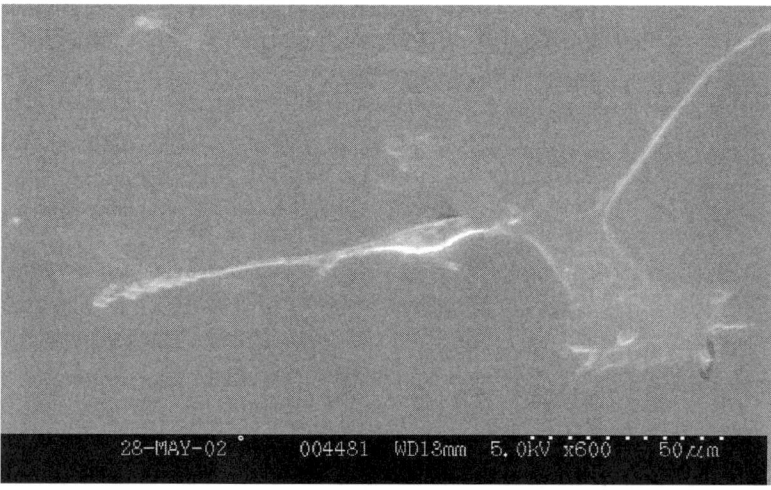

Fig. 11.23. Pericytes growing on the as deposited a-C:H after seeding for >12hrs [37]

there are no smooth muscles) [177]. Pericytes embrace capillary, and their nuclei bulge outward rather than inward like endothelial nuclei. Pericytes are thought to contribute to endothelial cell proliferation, via selective inhibition of endothelial cell growth and lack of pericytes has lead to endothelial hyperplasia and abnormal vascular morphogenesis in the brain [178]. They exhibit small, oval cell body with multiple processes extending for some distances along the vessel axis and these primary processes then give rise to orthogonal secondary branches which encircle the vascular wall. They function as macrophages and have multiple suggested functions. Pericytes have a close physical association with the endothelium. Gap junction communication between pericyte and endothelial cells, as well as at endothelial-endothelial junctions, has been shown *in vitro*.

This SEM image (Fig. 11.24) shows that bovine retinal pericytes grow well in silicon modified a-C:H thin films. On the whole these results show that bovine retinal pericytes grow well in both coated and uncoated TCPS tissue culture wells, thus the a-C:H and a-C:H:Si coatings are not toxic to these cells. Thus a-C:H and a-C:H:Si thin films could possibly function well in the eye compartment as materials for medical/optical prostheses. However, this study requires further

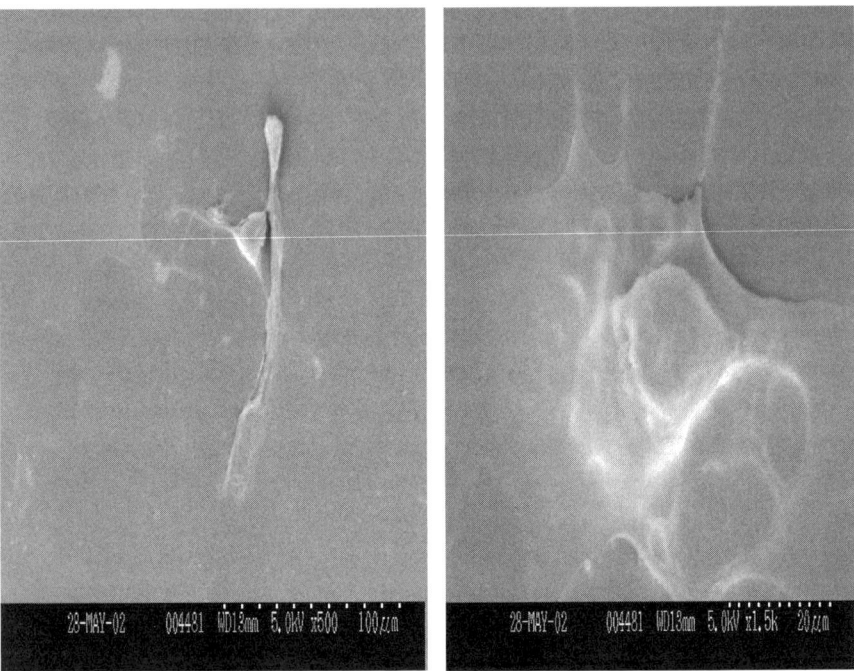

Fig. 11.24. Pericytes growing on the as deposited a-C:H:Si (SD10) after seeding for >12hrs

investigation possibly with human pericytes and possibly to assess the differential cell attachment in relation to micro-structural changes like the electrical properties, graphitisation and so on.

11.13.1.2 The Interaction of a-C:H and a-C:H:Si Thin Films with L132 Cell Line

The L132 cell line (human embryonic lung cells) originally purchased from ATCC (CCL-5), is epithelial and normal but with a Hela-characteristics. This cell line has Hela-contamination and thus proliferates in a very rapid rate characteristic of Hela-tumour cells. The results of MTT-assay of L132 cell lines on 96-well TCPS plates coated with a-C:H and a-C:H:Si (Fig. 11.25) show that L132 cell lines proliferate well on both the uncoated TCPS (control) and the a-C:H, a-C:H:Si coated wells.

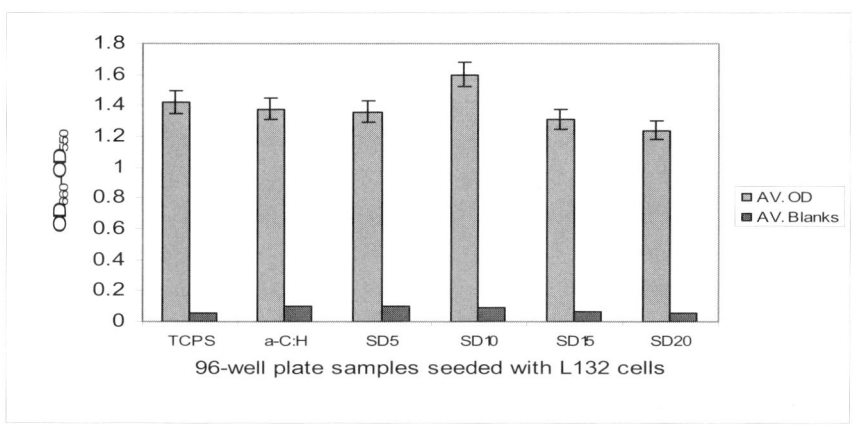

Fig. 11.25. MTT-assay of L132 cells seeded on standard tissue culture 96-well plates coated with a-C:H and a-C:H:Si (SD5-20), TCPS=control

Fig. 11.26. SEM image of L132 cell lines seeded on uncoated silicon wafer samples

In this SEM images (Fig. 11.26), the L132 cell lines is seen growing on silicon wafer but, the morphology of these cells are not as good as those of the coated samples (Fig. 11.27–30), indicating possibly less proliferation in these samples. This could be attributed to high surface energy of this uncoated silicon wafer substrate.

The L132 cell lines proliferate well on a-C:H samples as shown in Fig. 11.27. The morphology of these cells is clearly defined indicating a good proliferation in this sample.

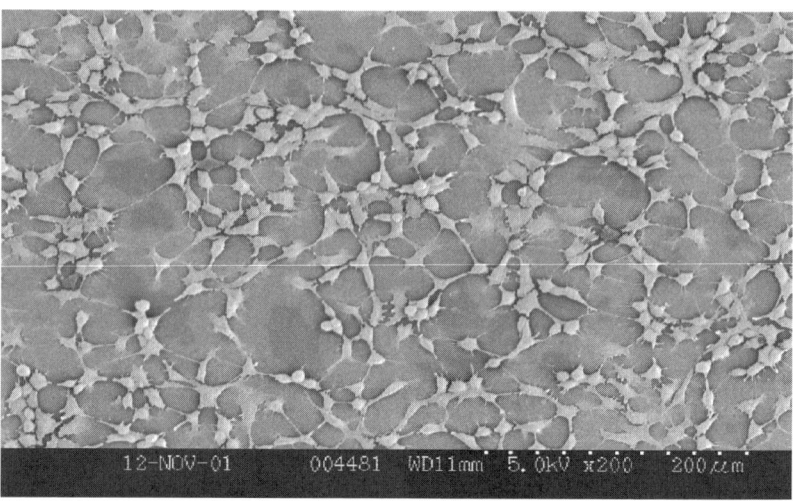

Fig. 11.27. SEM image of L132 cell lines seeded on as deposited a-C:H sample

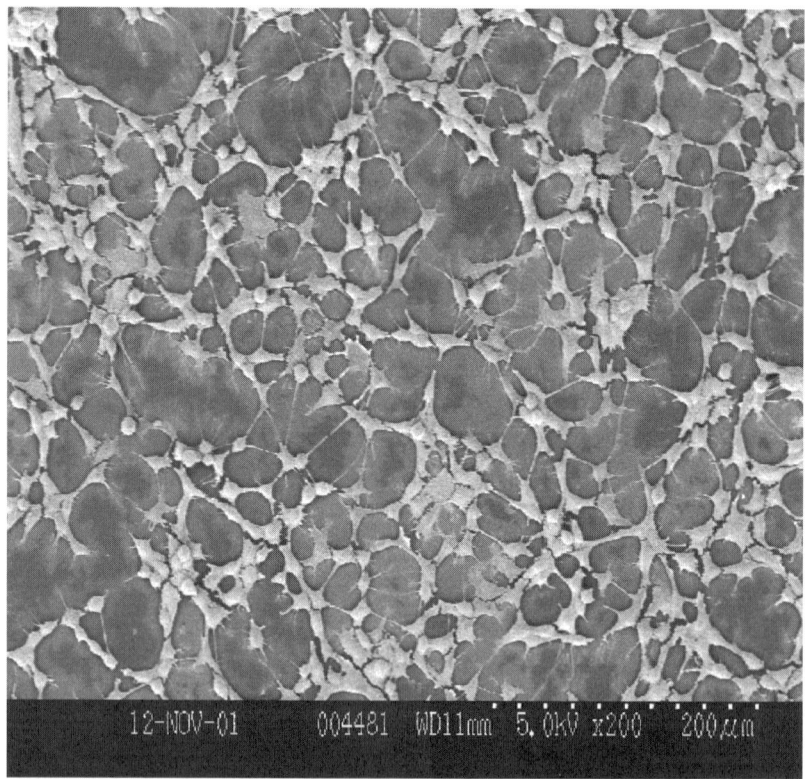

Fig. 11.28. SEM image of L132 cell lines seeded on as obtained a-C:H:Si (SD10) sample

Again the L132 cells are seen proliferating very well in this a-C:H:Si sample (Fig. 11.28). The morphology and spreading of these cells are better in this sample compared to the uncoated substrate. Compared to the a-C:H samples the morphology and spreading seems to be slightly better in these silicon doped samples.

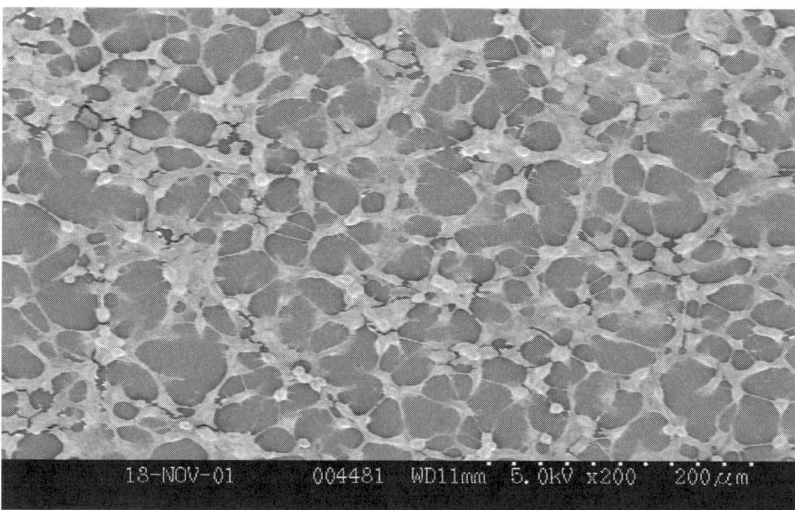

Fig. 11.29. SEM image of L132 cell lines seeded on a-C:H (thermally annealed at 400°C)

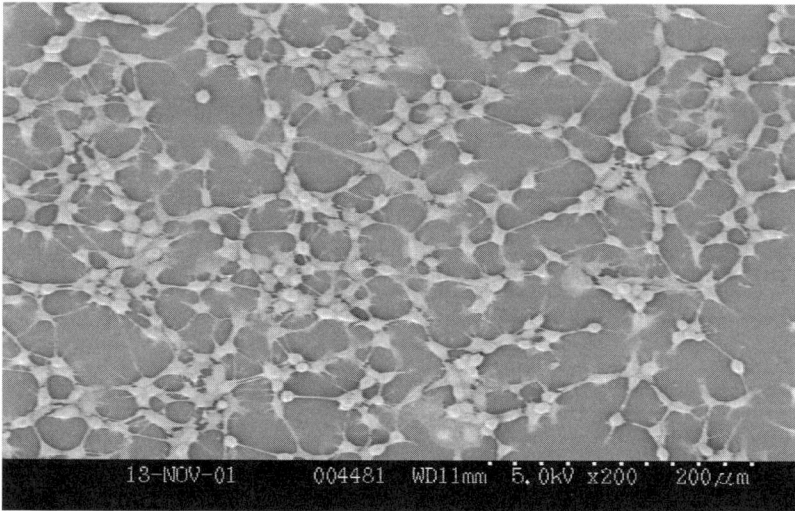

Fig. 11.30. SEM image of L132 cell lines seeded on a-C:H:Si (SD5, thermally annealed at 400°C)

In a-C:H (Fig. 11.29) and a-C:H:Si (Fig. 11.30) films thermally annea-led at 400°C the L132 cells are seen proliferating fairly well. Though the onset of graphitisation is expected at this temperature in the a-C:H film thermally annealed at 400°C, the proliferation of L132 cells on these samples does not seem to reflect this. This could be attri-buted to the nature of this Hela-contaminated cell line. These cell lines are contaminated with Hela-cells (high proliferation/cancerous) and are suitable for investigating cellular toxicity but may not be sen-sitive enough for assessing differential cell proliferation on closely varied sample surface properties due to this contamination on the original cell line.

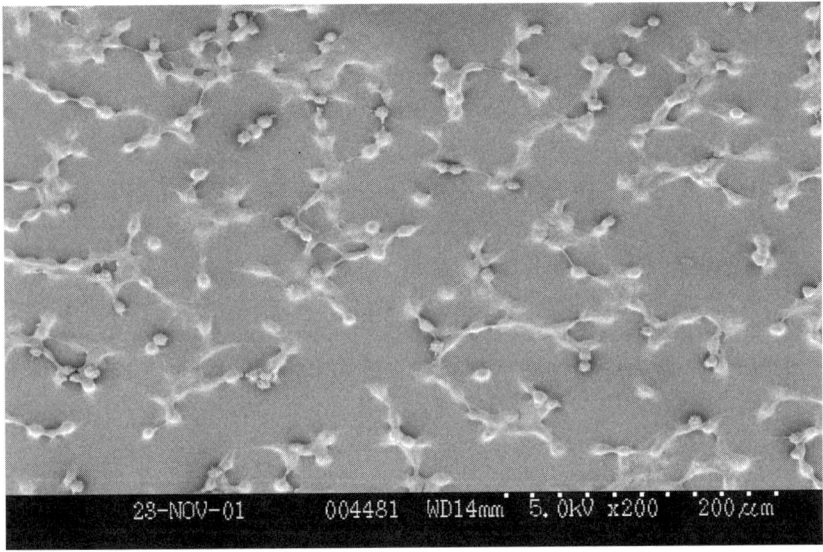

Fig. 11.31. SEM image of L132 cell lines seeded on a-C:H (thermally annealed at 600°C)

However, thermal annealing at higher temperature 600C, thought to produce high graphitic films does show a poor proliferating/ spread-ing L 132 cell lines (Fig. 11.31). This a-C:H sample thermally annealed at 600°C shows the L132 cells growing on the sample but with poorly defined morphological features similar to that observed on the un-coated sample (Fig. 11.26). This poorly defined morphology could be attributable to graphitisation in this film. Thus low contact angle (high surface energy) and graphitisation may lead to impaired L132 cell spreading and morphology.

11.13.1.3 *The Interaction of a-C:H and a-C:H:Si Thin Films with V79 Cell Line*

The figure (Fig. 11.32) shows the MTT-assay of Chinese Hamster cell line (V79) on uncoated TCPS (control), a-C:H and a-C:H:Si coated 96-well plate standard tissue culture plates. The cells proliferate well on the coated and uncoated control plates, but generally the cells seem to proliferate slightly better on the coated samples. This could be attributed to the change in surface energy in these as obtained thin film coatings. This result also agrees with reports in the literature using some other cell lines [74–75], a-C:H has been reported to do well in both organ [74] and cell culture [75] when compared to the materials conventionally used for this purpose.

In summary, it could therefore be inferred from this section that endothelial cells prevent platelet adhesion and aggregation. The a-C:H and a-C:H:Si thin films are not toxic to the bovine retinal pericytes, the L132 (human embryonic lung) cell line and the Chinese hamster (V79) cell lines. Based on these MTT-assay results it seems that these cells seem to proliferate slightly better in the a-C:H and a-C:H:Si coated wells compared to uncoated TCPS conventionally used for cell culture.

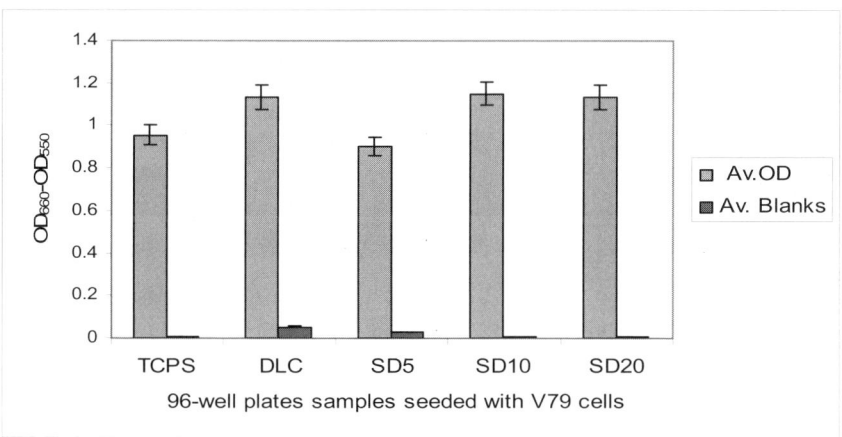

Fig. 11.32. MTT-assay of V79 cells (Chinese hamster) seeded on standard tissue culture 96-well plates (TCPS) coated with a-C:H and a-C:H:Si (SD)

11.13.2 Other Bioassays Techniques

Alamar Blue Assay

The alamar Blue dye contains a fluorometric/colorimetric growth indicator which detects metabolic activity in the cell. Cell growth causes a chemical reduction of the medium; this causes the Redox indicator to change from an oxidised, non-fluorescent blue colour, to a reduced, fluorescent red colour. The indicator has been shown to be minimally toxic to living cells and produces a clear, stable, distinct colour change. When the assay is completed the alamar Blue dye can be removed and the cells can be used for further experiments.

See samples with 1×10^5 cells per well in a 6 well plate with 3 ml of the RPMI medium. Incubate cells in standard incubator (37°C and 5% CO_2). Perform test a chosen/different day/time points. Prepare 1:20 dilution of the stock solution of alamar blue dye by adding 1 ml of alamar blue and 19 ml of Hank's Balanced Salt Solution (HBSS). Wash cells twice with HBSS and add the alamar blue dye to each well in 2.5 ml volumes. Incubate at 37°C and 5% CO_2 for 90 mins. Following incubation, volumes of 100 µl were taken from each well in duplicate and added to a black 96 well plate. Read the fluorescence on a fluorescence Reader at excitation 530 nm and emission 590 nm.

DNA Assay

The DNA content of cell suspensions can be measured by using the binding of bis-Benzamide to DNA. Prepare a stock solution of (Bis-Benzamide) 33258 by the addition of 1 mg of the dye to 1 ml deionised water, covered in foil and stored at 4°C. Prepare a 1:50 dilution of the 1 mg/ml Bis-Benzamide with TNE buffer, to give a working solution of 20 µg/ml. Lyse cells to release DNA, using a freeze thaw method by freezing at -80°C and allowed to thaw; repeat procedure three times. Add sample volumes of 50 µl to 50 µl of TNE buffer and 100 µl of bis Benzamide 33258. Read plate on the plate reader at excitation wavelength of 360 nm and an emission wavelength of 460 nm. Determined values against a calf thymus DNA standard curve and displayed as µg/ml.

Hydrogen Peroxide Production: Dichlorofluorescein Diacetate Assay

The dichlorofluorescein diacetate assay was employed to measure the amount of H_2O_2 peroxide production from cells eg. monocytes following growth on materials. The generation of reactive oxygen intermediates, such as H_2O_2, is usually increased in inflammatory, injury and repair processes.

Prepare a 1 mM stock solution of dichlorofluorocein diacetate by the addition of 4.87mg of dichlorofluorocein diacetate (DCFH-DA) to 10 ml of ethanol. Suspend monocytes in Hanks Balanced Salt Solution without phenol red. Seed samples with 5×10^5 cells in a 6 well plate, with 2 ml of Hanks Balanced Salt Solution. Incubated for 3–4 h in a standard incubator (37°C and 5% CO_2). As a positive control, 1 µg/ml PMA [179] should be added to cells to give a final concentration of 50ng/ml. After incubation, add 10.2 µl of 1000 µM DCFH-DA to each well and incubated for 1 h at 37°C and 5% CO_2. After incubation, transfer 100 µl to a 96 well plate. Read plate on a fluorescence plate reader at an excitation wavelength of 485 nm and an emission wavelength of 528 nm.

11.14 In-vivo Studies of Carbon-based Materials: Cell-Tissue Interactions In-situ

11.14.1 In-vivo Studies on the Biocompatibility and Hemocompatibility of DLC

Allen *et al* (2001) [110] implanted DLC-coated cobalt-chromium cylinders in the intramuscular locations in rats and in transcortical sites in sheep and their histological analysis of specimens retrieved 90 days after surgery showed that the DLC-coated specimens were well tolerated in both sites [110].

Fournier *et al* (2001) [180] has shown from their clinical and angiographic data that the ***hydrogenated silicon carbide coating*** of the Tenax coronary stent may indeed play a beneficial role in patient outcome, and should therefore be evaluated by prospective clinical trials. They implanted the prostheses (231 Tenax stents) in 206 patients (62

+/– 5 years) in the patients left anterior descending (51%) and right coronary arteries (36%). Their results show that revascularization was complete in 70%, elective in 80%, and the implantation was direct in 25% of the cases and that the procedure was successful in all the lesions, reducing stenosis from 62 +/– 16 to 16 +/– 10% and increasing the minimal luminal diameter from 0.81 +/– 0.40 to 2.61 +/– 0.59 mm. Also the TIMI flow was reduced in 30%, but normalized after the stent in all but one case. They also reported that the incidence of cardiac events was minimal: 1 acute thrombosis (0.5%) resolved by a new angioplasty and 1 non-Q myocardial infarction (0.5%) and finally at the 6-month clinical follow-up 10% of the patients presented complaints of angina greater than class II [180].

De Scheerder *et al* **(2000)** [181] investigated the in-vivo bio-interaction with one particular class of modified DLC coatings: diamond-like nanocomposite coatings (DLN or Dylyn, Bekaert, Kortrijk, Belgium). Either coated or non-coated stents were randomly implanted in two coronary arteries of 20 pigs so that each group contained 13 stented arteries. Pigs underwent a control angiogram at 6 weeks and were then sacrificed. They performed a quantitative coronary analysis before, immediately after stent implantation, and at 6 weeks using the semi-automated Polytron 1000 system (Siemens, Erlangen, Germany). They also performed a morphometry using a computerized morphometric program and their angiographic analysis showed similar baseline selected arteries and post-stenting diameters. At 6-week follow-up, they discovered no significant difference in minimal stent diameter and their histo-pathological investigation revealed a similar injury score in the 3 groups. According to De Scheerder *et al* (2000) [181] inflammatory reactions were significantly increased in the DLN-DLC coating group, thrombus formation was significantly decreased in both coated stent groups and neointimal hyperplasia was decreased in both coated stent groups; however, the difference with the non-coated stents was not statistically significant; and also area stenosis was lower in the DLN-coated stent group than in the control group ($41 \pm 17\%$ vs. $54 \pm 15\%$; $p = 0.06$). in their conclusion they indicated that the diamond-like nanocomposite stent coatings are compatible, resulting in decreased thrombogenicity and decreased neointimal hyperplasia and covering this coating with another diamond-like carbon film (DLC) resulted in an increased inflammatory reaction and no additional advantage compared to the single-layer diamond-like nanocomposite coating [181].

Tran *et al* **(1999)** [182] reviewed the mechanical heart valves' (MHV) thrombogenicity and pointed out that the application of surface modification technology to reduce the incidence of thrombus formation on MHV is a novel undertaking requiring the collaboration within the bioengineering and cardiothoracic surgery fields. From reviewing results of recent and past investigations, and their own preliminary study with diamond-like carbon coating (DLC) and plasma or glow discharge treatment (GDT) of MHV, they identified and discussed several potentially beneficial effects that may reduce the extent of valve-related thrombogenesis by surface modification: DLC and GDT may affect the surfaces of MHV in many ways, including cleaning of organic and inorganic debris, generating reactive and functional groups on the surface layers without affecting their bulk properties, and making the surfaces more adherent to endothelial cells and albumin and less adherent to platelets; therefore these different effects of surface modification, separately or in combination, may transform the surfaces of MHV to be more thromboresistant in the vascular system [182].

Dowling *et al* **(1997)** [69] implanted two DLC-coated and uncoated stainless steel cylinders into both cortical bone (femur) and muscle (femoral quadriceps) sites of six adult (>40kg) sheep, for a period of 4 weeks (three sheep) and the rest for 12 weeks. According to Dowling *et al* (1997) after explantation of the implants and the pathological/ histological examination of the implanted cylinders, no macroscopic adverse effect was observed on both the bone and the muscles of the used sheep [69].

Yang *et al* **(1996b)** [9–10] examined in-vivo interactions of dics coated with TiN, DLC (deposited on SS316L disc using PVD) and or Pyrolytic carbon (PyC) films, implanted into the descending aorta of anaesthesized sheep (6 animals) for 2 hours. They evaluated the three different samples simultaneously in each animal. After explantation they examined the thrombus free area on the disc with close-up photography and planimetry, and the test surfaces with SEM. Yang *et al* [9–10] found out that there were many leucocytes adherent, activated and spread onto PyC and DLC, but on TiN it was the erythrocytes that were mainly adherent [9–10].

Patients using implanted prostheses are faced with life threatening bleeding problems because they are kept under life-long anticoagulant therapies in order to reduce the risk of thrombo-embolism. In order to reduce the risk of platelet aggregation/thrombo-embolism and complications following the life-long course(s) of anticoagulants, the biomaterials need to be improved in order to achieve better haemo-compatibility. Platelet aggregation in the surfaces of these prostheses is the key factor in thrombus formation. The platelets is known to play a crucial role in blood clotting/thrombus formation which is indicative of the ability of a foreign body to trigger off clot forma-tion and or thrombosis which may impair free flow of blood and result in some damaging effect on the internal body organs.

11.14.2 Summary

In this chapter, the role of microstructure, electrical properties and surface energy of amorphous hydrogenated carbon (a-C:H) and silicon modified a-C:H (a-C:H:Si) in relation to their biocompatibility/haemo-compatibility interactions with human micro-vascular endothelial cells and human platelets have been investigated in full. Preliminary investi-gations on nitrogen modified films (a-C:H:N) and a-C:H, a-C:H:Si interactions with other cell lines have been carried out. The findings are summarised as follows.

Microstructure

- The microstructure was tailored by the films deposition para-meters, silicon and nitrogen doping and thermal annealing
- Surface roughness: a-C:H, a-C:H:Si and a-C:H:N are ultra-smooth thin films that can be used to improve the surface features necessary for improved biocompatibility in biomedi-cal implants and devices made with conventionally rougher metals or polymers for example. Silicon incorporation seems to slightly increase the surface roughness. Thermal annealing also seems to increase slightly the surface roughness.
- Intrinsic Compressive stress: stress reduction in the film seems to be important in improving biocompatibility. Stress reduction

was achieved by thermal annealing of the films. Silicon incorporation also seems to reduce the intrinsic stress in the film. Lowered intrinsic stress is associated with decreased surface energy.

- Graphitisation: excessive graphitisation seems to impair biocompatibility. Graphitisation occurs at annealing temperature greater than 400°C in a-C:H thin films and may occur as well in a-C:H:Si film at a higher annealing temperatures of ~600°C. Silicon seems to lower the rate and degree of graphitisation in a-C:H and thereby improve hemocompatibility.
- The sp^2/sp^3 ratio and the sp^2 cluster size in the sp^3 matrix and not the absolute sp^2 or sp^3 content play a role in hamocompatibility.
- Silicon doping of a-C:H thin films decreases the I_D/I_G ratio and on annealing the I_D/I_G ratio starts to increase above 300°C of annealing temperature.

Electrical properties

- Conductivity/resistivity: silicon incorporation lowers the resistivity (Fig. 11.33) of a-C:H thin film from 6.7×10^{10} Ω.cm (a-C:H) to 5.4×10^9 Ω.cm (a-C:H:Si) and thereby increased the conductivity. Increased conductivity (without graphitisation) is associated with an improved haemo-compatibility (Fig. 14.1). Resistivity behaviour is electric field dependent, and for high amount of silicon, the resistivity first increases and then later decreases as the field increases ($\geq 1.5–1.8 \times 10^4$ V/cm).
- Work Function: Lower work functions in the examined films seem to improve biocompatibility. Silicon doping and nitrogen doping seem to lower the work function. Silicon doping lowers the work function from 4.77eV (a-C:H) to 4.56–4.34eV (a-C:H:Si), a reduction of 0.21–0.43eV. Silicon atomic percentage concentration of up to ~7.61% (TMS = 10sccm) led to a rapid jump in decrease of work function (Fig. 11.34).
- Band gap: Optical band gap was determined by ellipsometry technique. Silicon addition seems to increase the optical band gap energy. The optical gap values depend on the type of transition assumed and the model used for the calculation.

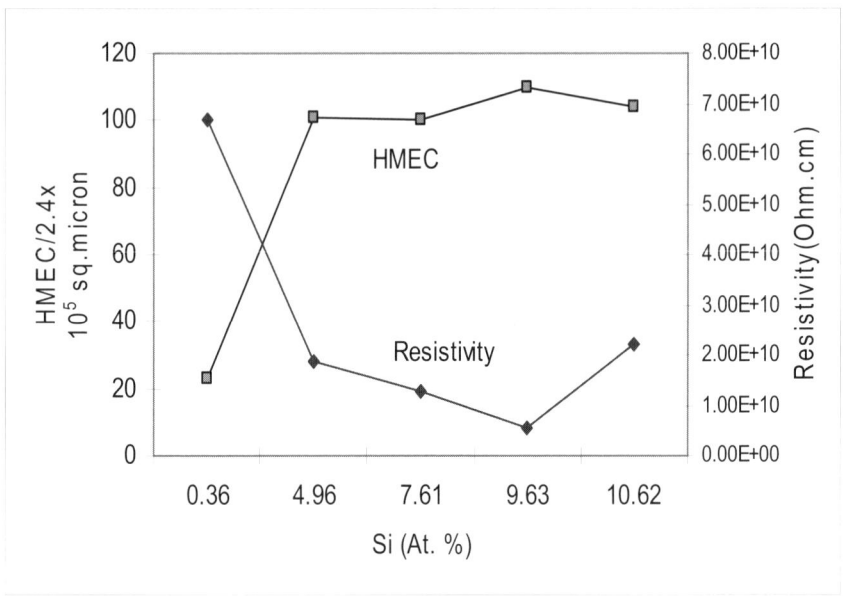

Fig. 11.33. HMEC adhesion related to resistivity as a function of percentage atomic concentration of silicon in a-C:H thin films

- The author suggests that the Density of states (DOS) and charge carriers could play a greater role compared to the optical band gap energy. Moderate amount of silicon (≤ 10 at.%) seems to increase the DOS in the sp^2 cluster region even though silicon does not form π-bond. There is however the need to establish the exact DOS distribution, at the extended states, at the mobility edge, the band tail or the defect states in the valence and conduction band within the band structure; and for the band gap determination there is need to ascertain the degree (percentage) of different electrons in (assumed different) transitions.

Surface energy/contact angle

- Contact angle: a higher contact angle seems to improve hemocompatibility. Silicon incorporation increases the contact angle (by up to $\sim 17^{\circ}$) while nitrogen incorporation decreases it (Fig. 11.35).

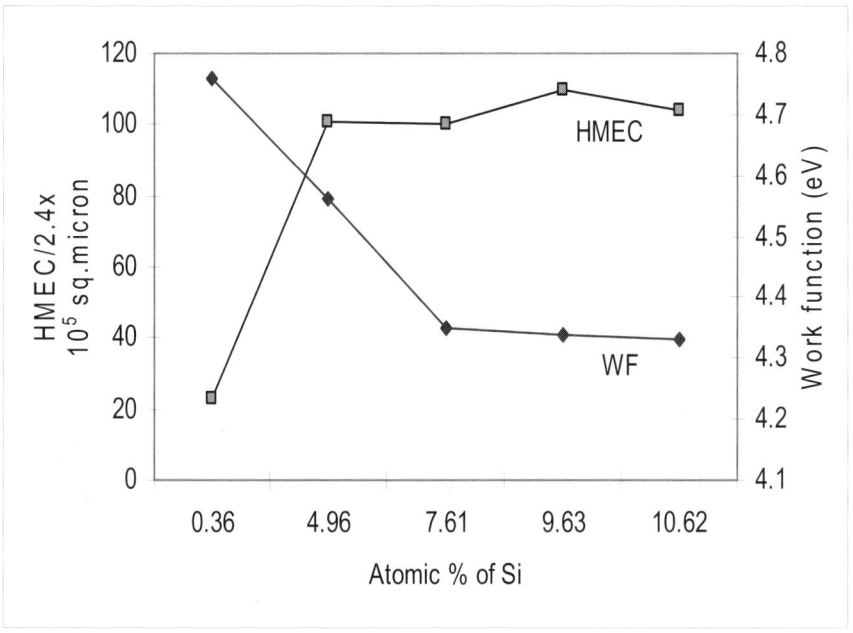

Fig. 11.34. HMEC adhesion related to the work function of a-C:H thin films as a function of percentage atomic concentration of silicon in the a-C:H thin films

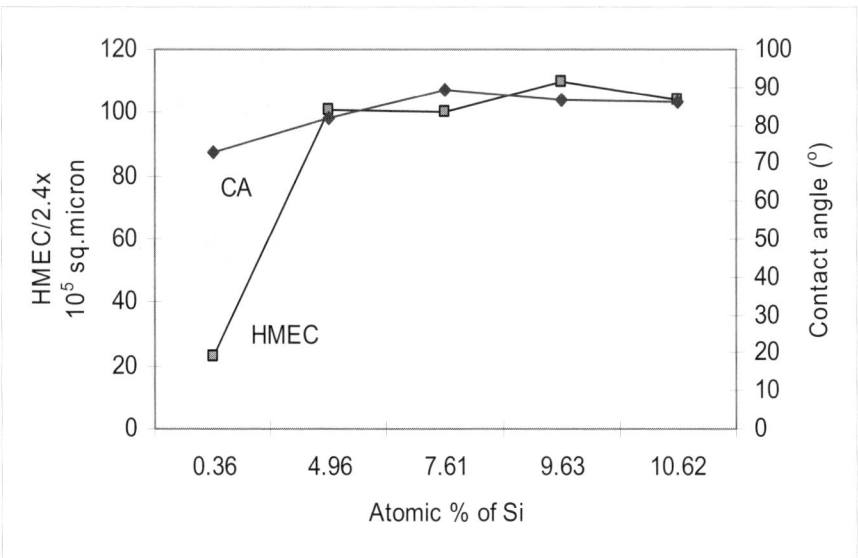

Fig. 11.35. HMEC adhesion related to the water contact angle of a-C:H films as a function of percentage atomic concentration of silicon in a-C:H thin films

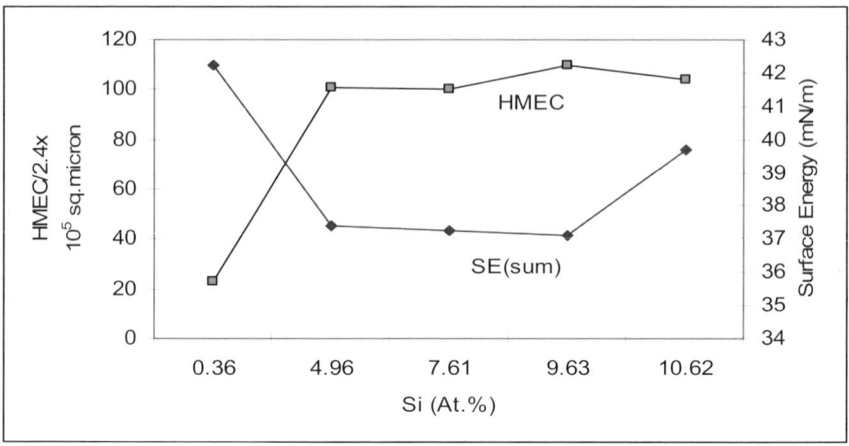

Fig. 11.36. HMEC adhesion related to the surface energy (sum) of a-C:H films as a function of percentage atomic concentration of silicon in a-C:H thin films

- Surface energy: About 99% of the surface energy components of a-C:H:Si and a-C:H are dispersive and seem to be responsible for the biocompatibility of these thin films. Silicon addition decreased the surface energy by ~5–10mN/m (Fig. 11.36–11.37).

Cellular interactions

- Cellular origin-specie differences: Human cells are relevant to biocompatibility studies for biomedical use in man and seem to be much more sensitive than cells from other species that may not be relevant to man.
- Cell specificity/function/sensitivity: Human cells should be carefully chosen from tissue of interest for biocompatibility studies in relation to those tissues for specific application in the particular tissue, organ and or system. Endothelial cells and platelets seem to be the most suitable cells in the blood compatibility studies.
- The results of these investigations have shown that microstructural changes in the films, especially the electrical conductivity, work functional changes, the contact angle/surface energy and the degree of graphitisation in these films play a key role in especially the haemo-compatibility of these thin films. These physical changes are easily detected when these thin films interact with micro-vascular endothelial cells and

human platelets in vitro. However, with other cell lines which may not be human and which may not play crucial roles like the endothelial cells and platelets, these changes may not easily be detected. Nevertheless these other cells may be good enough in investigations to say whether or not a material is just cyto-toxic.

- MTT-assay reveals that a-C:H and a-C:H:Si are generally not toxic to the cells, and may in some cases even encourage better cell proliferation when compared to tissue culture polystyrene (TCPS) conventionally used for cell culture.

Film adhesion in biological fluids

- Si-DLC coated stainless steel substrates were also immersed in various biological fluids (Saliva, PBS and FCS) similar to human body fluids incubated at body temperature. The adhe-sion properties of the films were tested using both the four-point bend test and the pull tensile substrate plastic straining techniques. With the use of scanning electron microscope (SEM) the crack initiation strain and the saturation crack spacing were determined. Statistical analysis of the data using two-parameter Weibull, lognormal probability density function models as well as Gamma function suggests long term relia-bility and good adhesion of Si-DLC to stainless steel and meta-llic substrates if used in biomedical implants and devices in continuous contact with body fluids.

Silicon doping of a-C:H lowers the work function, the resistivity, the surface energy of the films and the rate and degree of graphiti-sation in the films. Silicon doping also improves the adhesion of a-C:H thin films to its substrates. The a-C:H:Si thin films and conducting a-C:H thin films with less internal stress and less degree of graphiti-sation are good haemo-compatible materials suitable for biomedical applications as in heart valve prostheses, stents, etc.

Generally a-C:H and a-C:H:Si are non-toxic to cells in-vitro. Also a-C:H:Si coated stainless steel on immersion and incubation inside biological fluids have shown good shear strength and are therefore suitable for use for biomedical applications.

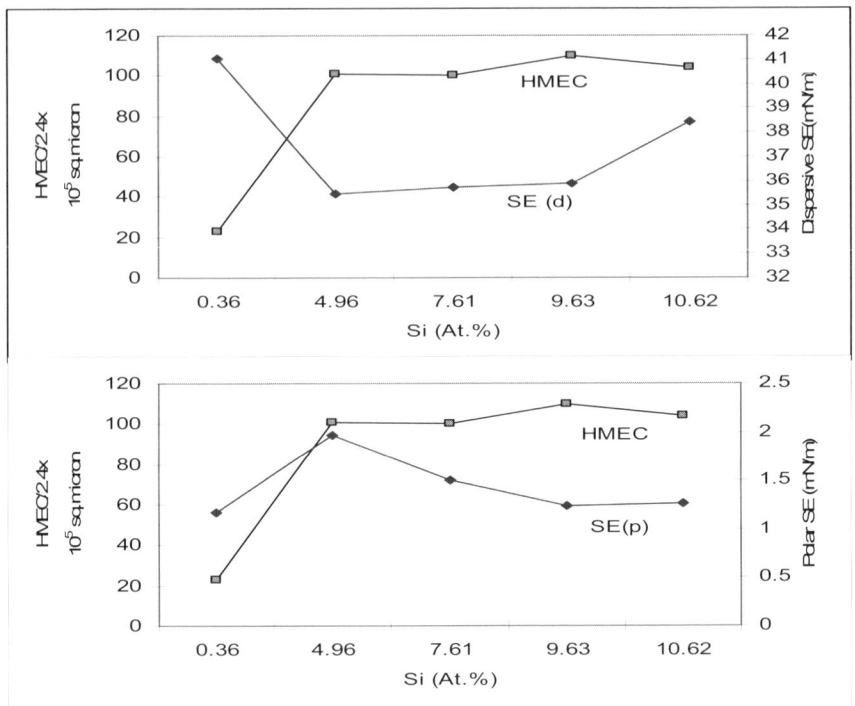

Fig. 11.37. HMEC adhesion related to the surface energy (dispersive and polar components) of a-C:H films as a function of percentage atomic concentration of silicon in a-C:H thin films

11.15 On-going and Future Investigations

- ^{13}C-NMR and the x-ray excited Auger electron spectroscopy studies of a-C:H:Si.
- Determination of exact amorphous-crystal structure, bond angles, bond angle disorders, number of and or distributions of dangling bonds, and dislocations in the films.
- Density of States (DOS) determination in a-C:H, a-C:H:Si, a-C:H:N, and a-C:H:Me
- Determination of the effective Band gaps of the a-C:H:Si - Protein-Cellular system and to relate the interactions with the DOS and or the band gap.
- The reported results and discussions termed 'preliminary investigations' are under further investigation.

- The interaction of a-C:H and a-C:H:Si with proteins with adhesive and non-adhesive functions needs to be investigated in order to see the exact role of these proteins in biocompatibility and haemo-compatibility. The role and the mechanism interaction of Cell Adhesion Molecules (CAM) should be investigated. Surface Plasmon Resonance (SPR) technique is envisaged to be useful in monitoring these interacting proteins in-situ.
- The effects of a-C:H, a-C:H:Si and a-C:H:N optical properties on cellular growth and interaction in-situ.
- Optical techniques to monitor the underside of cells, the mechanism of cell attachment, cell proliferation and density of distribution over the attached surfaces of a-C:H, a-C:H:Si and a-C:H:N coated transparent/glass substrates. By measuring the refractive indices of the glass, the medium (protein layer without cells), the cells and how the refractive index changes, the cell density and the cell cytoskeleton and how they change when the cell attaches to the surface can be monitored (Fig. 11.38).

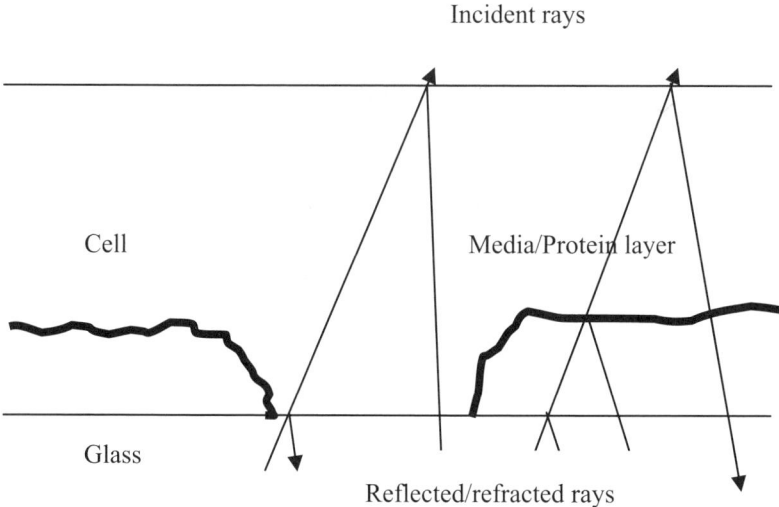

Incident rays

Cell

Media/Protein layer

Glass

Reflected/refracted rays

Fig. 11.38. Illustration of principles of optical method of cell characterization [183]

The relative intensity, R, of the reflected light at the interfaces of the glass and cell is given by:

$$R = \left(\frac{r_g - r_c}{r_g + r_c} \right)^2 \qquad (1)$$

$$r_c = r_g \times \frac{1 - \sqrt{R}}{1 + \sqrt{R}}$$

Where r_c = Refactive index of cell; r_g = refractive index of glass.

In this case, the refractive index of a-C:H, a-C:H:Si (SD) or a-C:H:N thin film coating has to be substituted in the above equation:

$$R = \left(\frac{r_{SD} - r_c}{r_{SD} + r_c} \right)^2 \qquad (2)$$

$$r_c = r_{SD} \times \frac{1 - \sqrt{R}}{1 + \sqrt{R}}$$

The cytoplasmic content of the cells follows the distribution of the cellular attachments on the substrates/samples. Therefore, regions of focal or more condensed cellular attachment imply heavy density of cytoplasm and a relatively higher refractive index. The cellular microfilaments and cytoskeleton are condensed in the region of attachment, thereby allowing less amount of light to pass through and thus a higher refraction occurring. Well spread cells on the other hand have more widely distributed ctoplasmic content since the volume of the cytoplasm is relatively the same (the larger the area, the lesser the thickness for the same volume of material inclusion). However, there has to be an assumption that the same thickness of protein interlayer exists between every cell and the substrate at the points of cell attachment, and at regions where there is no attached cells but just layer of protein/culture media. For further reading refer to [184–185].

- Further investigation on the role of surface charges is needed. By applying a small electric field (not to affect the cells) through the film while the cells are growing in culture will help to redistribute the charges (See Fig. 11.39).

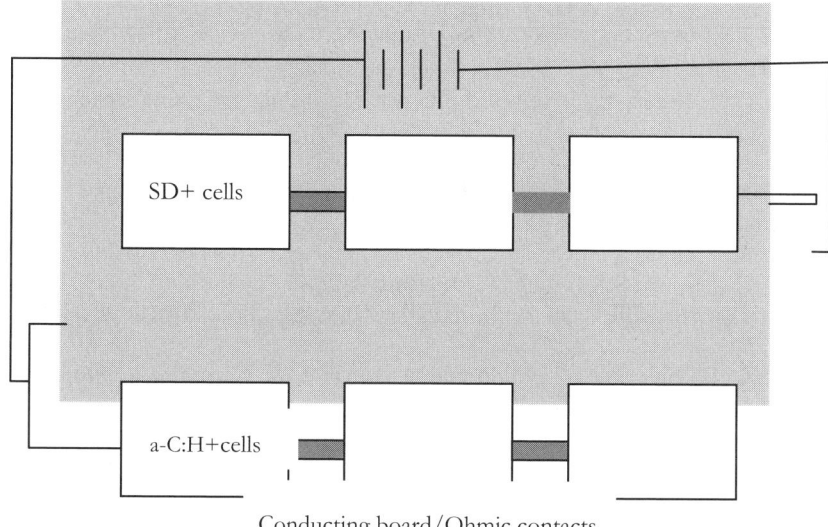

Fig. 11.39. Schematics of the layout for applying electric field on the coatings with seeded cells in culture

- Similarly impedance spectroscopy can be used to monitor the cell growth and distribution on the a-C:H, a-C:H:Si and a-C:H:N thin film coatings in-situ.
- For a-C:H, a-C:H:Si and a-C:H:N to find their way into routine use as biomedical prostheses, they need to be tested in-vivo with animals with close physiological relation with the humans, e.g., the baboons.
- Extensive clinical trials of a-C:H, a-C:H:Si and a-C:H:N coated implant.

References

1. Bittl J.A., Advances in coronary angioplasty. N. Engl. J. Med. 335 (1996) 1290-1302.
2. Gawaz M., Neumann FJ, Ott I, May A, Schomig A., Circulation 94 (1996) 279-285
3. Inoue T, Sakai Y, Fujito T, Hoshi K, Hayashi T, Takayanagi K, Morooka S, Sohma R; Circulation 94 (1996) 1518-1523
4. Lahann J., D. Klee, H. Thelen, H. Bienert, D.Vorwerk, H. Hocker; J. Mater. Sci. Mater. In Med. 10 (1999) 443-448.

5.   Haycox C.L., Ratner B.D.; J. Biomed. Mater Res. 27 (1993) 1181-1193.
6.   Courtney J.M., Lamba N.M.K., Sundaram S., Forbes C.D.; Biomaterials 15 (1994) 737-44.
7.   Klein C.L., Nieder P., Wagner M., Kohler H., Bittinger F., Kirkpatrick C.J., Lewis J.C.; J. Pathophysiol. 5 (1994) 798-807.
8.   Gutensohn K., Beythien C., Koester R., Bau J., Fenner T., Grewe P., Padmanaban K., Schaefer P., Kuehnl P.; Infusionstherapie und Transfusionmedizin, 27(4) (2000) p. 200-206.
9.   Yang Y., Franzen S.F., Olin C.L.; Cells and Materials, 6(4) (1996) 339-354.
10. Yang Y., Franzen S.F., Olin C.L.; J. Heart Valves Dis., 5 (1996) 532-537.
11. Bittl J.A., Subacute stent occlusion: thrombus horribilis. JACC 28 (1996) 368-370.
12. Mark K., Belli G., Ellis S., Moliterno D.; J. Am. Coll. Cardiol 27 (1996) 494-503.
13. Colombo A., Hall P., Nakamura S., Almagor Y., Maiello L., Martini G., Gaglione A., Goldberg S.L., Tobis J.M.; Circulation 91 (1995) 1676-1688.
14. Gott V.L., Koepke D.E., Daggett R.L., Zarnstorff W., Young W.P. The coating of intravascular plastic prostheses with colloidal graphite. Surgery 1961; 50: 382-389.
15. Haubold A. Annals N.Y. Aca. Sc. 283 (1977) 383.
16. Goodman S.L., Tweden K.S., Albrecht R.M. Platelet interaction with pyrolytic carbon heart-valve leaflets. J. Biomed. Mater Res. 1996; 32: 249-258.
17. Baier R.E., Bull N Y Aca. Med. Vol 48 (1972b) 273.
18. Williams D.F.; J. Biomed. Eng., Vol. 11 (1989) 185.
19. Salzman E. (Ed.); 1981; Interaction of blood with natural and artificial surfaces; Marcel Dekker, 1981.
20. Gordon J.L., 1986 in Blood-Surface Interactions: Biological Principles Underlying Hemocompatibility with Artificial Materials edited by Cazenave J.P, Davies J.A, Kazatchkine M.D, and W.G van Aken; Elsevier Science Publishers (Biomedical Division) 1986; pp. 5.
21. Cenni E., Arciola C.R., Ciapetti G., Granchi D., Savarino L., Stea S., Cavedagna D., Curti T., Falsone G., Pizzoferrato A.; Biomaterials 16 (1995) 973-976.
22. Herring M.B., Gardner A. and Gloves J.A.; Surgery 84 (1978) 498.
23. Remy M., Bordenave L., Bareille R., Rouais F., Baquey Ch., Gorodkov A., Sidorenko E.S., Novikova S.P.; Journal of Materials Science: Materials in Medicine 5 (1994) 808.
24. Pesakova V., Klezl Z., Balik K. and Adam M., Journal of Material Science: Materials in Medicine 11 (2000) p. 797.
25. Hallab N.J., Bundy K.J., O'Connor K., Clark R., Moses R.L., Journal of Long-term effects of medical implants 5 (1995) p. 209.
26. Ahluwalia A., Basta G., Chiellini F., Ricci D., Vozzi G., Journal of Material Science: Materials in Medicine 12 (2001) 613-619.
27. Bowlin G.L., Rittger S.E., Cell Transplantation 6 (1997) 623.
28. Altankov G., Grott T., J. Biomaterial Sci. Polymer Edn 8 (1997) 299.
29. Grinnell F., Int. Rev. Cytol. 53 (1978) p. 65.

30.   Van Wachem P.B., Schakenraad J.M., Feijen J., Beugeling T., van Aken W.G., Blaauw E.H., Nieuwenhuis P., Molenaar I.; Biomaterials 10 (1989) 532-539.
31.   Van Wachem P.B., Beugeling T., Feijen J., Bantjes A., Detmers J.P., van Aken W.G.; Biomaterials 6 (1985) 403-408.
32.   McLaughlin J., Meenan B., Maguire P., Jamieson N.. Properties of Diamond like carbon thin film coatings on stainless steel medical guidewires. Diamond Rel Mater 1996; 8: 486-491.
33.   Jones M.I., McColl I.R., Grant D.M., Parker K.G., Parker T.L.. Hemocompatibility of DLC and TiC-TiN interlayers in titanium. Diamond Rel Mater 1999; 8: 457-462.
34.   Okpalugo T.I.T., Ogwu A.A., Maguire P., McLaughlin J.A.D.; Technology and Health Care: Int. J. Health Care Engineering, 9(1-2) 2001, 80-82.
35.   Okpalugo T.I.T., Ogwu A.A., Maguire P.D., McLaughlin J.A., Hirst D.G. In-vitro blood compatibility of a-C:H:Si and a-C:H thin films; Diamond and Related Materials, 2004; 13 (4-8): 1088-1092.
36.   Okpalugo T.I.T., Ogwu A.A., Maguire P.D., McLaughlin J.A. Platelet adhesion on silicon modified hydrogenated amorphous carbon films; Biomaterials, 2004; 25 (3): 239-245.
37.   Okpalugo T.I.T., McKenna E., Magee A.C., McLaughlin J.A., Brown N.M.D. The MTT assays of bovine retinal pericytes and human microvascular endothelial cells on DLC and Si-DLC-coated TCPS; Journal of Biomedical Materials Research Part A, 2004; 71A (2): 201-208.
38.   Okpalugo T.I.T., Maguire P.D., Ogwu A.A., McLaughlin J.A. The effect of silicon doping and thermal annealing on the electrical and structural properties of hydrogenated amorphous carbon thin films; Diamond and Related Materials, 2004; 13 (4-8): 1549-1552.
39.   Okpalugo T.I.T, Ogwu A.A., Maguire P.D., McLaughlin J.A., McCullough R.W. Human micro-vascular endothelial cellular interaction with atomic N-doped compared to Si-doped DLC; Journal of Biomedical Materials Research Part B; Applied Biomaterials 2006; 78B (2): 222-229.
40.   Okpalugo T.I.T., 2002, G2c., Ph.D., Ulster, 53-4066. The hemocompatibility of ultra-smooth silicon and nitrogen doped hydrogenated amorphous carbon thin films - the role of the microstructure, electrical properties, and surface energy. (BL: DXN062999)
41.   Parker T.L., Parker K.L, McColl I.R., Grant D.M., Wood J.V.; Diamond and Related Materials 93 (1993) 118.
42.   Dion I., Roques X., Baquey C., Baudet E., Basse Cathalinat B., More N.; Biomed. Mater. Eng. 3 (spring 1993) 51-55.
43.   O'Leary A., Bowling D.P., Donnelly K., O'Brien T.P., Kelly T.C., Weill N., R.Eloy, Key Engineering Materials 99-100 (1995) 301-308.
44.   Freitas R.A., IMM report number 12, http://www.imm.org/reports/rep012.html
45.   Allen M., Law F.C, Rushton N., Clin. Mater. 17 (1994) p. 1-10.
46.   Allen M.J., Myer B.J., Law F.C., Rushton N., Trans. Orthop. Res. Soc. 20 (1995) 489.
47.   Szent-Gyorgyi A.; 'Bioenergetics', Academic Press, New York, 1957.
48.   Szent-Gyorgyi A.; Nature 157 (1946), 875.

49. Eley D.D., Parfitt G.D., Perry M.J., Taysum D.H.; Trans. Faraday Soc. 49 (1953) 79.
50. Postow E. and Rosenberg B., Bioenergetics, 1 (1970), 467.
51. Bruck S.D., Polymer 6 (1965), 319.
52. Bruck S.D., Polym. Sci. (C), 17 (1967) 169.
53. Bruck S.D., Intrinsic semiconduction, electronic conduction of polymers and blood compatibility. Nature 1973; 243: 416-417.
54. Bruck S.D., The role of electrical conduction of macromolecules in certain biomedical problems. Polymer 1975; 16: 25.
55. Van Oss C.J. Phagocytosis as a surface phenomenon. Ann Rev Microbiol 1978; 32: 19-39.
56. Kochwa S., Litwak R.S., Rosenfield R.E., Leonard E.F.; Annals of New York Academy of Sciences, 1977, 283, p. 37.
57. Lettington A.H.; Applications of Diamond Films and Related Materials, Materials Science Monographs, 73 (1991) edited by Tzeng Y., et al; Elsevier. New York, p. 703.
58. Evans A.C., Franks J., and Revell P.J.; Surf. Coat. Technol. 47 (1991) 662-667.
59. Grill A., Diamond and related materials 8 (1999) 428.
60. Gutensohn K., Beythien C., Bau J., Fenner T., Grewe P., Koester R., Padmanaban K., Kuehnl P.; Thrombosis Research 99 (2000) 577-585.
61. Gutensohn K., Beythien C., Koester R., Bau J., Fenner T., Grewe P., Padmanaban K., Schaefer P., Kuehnl P.; Infusionstherapie und Trans-fusionmedizin, 27(4) (2000) p. 200-206.
62. Zheng C., Ran J., Yin G., Lei W.; in Applications of Diamond Films and Related Materials, Materials Science Monographs, Vol. 73 (1991) edited by Tzeng Y., et al., Elsevier. New York, p. 711.
63. Jones M.I., McColl I.R., Grant D.M., Parker K.G., Parker T.L.; J. Biomed Mater Res, 52 (2) (2000) 413-421.
64. Alanazi A., Nojiri C., Noguchi T., et al; ASAIO Journal, 46 (4) 2000, p. 440-443.
65. Alanazi A., Nojiri C., Noguchi T., Ohgoe Y., Matsuda T., Hirakuri K., Funakubo A., Sakai K., Fukui Y; Artificial Organs 24(8) (2000) 624-627.
66. Bangali and Shea; MRS Bulletin 2005; 30 (9): 659.
67. Morrison M.L., Buchanan R.A., Liaw P.K., Berry C.J., Brigmon R.L., Riester L., Abernathy H., Jin C. and Narayan R.J. Electrochemical and antimicrobial properties of diamond like carbon-metal composite films. Diamond and Related Materials, Volume 15, Issue 1, January 2006, Pages 138-146.
68. Maizza G., Saracco G., Abe Y., In Vincenzini P editor. Advances in Science and Technology, 9th Cimetec-World forum on New Materials, Faenza: 1999, p. 75-82.
69. Dowling D.P., Kola P.V., Donnelly K., Kelly T.C., Brumitt K., Lloyd L., Eloy R., Therin M., Weill N.; Diamond and Rel. Mater. 6 (1997) 390-393.
70. Veli-Matti Tiainen; Diamond and Rel. Mater. 10 (2001) 153-160.
71. Butter R.S., Lettington A.H.; DLC for biomedical applications (Reviews); J. Chem. Vapour Depo. 3 (1995) 182-192.

72. Higson S.P.J and Pankaj M. Vadgama; Analytica Chemica Acta 300 (1995) 77-83.
73. Higson S.P.J. and Pankaj M. Vadgama, Biosensors and Bioelectronics 10 (5) 1995, p. viii.
74. Du C., Su X.W., Cui F.Z., Zhu X.D.; Biomaterials 19 (1998) 651-658.
75. Cui F.Z. and Li D.J., Surface Coatings Technol. 131 (2000) 481-487.
76. Ivanov-Omskii V.I., Panina L.K., Yastrebov S.G.; Carbon 38 (2000) 495-499.
77. Dyuzhev G.A.; Ivanov-Omskii V.I., Kuznetsova E.K., Rumyantsev V.D., et al; Mol Mat 8 (1996) 103-106.
78. Ivanov-Omskii V.I., Tolmatchev A.V., Yastrebov S.G., Phil Mag B 73 (4) 1996; 715-22.
79. Andrade J.D., ed. Surface and interfacial aspect of biomedical polymers Vol. 2, Protein Adsorption, New York: Plenum, 1988.
80. William D.F., Physiological and microbiological corrosion CRC Crit.Rev; Biocompatibility, Vol. 1 (1985), 1-30.
81. William D.F. (ed.) Definitions in Biomaterials, Elsevier, 1987.
82. William D.F., Systemic Aspects of Biocompatibility, Vol. 1-2. Boca Raton: CRC Press 1981.
83. Peppas N.A., Adv. in Chemistry Series 199, American Chemical Society (1982).
84. Martini F.C., 2001; Fundamentals of Anatomy and Physiology; 5th Edition; Prentice Hall (2001), New Jersey, USA.
85. Hoffman A.S., Advances in Chemistry Series 199 (1982) 3.
86. Vroman L.; in Vroman L. and Leonard E.F. (Eds) Ann. N.Y. Ac. Sc. 283 (1977) 65.
87. NHLBI (USA) Guidelines, 1980.
88. Neumann A.W., Absolom D.R., Francis D.W., Omenyi S.N., Spelt J.K., Policova Z., Thomson C., Zingg W., van Oss C.J.; Ann. N.Y. Ac. Sci., 416 (1983) 276.
89. Srinivasan S. and Sawyer P.N., J. Colloid and Interface Sci., Vol. 32 (3) (1970) 456.
90. Sawyer P.N. and Pate J.W., Am. J.Physiol. 1953; 175: 113.
91. Sawyer P.N. and Srinavasan S., Am. J. Surgery 1967; 114: 42.
92. Srinivasan, S., and Sawyer P.N., JAAMI, 3 (1969) 116.
93. Martin J.G., Afshar A., Kaplitt M.J., Chopra P.S., Srinivasan S., Sawyer P.N. Implantation studies with some non-metallic prostheses; Trans. Amer. Soc. Artif. Int. Organs 14 (1968) 78.
94. Wilcox C.D., Dove S.B., McDavid W.D. and Greer D.B. Imagetool. http://ddsdx.uthscsa.edu/dig/itdesc
95. Baier R.E., Bull N Y Aca. Med. Vol. 48 (1972) 273.
96. Baier R.E., Loeb G.I., Wallace G.T.; Fed. Proc. 30 (1971) 1523-1538.
97. Chen J.Y., Leng Y.X., Tian X.B., Wang L.P., Huang N., Chu P.K., Yang P. Antithrombogenic investigation of surface energy and optical bandgap and hemocompatibility mechanism of Ti (Ta+5)O2 thin films. Biomaterials 2002; 23: 2545.

434   Surface Engineered Surgical Tools and Medical Devices

98. Curtis, A. Tutorial on the Biology of Nanotopography. IEEE Transactions on Nanobioscience, 2004; 3 (4): 293-295.
99. Matsuda T., Kurumatani H. Surface induced in vitro angiogenesis: surface property is a determinant of angio-genesis. ASAIO Trans 1990; 36: M565-568.
100. Hubbell J.A., Massia S.P., Drumheller PD. Surface-grafted cell-binding peptides in tissue engineering of vascular graft. Ann. N.Y. Aca. Sci. 1992; 665: 253-258.
101. Goodman S.L., Lelah M.D., Lambrecht L.K., Cooper S.L., Albrecht R.M.; Scanning Electron Microscopy 1 (1984) 279.
102. Dowling D.P., Kola P.V., Donnelly K., Kelly T.C., Brumitt K., Lloyd L., Eloy R., Therin M., Weill N.; Diamond and Rel. Mater. 6 (1997) 390-393.
103. Allen M., Law F.C., Rushton N., Clin. Mater. 17 (1994) p. 1-10.
104. Hauert R., Muller U., Francz G., Birchler F., Schroeder A., Mayer J., Wintermantel E.; Thin Solid Films 308-309 (1997) 191-194.
105. Allen M., Butter R., Chandra L., Lettington A., N. Rushton; Biomed. Mater. Eng. 5 (3) (1995) 151-159.
106. McColl I.R., Grant D.M., Green S.M. et al; Diamond and Rel. Mater. 3 (1993) 83.
107. Parker T.L., Parker K.L., McColl I.R., Grant D.M., Wood J.V.; Diamond and Related Materials 93 (1993) 118.
108. Parker T.L., Parker K.L., McColl I.R., Grant D.M., Wood J.V.; Diamond and Related Materials 3 (1994) 1120-1123.
109. Thomson L.A., Law F.C., Rushton N., Franks J.; Biomaterials 12 (1991) 37-40.
110. Allen, M., Myer B., Rushton N.; J. Biomed. Mater Res. 2001; 58 (3) 319-328.
111. Schroeder A., Gilbert Francz, Arend Bruinink, Roland Hauert, Joerg Mayer, Erich Wintermantel; Biomaterials 21 (2000) 449-456.
112. Lu L., Jones M.W., Wu R.L.C.; Biomed. Mater. Eng. 3 (1993) 223.
113. Evans A.C., Franks J., and Revell P.J.; Surf. Coat. Technol. 47 (1991) 662-667.
114. Ames B.N., McCann J., Yamasaki E.; Mutat. Res. 31 (1975) 347-367.
115. Bruck S.D., Biomat., Med. Dev. Art. Org. 5 (1) 1977.
116. McHargue C. J in: Tzeng Y et al (eds), Applications of Diamond Films and Rel. Mater., Mater. Sci. Monographs, Elsevier, New York, 1991, p. 113.
117. Devlin D et al in: Simons B (ed) ASME International Mechanical Engineering Congress and Exposition, Proceeding (1997), ASME, Bioengineering Division, Fairfield, NJ, USA, 1997, p. 265.
118. Gordon J.L., 1986 in Blood-Surface Interactions: Biological Principles Underlying Hemocompatibility with Artificial Materials edited by Cazenave J.P, Davies J.A, Kazatchkine M.D, and W.G van Aken; Elsevier Science Publishers (Biomedical Division) 1986; pp. 5.
119. Moncada, S. and Vane, J.R. (1982) The role of prostaglandins in platelet-vessel wall interactions. In Pathobiology of endothelial cells (H.L Nossel and H.J Vogel, eds) pp.253-285. New York Academic Press.
120. Gimbrone M.A., Jr, in M.A Gimbrone Jr. Ed. Vascular Endothelium in Hemostasis and Thrombosis, 1986, Churchill Livingstone, Edinburgh, pp. 1-13.

121. Gimbrone M.A., Jr., Annals of New York Acad. Sci. 516 (1987) 5-11.
122. Chan, T.K. and Chan, V. (1981): Antithrombin III, the major modulator of intravascular coagulation is synthesised by human endothelial cells. Thrombosis and Haemostasis 46 (1981) 504-506.
123. Busch, C., Ljungman, C., Heldin, C-M., Waskson, E. and Obrink, B. (1979): Surface properties of cultured endothelial cells. Haemostasis, 8 (1979) 142-148.
124. Jaffe, E.A. (1982): Synthesis of factor VIII by endothelial cells. Annals of New York Academy of Sciences, 401 (1982) 163-170.
125. Mosher, D.F., Doyle, M.J., and Jaffe, E.A. (1982): Secretion and synthesis of thrombospondin by cultured human endothelial cells. Journal of cell biology, 93 (1982) 343.
126. Folkman J., Haudenschild C.: Angiogenesis in vitro; Nature 288 (1980) 551-556.
127. Tonnesen M.G., Smedly L.A., Henson P.M.; J. Clin. Invest. 74 (1984) 1581-1592.
128. Kubota Y., Kleinman H.K., Martin G.R., Lawley T.J.; J. Cell Biol. 107 (1988) 1589-98.
129. Pauli B. and Lee C., Lab Invest. 58 (1988) 379-387.
130. Picker L.J., Nakache M., Butcher E.C.; Monoclonal antibodies to human lymphocyte homing receptors define a novel class of adhesion molecules on diverse cell types; J. Cell Biol. 109 (2) (1989) 927-937.
131. Pober J., Am. J. Pathol. 133 (1988) 426-433.
132. Berg E.L., Goldstein L.A., Jutila M.A., Nakache M., Picker L.J., Streeter P.R., Wu N.W., Zhou D., Butcher E.C.; Immunol. Rev. 108 (1989) 1-18.
133. Rice G.E. and Bevilacqua M.P., Science 246 (1989) 1303-1306.
134. Springer T., Nature 346 (1990) 425-433.
135. Hynes R., Cell 69 (1992) 11-25.
136. Folkman J., Haudenschild C., Zetter B.R.; Proc. Natl. Acad. Sci. USA 76 (1979) 5217.
137. Keegan A., Hill C., Kumar S., Phillips P., Schof A., Weiss J.; J. Cell Sci. 55 (1982) 261.
138. Charo I., Karasek M.A., Davison P.M., Goldstein I.M., J. Clin. Invest. 74 (1984) 914.
139. Gerritsen M.E., Biochem. Pharmacol. 36 (1987) 2701-2711.
140. Fujimoto T. and Singer S.J., J. Histochem. Cytochem. 36 (1988) 1309-1317.
141. Kubota Y., Kleinman H.K., Martin G.R., Lawley T.J.; J. Cell Biol. 107 (1988) 1589.
142. Ades E.W., Candal F.J., Swerlick R.A., George V.G., Summers Susan, Bosse D.C, Lawley T.J; J. Invest. Dermatol. 99 (1992) 683-690.
143. Van Wachem P.B., Beugeling T., Feijen J., Bantjes A., Detmers J.P., van Aken W.G.; Biomaterials 6 (1985) 403-408.
144. Van Wachem P.B., Schakenraad J.M., Feijen J., Beugeling T., van Aken W.G., Blaauw E.H., Nieuwenhuis P., Molenaar I.; Biomaterials 10 (1989) 532-539.
145. Kaukonen M., Nieminen R.M., Poykko S., Settsonen A. Nitrogen Doping of Amorphous Carbon Surfaces. Phys Rev Lett 1999; 83 (25): 5346-5349.

146. Ganong W.F., Rev. of Med. Physiol., 17th. Ed., Appleton & Lang (1995).

147. Chen J.Y., Wang L.P., Fu K.Y., Huang N., Leng Y., Leng Y.X., Yang P., Wang J., Wan G.J., Sun H., Tian X.B., Chu P.K.; Surface and Coatings Technology 156 (2002) 289-294.

148. Krishnan L.K, Varghese N., Muraleedharan C.V., Bhuvaneshwar G.S., Derangere F., Sampeur Y., Suryanarayanan R.; Biomolecular Engineering (2002) 1-3.

149. Gutensohn K., Beythien C., Bau J., Fenner T., Grewe P., Koester R., Padmanaban K., Kuehnl P.; Thrombosis Research 99 (2000) 577-585.

150. Ogwu A.A., Lamberton R.W., McLaughlin J.A., Maguire P.D.; J. Phys. D: Appl. Phys. 32 (1999), 981.

151. Jiu J-T., Hao Wang, Chuan-Bao Cao, He-Sun Zhu; J. Mater. Sci. 34 (1999) 5205-09.

152. Dementjev A.P., Petukhov M.N., Baranov A.M.; Diamond and Related Materials 7 (1998) 1534-1538.

153. Dementjev, A.P. and Petukhov M.N., Diamond and Related Materials 6 (1997) 486.

154. Grill A., Meyerson B., Patel V., Reimer J.A., and Petrich M.A.; J. Applied Physics 61 (1987) 2874.

155. Miyake S., Kaneko R., Kikuya Y., Sugimoto I.; Trans. ASME J. Tribol. 113 (1991) 384.

156. Baker M.A. and Hammer P., Surface and Interface Analysis, 25 (1997), 629-642.

157. Demichelis F., Pirri C.F., Tagliaferro A.; Mater. Sci. Eng. B 11 (1992) 313-316.

158. Li D.J., Cui F.Z., Gu H.Q.; J. Adhesion Sci. Technol. 13 (1999) 169.

159. Linder Stefan, Wolfhard Pinkowski, Martin Aepfelbacher; Biomaterials 23 (2002) 767-773.

160. Goodman S.L., Cooper S.L., Albrecht R.M.; J. Biomater. Sci. Polymer Edn. 2(2) (1991) 147-159.

161. Tangen D., Berman H.J., Marfey P.; Throm. Diath. Haemorrh. 25 (1971) 268.

162. Schakenraad J.M., Busscher H.J., Ch. R.H Wildevuur, Arends J.; Cell Biophys. 13 (1988) 75.

163. Goodman S.L., Cooper S.L., Albrecht R.M.; Progress in Artificial Organs, pp. 1050-1055; Y. Nose, C. Kjellstrand, P. Ivanovich (Eds.) ISAO Press, Cleveland, OH, 1985.

164. Schakenraad J.M, Busscher H.J., Ch. R.H Wildevuur, Arends J.; J. Biomed. Mater. Res. 20 (1986) 773.

165. Grinnell F.; Ann. NY Acad. Sci. 516 (1987) 280.

166. Grinnell F.; J. Cell Biol. 103 (1986) 2697.

167. Feuerstein I.A.; Ann. N.Y. Acad. Sci. 516 (1987) 484.

168. Park K., Park H.; Scanning Microsc. (Suppl.) 3 (1989) 137.

169. Pitt W.G., S.H. Spiegelberg, S.L Cooper; Trans. Soc. Biomater. 10 (1987) 59.

170. Park K., Mosher D.F., Cooper S.L.; J. Biomed. Mater. Res. 20 (1985) 589.

171. Brash J.L., Macromol. Chem. Suppl. 9 (1985) 69.

172. Lambrecht L.K., Young B.R., Stafford R.E., Park K., Albrecht R.M., Mosher D.F., Cooper S.L.; Thrombosis Res. 41 (1986) 99.

173. Wildner O., Lipkow T., Knop J.; Increased expression of ICAM-1, E-selectin and V-CAM-1 by cultured endothelial cells upon exposure to haptens; Exp Dermatol 1 (1992) 191.
174. Klein C.L., Nieder P., Wagner M., Kohler H., Bittinger F., Kirkpatrick C.J., Lewis J.C.; J. Pathophysiol. 5 (1994) 798-807.
175. Albelda S., Smith C., Ward P.; Adhesion molecules and inflammatory injury; FASEB J. 8 (1994) 504-512.
176. Gerdes J., Schwab U., Lemke H., Stein H. Production of a mouse monoclonal antibody reactive with a human nuclear antigen associated with cell proliferation. Int J Cancer 1983; 31 (1): 13-20.
177. Thomas W.E.; Brain Res. Brain Res. Rev. 31 (1) 1999, 42-57.
178. http://users.ahsc.arizona.edu/davis/bbbpericytes.htm
179. Chen X., Zuckerman S.T., Weiyuan John Kao. Intracellular protein phosphorylation in adherent U937 monocytes mediated by various culture conditions and fibronectin-derived surface ligands, Biomaterials 2005; 26 (8): 873-882.
180. Fournier, J.A., Calabuig J., Merchán A., Augé J.M., Melgares R., Colman T., Martín De Dios R., Insag L., Santos I., Revista Espanola de Cardiologia, Volume 54, Issue 5, May 2001, Pages 567-572.
181. De Scheerder I., Szilard M., Yanming H., Ping X.B., Verbeken E., Neerinck D., Demeyere E., Coppens W., Van de Werf F.; The Journal of Invasive Cardiology, Volume 12, Issue 8, August 2000, Pages 389-394.
182. Tran H.S., Puc M.M., Hewitt C.W., Soll D.B., Marra S.W., Simonetti V.A., Cilley J.H., DelRossi A.J.; Journal of Investigative Surgery: the Official Journal of the Academy of Surgical Research, Volume 12, Issue 3, May - June 1999, Pages 133-140.
183. http://www.tiem.utk.edu/~gross/bioed/webmodules/cellattach.htm
184. Izzard C.S., and Lochner L.R.; Cell-to-substrate contacts in living fibroblasts: An interference reflection study with an evaluation of the technique; J. Cell Sci. 21 (1976) 129.
185. Bereiter-Hahn J., Fox C.H., Thorell B.; Quantitative reflection contrast microscopy of living cells; J. Cell Biology 82 (1979) 767-779.

12. Applications of Carbon Nanotubes in Bio-Nanotechnology

12.1 Introduction

Patients today are seeking for better health care, while healthcare providers and insurance companies are calling for cost-effective diagnosis and treatments. The biomedical industry thus faces the challenge of developing devices and materials that offer benefits to both patients and healthcare industry. The combination of biology and nanotechnology, is expected to revolutionize biomedical research by exploiting novel phenomena and properties (physical chemical and biological) of material present at nanometer length (10^{-9}m) scale [1–5]. This will lead to the creation of functional materials, devices and systems through control of matter on the nanometer meter scale and the direct application of nano-materials to biological targets.

The nano-materials were existed in nature, long before mankind was able to identify forms at the nanoscale level. Today, nano-materials have been designed for a variety of biomedical and biotechnological applications, including biosensors, enzyme encapsulation, neuronal growth, drug delivery and bone growth [6–10]. The advances in bio-nanotechnology is based on the introduction of novel nano-materials which can result in revolutionary new structures and devices using extremely biological sophisticated tools to precisely position molecules and assemble hierarchal structures and devices. The application of the principles of biology to nanotechnology provides a valuable route for further miniaturization and performance improvement of artificial devices. The feasibility of the bottom-up approach that is based on molecular recognition and self-assembly properties of bio-molecules has already been proved in many inorganic-organic hybrid systems and devices [11]. Nanodevices with bio-recognition properties provide tools at a scale, which offers a tremendous opportunity to study bio-chemical processes and to manipulate living cells at the single molecule level. The synergetic future of nano-and bio-technologies holds great promise for further advancement in tissue engineering, prostheses, pharmacogenomics, surgery and general medicine.

In this chapter, we discuss about various aspects of carbon nanotubes that have been successfully applied to bio-nanotechnology. We focus particularly on biological applications of carbon nanotubes, and take a comprehensive look at the advances in this fast-moving and exciting research field. We review the results on modifications of carbon nanotubes, and highlight some of the recent achievements in the fabrication and evaluation of carbon nanotube-based biological devices and implants.

12.2 Bio-Nanomaterials

Many nanomaterials have novel chemical and biological properties and most of them are not naturally occurring [12]. Carbon nanotubes (CNTs) are in the top list of artificial bio-nanomaterials [16–20], which has won enormous popularity in nanotechnology for its unique properties and applications. CNTs have highly desirable physicochemical properties for use in commercial, environmental and medical sectors. The inclusion of CNTs to improve the quality and performance of many widely used products, as well as potentially in medicine, will dramatically affect occupational and public exposure to CNT-based bio-nanomaterials in the near future.

Even since the discovery of carbon nanotubes, researchers have been exploring their potential in bio applications [21]. One focal point has been the development of nanoscale biosensor [22] and drug delivery systems [23] based on carbon nanotubes, which has been driven by the experimental evidence that biological species such as proteins and enzymes can be immobilized either in the hollow cavity or on the surface of carbon nanotubes [24]. Recently, hopes have been raised for the use of carbon nanotubes as superior biosensor materials in light of the successful fabrication of various electroanalytical nanotube devices, especially those modified by biological molecules [25]. These prototype devices, sometimes prepared as ordered arrays or single-nanotube transistors, have shown efficient electrical communications and promising sensitivities required for such applications as antigen recognition, [26] enzyme-catalyzed reactions [27] and

DNA hybridizations [28]. The CNT/hydroxyapatite composite coated [29] bio-implants has also received much attention recently, for the surface modification of implant materials to promote interaction with living bone tissues owing to its similar chemical composition and crystal structure as natural apatite in the human skeleton.

12.3 Carbon Nanotubes

12.3.1 Introduction

The discovery of carbon nanotubes [30] in 1991 has stimulated significant scientific interest and research leading to rapid progress in the field. Since their discovery, enormous research have been focused on the problems of synthesizing nanotubes, on their physical properties and on possible applications in nanoelectronics [31–35], catalysis [36–38] and other fields including bio-applications [39–45]. The highly impressive structural, mechanical, and electronic properties such as small size and mass, high strength, higher electrical and thermal conductivity, etc. are some of the fascinating properties of this remarkable material that are ideal for various potential applications.

12.3.2 Synthesis

Carbon nanotubes can be manufactured using a variety of methods that includes (Figure 12.1): (i) Laser ablation [46] uses a high-power laser to vaporise a graphite source loaded with a metal catalyst (Figure 12.1a). The carbon in the graphite reforms as predominantly single-wall nanotubes on the metal catalyst particles. (ii) Arc discharge [47] involves an electrical discharge from a carbon-based electrode in a suitable atmosphere to produce both single and multi-wall tubes of high quality but in low quantities (Figure 12.1b). (iii) Chemical vapour deposition (CVD) [48], where a hydrocarbon feedstock is reacted with a suitable metal-based catalyst in a reaction chamber to grow CNTs (Figure 12.1c) which are subsequently removed from the substrate and catalyst by a simple acid wash.

The laser-vaporization method is widely used for the production of single walled (SW) CNTs. The laser is suitable for materials with a high boiling temperature, such as carbon, as the energy density of lasers is much higher than that of other vaporization devices. The basic principle of this method is as follows: a CO_2 laser beam is introduced onto the target (carbon composite doped with catalytic

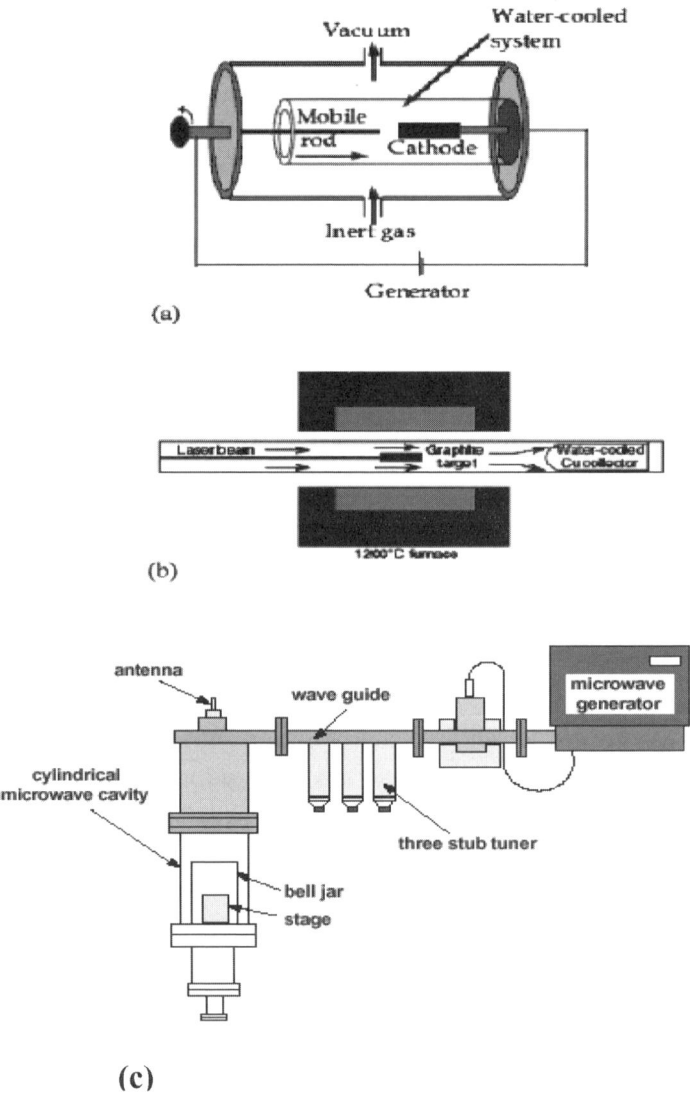

(a)

(b)

(c)

Fig. 12.1. Methods of manufacture for carbon nanotubes: (a) laser ablation; (b) arc discharge; and (c) microwave chemical vapor deposition

metals) located in the center of a quartz tube furnace; the target is vaporized in a high-temperature argon atmosphere and SWCNTs are formed; and, the SWCNTs produced are conveyed by the gas to a special collector. The method has several advantages, such as the high quality of the diameter and controlled growth of the SWCNTs. The change of the furnace temperature, catalytic metals and flow rate directly affect the SWCNT diameter [49].

Large-scale synthesis of Multiwalled (MW) CNTs by arc-discharge was reported [50] in a helium, argon, and methane atmosphere. It was found that methane is the best gas for the synthesis of MWCNTs. This is due to the thermal decomposition of methane producing hydrogen that achieves higher temperature and activity compared to inert gases, such as Helium or Argon. The hydrogen is also found to be an effective factor in the synthesis of MWCNTs [51–53]. The drawback of arc-discharge method is purification of CNTs. Removal of non-nanotube carbon and metal catalyst material in as-grown CNTs is much more expensive than production itself.

The first two methods, arc-discharge and laser furnace, also have the drawback that they do not allow control of the location and the alignment of the synthesized CNTs. CVD is suggested as an alternate method which uses hydrocarbon vapor (e.g., methane) that is thermally decomposed in the presence of a metal catalyst and CNTs are deposited directly on desired substrate.

12.3.3 Structure and Properties

A CNT can be regarded as one gigantic carbon molecule obtained by folding graphite planes into a cylinder (Figure 12.2) whose diameter is measured in nanometers and whose length can reach macroscopic dimensions [54].

This linear structure determines the extremely high mechanical strength of CNT [55] whereas their electrical conductivity depends strongly on the diameter and the helicity that is the angle between the most highly packed chains of atoms and the axis of the cylinder [56].

Two types of CNTs exist: (i) whose walls contain a single layer of carbon atoms, SWCNTs [57] and (ii) nanotubes with walls consisting of several concentric cylindrical graphite layers, MWCNTs [58] as illustrated in Figure 12.3. The hexagonal lattice structure of

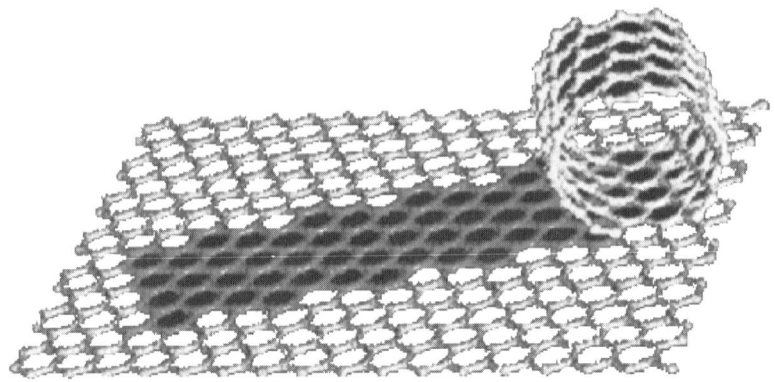

Fig. 12.2. Folding of graphite sheets to form a carbon nanotube

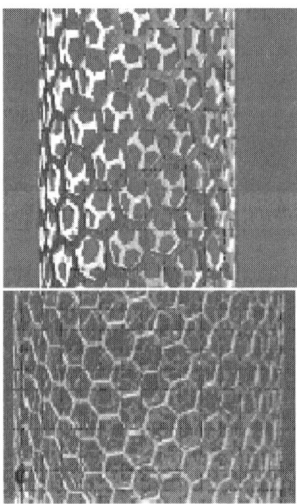

Fig. 12.3. Nanotubes with walls consisting of several concentric cylindrical graphite layers known as multi-walled carbon nanotubes (MWCNTs)

CNTs, gives rise to three types of SWCNTs and their diameter varies between 0.4 and 2 nm. Based on the unit cell of a CNT (Figure 12.4), it is possible to identify armchair nanotubes, formed when n=m and the chiral angle is 30; zig-zag nanotubes, formed when either n or m are zero and the chiral angle is 0; and chiral tubes, with chiral angles intermediate between 0 and 30. SWNTs are either metallic or semi-conducting depending on their diameter and helicity. All armchair nanotubes are metallic, while zig-zag and chiral nanotubes can be metallic or semiconducting.

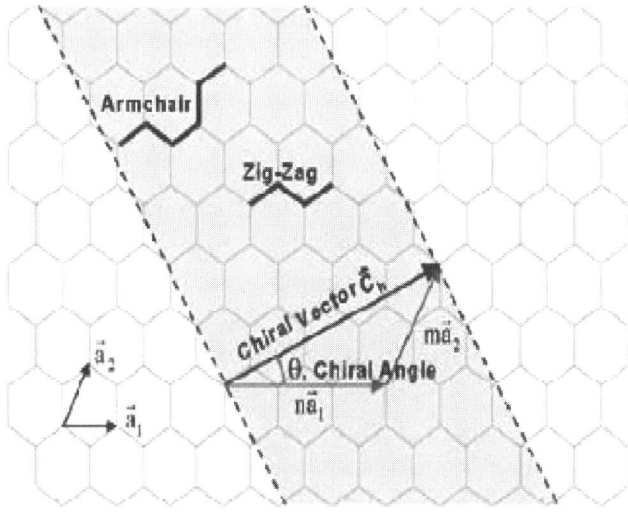

Fig. 12.4. Unit cell of a carbon nanotube

12.3.4 Applications

There is a wealth of potential applications for CNTs [59–62] due to their extraordinary properties [63–66]. They are probably the best electron field-emitter possible. They are polymers of pure carbon and can be reacted and manipulated using the tremendously rich chemistry of carbon. This provides opportunity to modify the structure and to optimize solubility and dispersion. Very significantly, CNTs are molecularly perfect, which means that they are free of property-degrading flaws in the nanotube structure. These extraordinary characteristics give CNTs potential in numerous applications including electronic, mechanical, chemical, thermal and biological applications. The electrical properties of single wall carbon nanotubes are highly sensitive to surface charge transfer and changes in the surrounding environment as the walls of nanotubes constitute a monolayer of atomic arrangement. Due to their surface sensitivity, surface charge mechanisms can cause covalent/non-covalent interactions and van der Waals forces to induce sufficient change in their electronic properties and local density of states. The diversity of available chemistries and easiness of modification makes CNTs viable candidates for biological applications. The biological assembly of CNTs can be attained by their manipulation, dispersion and separation.

12.3.4.1 CNTs as Biosensors

The first application of CNTs for biosensors was proposed in 1996 by Britto et al. [67]. Later, the study by A.Star et alia, on SWNTs have shown to exhibit a significant change in response to the presence of small biomolecules and proteins [68]. The adsorption of cytochrome c, a redox catalyst in the respiratory chain of mitochondria, has been detected in situ using a SWCNT device [69]. Biotin-modified SWCNTs have been used to electronically detect biotin-streptavidin binding [68]. It has been demonstrated that the binding of streptavidin to biotin-functionalized SWCNTs results in a reduced conductance of the carbon nanotubes. Although the mechanism of chemical sensing exhibited by SWNTs has not been unequivocally identified, it seems probable that the resistance changes experienced by these devices originate from the doping of the carbon nanotubes as a result of charge transfer processes that are associated with interactions between the SWCNTs and the analyte. Nevertheless, the interpretation of the electrical responses in thin film devices is complicated by the nature of carbon nanotube networks that are a mixture of bundled semiconducting and metallic SWCNTs.

In some cases the conductance change originates from electronic effects occurring at the metal-nanotube contacts during adsorption. Despite the absence of a definitive understanding of the sensing mechanism, remarkable achievements in electrical biosensing have been reported [70,71]. Covalent coupling of the alkaline phosphatase (ALP) enzyme to CNTs has lead to the highest sensitivity (detection limit of 1 pg L^{-1}) reported thus far for electrical detection of DNA. This CNT-ALP-linked assay can be modified for antigen detection by using specific antibody-antigen recognition. Thus, it could provide a fast and simple solution for molecular diagnosis in pathologies where molecular markers exist, such as DNA or protein [72].

Demand for the reliable monitoring of blood glucose has stimulated further research on the development of biosensors based on CNTs. The voltametric behavior of oxidized SWCNTs with physically adsorbed glucose oxidase has been investigated [73]. The magnitude of the catalytic response to the addition of D-glucose was 10-fold greater than that observed with a glassy carbon electrode. Further improvements in sensitivity and temporal resolution were made by using glucose oxidase-functionalized individual SWCNTs in a (Field emission transistor) FET configuration which allowed for the measurement of enzymatic activity at the level of a single molecule.

Enzyme immobilization is central to bioreactor and biosensor technologies. The current immobilization methods include covalent binding and physical adsorption of enzymes on high surface area materials (carbon silica and polymers). The first step in enzyme immobilization on CNTs is to create active sites on their stable walls. The immobilization of antibodies on the sensor platform to convert a non-electrical, physical or chemical, quantity into an electrical signal is the key for the control and the improvement of the performance of such a biosensor. Several immobilization methods have been reported for the improvement of the anti- body –antigen binding to increase detection sensitivity or for covalent binding of antibody or protein on solid surface [74]. T.S. Huang et alia studied the antibody immobilization on the surfaces of various oxidation processed nanodiamond and carbon nanotubes [75].

Electron transfer in biological systems is one of the leading areas of biochemical and biophysical sciences, and has received more and more attention [76–83]. The direct electron transfer of enzymes with electrodes can be applied to the study of enzyme-catalyzed reactions in biological systems and the development of an electro- chemical basis for the investigation of the structure of enzymes, mechanisms of redox transformations of enzyme molecules and metabolic processes involving redox transformations. Enzyme-modified electrodes provide a basis for constructing biosensors, biomedical devices, and enzymatic bioreactors. If an enzyme immobilized on an electrode surface is capable of direct electron transfer and keeping its bioactivity, it can be used in biosensors without the addition of mediators or promoters onto the electrode surface or into the solution. Unfortunately, it is difficult for an enzyme to carry out a direct electrochemical reaction due to several factors. For example, enzymes would be adsorbed on the electrode surface, resulting in the de-naturation, and loss of their electrochemical activities and bioactivities. In addition, usually, the larger three-dimensional structure of enzymes and the resulting inaccessibility of the redox centers have made it generally difficult to obtain direct electron transfer between enzymes and electrode surfaces, so that promoters and mediators are needed to obtain their electrochemical responses. Therefore, suitable electrode materials and immobilization methods of enzymes onto the electrode surface are important for obtaining their direct electrochemical reaction and keeping their bioactivities.

For the immobilization and/or modification of cells, there are mainly two types of interface interactions being used between a substrate and cells; one is chemical modification of substrate surface to have high affinity to cells, and the other is attaching biomolecules on substrate to recognize the cells [84]. Since the pioneering work of Decher [85], there has been great interest in using the layer-by-layer immobilization of polyelectrolytes for the development of biosensors [86].

Since carbon nanotubes show good electric conductivity, they have been used to modify electrodes and catalyze various biomolecules electrochemically [87]. Direct electrochemistry of redox proteins may provide a model for the study of electron transport of enzymes in biological system [88] and establish a foundation for fabricating a new generation of electrochemical biosensors without using mediators [89]. The CNT has been demonstrated as biochemically compatible, electrically conductive nano-scale interface between redox enzymes and macro-scale electrodes [90]. CNTs are capable of maintaining the functional properties of redox enzymes while linking biomolecules into nanoelectronic platforms. Guiseppi-Elie et al. [91] have developed biosensors of exceptional sensitivity by exploiting the efficiency and specificity inherent in redox proteins. The direct electron transfer of redox enzymes to an electrode surface of CNT has been reported by Cai and Chen also [92]. A relatively new approach to realize direct electrochemistry of proteins is to incorporate proteins into films modified on surface of solid electrodes [93]. Li and co-workers [94] reported direct electrochemistry of cytochrome c (Cyt c) in SWCNT films cast on glassy carbon (GC) electrodes.

Several other important investigations have also been directed to the attachment of natural proteins and DNA immobilization onto MWCNT to construct biosensors. Davis et al. [95] have reported the high surface area possessing multiply acidic sites that may make an offer of special opportunities for the immobilization of enzymes. MWNTs can be activated in acid oxidation conditions due to the residues such as –COOH, –COH, and –OH introduced on the surface of MWNTs [96,97]. The schematic of self-assembly of DNA probes onto MWCNTs described by them are illustrated in Figure 12.5. Shim et al. [98] have described that BSA can be covalently attached to SWNTs and MWNTs by way of diimide-activated amidation under ambient conditions, while the majority of the protein in the nanotube–BSA conjugates remain bioactive. CNTs have been used as modified electrodes to catalyze the electrochemical reaction of some biomolecules, such as dopamine, β-nicotinamide adenine dinucleotide (NADH), cytochrome c, etc., [99].

GC electrodes, in which they dipped the MWCNT electrodes into Mb solution and Mb was absorbed into MWCNT films. A reversible CV peak pair of catalase in SWCNT films cast on gold (Au) electrodes and used the films to electrochemically catalyze reduction of hydrogen peroxide was also observed by them. Cai et al. [104] coated a mixture of hemoglobin (Hb), glucose oxidase (GO_X), or horseradish

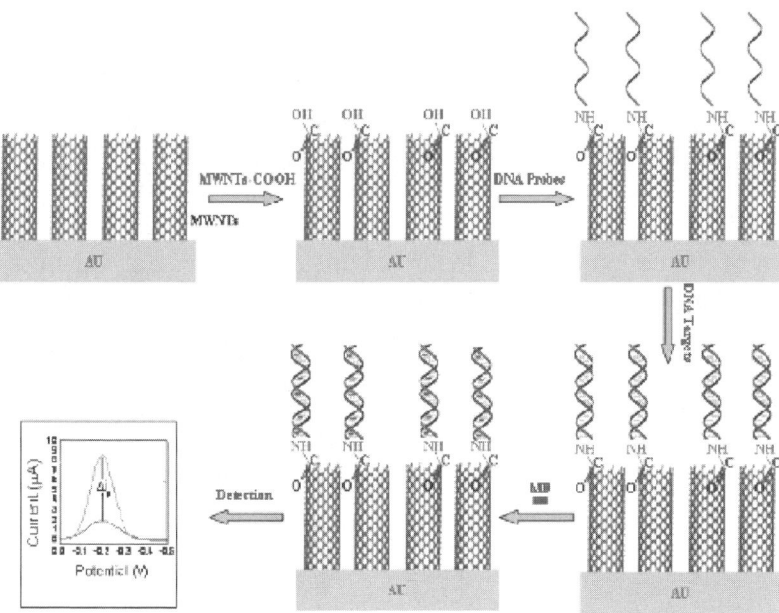

Fig. 12.5. Schematic diagram of self-assembly of DNA probes to MWCNTs

peroxidase (HRP) solution with carbon nanotube dispersions onto GC electrodes. After drying, the films demonstrated good direct electrochemistry of Hb, GO_X, or HRP in blank buffers. Rusling and coworkers [105] covalently attached Mb or HRP onto the ends of vertically oriented SWCNT forest arrays assembled on pyrolytic graphite (PG) electrodes. Quasi-reversible heme Fe(III)/Fe(II) CV response of Mb or HRP at this electrode was observed. There have also been suggestions of using nano-diamond [106] as the signal transducer for glucose sensing enzyme due to the biocompatibility and chemical robustness of these systems.

12.3.4.2 Processibility

A major drawback of CNTs particularly relevant to biological applications is their complete insolubility in all types of solvents. However, CNTs have been found to exhibit a certain degree of chemical reactivity towards many reagents, thus leading to increased solubility and processability. An aqueous medium is highly essential in order to study CNTs in the presence of live cells and therefore the solubilization of CNTs in aqueous solutions is the focus of biological research. Several strategies have been developed to introduce carbon nanotubes into solvent systems, including dispersion and suspension under special experimental conditions and the chemical modification and functionalization. The well-dispersed and solubilized carbon nanotubes make it possible to characterize and study the carbon nanotubes by using solution-based techniques, to realize some of the unique properties of the nanotubes, and to carry out further chemical transformations. The recent bloom of chemical modification and functionalization methods has made it possible to solubilize and disperse carbon nanotubes in water, thus opening the path for their facile manipulation and processing in physiological environments. Equally important is the recent experimental demonstration that biological and bioactive species such as proteins, carbohydrates, and nucleic acids can be conjugated with carbon nanotubes. These nanotube bioconjugates will play a significant role in the research effort toward bioapplications of carbon nanotubes.

Chemical functionalization of CNTs has been shown to impart solubility in a variety of solvents, to modify their electronic properties and to cause significant de-bundling. The chemical reactivity of CNTs arises from the curvature-nduced strain due to misalignment of the π-orbitals of adjacent conjugated carbon atoms. The induced strain is higher at the carbon atoms that comprise the CNT caps because they are curved in two-dimensions, and therefore the caps are more reactive than the sidewalls. Hence, treatment in strong oxidizing agents such as HNO_3 or H_2SO_4 preferentially disrupts the aromatic ring structure at the caps of CNTs and introduces carboxylic acid groups that undergo further chemical reactions. Thus, SWNT-COOH are produced by refluxing arc discharge produced SWCNTs in HNO_3 or H_2SO_4. Numerous amidation and esterification reactions of acid functionalised SWCNTs have been reported [107]. In addition to the chemistry that occurs at the oxidized open ends of SWCNTs, it is also possible to react the side- wall carbon atoms with highly reactive reagents, such as carbenes, fluorine, aryl radicals and azomethineylides. Furthermore, the surface chemistry developed for SWCNTs has been applied to MWCNTs in specific cases. End [108] and/or sidewall [109] functionalization, use of surfactants with sonication [110], polymer wrapping of nanotubes [111], and protonation by super-acids [112] have been reported. Although acid treatment methods for CNT functionalisation are quite successful, they often indicate cutting the CNTs into smaller pieces (sonication and/or functionalization), thus partly losing the high aspect ratio of SWCNTs. J.E. Riggs et al., have shown, it is possible to solubilize carbon nanotubes in aqueous solutions by covalently attaching water soluble linear polymer [114]. By applying the preceding functionalization scheme, poly-m-aminobenzene sulphonic acid (PABS), has been covalently linked to SWCNTs to form a water soluble nanotube-graft copolymer [115], which could be used for future biological applications.

Titus et al., reported the effective functionalisation of CNT using a novel surfactant and its performance of dispersion of CNT in Poly Vinyl Alcohol (PVA) medium [116]. The as-grown CNTs synthesized using microwave plasma (MP) CVD method was in bundle form (Figure 12.6) and bundles were dispersed effectively in nanodisperse surfactant by ultrasonication (Figure 12.7). The surfactant functionalisation promoted further unbundling of CNT in PVA medium. Titus et al. have also reported the attachment of COOH group onto CNT in non-aqueous medium by metal CVD process [117]. The advantage

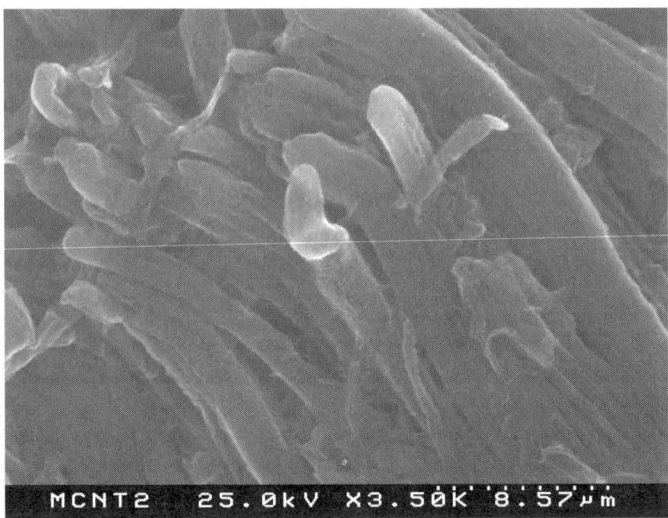

Fig. 12.6. As-grown CNTs in bundle form synthesized using the microwave plasma (MP) CVD method

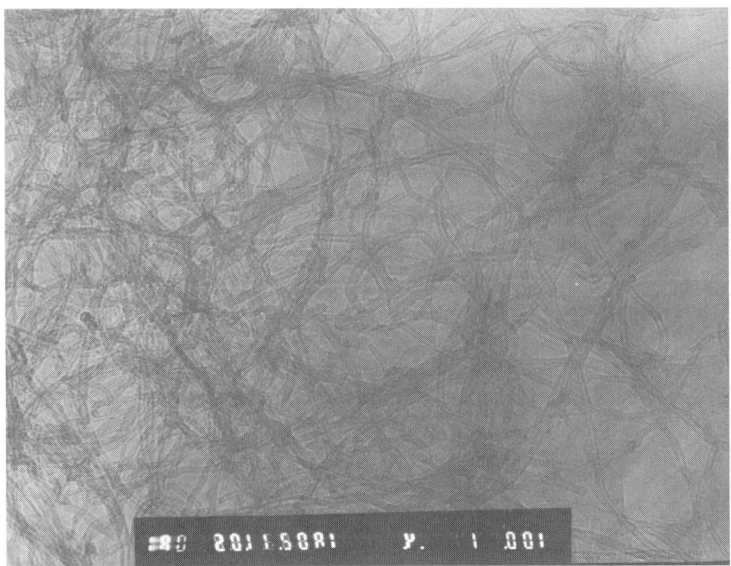

Fig. 12.7. Bundles of CNTs dispersed effectively using a nanodisperse surfactant by ultrasonication

of this method is the formation of high population density of carboxyl groups along the walls of CNT compared to the aqueous method.

A carboxylic acid groups forms an overall electrophylic surface that can minimizes hydrophobic adsorption of biological molecules like protein. In addition, the wall thickness (20–30Å) of high-curvature MWCNT, is dimensionally compatible with typical dimensions of proteins (e.g., GO_x, 60 Å × 52 Å × 77Å). This presumably reduces potential non-specific binding regions, minimizing surface denaturation effects [119].

Biological functionalization of CNTs has come to be of significant interest also due to the possibility of developing sensitive and ultra-fast detection systems that can be addressed using electronic or optical techniques [120]. Most often, biological sensing techniques depend on optical signals derived from the analytes in use, thus involving a series of steps for preparation, varying reagents to differentiate components, and a relatively large sample size. Although these techniques are relatively sensitive, they result in complex data analysis involving unnecessary time consumption and expensive examination techniques. Miniaturizing processes in biological sensing could result in lowering sample size, time and expenses related to detection and sensing. CNTs functionalized with biological assays could be the key to novel nano-biosensing techniques. Several issues are important regarding functionalization of biomaterials on solid-state nanomaterials such as biocompatibility, specificity to the target biomolecule, extent of functionalization, interface effects and the corresponding sensor performance.

12.3.4.3 Fabrication

One of the key focuses of biosensor research is the need to fabricate bio-electrodes which exhibit high selectivity, high-sensitivity and long-term stable response to bioanalytes. Researchers have demonstrated that CNTs have a high electrocatalytic effect, a fast electron-transfer rate, and a large working surface area [121]. CNTs have been used for preparing biosensors employing different strategies: by dispersing them in acidic solutions [122], N,N_-dimethylformamide [123], Nafion [124] and chitosan [125] among others; by incorporating in composites matrices using different binders like Teflon [126], bromoform [127], mineral oil [128] and inks [129]; by immobilizing on pyrolitic graphite electrodes [130]. Wang et al. [131] have demonstrated the ability of the perfluorosulfonated polymer Nafion to disperse single wall (SWCNTs) and multi-wall (MWCNTs) carbon nanotubes. They reported a dramatic decrease in the overvoltage for hydrogen peroxide oxidation and reduction as well as highly selective glucose quantification after immobilization of glucose oxidase (GO_x) by cross-linking with glutaraldehyde.

Various chemical sensors and biosensors based on CNTs have been developed to detect some important species that are related to human health, such as glucose, NADH, ascorbic acid, and cytochrome C [132]. Significant research and development efforts have been devoted to producing CNT-based glucose chemical-and biosensors for in vitro or in vivo applications because of the importance of monitoring blood glucose for the treatment and control of diabetes [133–136]. The measurement principle of CNT-based electrochemical glucose sensors relies on the direct measurement of the oxidation current of glucose on the CNT surface [137] or the immobilization of glucose oxidase on the CNT surface to detect the redox current produced by the enzymatic product H_2O_2 [138]. Ye et al. [139] reported non-enzymatic glucose detection using a well-aligned multi-wall CNT electrode in an alkaline medium. The oxidation over-voltage of the glucose on CNT electrodes was reduced 400 mV compared to that on a glassy carbon electrode. The direct oxidation of glucose on the CNT surface avoids the use of glucose oxidase and over comes the problem of the sensors' life and stability; however, the interference from other electroactive species, such as uric acid and ascorbic acid, still exists. Most of the CNT based electrochemical glucose biosensors are based on glucose oxidase (GO_x), which catalyzes the oxidation of glucose to gluconolactone:

$$\text{Glucose} + O_2 \xrightarrow{\;GO_x\;} \text{Gluconolactone} + H_2O_2$$

The quantification of glucose can be achieved via electro- catalytic redox detection of the enzymatic product H_2O_2 on the CNT transducer at reduced oxidation or reduction over voltage. Here, CNTs play multiple roles: (1) a substrate to immobilize GO_x, (2) electrocatalytic oxidation or reduction of H_2O_2 at the CNT surface to reduce over-voltage and avoid interference from other co-existing electroactive species, and (3) an enhanced signal because of its fast electron transfer and large working surface area.

The approaches, such as covalent binding [140], direct adsorption [141], and entrapment [142] has also been widely used to construct GO_x/CNT biosensors. A drawback to physical adsorption and entrapment is that the distribution of enzyme molecules is not uniform, is sometimes unstable, and tends to leach with time. Covalent binding of GO_x on the functionalized CNT needs a relatively longer reaction time. The ideal immobilization method should employ mild chemical conditions and a short immobilization time to allow for large quantities of enzyme to be immobilized, provide a large surface area for enzyme –substrate contact within a small total volume, minimize barriers to mass transport of substrate and product, and provide a chemically and mechanically robust system.

12.3.4.2 Carbon Nanotubes for Neuronal Growth

Neurons are electrically excitable cells that on network formation serve as conduits for information transfer. A vast amount of information is transferred through the cells in the spinal cord via synaptic and gap junctions in an electroionic fashion mediated by neurotransmitters. The growth of neurites and formation of synapses during development and regeneration is controlled by a highly specialized motile structural specialization at the tip of the neurite called the growth cone. Carbon nanotubes (CNT) are strong, flexible and conduct electrical current. They are biocompatible and non-biodegradable. They can be functionalized with different biomolecules like neuron growth factors and adhesion agents. These properties are useful in the formation of neuron hybrids [143]. These capabilities of carbon nanotubes make them potentially successful candidates to form scaffolds to guide neurite outgrowth.

Xuan Zhang et al., established the ability of the growth cone to grasp onto carbon nanotube matrices functionalized with neuron growth factors [144]. The need, however, is the ability to guide the formation of neuronal networks and establishment of synaptic connections essential for signal transmission leading to re-generation. Their latest research shows the ability of forming highly directed neural networks in vitro over patterned nanotube substrates (Figure 12.8). The nanotubes not only function as scaffolds for the neurons, but the patterned

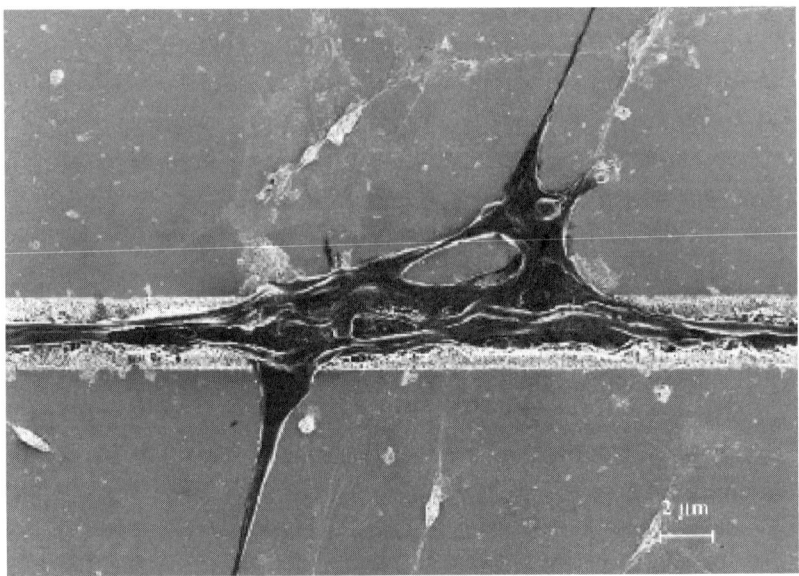

Fig. 12.8. Highly directed neural networks in-vitro over a patterned nanotube substrate

boundary serves as markers for directing the growth and network formation. A number of patterns of vertically aligned multiwalled carbon nanotube (MWNT) substrates are used to determine the geometry and NT length most suitable for scaffolding purposes. Surface characterization is performed using scanning electron microscopy. The interaction between the neuron membrane and the CNT scaffold is also visually analyzed to obtain an insight into network formation in in vitro conditions. This is an essential pre-requisite in forming three-dimensional scaffolds.

The growth of cells and neurons on carbon nanotube films also have been reported [145]. The two-dimensional network of aligned nanotube arrays are proposed to be a good substrate for the cell growth. Elena Bekyarova et al. [146], demonstrated the growth of cultured hippocampal neurons on MWCNTs deposited on poly- ethylenei-mine-coated coverslips. SEM was used to identify the morphological changes of neuron growth brought about by the presence of the MWCNTs. The neuronal bodies were found to adhere to the surface of the MWCNTs with their neurites extending through the bed of CNTs and elaborating into many small branches (Figure 12.9). The neurons remained alive on the nanotubes for at least 11 days, and it was shown that the physisorption of 4-hydroxynonenal on the MWNTs enhanced both neurite outgrowth and branching.

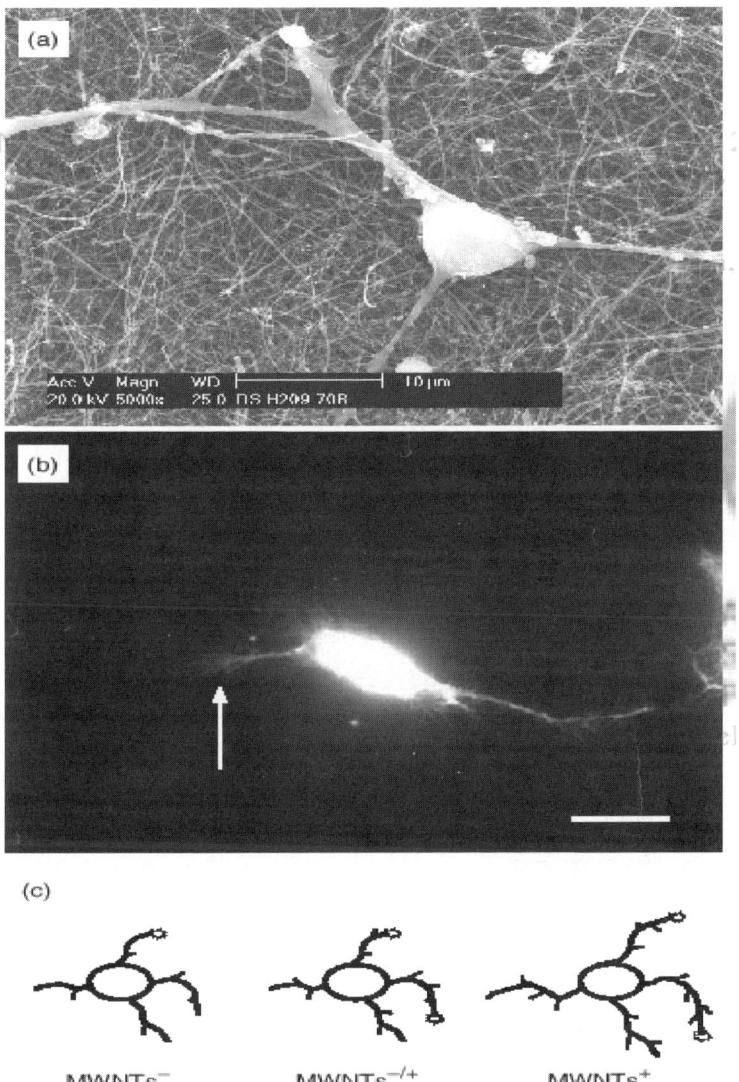

Fig. 12.9. (a) Neuronal bodies found to adhere to the surface of the MWCNTs with their neurites extending through the bed of CNTs (b); and (c) elaborating into many small branches

12.3.4.3 Drug Delivery by CNTs

The diversity of available chemistries and cell-penetrating structures makes CNTs viable candidates as carriers for the delivery of drugs, DNA, proteins and other molecular probes into mammalian cells [147]. An important issue in intracellular drug delivery is the poor permeability of the plasma membrane to many drugs. Thus, various carriers, including polyethylene glycol, peptides and lipids, have been developed to facilitate the cellular entry of drugs. One of the prerequisites for such a task is therefore the ability of the carrier to bind to biologically relevant molecules [148].

The feasibility of using SWCNTs for intracellular drug delivery has been demonstrated [149]. Water soluble SWCNTs were functionalized with a fluorescent probe, FITC, to allow tracking of SWCNTs. When murine and human fibroblast cell lines were exposed to SWCNT-FITC, the nanotubes could be shown to accumulate within the cells. Similarly, SWCNTs, covalently functionalized with biotin and reacted with streptavidin, were internalized within human promyelocytic leukemia (H60) cells, human T cells, Chinese hamster ovary (CHO) and 3T3 fibroblast cell lines.

While the mechanism of the CNT cell entry remains undelined, these experiments suggest the viability of CNTs as carriers for delivering relatively large molecules to mammalian cells.

MWCNTs are also demonstrated in drug delivery. F. Balavoine et al., reported the interactions between MWCNTs and proteins and revealed the self-organization of streptavidin molecules and the growth of its helical crystals on the CNT surface [150]. Similarly, DNA molecules may be adsorbed on MWCNTs, and small protein molecules, such as cytochrome c and -Lactase I, can be inserted within the interior cavity of open CNT. CNTs have also been used to deliver proteins and peptides inside a cell [151] by the direct covalent bonds between a CNT and biomolecules formed by attaching functional groups via acidic treatments [152]. As an alternative to the binding of molecules to the outside of the CNTs, it would also be convenient to fill the interior cavity of tubes, whose open ends might be capped to generate a nanopill containing a drug for delivery to the cell. Towards that end, a template method has been used to synthesize nano test tubes, which are CNTs with one end closed and the other open. Such an approach might be construed as a first step toward the development of a nanopill, in which the substance to be delivered is introduced into the interior of the nanotube and then bottled by resealing the open end.

12.3.4.4 Biomedical Implant Applications of CNT

The primary issues in materials science of new bone biomaterials are mechanical properties and biocompatibility. Although mechanical properties of biomaterials have been well characterized, the term biocompatibility is only a qualitative description of how the body tissues interact with the biomaterial within some expectations of certain implantation purpose and site [153]. Materials scientists have investigated metals, ceramics, polymers and composites as biomaterials. The general criteria for materials selection for bone implant materials are:

- It is highly biocompatible and does not cause an inflammatory or toxic response beyond an acceptable tolerable level.
- It has appropriate mechanical properties, closest to bone.
- Manufacturing and processing methods are economically viable.

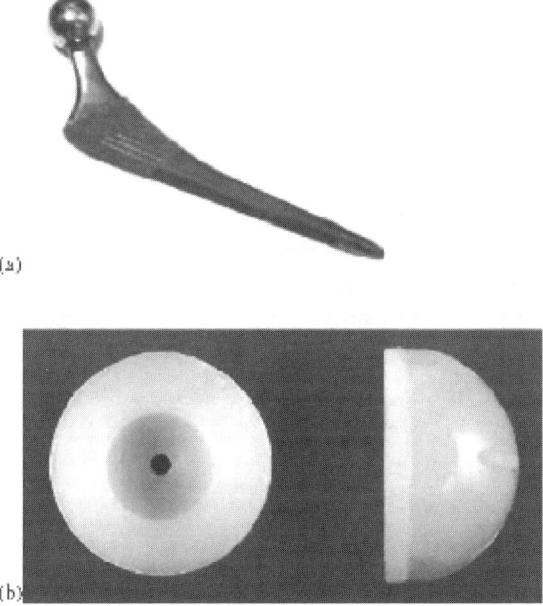

Fig. 12.10. Artificial hip implant fabricated from titanium alloy: (a) artificial hip implant; and (b) femoral head manufactured from UHMWPE

The hip joint consists of two complementary articular surfaces separated by articular cartilage and the synovial fluid that has a pH between 7.29 and 7.45. Excessive wear of the interfaces due to degenerative disease (such as osteoarthritis) or injury requires a replacement of the entire hip joint. Historically, a total hip replacement the articulation of a human hip is simulated with the use of two components, a cup type and a longfemoral type element [154 K.S. Katti, Colloids and surfaces]. A typical hip implant fabricated from titanium is shown in Figure 12.10.

The head of the femoral element fits inside the cup to enable the articulation of the human joint. These two parts of the hip implant have been made using a variety of materials such as metals, ceramics, polymers and composites. Typically polymeric materials alone tend to be too weak to be suitable for meeting the requirement of stress deformation responses in the THR components. Metals typically have good mechanical properties but show poor biocompatibility, cause stress shielding and release of dangerous metal ions causing eventual failure and removal of implant. Ceramics generally have good biocompatibility but poor fracture toughness and tend to be brittle. A hip implant therefore should be such that it exhibits an identical response to loading as real bone and is also biocompatible with existing tissue. The average load on a hip joint is estimated to be up to three times body weight and the peak load during other strenuous activities such as jumping can be as high as 10 times body weight. In addition hipbones are subjected to cyclic loading as high as 10^6 cycles in 1 year [154]. The compatibility issue involves surface compatibility, mechanical compatibility and also osteocompatibility. These materials are also classified as bioactive (illicit a favorable response from tissue and bond well), bioinert and biodegradable. The commercial metallic total hip replacement (THR) implants are five to sixtimes stiffer than bone and result in significant problems associated with stress shielding. Titanium (Ti) alloys in the femoral elements of the THR have shown improvement in wear properties [155]. The regenerative and remodeling processes in bone are directly triggered by loading, i.e., bone subjected to loading or stress regenerates and bone not subjected to loading results in atrophy. Thus, the effect of a much stiffer bone implant is to reduce the loading on bone resulting in the phenomenon called as stress shielding. The key problems associated with the use of these metallic femoral stems are thus release of dangerous particles from wear debris, detrimental effect on the bone remodeling process due to stress shielding and also loosening of the implant tissue interface. It has been shown that the degree of stress shielding is directly related to the difference in stiffness of bone and implant material [156]. Ti alloys are favorable materials for orthopedic implants due to their good mechanical properties. However, Ti does not bond directly to bone resulting in loosening of the implant. Undesirable movements at the implant-tissue interface results in failure cracks of the implant.

One approach to improving implant lifetime is to coat the metal surface with a bioactive material that can promote the formation and adhesion of hydroxyapatite, the inorganic component of natural bone [157]. The application of bioactive coatings to Ti-based alloys enhance the adhesion of Ti-based implants to the existing bone, resulting in significantly better implant lifetimes than can be achieved with materials in use today. Typically, several silicate glasses are used as bioactive coatings [158]. Some ceramic coatings are known to be bioactive and have also been tested on Ti implants. As compared to metals, ceramics often cause reduced osteolysis and are regarded as favorable materials for joints or joint surface materials. Several ceramics due to their ease of processing and forming and superior mechanical properties were investigated as bone substitute materials [159].

Conventional ceramics such as alumina were evaluated due to their excellent properties of high strength, good biocompatibility and stability in physiological environments [160]. Alumina, because of the ability to be polished to a high surface finish and its excellent wear resistance, is often used for wear surfaces in joint replacement prostheses. Femoral heads for hip replacements and wear plates in knee replacements have been fabricated using alumina. In year 2003, the United States Food and Drug Administration (FDA) has approved alumina ceramic-on-ceramic articulated hips for marketing in the United States of America. Other ceramic materials have also been investigated for potential applications in orthopedics. Considerable research has focused on zirconia and yttria ceramics that are characterized by fine-grained microstructures. These ceramics are known as tetragonal zirconia polycrystals (TZP). TZP in the body have been limited by the low strength and low fracture toughness of the synthetic phosphates. Alumina and titanium dioxide have been used as nano-ceramics separately or in nanocomposites with polymers such as poly-lactic acid or polymethlyl methacrylate. The nanoceramic formulations promote selectively enhanced functions of osteoblasts (bone-forming cells). These functions include cell adhesion, proliferation, and deposi-tion of calcium-containing minerals, an indication of new bone for-mation in a laboratory setting. Despite of many advantages, the lack of chemical bonding between sintered alumina and tissue, however limited its applications as a potential bone substitute to a certain extent. The other problems arise when attempting to coat metals with cera-mics are: the thermal expansion coefficients of the ceramic and metal are usually different, and as a result, large thermal stresses are gene-rated during processing. These stresses lead to cracks at the interface and compromise coating adhesion. In addition, chemical reactions between the ceramic and metal can weaken the metal in the vicinity of the interface, reducing the strength of the coated system.

Calcium hydroxyapatite $[Ca_{10}(PO_4)_6(OH)_2]$ is the principal calcium phosphate commonly used for biomedical implant applications. How-ever, the high brittleness and poor strength of sintered hydroxyapatite (HA) restricts its clinical applications under load-bearing conditions and therefore coating of HA on metal implants is an alternate option. The excellent biocompatibility and osteointegration are the key char-acteristics of existing bioceramic hydroxyapatite coatings. Synthetic HA elicits a direct chemical response at the interface and forms a very tight bond to tissue [161]. Attempts have been made to form high strength consolidated HA bodies [162]. However, its poor mecha-nical properties such as low strength and limited fatigue resistance restrict its applications. Bending strength as high as 90 MPa has been achieved by colloidal processing of HA [163].

The first priority for the development of a better coating is there-fore improvement of the interfaces (metal-coating and coating-bone interfaces) so that the coating binds well with both metal and bone. Thermal stresses, chemical reactions between coating and metal, and biocompatibility are the key issues to be considered. Although HA coverings are able to enhance bone ingrowth and reduce early loosen-ing of hip and knee prostheses, the optimum coating quality and surface texture are still a matter open to debate. Moreover the significance of coating resorption is controversial. It has been suggested that resorption disintegrates the coating and reduces the bonding strength between the implant and bone, and the strength of coating implant interface, which might lead to implant loosening, coating delamination and acceleration of third body wear process.

The extremely light-weight, extra strong and nano sized carbon nanotubes (CNTs) are highly recommended as an additional ingredient for the synthesis of HA coating. Carbon materials are known to be inert to cells and tissues because of their pure carbon composition. Some recent investigations indicated that carbon nanotubes may have promising potentials in biomedical applications both at molecular and cellular levels [164]. Functionalisation of CNTs are important in HA/CNT coating since chemical reactions can take place at functionalised sites on CNT in a colloidal state and it play a potential role

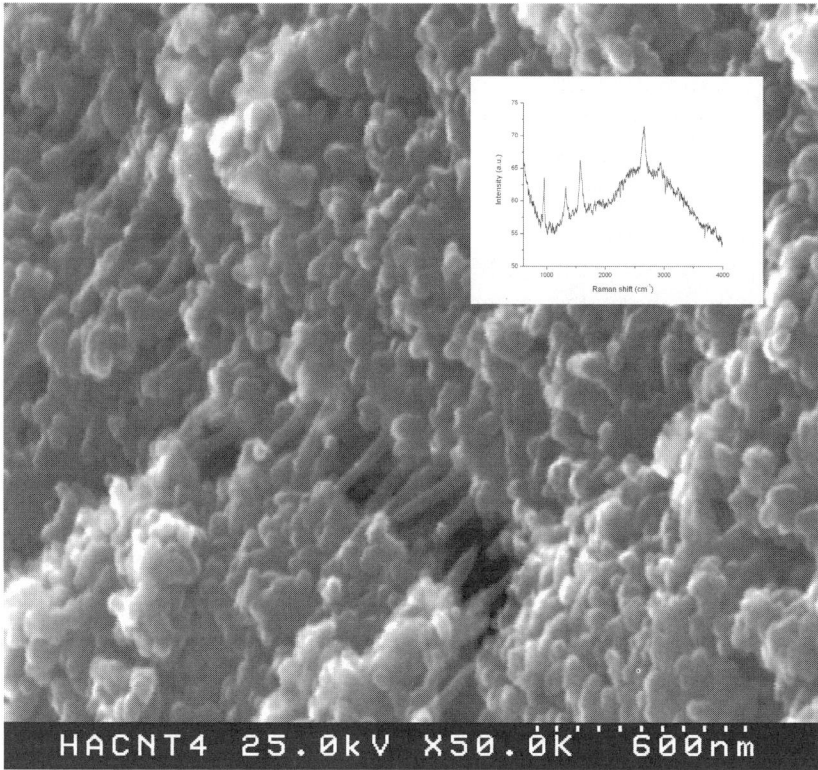

Fig. 12.11. SEM image of uniform distribution of CNT in HA matrix. The uniform distribution of CNT in HA matrix was confirmed using Raman analysis (inset)

in its uniform dispersion in composite media. E. Titus et al., achieved, uniform dispersion of CNT into HA matrix by ultrasonication and other novel methods. Figure 12.11 shows the SEM image of uniform distribution of CNT in HA matrix. The uniform distribution of CNT in HA matrix also was confirmed using Raman analysis (inset of Figure 12.11).

12.4 Analysis

The determination of structure function relationships in biological macromolecules is central to elucidating biochemical pathways, and thereby designing new drugs and understanding their mode of action. Structural biology has played and will continue to play a key role in these endeavors because bio-molecule function is closely tied to three-dimensional structures. The workhorse tools of structural biology, X-ray diffraction, electron diffraction and NMR, can almost routinely be used to determine atomic resolution structure of single bio-molecule. Continued advances in these methods are pushing the limits of the size and complexity of systems that can be characterized [165], although in the future, expanded needs for biomolecular structure analysis are expected in several areas, including: (i) increased throughput to characterize new gene products discovered by genomic DNA sequencing [166], (ii) routine analysis of multimeric protein, protein and protein-nucleic acid structures involved in, for example, signaling and gene regulation; and (iii) elucidation of dynamic processes in these multimeric systems. It is unlikely that the conventional structural tools will meet all of these needs, both because of the increased difficulty of crystallizing large signaling and regulatory protein complexes, which will limit diffraction methods, and the challenges of using solution NMR for large biomolecular systems [167]. Future progress in understanding complex processes in biological systems will therefore clearly require additional, perhaps revolutionary techniques for structural analysis.

AFM [168] is one technique with the potential to probe both structure and dynamics of large macromolecular systems, since it permits direct visualization of individual biological structures in vitro [169]. The potential for AFM to impact structural biology has been suggested by beautiful images of, for example, two-dimensional arrays of proteins with 'sub-molecular resolution' [170], although such captivating data are not without limitations. These shortcomings, which if overcome could dramatically extend the applicability of AFM to structural biology, can be understood by reviewing the key features of an AFM (Figure 12.12).

Common to all AFMs are an integrated cantilever-tip assembly, a detector to measure cantilever displacement as the sample is scanned, and electronics to acquire and display images. The basic features of AFM and methods of imaging have been reviewed recently [171]. Central to reproducible high-resolution characterization of biological

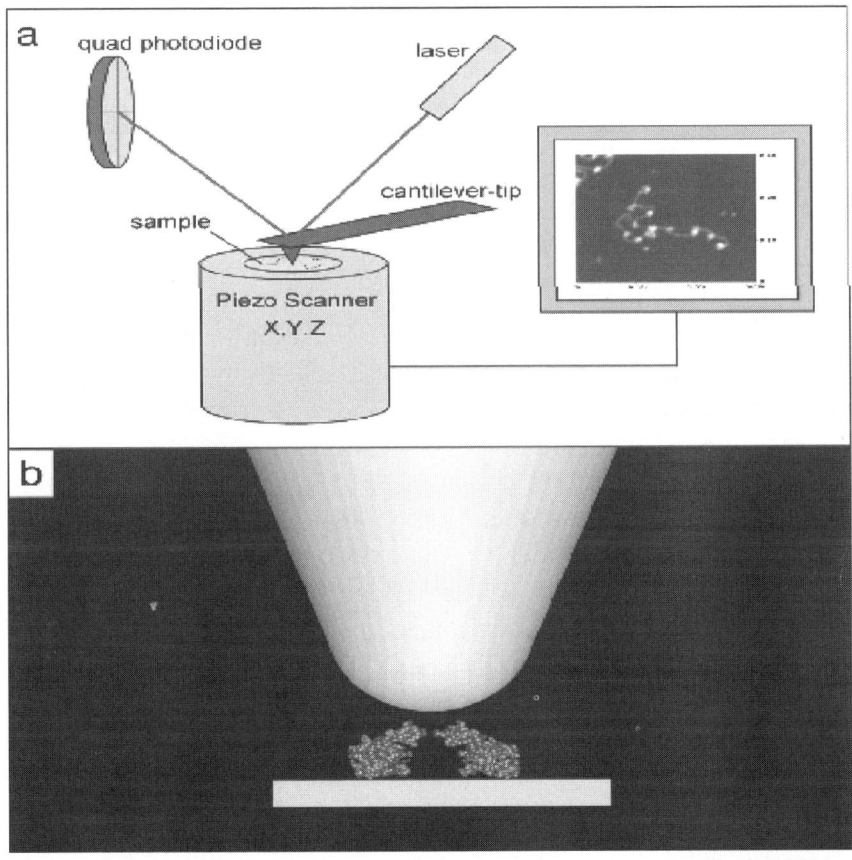

Fig. 12.12. Schematic diagram of the atomic force microscope

macromolecules with AFM especially new systems not crystallographically characterized is the size and shape of the probe tips used for imaging.

An ideal AFM probe tip should have (i) a sub-nanometer radius, (ii) a zero degree cone angle, (iii) mechanical and chemical robustness and, (iv) the potential for molecularly precise modification of the tip end. Moreover, it should be possible to prepare such tips reproducibly with the same features, such that the resolution and other imaging characteristics are predictable, as is the case in diffraction experiments.

The first demonstration of nanotube probes used mechanical mounting of bundles of MWCNTs onto standard AFM tips [172]. AFM studies with mechanically mounted MWCNT probes yielded only modest improvements in resolution on amyloid fibrils, protofibrils, and gold nanocluster standards with respect to standard Si tips [173]. In contrast, probes fabricated from etched SWCNT bundles, which occasionally have just a few SWCNTs protruding from the end, demonstrated up to five fold better resolution than conventional probes on inorganic nanostructures and DNA [174]. While these results indicated the potential for SWCNTs to enhance AFM resolution, tip radii were still 10 times larger than what would be obtained with a single 0.25 nm radius nanotube [175]. Moreover, the conceptual simplicity of mechanical nanotube tip fabrication is hampered by its difficulty in scale-up and by its intrinsic selectivity towards thicker nanotube bundles. Mechanical nanotube tip assembly in a scanning electron microscope (SEM) [176] allows assembly of somewhat smaller 10 nm diameter tips, but this method is even slower than mounting in an optical microscope. Thus, a different approach is needed for reproducible and scaleable fabrication of ultrahigh resolution nanotube tips. All of the problems associated with manual assembly can be solved by directly growing nanotubes on AFM tips using metal-catalyzed chemical vapor deposition (CVD). By carefully manipulating CVD reaction conditions and the catalyst, one can selectively produce SWCNTs [177] with radii as small as 0.35 nm [178]. Since carbon nanotube probes can provide enhanced imaging in diverse areas, they are particularly suited to the field of biology. The use of SWCNT probes will therefore improve the imaging resolution of small proteins, such as antibodies in DNA analysis. The widespread availability of nanosurgery involves the manipulation of single cells or cell structures. The advantage of CNT AFM probes is that they can be easily functionalized, and the range of chemical groups that can be added specifically to the tip of a CNT make them an ideal high resolution probe for mapping chemical domains by using chemical force microscopy. Thus probes can be developed to sense polarity, pH and many other chemical characteristics of the sample by adding different residues to the CNT, as shown by studies using CNT AFM tips patterned with terminal carboxylic groups in which it was possible to chemically map the substrate by using tapping mode AFM (Figure 12.13). Figure 12.14 shows the oriented isolated CNTs grown directly on copper substrate by CVD method [Gil Cabral et al.]. The orientation and size of the CNTs were controlled by maintaining growth conditions.

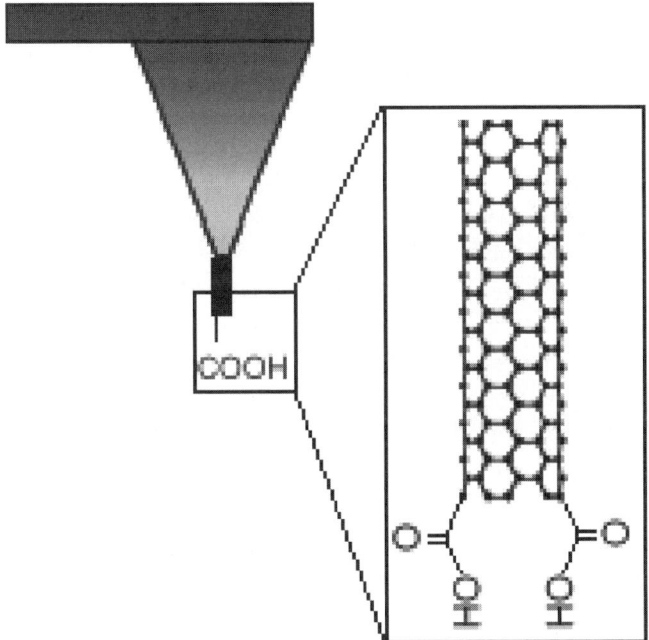

Fig. 12.13. AFM probes are developed to sense polarity, pH and many other chemical characteristics of the sample by adding different residues to the CNT. It is possible to chemically map the substrate by using tapping-mode AFM

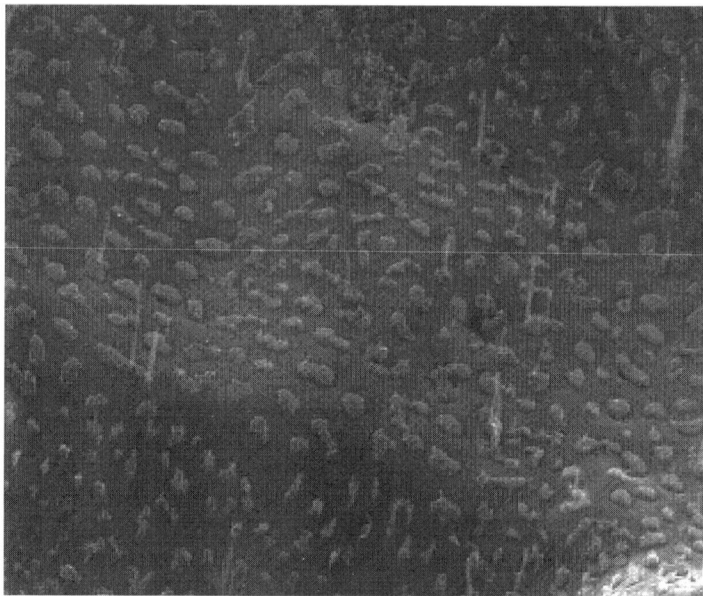

Fig. 12.14. Oriented isolated CNTs grown directly on a copper substrate by the CVD method. Maintaining strict growth conditions controls the orientation and size of the CNTs

12.5 Toxicity of Carbon Nanotubes

Toxicity is one of the important issues regarding the use of CNTs in biology and medicine [179]. Currently, CNTs are under investigation in various laboratories, and therefore, the widespread commercialization and exposure of the general populace to this material must occur only after adequate testing. Detailed toxicological studies are required in this regard, and few of the reported studies have shown negative effects on human health. For example, the exposure of cultured human skin cells to SWCNTs caused oxidative stress and loss of cell viability, indicating that dermal exposure may lead to skin conditions [180]. This is perhaps to be expected, since graphite and carbon materials have been associated with increased dermatitis and kerato sis. Additional studies have investigated the pulmonary toxicity of SWCNTs, and it was shown that exposure to SWCNTs lead to the development of granulomas in rodents. Since these studies used very high concentrations of SWCNTs, which were directly exposed to skin and instilled into the lungs of the animals, further testing is required to establish their toxicity. Also the toxicity is expected to be less in functionalized CNTs and CNT composites.

12.6 Conclusions

The multidisciplinary field of Bio-nanotechnology holds the promise of delivering the technological breakthrough and is moving very fast from concept to reality. The flexibility to modify or adapt bio-nanotechnology to meet the needs of pathologic conditions either for therapeutic applications or as a diagnostic tool is the important characteristic of the technology.

The CNTs represent one of the most promising materials for application in bio-nanotechnology due to their amazing electronic and mechanical properties. CNTs provide a new generation of bio-compatible nanomaterials for sensors and probes, implants, electro-chemical devices, reinforcements in composites and nanometer-sized electronics that could revolutionize the world. It provides a wide range of new technologies for developing customized solutions that optimize the delivery of pharmaceutical products. It involves the creation and use of materials and devices at the atomic and molecular level. The scope of nano-material is vast and the potential for breakthroughs is enormous and is being pursued on multiple fronts.

References

1. Y. Xiao, et al., Science 299 (2003)1877.
2. R. Elghanian, et al., Science 277 (1997) 1078.
3. R.D. Averitt, et al., Phys. Rev. Lett. 78 (1997) 4217.
4. A.R. Clapp, et al., J. Am. Chem. Soc. 126 (2004) 301.
5. W. Wang, et al., Nano Lett. 2 (2002) 817.
6. W.C.W. Chan and S. Nei, Science 281 (1998) 2016.
7. Y.W.C. Cao, R.C,. Jin C.A. Mirkin, Science 297 (2002) 1536.
8. J.D. Hartgerink, E. Beniash S.I. Stupp. Science 294 (2001) 1684.
9. I. Koltover, T. Salditt, J.O. Radler C.R. Sa, Science 281 (1998) 78 .
10. K.E. Ulrich. S.M. Cannizzaro, R.S. Langer, K.M. Shakesheff, Chem. Rev. 99 (1999) 3181.
11. M. Fandrich, M.A. Fletcher, C.M. Dobson, Nature 410 (2001) 165.

12. N.C. Seeman, and A.M. Belcher, (2002). Proc. Natl. Acad. Sci. USA 99, 6451.
13. R. Garjonyte, A. Malinauskas, Biosensor & Bioelectronics 15 (2001) 445.
14. M.D. Rubianes, G.A. Rivas, Electrochemistry Communications 5 (2003) 689.
15. J. Wang, A.N. Kawde, M.R. Jan, Biosensors & Bioelectronics 20 (2004) 995.
16. S. Ravindran, S. Chaudhary, B. Colburn, M. Ozkan, C.S. Ozkan, Nano Lett. 3 (2003) 447.
17. J.M. Haremza, M.A. Hahn, T.D. Krauss, S. Chen, J. Calcines, Nano Lett. 2 (2002) 1253.
18. M. Hazani, R. Naaman, F. Hennrich, M.M. Kappes, Nano Lett. 3 (2003)153.
19. C. Dwyer, M. Guthold, M. Falvo, S. Washburn, Nanotechnology 13 (2002) 601.
20. A.V. Ellis, K. Vijayamohanan, R. Goswami, N. Chakrapani, L.S. Ramanathan, P.M. Ajayan, G. Ramanath, Nano Lett. 3 (2003) 279.
21. J.M. Wilkinson, Med. Device Technol. 14 (2003) 29.
22. M. Shim, N.W.S. Kam, R.J. Chen, Y. Li, H. Dai, Nano Letters 2 (2002) 285.
23. D. Pantarotto, J.-P. Briand, M. Prato, A. Bianco, Chem. Commun. 16 (2004).
24. N.W.S. Kam, T.C. Jessop, P.A. Wender, H. Dai, J. Am. Chem. Soc. 126, 6850 (2004).
25. J.J. Gooding, R. Wibowo, J. Liu, W. Yang, D. Losic, S. Orbons, F.J. Mearns, J.G. Shapter, D.B. Hibbert, J. Am. Chem. Soc. 125 (2003) 9006.
26. J. Wang, G. Liu, M.R. Jan. J. Am. Chem. Soc. 126 (2004) 3010.
27. J. Wang, M. Li, Z. Shi, N. Li, Z. Gu, Electroanalysis 14 (2002) 225.
28. Joseph Wang, Abdel-Nasser Kawde, M. Rasul Jan, Biosensors and Bioelectronics, 20 (2004) 995.
29. Y. Chen, Y.Q. Zhang, T.H. Zhang, C.H. Gan, C.Y. Zheng, G. Yu, Carbon, 44 (2006) 37
30. S. Iijima, Nature 354 (1991)56.
31. A. Thess, et al., Science 273 (1996) 483.
32. S.J. Tans, et al., Nature 386 (1997) 474.
33. C. Joachim, and J.K. Gimzewski, Chem. Phys. Lett. 265 (1997) 353.
34. D. Tomanek, and R.J. Enbody (Eds.), Science and Application of Nanotubes, Kluwer Academic/Plenum Publishers, New York, 2000.
35. J.E. Fischer, et al., Phys. Rev. B 55 (1997) R4921.
36. R.S. Lee, H.J. Kim, J.E. Fischer, A. Thess, R.E. Smalley, Nature 388 (1997) 255.
37. A.S. Claye, J.E. Fischer, C.B. Huffman, A.G. Rinzler, R.E. Smalley, J. Electrochem. Soc. 147 (2000) 2845.
38. V.G. Gavalas, R. Andrews, D. Bhattacharyya, L.G. Bachas, Nano Lett. 1 (2001) 719.
39. R. Garjonyte, A. Malinauskas, Biosensor & Bioelectronics 15 (2001) 445.
40. M.D. Rubianes, and G.A. Rivas, Electrochemistry Communications 5 (2003) 689.
41. J. Wang, A.N. Kawde, M.R. Jan, Biosensors & Bioelectronics 20 (2004) 995.
42. J.S. Ye, X. Liu, H.F. Cui, W.D. Zhang, F.S. Sheu, Tit Meng Lim, Electrochemistry Communications 7 (2005) 249.
43. M. Shim, N.W.S. Kam, R.J. Chen, Y. Li, H. Dai, Nano Letters 2 (2002) 285.

44. W. Huang, S. Taylor, K. Fu, Y. Lin, D. Zhang, T.W. Hanks, A.M. Rao, Y.-P. Sun, Nano Letters 2 (2002) 311.
45. S. Sotiropoulou, V. Gavalas, V. Vamvakaki, N.A. Chaniotakis, Biosensors & Bioelectronics 18 (2003) 211.
46. L.P. Biro, Z.E. Horvath, L. Szlamas, K. Kertesz, F. Weber, G. Juhasz, et al., Chem Phys Lett 399 (2003) 402.
47. C. Journet, P. Bernier, Appl. Phys. A 1 (1998) 9.
48. Y. Zhang, Applied Physics Letters 79 (2001) 3155.
49. M. Chiang, K. Liu, T. Lai, C. Tsai, H. Cheng, I. Lin, J. Vac. Sci. Technol. B 19 (2001) 1034.
50. L.P. Biro, Z.E. Horvath, L. Szlamas, K. Kertesz, F. Weber, G. Juhasz, et al., Chem Phys Lett, 399 (2003) 402.
51. D. Park, Y.H. Kim, J.K. Lee, Carbon 41, (2003) 1025.
52. L.C. Qin, D. Zhou, A.R. Krauss, D.M. Gruen, Appl. Phys. Lett. 72 (1998) 3437.
53. M. Meyyappan, L. Delzeit, A. Cassell, D. Hash, Plasma Sources Sci. Technol. 12 (2003) 205.
54. H. Dai, Surface Sci. 500 (2002) 218.
55. V. Popov. Mater Sci. Eng. R. Rep. 43 (2004)
56. M. Dresselhaus, G. Dresselhaus, R. Saito, Carbon 33 (1995) 883
57. W.Z. Liang, G.H. Chen, Z. Li, Z.K. Tang, Appl. Phys. Lett. 80 (2002) 3415.
58. P.M. Ajayan, Chem. Rev. 1999; 99:1787.
59. R.H. Baughman, A.A. Zakhidov, W.A. Heer, Science 297 (2002) 787.
60. E.T. Thostenson, Z. Ren, T.W. Chou, Compos Sci. Technol. 61 (2001) 1899.
61. F. Lupo, R. Kamalakaran, C. Scheu, N. Grobert, M. Ruhle. Carbon, 42 (2004) 1995.
62. J.W. Mintmire, B.I. Dunlap, C.T. White, Phys. Rev. Lett. 68 (1992) 631.
63. N. Hamada, S.I. Sawada, and A. Oshiyama, Phys. Rev. Lett. 68 (1992) 1579.
64. R. Saito, M. Fujita, G. Dresselhaus, and M.S. Dresselhaus, Appl. Phys. Lett. 60 (1992) 2204.
65. H. Kataura, Y. Kumazawa, Y. Maniwa, I. Umezu, S. Suzuki, Y. Ohtsuka, Y. Achiba, Synth. Met. 103 (1999) 2555.
66. J. Chen, M.A. Hamon, H. Hu, Y. Chen, A.M. Rao, P.C. Eklund, R.C. Haddon, Science 282 (1998) 95.
67. P.J. Britto, K.S.V. Santhanam, P.M. Ajayan, Bioelectrochem. Bioenerg. 41, (1996) 121.
68. A. Star, J.-C.P. Gabriel, K. Bradley, G. Gruner, Nano Lett. 3 (2003) 459.
69. S. Boussaad, N.J. Tao, R. Zhang, T. Hopson, L.A. Nagahara, Chem. Commun. 1502 (2003)
70. K. Besteman, J.O. Lee, F.G. Wiertz, H.A. Heering, C. Dekker, Nano Lett. 3 (2003) 727.
71. R.J. Chen, S. Bangsaruntip, K.A. Drouvalakis, N.W.S. Kam, M. Shim, Y. Li, W. Kim, P.J. Utz, H. Dai, Proc. Nat. Acad. Sci. U.S.A. 100 (2003) 4984.
72. J. Wang, G. Liu, M.R. Jan, J. Am. Chem. Soc. 126 (2004) 3010.

73. B.R. Azamian, J.J. Davis, K.S. Coleman, C.B. Bagshaw, M.L.H. Green, J. Am. Chem. Soc. 124 (2002) 2664.

74. S.T. Pathirana, J. Barbaree, B.A. Chin, M.G. Hartell, W.C. Neely, V. Vodyanoy, Biosens. Bioelectron. 15 (2000) 135.

75. T.S. Huang, Y. Tzeng, Y.K. Liu, Y.C. Chen, K.R. Walker, R. Guntupalli, C. Liu, Diamond and Related Materials 13 (2004)1098.

76. R. Bandyopadhyaya, E. Nativ-Roth, O. Regev, R. Yerushalmi-Rozen, Nanoletters 2 (2002) 25.

77. M. Bodanszky, A.J. Bodanszky, Am. Chem. Soc. 126 (1994) 12750.

78. C. Bourdillon, C. Demaille, J. Gueris, J. Moirourx, J.M. Saveant, J. Am. Chem. Soc. 115 (1993) 12264.

79. C. Cai, J. Chen, Anal. Biochem. 332 (2004) 75.

80. J.J. Gooding, R. Wibowo, J. Liu, W. Yang, D. Losic, S. Orbons, F.J. Mearns, J.G. Shapter, D.B. Hibbert, J. Am. Chem. Soc. 125 (2003) 9006.

81. A. Guiseppi-Elie, C. Lei, R.H. Baughman, Nanotechnology 13 (2002) 559.

82. H.J. Hecht, H.M. Kalisz, J. Hendle, R.D. Schmid, J. Shomburg, Mol. Biol. 229 (1993) 153.

83. J. Li, H.T. Ng, A. Cassell, W. Fan, H. Chen, Q. Ye, J. Koehne, J. Han, M. Meyyappan, Nanoletters 3 (2003) 597.

84. E. Domynguez, O. Rincon, A. Narvaez, Anal. Chem. 76 (2004) 3132.

85. G. Decher, Science 277 (1997) 1232.

86. G. Narvaez, I. Suarez, I. Popescu, E. Katakis, Domynguez, Biosens. Bioelectron. 15 (2004) 43.

87. M. Musameh, J. Wang, A. Merkoci, Y. Lin, Electrochem. Commun. 4 (2002) 743.

88. G. Dryhurst, K.M. Kadish, F. Scheller, R. Rennerberg, Biological Electrochemistry, Academic Press, New York, 1982.

89. L. Gorton, A. Lindgren, T. Larsson, F.D. Munteanu, T. Ruzgas, I. Gazaryan, Anal. Chim. Acta 400 (1999) 91.

90. N. Hu, Pure Appl. Chem. 73 (2001) 1979.

91. J. Wang, M. Li, Z. Shi, N. Li, Z. Gu, Anal. Chem. 74 (2002) 1993.

92. A. Guiseppi-Elie, C. Lei, R.H. Baughman, Nanotechnology 13 (2002) 559.

93. C. Cai, J. Chen, Anal. Biochem. 332 (2004) 75.

94. M. Li, J. Wang, Z. Shi, N. Li, Z. Gu, Anal. Chem. 74 (2002) 1993.

95. J.J. Davis, R.J. Coles, H.A.O. Hill, J. Electroanal, Chem. 440 (1997) 279.

96. S.G. Wang, Ruili Wang, P.J. Sellin, Qing Zhang, Biochemical and Biophysical Research Communications 325 (2004) 1433.

97. S.C. Tsang, J.J. Davis, M.L.H. Green, H.A.O. Hill, Y.C. Leung, P.J. Sadler, J. Chem. Soc. Chem. Commun. (1995) 1803.

98. B.R. Azamian, J.J. Davis, K.S. Coleman, C.B. Bagshaw, M.L.H. Green, J. Am. Chem. Soc. 124 (2002) 664.

99. M. Shim, N.W.S. Kam,, R.J. Chen, Y. Li, H. Dai, Nanoletters 2 (2002) 285.

100. J. Wang and M. Musameh, Anal. Chem. 75 (2003) 2075.

101. Z. Wu, Z. Chen, X. Du, J.M. Logan, J. Sippel, M. Nikolou, J.J. Davis, M.L.H. Green, H.A.O. Hill, Y.C. Leung, P.J. Sadler, J. Sloan, A.V. Xavier, and S.C. Tsang, Inorg. Chem. Acta 27 (1998) 261.

102. D.R. Maria, and A.R. Gustavo, Electrochem. Commun. 5 (2003) 689.

103. M. Rubianes, and G. Rivas, Electrochem. Commun. 5 (2003) 689.

104. G. Zhao, L. Zhang, X. Wei, Z. Yang, Electrochem. Commun. 5 (2003) 825.

105. C. Cai, J. Chen, T. Lu, Sci. China Ser. B 47 (2004) 113.

106. X. Yu, D. Chattopadhyay, I. Galeska, F. Papadimitrakopoulos, J.F. Rus-ling, Electrochem. Commun. 5 (2003) 408.

107. W.C.W. Chan, and S. Nei, Quantum dot bioconjugates for ultra- sensitive nonisotopic detection. Science 281 (1998) 2016.

108. M.A. Hamon, J. Chen, H. Hu, Y. Chen, M.E. Itkis, A.M. Rao, P.C. Eklund, R.C. Haddon, Adv.Mater. 11 (1999) 834.

109. M.S. Strano, C.A. Dyke, M.L. Usrey, P.W. Barone, M.J. Allen, H. Shan, C. Kittrell, R.H. Hauge, J.M. Tour, R.E. Smalley, Science 301 (2003) 1519.

110. A. Koshio, M. Yudasaka, M. Zhang, S. Iijima, Nano Lett. 1 (2001) 361.

111. H. Hiura, T.W. Ebbesen, K. Tanigaki, Adv. Mater. 7 (1995) 275.

112. J.X. Wang, M.X. Li, Z.J. Shi, N.Q. Li, Z.N. Gu, Electrochim. Acta 47 (2001) 651.

113. J.X. Wang, M.X. Li, Z.J. Shi, N.Q. Li, Z.N. Gu, Anal. Chem. 74 (2002) 1993.

114. Y. Lin, S. Taylor, H.P. Li, K.A.S. Fernando, L.W. Qu, W. Wang, L.R. Gu, B. Zhou, Y.-P. Sun, J. Mater. Chem. 14 (2004) 527.

115. J.E. Riggs, Z.-X. Guo, D.L. Carroll, Y.P. Sun, J. Am. Chem. Soc. 122, (2000) 5879.

116. B. Zhao, H. Hu, R.C. Haddon, Adv. Func. Mater. 14, (2004) 71.

117. E. Titus, N. Ali, G. Cabral, P. Ramesh Babu, J. Gracio, Journal of Materials Engineering and Performance, 2 (2006).

118. E. Titus, N. Ali, G. Cabral, P. Ramesh Babu, J. Gracio, Proceedings of the MP3 conference, Tsukuba, Japan, 2006-07-23

119. S. Sotiropoulou, N.A. Chainiotakis, Anal. Bioanal. Chem. 375 (2003) 103.

120. J.M. Xu, 2003. Nanotube electronics: non-CMOS routes. In: Proceedings of the IEEE, Special Issue on Nanoelectronics and Giga-scale Systems 91, pp. 1819–1829.

121. K. Fu, W. Huang, Y. Lin, D. Zhang, T.W. Hanks, A.M. Rao, Y.-P. Sun, Nanotechnology 2 (2002) 457.

122. Dong-sheng Yao, Hong Cao, Shengmei Wen, Da-ling Liu, Yan Bai, Wen-jie Zheng, Bioelectrochemistry 68 (2005) 131.

123. M. Musameh, J. Wang, A. Merkoci, Y. Lin, Electrochem. Commun. 4 (2002) 743.

124. J. Wang, M. Li, Z. Shi, N. Li, Z. Gu, Anal. Chem. 74 (2002) 1993.

125. K. Gong, Y. Dong, S. Xiong, Y. Chen, L. Mao, Biosens. Bioelectron., 20 (2004) 253.

126. M. Zhang, A. Smith, W. Gorski, Anal. Chem. 76 (2004) 5045.

127. J. Wang, M. Musameh, Anal. Chem. 75 (2003) 2075.
128. J. Davis, R. Coles, H. Hill, Electroanal. Chem. 440 (1997) 279.
129. M. Rubianes, G. Rivas, Electrochem. Commun. 5 (2003) 689.
130. J. Wang, Musameh, Analyst 129 (2004) 1.
131. M. Guo, J. Chen, D. Liu, L. Nie, S. Yao, Bioelectrochemistry 29 (2004) 29.
132. J. Wang, M. Musameh, Y. Lin, J. Am. Chem. Soc. 125 (2003) 2408.
133. Z. Wang, J. Liu, Q. Liang, Y. Wang, G. Luo, Analyst 127 (2002) 653.
134. P. Joshi, S.A. Merchant, Y. Wang, D. Schmidtke, Anal. Chem. 77 (2005) 3183.
135. Y. Lin, F. Lu, Y. Tu, Z. Ren, Nano. Lett. 4 (2004) 191.
136. W. Guan, Y. Li, Y. Chen, X. Zhang, G. Hu, Biosens. Bioelectron. 21 (2005) 508
137. H. Tang, J. Chen, S. Yao, L. Nie, G. Deng, Y. Kuang, Anal. Biochem. 331 (2004) 89.
138. M. Gao, L. Dai, G. Wallace, Synthetic Met. 137 (2003) 1393.
139. S. Lim, J. Wei, J. Lin, Q. Li, J. Kua You, Biosens. Bioelectron. 20 (2005) 2341.
140. J. Ye, Y. Wen, W. Zhang, L. Gan, G. Xu, F. Sheu, Electrochem. Commun. 6 (2004) 66.
141. M. Yang, Y. Yang, Y. Liu, G. Shen, R. Yu, Biosens. Bioelectron. 27 (2006) 246.
142. P. Joshi, S.A. Merchant, Y. Wang, D. Schmidtke, Anal. Chem. 77 (2005) 3183.
143. H. Sun, N. Hu, Analyst 130 (2005) 76.
144. M.P. Mattson, R.C. Haddon, A.M. Rao, J. Mol. Med. 14 (2000) 175.
145. Xuan Zhang, Shalini Prasad, Sandip Niyogi, Andre Morgan, Mihri Ozkan Cengiz S. Ozkan, Sensors and Actuators B 106 (2005) 843.
146. H. Hu, Y. Ni, V. Montana, R.C. Haddon, V. Parpura, Nanoletters 4 (3) (2004) 507.
147. Elena Bekyarova, Yingchun Ni, Erik B. Malarkey, Vedrana Montana, Jared L. McWilliams, Robert C. Haddon, Vladimir Parpura, Journal of Biomedical Nanotechnology, 1 (2005) 17.
148. N.W.S. Kam, T.C. Jessop, P.A. Wender, H. Dai, J. Am. Chem. Soc.126, 6850 (2004).
149. F. Balavoine, P. Schultz, C. Richard, V. Mallouh, T.W. Ebbesen, C. Mioskowski, Angew. Chem. Int. Ed 38, 1912 (1999).
150. D. Pantarotto, J.-P. Briand, M. Prato, A. Bianco, Chem. Commun. 16 (2004).
151. F. Balavoine, P. Schultz, C. Richard, V. Mallouh, T.W. Ebbesen, C. Mioskowski, Angew. Chem. Int. Ed 38 (1999) 1912.
152. Z. Guo, P.J. Sadler, S.C. Tsang, Adv. Mater. 10 (1998) 701.
153. J. Sloan, A.V. Xavier, S.C. Tsang, 272 (1998) 261.
154. Von Recum (Ed.), Handbook of Biomaterials Evaluation, Scientific, Technical and Clinical Testing of Implant Materials, second ed., Taylor and Francis, PA, (1999) 915.
155. J. Black, Biological Performance of Materials: Fundamentals of Biocompatibility, Marcel Deckker, New York, 1992.
156. P. Christel, A. Meunier, A.J.C. Lee (Eds.), Biological and Biomechanical Performance of Biomaterials, Elsevier, Amsterdam, The Netherlands, 1997, pp. 81.

157. K.A. Khor, L. Fu, V.J.P. Lim, P. Cheang, The effects of ZrO_2 on the phase compositions of plasma sprayed HA/YSZ composite coatings, Mater. Sci. Eng. A 276 (2000) 160.
158. C.A. Van Blitterswijk, J.J. Grote, W. Kuijpers, W.T. Daems, K.A. de Groot, Biomaterials 7 (1986) 553.
159. D.C. Tancred, B.A.O. McCormack, A.J. Carr, Biomaterials 19 (1998) 1735.
160. M. Wang, S. Deb, K. Tanner, W. Bonfield, in: Proceedings of the 7th European Conference on Composite Materials, London, 1996, 455.
161. S.F. Hulbert, L.L. Hench, in: P. Vineenzini, High Technology Ce-ramics, Elsevier, Amsterdam, 1987, 3.
162. F.B. Bagambisa, U. Joos, W. Schilli, J. Biomed. Res. 27 (1993) 1047.
163. L.M. Rodriguez-Lorenzo, M. Valler-Regi, J.M.F. Ferreira, Bioma-terials 22 (2001) 583.
164. H.Y. Yasuda, S. Mahara, Y. Umakoshi, S. Imatazo, S. Ebisu, Bio-materials 21 (2001) 2045.
165. F. Lupo, R. Kamalakaran, C. Scheu, N. Grobert, M. R. uhle, Carbon 42 (2004) 1995.
166. G. Siegal, J. van Duynhoven, M. Baldus, Chem. Biol. 3 (1999) 530.
167. S.H. Kim, Nature Struct. Biol. 5 (1998). 643.
168. K. Wuthrich, Nature Struct. Biol. 7, (2000) 188.
169. C. Bustamante, C. Rivetti, D.J. Keller, Curr. Opin. Struct. Biol. 7 (1997) 709.
170. H.G. Hansma, and L. Pietrasanta, Curr. Opin. Chem. Biol. 2 (1998) 579.
171. Adam T Woolley, Chin Li Cheung, Jason H Hafner, Charles M Lieber, Chemistry & Biology 2000, 7:R193^R204.
172. S. Kasas, N.H. Thomson, B.L. Smith, P.K. Hansma, J. Miklossy, H.G. Hansma, Int. J. Imaging Syst. Technol. 8 (1998) 151.
173. H. Dai, J.H. Hafner, A.G. Rinzler, D.T. Colbert, R.E. Smalley, Nature 384, (1996)147.
174. S.S. Wong, J.D. Harper, P.T. Lansbury Jr., C.M. Lieber, J. Am. Chem. Soc. 120 (1998), 603.
175. S.S. Wong, A.T. Woolley, T.W. Odom, J.-L. Huang, P. Kim, D.V. Ve- zenov, C.M. Lieber, Appl. Phys. Lett. 73 (1998) 3465.
176. L.F. Sun, S.S. Xie, W. Liu, W.Y. Zhou, Z.Q. Liu, D.S. Tang, G. Wang, L.X. Qian, Nature 403 (2003) 384.
177. H. Nishijima, S. Kamo, S. Akita, Y. Nakayama, K.I. Hohmura, S.H. Yoshimura, K. Takeyasu, Appl. Phys. Lett. 74 (1999) 4061.
178. J.H. Hafner, M.J. Bronikowski, B.R. Azamian, P. Nikolaev, A.G. Rin zler, D.T. Colbert, K.A. Smith, R.E. Smalley, Chem. Phys. Lett. 296 (1998) 195.
179. P. Nikolaev, M.J. Bronikowski, R.K. Bradley, F. Rohmund, D.T. Col- bert, K.A. Smith, R.E. Smalley, Chem. Phys. Lett. 313 (1999) 91.
180. Gil Cabral et al.,
181. R.F. Service, Nanomaterials show signs of toxicity. Science 300 (2003) 243.
182. A.A. Shvedova, V. Castranova, E.R. Kisin, D. Schwegler-Berry, A.R. Murray, V.Z. Gandelsman, A. Maynard, P. Baron, J. Toxicol. Environ. Health A 66 (2003) 1909.

13. Bonelike® Graft for Regenerative Bone Applications

13.1 Introduction

13.1.1 Bone physiology

Bone is a complex mineralized living tissue, exhibiting the property of marked rigidity and strength whilst maintaining some degree of elasticity. In general, there are two types of bones in the skeleton, namely, the flat bones, i.e., skull bones, scapula, mandible, ilium, and the long bones, i.e., tibia, femur and humerus. In principle, bone serves the following three main functions in human bodies: (i) acts as a mechanical support; (ii) is the site of muscle attachment for locomotion, protective, for vital organs and bone marrow; and (iii) to assist metabolism, it acts as a reserve of ions for the entire organism, especially calcium and phosphate.

Bone consists of metabolically active cells that are integrated into a mineralized extracellular matrix. The cellular components consist of osteogenic precursor cells, osteoblasts, osteoclasts, osteocytes, bone lining cells and the hematopoietic elements of bone marrow. The matrix, essentially collagen type I (about 90%), noncollagenous proteins and proteoglycans, is calcified with the deposition of highly substituted hydroxyapatite.

During a lifetime, bone mass is continuously involved in the remodelling process, which is responsible for the renewal of the skeleton, necessary for the maintenance of bone tissue integrity and mineral homeostasis. Bone remodelling involves the coordination of activities of cells from two distinct lineages, the osteoblasts and the osteoclasts, which form and resorb the mineralized tissue.

Bone exists in two main forms, woven bone and lamellar bone. Woven bone is a primitive and immature bone that is formed during bone development, fracture healing, tumours and metabolic diseases. It forms very rapidly and is characterized by a random organization of the collagen fibers, irregularly shaped vascular spaces and calcification occurring as irregularly distributed patches. Woven bone is gradually replaced by organized lamellar bone, which is the form that constitutes most of the mature skeleton. In lamellar bone, the collagen fibers form highly organized sheets in which successive layers of fibers are oriented perpendicular to each other with little interfibrillar space, and calcification occurs in an orderly manner. Lamellar bone may be formed as cortical and trabecular bone. The external parts of the bones are formed by a thick and dense layer of calcified tissue – the cortical bone, in which 80–90% of the volume is calcified. The internal space of the bones is filled with a network of thin calcified trabeculae - the trabecular bone, with 15–25% of the volume being calcified. The cortical bone fulfils mainly mechanical and protective functions and the trabecular bone, which is filled with bone marrow, is mainly involved in metabolic functions.

Osteoblast and Bone Formation

Cells of the osteoblast lineage originate from multipotent mesenchymal stem cells, adjacent to all bone surfaces. Osteoblastic differentiation proceeds along an osteogenic pathway progressing from osteoprogenitors, preosteoblasts to fully mature osteoblasts and then, to lining cells or osteocytes. This process is controlled by a cascade of events that involves a combination of genetic programming and gene regulation by various hormones, cytokines, growth factors and other soluble factors [1].

The osteoblasts are cuboidal or slightly elongated cells that are located at the bone surfaces undergoing remodelling. Their primary function is to lay down the extracellular matrix and regulate its mineralization. The osteoblast is highly anchorage dependent and is a typical protein-producing cell having an extremely well-developed rough endoplasmatic reticulum and a large circular Golgy complex. Teams of 100 – 400 osteoblasts per bone-forming site produce a highly organized organic extracellular matrix called the osteoid. The lifespan of an osteoblast is around one month, during which it lays down 0.5 – 1.5 m^3 osteoid per day [2]. Eventually, some osteoblasts may become "trapped" in the newly formed matrix, thus they are converted into osteocytes, whereas, others remain on the bone surface as bone-lining cells or undergo apoptosis. The osteoid is a complex mixture of a variety of proteins secreted in a certain and specific order. It includes type I collagen (about 90%) and a number of other matrix proteins, i.e., osteocalcin (bone GLA protein), osteopontin, osteonectin, bone sialoprotein and fibronectin, proteoglycans, carbohydrates and lipids. The initial mineralization of the osteoid typically occurs within a few days of secretion but is completed over the course of several months [2]. Matrix vesicles, lipid bilaminar organelles which bud from the osteoblasts, appear to be the initiators of this process. These structures contain phosphatases (which hydrolyse organic phosphates providing high local levels of phosphate ions, i.e., alkaline phosphatase, ALP), phospholipds and calcium ions. At a point of supersaturation, mineral deposition begins. As the matrix vesicles disintegrate, the mineral is exposed to the matrix, where the mineralization proceeds in a self-perpetuating manner [2]. The inorganic content of bone consists primarily of calcium phosphate and calcium carbonate, with small quantities of magnesium, fluoride and sodium. Bone mineral is referred as hydroxyapatite $[Ca_{10}(PO_4)_6(OH)_2]$, a plate-like crystal $20 - 80$ nm in length and $2 - 5$ nm thick. It is four times smaller than naturally occurring apatites and less perfect in structure, being more reactive and soluble which facilitates chemical turnover [3].

Osteoclast and Bone Resorption

The osteoclast is the bone cell responsible for bone resorption. Osteoclasts originate from hematopoietic stem cells that undergo proliferation and differentiate into preosteoclasts and osteoclasts. Osteoclasts are found in contact with a calcified bone surface and within lacunae, which is the result of its own resorptive activity. Usually, there are only one or two osteoclasts per resorptive site. They are highly migratory, multinucleated and polarized cells that present abundance of Golgy complexes, mitochondria and transport vesicles loaded with lysosomal enzymes. The most prominent feature of the osteoclast is the existence of deep foldings of the plasma membrane in the area facing the bone matrix, the ruffled border that is surrounded by a ring of contractile proteins (actin, vinculin and talin) serving to attach the cell to the bone surface and sealing off the resorbing compartment (the sealing zone). Protons and lytic enzymes are secreted across the ruffled border to the resorbing compartment. The low pH (between 2 and 4) allows the dissolution of the mineral, exposing the organic matrix. Subsequently, a variety of enzymes (including tartrate-resistant acid phosphatase, TRAP), at low pH, degrades the matrix components. The resultant residues are either internalised or transported by transcytosis and released at the basolateral membrane. An activated osteoclast is able to resorb 200,000 m^3 per day, an average amount of bone formed by seven to ten generations of osteoblasts with an average lifespan of 15 – 20 days [2].

Cellular Organization Within the Extracellular Matrix –
The Osteocyte

Osteocytes are the most differentiated cells of the osteoblast lineage and may persist in bone matrix for prolonged time periods. They are derived from the osteoblasts but are different in morphology and function i.e., smaller in size, with less cell organelles such as ribosomes and endoplasmatic reticula and an increased nucleus to cytoplasm ratio. Osteocytes are the most abundant cells in bone (about 90%, 25,000 cells/mm^3) and are found embedded deep within the mineralized matrix in small lacunae. They present an extensive network of cellular processes that are enclosed within fine tubules (canaliculi) permeating the entire bone matrix.

The structural design of the osteocytes in the bone matrix allows for a tremendous cell-bone surface contact area, an extensive communication network between neighbouring osteocytes, cells lining the bone surface and blood vessels and, also, diffusion of solutes and gases (via the canaliculi). This structure has the characteristics of a three-dimensional sensor and communication system in bone. Osteocytes have a critical role in the maintenance of bone mass, with an important function in the local response of bone to stress, mechanical deformation and fluid flow [1].

Bone Re-modeling

Living bone is continually undergoing remodelling and the turnover rate is around 10% a year in adult bone. Remodelling is required to replace dead or damaged tissue and to give bone the capacity to adapt to changes in loading and to respond to nutritional and/or metabolic changes. It is an orderly sequence of events achieved by the concerted actions of different bone cells – the basic multicellular unit (BMU) [4]. It includes four different phases – activation, resorption, reversal and formation – the ARRF sequence, which takes about 3 – 6 months for completion.

The process begins with activation of bone lining cells, to uncover bone surface locally allowing for osteoclast adhesion. In the resorption phase, which lasts from one to two weeks, the osteoclasts degrade the mineralized matrix with the formation of resorption lacunae. Subsequently, in the reverse phase, believed to be critical for the coupling between bone resorption and bone formation, a heterogeneous population of mononuclear cells appears in the resorption lacunae and secretes a cement line substance. The formation phase, performed by the osteoblasts, includes the production of a highly organized collagenous matrix and its mineralization; these events continue to occur until the cavity becomes filled, i.e., taking up to several months. Complete refilling of the resorption lacunae is of great importance in maintaining the constant level of bone mass.

Regulation of bone remodelling is a complex process with several regulatory pathways operating and acting on the generation and activity of differentiated bone cells, namely circulating hormones, local growth factors and mechanical stress. Hormones, which are produced and secreted by endocrine glands at different locations in the body, enter the systemic circulation and reach the bone microenvironment, thus affecting local cells either directly or indirectly by inducing them to produce growth factors, peptide mediators with autocrine/paracrine effects. The regulation of bone remodelling at the molecular level remains to be inadequately understood. Specific factors are believed to regulate each step in the remodelling process and to integrate the development of osteoblasts and osteoclasts and their activities, as well as modulate control that is exerted through the endocrine system [1].

Bone Healing Process

A wound represents an anatomical or functional interruption in the continuity of a tissue that is accompanied by cellular damage and death. Every injury initiates a series of coordinated events directed towards restoring the injured tissue to as near normal as possible - the healing process, which can be accomplished by regeneration or repair.

Bone healing can be considered as regeneration rather than repair because it restores the tissue to its original physical, mechanical and functional properties. Healing is regulated by a complex interplay of systemic and local factors and occurs in three distinct but overlapping stages: the early inflammatory stage, the repair stage and the late remodelling stage [5,6].

The initial biological response to a disruption in the continuity of bone is bleeding from ruptured vessels. A hematoma develops within the injury site during the first few hours and days and is accompanied by a typical inflammatory response. The hypoxic tissue environment stimulates the migration of inflammatory cells and fibroblasts to the injured zone, and the initial hematoma is replaced by granulation tissue. Ingrowth of vascular tissue favours the migration of mesenchymal stem cells, which contributes to the establishment of the repair stage. Fibroblasts begin to lay down a stroma that helps supporting vascular ingrowth, while cells of the osteoblastic lineage secrete osteoid, which is subsequently mineralized (formation of a callus around the repair site). This early immature bone consists of arrays of collagen fibers and randomly oriented spicules of bone (woven bone). Bone healing is completed during the remodelling phase in which various cellular molecular and functional regulators modify woven bone to organized lamellar bone, restoring the original shape, structure and mechanical strength. Remodelling occurs slowly over months to years, and is strongly influenced by local mechanical stress placed in the bone.

The process of bone graft incorporation is similar to the bone healing process that occurs in fractured long bones [6]. The most critical period is the first one to two weeks in which inflammation and revascularization occur. Incorporation and remodelling of a bone graft requires that the precursor bone cells have vascular access to the graft to differentiate into osteoblasts and osteoclasts. Bone grafts are also strongly influenced by local mechanical forces during the remodelling phase. Mechanical demands modulate the density, geometry, thickness and trabecular orientation of bone, allowing to optimize the structural strength of the graft [6].

13.1.2 Regenerative Graft Procedures

Regenerative graft procedure refers to technologies that repair or replace any defective, diseased tissues or organs by trauma, ageing etc. Bone grafting is commonly used in the reconstruction of surgical procedures to swift a *de novo* bone formation *in vivo* with the aim of providing a rigid structure and supports other parts, in which the host bone can regenerate and heal in a proper way at particular defined time periods [7]. There are generally four types of bone grafts, namely autograft, allograft, xenograft and synthetic graft that have been widely used in regenerative surgery [8].

Autografts are those where the bone to be grafted is from another site in or on the body of the same individual. It is immunologically safe and thus limiting rejection concerns. The harvest of the autograft implies an extra and invasive surgical procedure coupled with the post-operative pain. Another disadvantage is the limited quantity of bone available for harvesting. Allografts are taken from human donors such as organ, tissues or cells donated from genetically distinct individual of the same specie. The use of this graft can solve some of the drawbacks related with autologous bone grafting since the second surgical procedure is eliminated and the quantity of tissue is available in large amounts. However, the risk of postoperative infection and disease transmission etc., are higher than with autograft. Xenografts are harvested from animals to human. The animal bone, most commonly bovine (cow) is specially processed to make it biocompatible and sterile. It acts like filler, which in time, the body will replace with natural bone. Synthetic graft substitutes have been developed to provide an alternative to autografts and allografts [7–11].

Synthetic bone graft substitutes offer many advantages compared to autografts including a lower probability of rejection or risk of morbidity, patient pain and recovery time. Hence, synthetic substances are gaining growing interest for use as bone graft materials. For successful bone grafting, there are three basic criteria namely, osteogenesis, osteoinduction and osteoconduction [8]. Osteogenesis is the process to produce a direct bone formation by transplanted living cells (osteoblasts and osteoblast precursors). To date, the only material that displays true osteogenic properties is autograft. Osteoinducion is the process, which stimulates new bone production in bone-forming cells [7]. Blood-borne proteins, peptides, growth factors and a specific group of named cytokines provide this stimulation. Osteoconduction is the process, which provides a structural framework and environment that supports the migration, attachment and growth of osteoblasts and osteopregeneitor cells into the graft. Autografts, allografts, many of mineral bone graft substitutes such as hydroxyapatite (HA) and bioactive glass provide this property.

Hence, osteoconductive materials, such as synthetic calcium phosphate ceramics are of special interest for bone repair because of the occurrence of biological apatites in normal calcified tissues, e.g. enamel, dentine and bone. Therefore, calcium phosphates act as motivators because they are the main inorganic constituents of hard tissues in the body. They are generally used in dense, granular or porous form as well as coatings of metal prostheses and implants or in the form of composites. The benefits of synthetic grafts include availability, sterility, cost-effectiveness, and reduced morbidity. However, the selection of grafting procedure to use is purely dependent on the nature and complication of the bone defects as well as the choice of available bone grafts.

Ideally, synthetic bone graft substitutes should be biocompatible, show minimal fibrotic reaction, undergo remodelling and support ossification. Hydroxyapatite (HA), $Ca_{10}(PO_4)_6(OH)_2$, is one of the most biocompatible material used today and has been used as bone graft for a long time. HA implant materials are osteoconductive, however they are very slow resorbable materials. Hence, different approaches have been used to overcome this hindrance. For example, HA can be modified or combined with other materials to improve its functionality and enable faster resorption. Tricalcium phosphate, $Ca_3(PO_4)_2$, in their allotropic forms β and α-TCP has higher solubility and resorption rate than HA. Hence due to their relative solubility TCP is generally used in circumstances where structural support is less important.

Glass-based materials are considered as a surface reactive ceramics. These types of materials when implanted undergo surface dissolution and release ions into the surrounding environment with consequent change of pH environment. The composition of the glasses controls surface reactivity and some are known to closely adhere to the surrounding living bone tissues.

In Table 13.1 are listed some of the most widely used and commercially available synthetic bone graft materials. These materials are available in different forms such as blocks, granules, cements, gels and strip products. The block and granule types are the most commercialised on the market [11]. Blocks are normally used in situations of trauma, interbody spinal fusion and non-unions; they stay in one place without migrating and can be shaped to fit the defect. Granules are generally used for posterior/lateral spinal fusion, filling cystic voids as well as for hip and knee revisions. Among other indications, cement is used for the augmentation of pedicle screw fixation, whereas gels can be used percutaneously and injected into closed fractures. Strips are less commonly used, but could be utilized in acetabular reconstructions [11].

According to the statistical and published data on bone graft substitutes, it is estimated that 500,000 to 600,000 bone grafting procedures are performed annually in the United States. Approximately half of these surgeries involve spinal arthrodesis whereas 35% – 40% are used for general orthopaedic applications [7,12]. In Europe the number of grafting procedures was reported to be 287,300 in the year 2000, with a predicted increase to 479,079 in the year 2005 [11,13]. Synthetic bone graft substitutes currently represent only 10% of the bone graft market, but their share is increasing day-by-day as experience and confidence accumulated [12].

Today, tissue-engineering methods are being increasingly used to optimise actual surgical treatments and to develop new treatment methods. The objectives are accelerated towards the healing of bone or soft-tissue defects. The properties of the carrier play an important role in the effect of biological modulators. On one hand they act as a delivery system for the morphogenic factors and on the other hand they have to provide a stable matrix for cell adhesion and growth. Recent approaches which are being implemented include (a) platelet-rich (PRP) harvested from the patient's blood, (b) genetically engineered growth factors (BMP's), (c) autogenous cells enriched millions of times in the laboratory and (d) autogeneous bone marrow (transplants), already containing cells and growth factors [14].

13.2 Synthetic Bone Graft Material - Bonelike®

13.2.1 Bonelike® Development and Preparation

Bone graft substitutes have been developed to provide an alternative to autografts and allografts. Nowadays, it is possible to prepare synthetic bone substitutes that have very similar composition to the mineral osseous tissue. This is important aspect to increase the regeneration of bone, since bone graft should promote an ideal microenvironment where it is possible for cellular adhesion, differentiation and mitoses to occur. Some of these biomaterials have the ability of being reabsorbable in a time-controlled way, in order to permit the correct process of natural reconstruction of involved bone tissue. Therefore, in order to design a scaffold that supports bone formation while gradually being replaced by bone, an optimum balance between the more stable phase like HA and the more soluble phases like TCPs' must be essentially required.

J. D. Santos and co-authors developed [15–18] glass-reinforced HA (GR-HA) composites with the incorporation of a $CaO-P_2O_5$ based glass in the HA matrix. This patented material has been recently registered and marketed as Bonelike®. This bone graft displays two distinctive advantageous characteristics: (a) enhanced bioactivity, by reproducing the inorganic phase of HA in bone which contains several ionic substitution which modulates its biological behaviour, (b) improved mechanical properties by using $CaO-P_2O_5$ based glasses that act as liquid phase sintering process of HA which reduces porosity and grain size [19]. Due to the chemical composition, Bonelike® induces bone formation through specific activation of the bone cells with controlled release of ions such as F^-, Mg^{2+}, Na^+ etc., from its surface to the surrounding medium, which can be schematically described and as shown in Fig. 13.1.

The preparation of Bonelike® can be briefly described as follows: $CaO-P_2O_5$ based glasses are prepared from reagent grade chemical using conventional glass making techniques, which are then crushed to fine particles and mixed to phase pure HA powder. Usual sintering temperature for HA in the range of 1200–1350°C the glass melts, and diffuses into the HA structure, leading to several ionic substitutions in the lattice [20].

13.2.2 Physico-Chemical Characterisation

The first step in a complex series of biophysical/biochemical processes related to the interaction of an implanted material with a host biological tissue consists of the spontaneous formation of a proteinaceous layer absorbed onto the implanted materials. Therefore, surface properties are critical in biomaterial biocompatibility and must be considered as in their selection for medical applications, together with bulk property characteristics [21–24]. Chemical composition, hydrophobicity and electrical surface charge are well known parameters that influence the complex process of protein adsorption onto biomaterials, and their subsequent influence on the attachment and spreading of cells that ultimately determine the success or failure of the implant during service [25].

Phase transformations and interstitial and/or substitution of trace elements during the liquid phase sintering process of Bonelike® were examined by X-ray diffraction and FTIR analyses [16,26].

Table 13. 1. List of the most commercially available synthetic graft material on the market

Name of the product	Characteristics	Company
Hydroxyapatite (HA)		
Ceros 80	Dense polycrystalline HA	Straumann Ltd, UK
Biocoral	Natural mineral skeletons of Scleractinian corals	Inotebm,France
Pro Osteon	Coralline macroporous HA	Interpore Int., USA
Osteograft	Crystalline HA	CeraMed, USA
Calcitite	Dense crystalline HA	Calcitek, USA
Tricalcium Phosphates (TCP)		
Chronos/Ceros 82	Tricalcium phosphate	Mathys Suisse
Biosorb	Tricalcium phosphate	SBM S.A, Lourdes
Vitoss	Ultraporous TCP	Orthovita, USA
Cerasorb	Tricalcium phosphate	Curasan AG, Germany
Bioactive glasses		
Bioglass®	Bioactive glass	Novabone , USA
Ceravital	Group of glasses and glass ceramics of various compo-	E. Pfeil & H. Bromer, Germany
Biogran	sitions Bioactive glass	Orthovita, USA

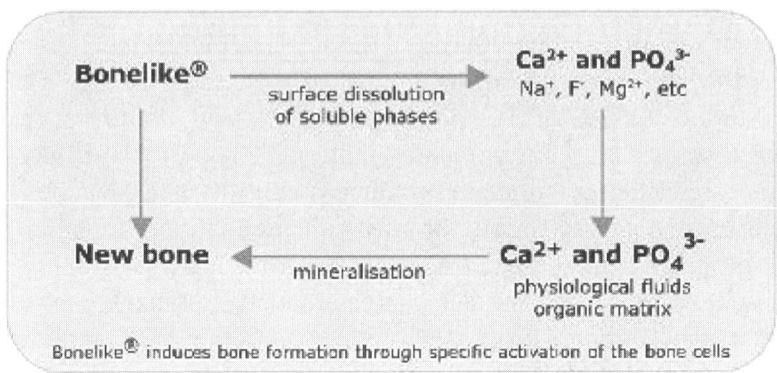

Fig.13.1. Bone regeneration mechanism of Bonelike®

The characterisation of Bonelike® using FTIR shows that carbonate ions CO_3^{-2} that are present in the samples at 1200°C (Fig. 13.2a), as indicated by bands in the region of 880cm^{-1}, systematically decreases until at 1350°C (Fig. 13.2b), the bands are almost indistinguishable. Furthermore, there is a significant loss in the hydroxyl groups OH$^-$, denoted by the band at 3570cm^{-1}. The bands at 880 cm^{-1} have been assigned to $\square_2$ vibrational mode [27].

XRD analysis showed that, depending on the glass amount and the glass composition, as well as the sintering temperatures, the microstructure of Bonelike® was composed of the HA matrix, and α- and β-TCP secondary phases (Fig. 13.3).

The HA is decomposed by the presence of a reactive glass, which enters into the HA structure and causes the OH$^-$ groups to be driven out and also alters the Ca: P ratio. However, the β-TCP formed in the presence of high amount of glass is likely to contain more residual ions from the glass and thus it becomes unstable, and therefore α-TCP is formed.

The glasses can give rise to formation of between approximately 30 and 60 % TCP. At high temperatures, the β-TCP inverts to α-TCP, without further decomposition of the residual HA. For example, the Mg^{2+} containing glasses induced the α-TCP phase formation in the structure of Bonelike® and retarded the α-TCP into β-TCP transformation at higher temperatures [16,28–30]. The chemical composition

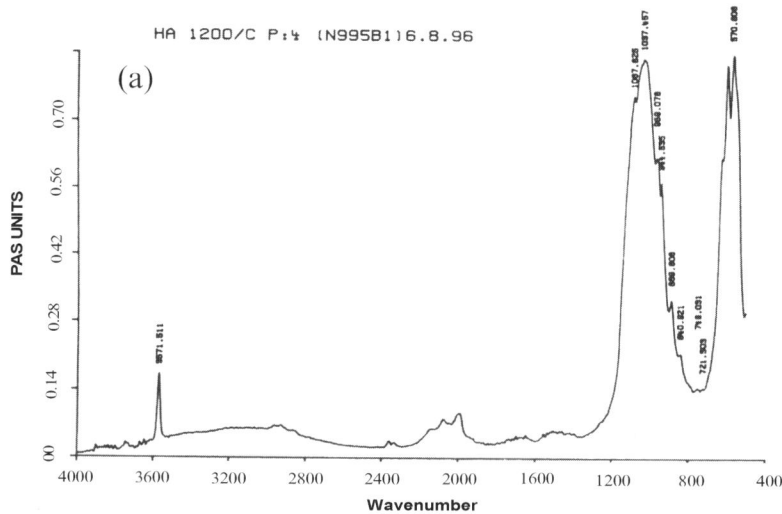

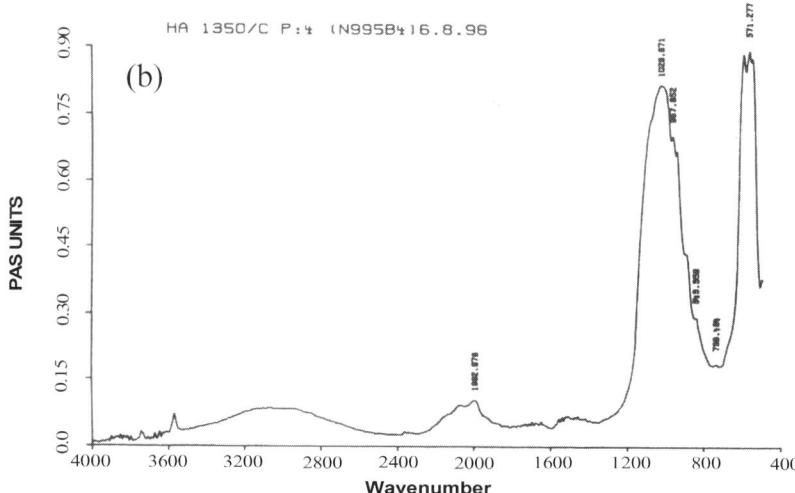

Fig. 13.2. FTIR spectra for Bonelike® with 4wt% glass addition sintered (a) at 1200°C and (b) at 1350 °C

of the glasses also induces modifications in the lattice parameters of the crystallographic phases present in the microstructure of the composites [16].

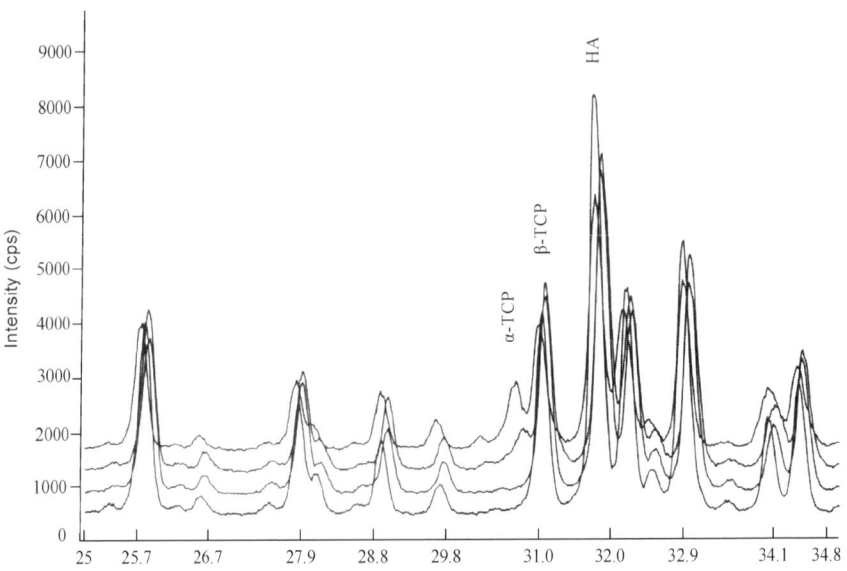

Fig. 13.3. Overlaid traces of XRD data sets from 1200°C (bottom) to 1350°C (top) for Bonelike® containing 4wt% glass

The mechanism and the extent of any ion substitutions and/or interstitial in the calcium phosphate phases lattice depends on the size and content of the ion as well as on the temperature of the liquid phase sintering process [19].

XRD studies highlighted the effect glass had on the crystallographic structure of the main phases of Bonelike® [16,26]. It was proposed that the larger addition of glass gives a larger number of ions present, and it is these ions that enter into phase structures and may become interstitial ions. Wettability and zeta potential studies have been performed to characterize the hydrophobicity and surface charge of Bonelike® [31]. Despite the glass composition used and with the glass amount addition between 2.5 and 4.0 wt%, zeta potential values in between pure phase HA and α-TCP values were observed, i.e., between –28.7 mV and –18.1 mV. This intermediate behaviour was also observed regarding the wettabillity studies performed by static contact angle measurements using water. The presence of the secondary phases namely α-TCP, influenced surface charge and hydrophobicity of the Bonelike®, and therefore indirectly affects cell-biomaterial interactions [31].

13.2.3 Mechanical Characterisation

Mechanical properties of bioactive ceramic materials that are known to form a direct bond with natural bone have been extensively studied to improve their reliability when used in medical applications [32,33]. To assess the mechanical behaviour of Bonelike®, namely the bending strength, elastic properties and fracture toughness, a concentric ring-on-ring testing, an impulse excitation method and Vickers indentation technique respectively have been performed. The Duckworth-Knudsen model was also applied to determine the dependence of bending strength as well as Young's and shear modulus on microstructural parameters of Bonelike®.

Bending Strength

Flexural bending strength (σ) of Bonelike® (Glass reinforced hydroxyapatite composites) was found [34] to be about two or three times higher than that of HA, as seen in Fig. 13.4. The level of reinforcement obtained is exemplified in Fig. 13.5 where, using Weibull statistics analysis, failure probability is plotted versus bending strength for HA and Bonelike® sintered at 1350°C.

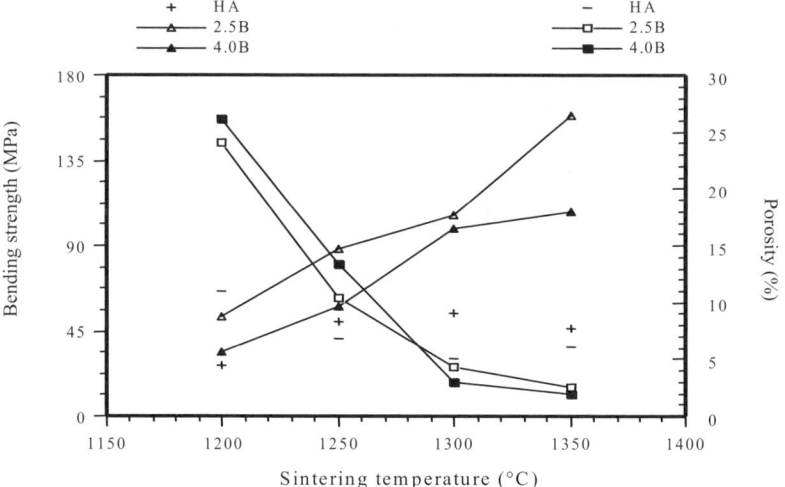

Fig. 13.4. Flexural bending strength (MPa) and Porosity (%) vs. sintering temperature (°C) for HA and Bonelike®

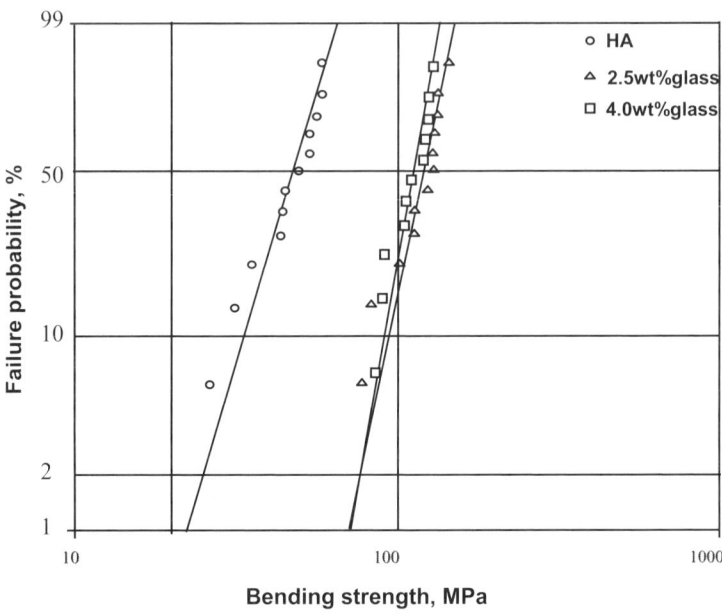

Fig. 13.5. Weibull distribution function (failure probability vs. flexural bending strength) for HA and Bonelike®

The strength of both HA and Bonelike® seems to be strongly dependent on porosity and tended to increase with densification, following the expected pattern for the sintering process (Fig. 13.4). However, for the Bonelike®, the secondary phases (β- and α-TCP) present in the microstructure and these phases are strongly influence the mechanical behaviour of the material.

The Duckworth-Knudsen exponential model is widely used to characterise the mechanical behaviour of ceramic matrix composites [35] and by using the rule of mixture it was possible to estimate the zero porosity rupture modulus for each phase (HA, β and α- TCP) and the porosity correction factor, b. Zero porosity rupture modulus for HA, β- TCP and α-TCP were estimated as being 83.1, 179.9 and 233.5 MPa, respectively and the strength vs porosity equation of Bonelike® was obtained using the relationship:

$$\sigma = (83.1 \, X_{HA} + 179.9 X_\beta + 233.5 \, X_\alpha) \, e^{-4.02P} \qquad (1)$$

Elastic Properties

The elastic properties such as dynamic Young's modulus (E) and shear modulus (G) of Bonelike® have been determined using an impulse excitation method [18] of vibration according to ASTM C 1256-96 using the following equation:

$$E = 0.9465 \ \{mf^2_f / b\} \ \{l^3/t^3 \} T_1 \tag{2}$$

Where E is the Young's modulus (Pa), *m* the mass of the bar (g), *b* the width of the bar (mm), *l* the length of the bar (mm), *t* the thickness of the bar (mm), f_f the fundamental resonant frequency of bar in flexure (H_z), and T_1 the correction factor for fundamental flexural mode.

For the shear modulus (G) determination the following equation was used:

$$G=\{4 \ l \ mf_t^2 / bt\} \ \{b / 1 +A\} \tag{3}$$

Where *G* is the dynamics shear modulus (Pa), f_t the fundamental resonant frequency of bar in torsion (Hz), *A* the empirical correction factor.

The Young's and shear moduli were experimentally determined and the results were correlated with respect to porosity and presence of secondary β- and α-TCP phases in the microstructure of Bonelike® using the Duckworth-Knudsen model. Zero porosity values determined for E and G were as follows:

$E_0 = 120$ GPa   and $G_0 = 41.8$ GPa for   HA, and $E_0 = 64.1$ GPa   and $G_0 = 27.4$ GPa for β- TCP and the obtained general equations were:

$$E = (120.0 \ X_{HA} + 64.1 X_\beta + 188.6 \ X_\alpha) \ e^{-4.04P} \tag{4}$$

$$G = (41.8 \ X_{HA} + 27.4 \ X_\beta + 91.3 \ X_\alpha) \ e^{-4.11P} \tag{5}$$

Fracture toughness

Fracture toughness (K_{IC}) of Bonelike® graft has been assessed using an indentation technique and results calculated according to Laugier and Evans' equations [17]. A 3 N load was applied for 10 s using a pyramid shaped diamond indenter. Each specimen was indented 15 times and at least three samples were used to check for reproducibility of results. Statistical analysis was performed using Student's *t-test*. The highest value was obtained for Bonelike® $K_{1C} = 1.45 \pm 0.25$ % MPa m$^{1/2}$ compared to 0.70 ± 0.09 % MPa m$^{1/2}$ for HA. However, the fracture toughness of Bonelike® strongly depended on the chemical composition and the percentage of the added glass and on the sintering temperature [17].

The combination of crack bridging, crack deflection and transformation –toughness mechanism were used to explain the enhanced fracture toughness of Bonelike® when compared to HA [17,36].

13.2.4 Biological Evaluation

Biological evaluation is of great importance in assessing the potential benefit of implantable materials for human applications. Biocompatibility testing is concerned with "biosafety" and "the ability to perform with appropriate host response in a specific application". *In vitro* and *in vivo* studies form an integral part of biological testing for potential implant materials.

13.2.4.1 In-vitro Studies

As a result of time, effort, cost, increasing restriction and lack of precise basic data derived from animal experimentation, *in vitro* testing has become a major tool for evaluating the basal and specific cytocompatibility of new and modified materials for biomedical applications. *In vitro* procedures are the basic starting point whereby biological responses to materials are determined initially, as required by a number of standardisation agencies [37].

The biological performance of Bonelike® composite scaffolds was assessed by using human bone osteoblastic cell cultures, namely the osteosarcoma cell line MG63 and osteogenic-induced bone marrow cells. The cell response was evaluated by a direct assay, i.e., culturing the cells on the material's surface, and also using an indirect assay, with the cultures being performed in the presence of Bonelike® extracts. The major advantage of the direct assay is that it closely mimics physiological conditions by establishing a concentration gradient of chemicals, which diffuse away from the cells, as it would occur in intact tissues *in situ*. Also, in studying such direct cell-surface interactions, the surface charge, surface topography and surface free energy play a fundamental part in cell adhesion, spreading, growth and, ultimately, function of the cell [37,38]. On the other hand, the indirect method, in which test materials are first "extracted" by immersion in a physiological solution, which is subsequently added to the target cells, has the advantage of being able to examine whether substances are "leached" from the test materials and exert a deleterious, or induce a beneficial effect [37,38].

Osteoblastic cell cultures were maintained for appropriate time periods and culture conditions, depending on the culture system used [39–41]. Cell behaviour was assessed in terms of attachment and spreading, morphology, cell viability/proliferation and functional activity regarding the expression of key extracellular matrix antigens and, in the case of human bone marrow cells, the ability to form a cell-mediated mineralized matrix. Bonelike® composites were prepared in several compositions in order to incorporate in its microstructure, ions commonly found in human bone tissue. Table 13.2 shows the compositions selected for *in vitro* testing. HA was always used as the control material.

Table 13.2: Bonelike® compositions subject to *in vitro* testing

Bonelike®	Glass (in % mol)					Glass added to HA (wt %)
	P_2O_5	CaO	MgO	Na_2O	CaF_2	
A	50	16.5	33.5	–	–	2.5
B	50	16.5	33.5	–	–	4.0
C	75	15.0	–	–	10	4.0
D	45	28	–	27	–	2.5
E	65	15	–	10	10	4.0

Biocompatibility of composites A, B, and C was assessed by monitoring their effects on the proliferation and function of osteoblast-like MG63 cells. Flow cytometry technique was used to evaluate cell-cycle progression, cell size and granularity and expression of collagen type I, fibronectin and osteocalcin. In addition, cultures were observed by scanning electron microscopy (SEM). Evaluation of DNA synthesis and content was also performed regarding composites B and C. Data showed that MG63 cells growing on the surface of the three composites presented a normal morphology, a high growth rate and were able to express collagen type I, fibronectin and osteocalcin, proteins known to have major roles in connective tissue integrity, adhesion and bone structure and differentiation [42]. In addition, cells cultured on the plastic surface in the presence of the extracts of composites B and C did not shown any significant deleterious effect on cell morphology and function, and data yielded information similar to that observed in the direct testing assay [40].

Bonelike® samples with the compositions D and E (Table 13.2) were evaluated regarding the proliferation/differentiation response of human bone marrow osteoblastic cells, using a direct assay. Bone marrow cells (first subculture) were plated on the surface of the composites and the colonized materials were incubated for 35 days in an osteogenic-inducing medium.

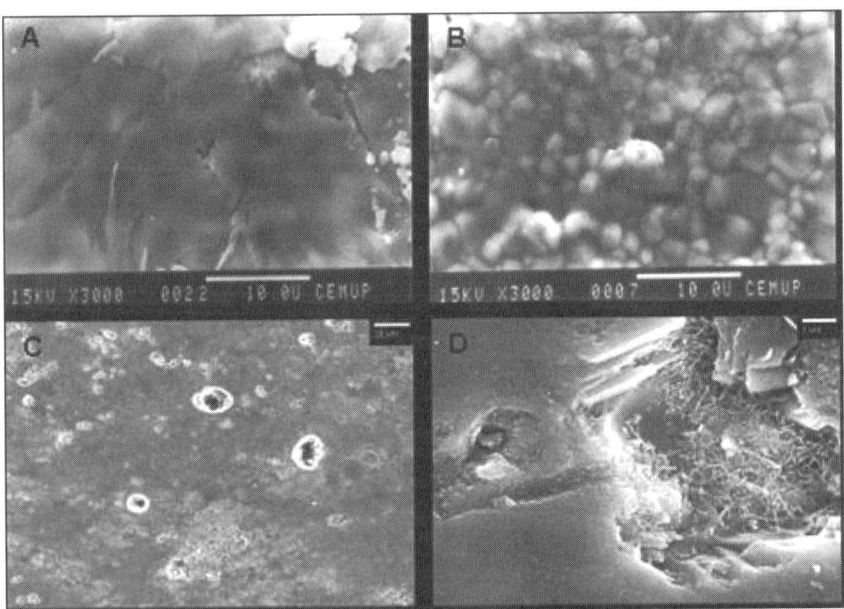

Fig. 13.6. SEM micrographs of the dissolution/deposition reactions on the surface of Bonelike® after incubation with culture medium. Composite D: "as-received" condition (A); 21 days incubation (B); Composite E: 21 days incubation: (C) Bar, 20 μm; (B) Bar, 1 μm

SEM observation of Bonelike® samples incubated in the absence of cells (Fig. 13.6) showed significant surface modifications resulting from the dissolution of resorbable phases, with the formation of different size microcavities that became progressively larger with the incubation time; simultaneously, deposition of an apatite layer was also observed.

Biochemical evaluation of seeded samples of Bonelike® showed that bone marrow cells presented a high proliferation rate and produced significant levels of alkaline phosphatase. HA had a slightly lower biological profile regarding these parameters (Fig. 13.7).

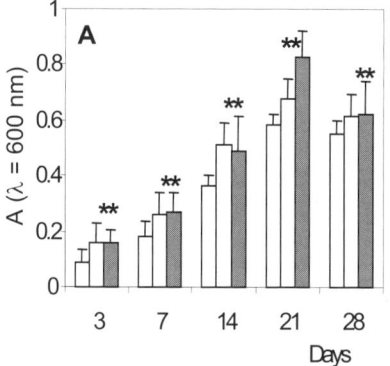 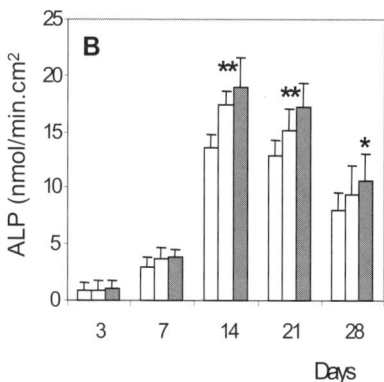

Fig. 13.7. Cell viability/proliferation (A, MTT assay) and alkaline phosphatase activity (B) of human bone marrow osteoblastic cells cultured on the surface of composites D (medium shaded column) and E (dark shaded column). HA (White column) presented a slightly lower biological profile

Cells growing on the material surface adapted successfully to the surface irregularities and were able to grow towards the forming cavities. Cell-mediated matrix mineralization began to occur during the third week, as evident by the presence of abundant globular mineralized structures incorporated into a network of fibers in 28-day cultures (Figs. 13.8 and 13.9). It is noteworthy that with composite E, the formation of the mineralized matrix was especially associated with the surface cavities resulting from the dissolution of resorbable phases. High magnification SEM micrographs showed exuberant cell-mediated mineralization inside these cavities (Fig. 13.9). In general, HA presented a slightly lower biological profile, reflected by a delayed and relatively poor matrix mineralization (Fig. 13.8).

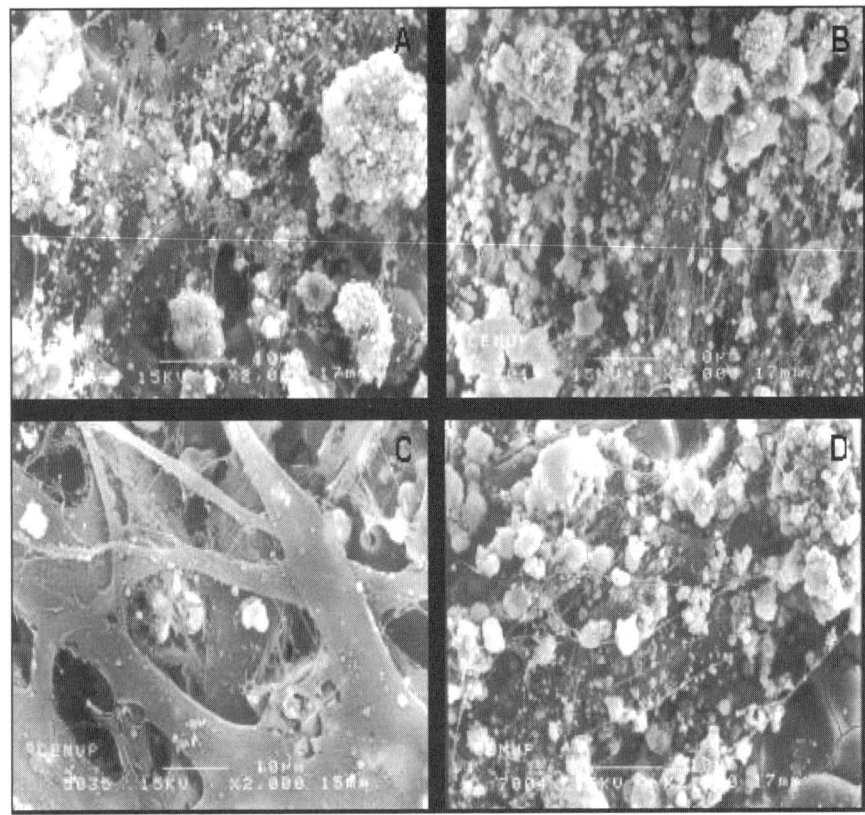

Fig. 13.8. SEM micrographs showing composite D, cultured with human bone marrow osteoblastic cells at days 21 (A) and 28 (B). For comparison, results regarding the cell response to HA at the same time points are also shown: C, 21 days; D, 28 days. Bonelike® presented an earlier onset of matrix mineralization

Results regarding the response of human bone marrow osteo-blastic cells to Bonelike® with the compositions D and E (Table 13.2) showed, in general, that, in comparison with HA, the composites had a positive effect regarding cell proliferation, synthesis of alkaline phosphatase and formation of a mineralised matrix. The improved biological performance of Bonelike® is most probably related with its chemical composition.

This biomaterial is composed of an HA matrix with bioresorbable α- and β-tricalcium phosphate phases, which are more soluble than single HA and liberate Ca and P ionic species to the local environment. Surface reactions occurring as a result of ongoing dissolution/

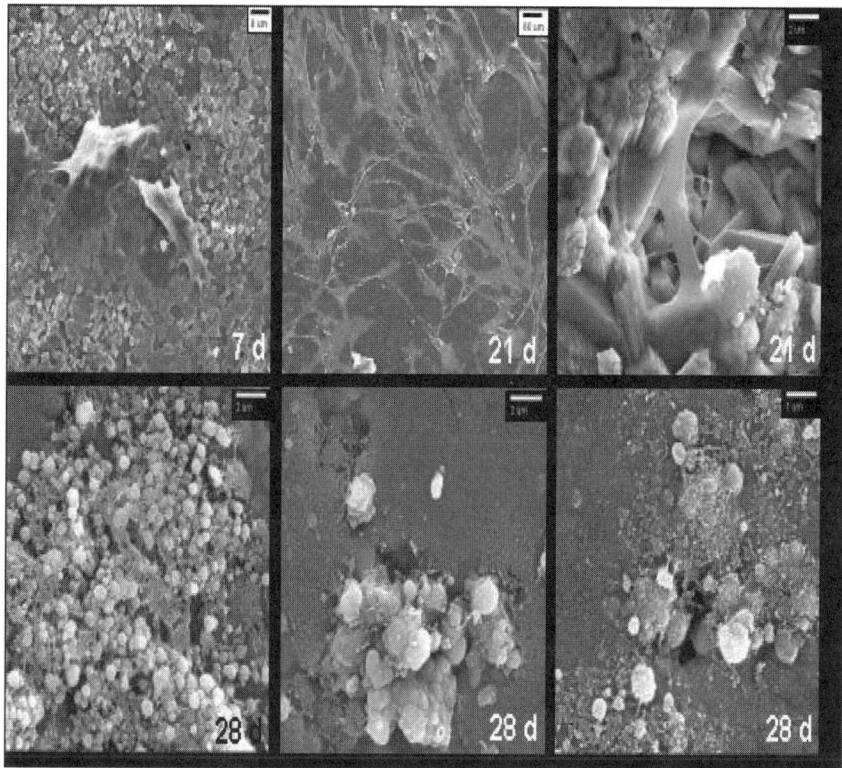

Fig. 13.9. SEM micrographs showing composite E, cultured with human bone marrow osteoblastic cells for 7 days (Bar: 6 μm), 21 days (Bar: 60 μm and 2 μm) and 28 days (Bar: 2 μm, 2 μm and 1 μm). Matrix mineralization occurred preferentially in the surface cavities resulting from dissolution of the resorbable phases

deposition events appeared to induce osteoblastic growth and differentiation. In addition, the presence of fluoride ions in the composite E may also have a positive contribution, as this ion is known to have a potent influence on cell proliferation [43].

13.2.4.2 In-vivo Experimentation

The *in vivo* animal testing of Bonelike® were performed in rabbits and sheep models and all procedures carried out with the approval of the National veterinary authorities and local Ethic Commission, and in accordance with the European Communities Council Directive 86/609/EEC [44,45]. For several medical applications of bone

regeneration, the use of a vehicle to carry the bone graft is considered as being a very relevant issue. In fact, this association not only
facilitates the medical application of the bone graft but also will
open-up new areas of application in medicine, namely those related
to: (i) minimal invasive surgery and (ii) the possibility of associating
therapeutic molecules that have crucial function in bone regeneration [46,47].

Rabbit and Sheep Testing

For the *in vivo* testing of Bonelike® associated to several vehicles,
healthy skeletally mature male New Zealand White rabbits [46,47] and
adult Merino sheep were used as experimental models. For surgery,
rabbits and sheep were placed prone under sterile conditions and under
deep anaesthesia a longitudinal incision was made on the lateral surface
exposing the femur. In each femur, several holes were drilled through
the cortex and into medulla using a micro-burr continuously flushed
with a saline solution (NaCl 0.9%, Braun) to minimize thermal damage
and to remove any residual bone [46,47] (Fig.13.10). Several vehicles
have been tried in association with Bonelike®, such as Floseal®,
Normal Gel 0.9% NaCl® and chitosan-derived gel that were mixed
with Bonelike®granules and implanted into the holes. These vehicles
were associated with two types of therapeutic molecules, raloxifene
hydrochloride and bone morphogenic proteins (BMPs).

FloSeal® is easily used and it can be extruded from a syringe and
applied topically to the bleeding area. This haemostatic agent has the
ability to acquire irregular shapes fitting the wounded site [48,49].

When the FloSeal® is in contact with blood, the collagen particles
are hydrated and swell. The thrombin present converts the patient
fibrinogen into a fibrin polymer, originating a clot around the granules
[48,49]. Normal Gel 0.9% NaCl® (Moneylycke, Portugal) is a polymeric vehicle [46] and the chitosan-derived materials are known for
their biocompatibility and bioactivity [50].

The raloxifene hydrochloride is a known selective estrogen receptor modulator (SERM) and acts as an estrogen agonist on bone and
liver, also to increase bone mineral density [51–53], being used for
prevention of osteoporosis in postmenopausal women [51]. Raloxifene

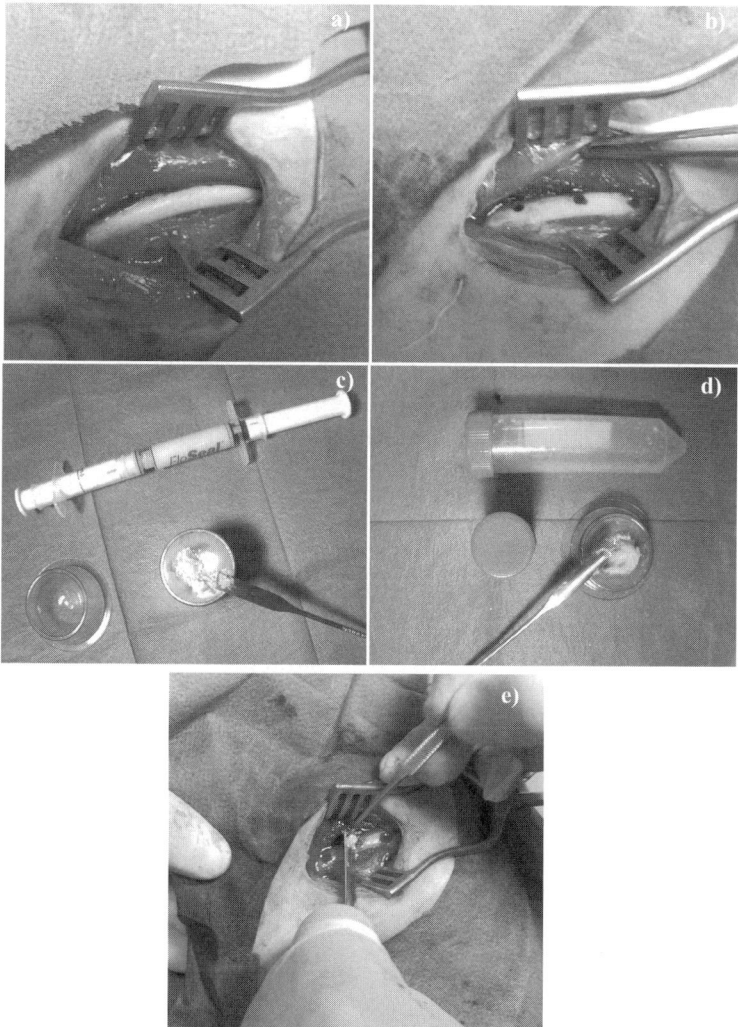

Fig. 13.10. Surgical procedures: (a) exposed rabbit femur; (b) rabbit femur showing 3 holes of 3 mm diameter; (c) mixture of Floseal ®and Bonelike ®granules; (d) mixture of Bonelike ®granules with raloxifene hydrochloride and (e) implantation of Floseal®, Bonelike® granules and raloxifene hydrochloride in the bone defects

hydrochloride also inhibits *in vitro* mammalian osteoclast differentiation and bone resorption in the presence of interleukin–6 (IL–6) [51–53]. The demonstration that recombinant BMPs have the potential to induce bone formation suggests that this may have enormous therapeutic potential in the management of numerous clinical conditions in which there is a requirement for new bone formation [54]. A number of studies have demonstrated that BMP–2, –3, –4, and –7 can up-regulate features of the mature osteoblast phenotype, including alkaline phosphatase (ALP) activity, collagen synthesis, and osteocalcin expression [54].

Through sequential x-ray images, it was possible to follow the healing process every week in both species. X-ray analysis of rabbit femurs revealed increased osteointegration and defect healing for both particle size ranges of implanted Bonelike® associated to both vehicles, Normal Gel 0.9% NaCl® and FloSeal® [46,47]. During the healing period, rabbits and sheep easily recovered and no rejection symptoms were observed in the implantation site for all implanted samples.

Rabbits and sheep were sacrificed 12 and 16 weeks after implantation, respectively, and the retrieved samples analyzed by scanning electron microscopy (SEM) and Solo Chrome R/Haematoxylin-Eosin stained for histological studies. SEM characterization of unstained slices was performed to quantify the contact percentage of new bone formed with implanted granules and assess the *in vivo* degradation process. The interface layer implanted material/new bone formed was evaluated by SEM-EDX (energy dispersion x-ray microanalyser) [46,47]. Both SEM and histological analyses confirmed the osteo-integration of Bonelike® granules and the new bone formation, with almost complete regeneration of the bone defect [46,47]. Bonelike® associated with Floseal® and raloxifene hydrochloride showed that new bone was rapidly apposed on implanted granules after 12 weeks of implantation in rabbits. Bonelike® granules are completely surrounded by *de novo* mature bone (Fig. 13.11) and in the SEM image (Fig. 13.11a) it is possible to observe the complete osteointegration of the Bonelike® granules with bone tissue formed among them with the presence of new osteon. Additionally, an extensive surface dissolution of Bonelike® granules could be observed [46,47]. No evidence of osteoclasts activity seems to have taken place which may be explained by the presence of raloxifene hydrochloride that is known to inhibit osteoclast activity [51–53] and BMPs, known to up-regulate osteoblast functions [54]. Similar results were observed on the histological slices in Fig. 13.11(b) where the granules were completely surrounded by new bone (fibroreticular) with vascular structures and cement lines indicating active bone regeneration. Several blood channels without signs of inflammation throughout the osteoid matrix have been observed and no inflammatory cells and fibrous tissue have

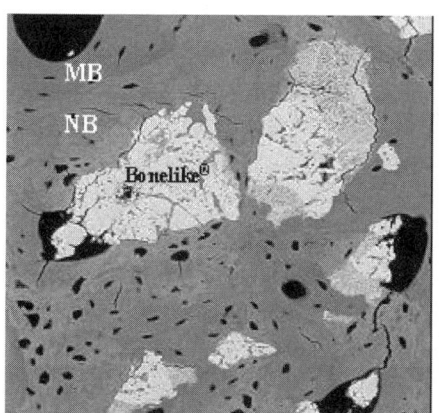
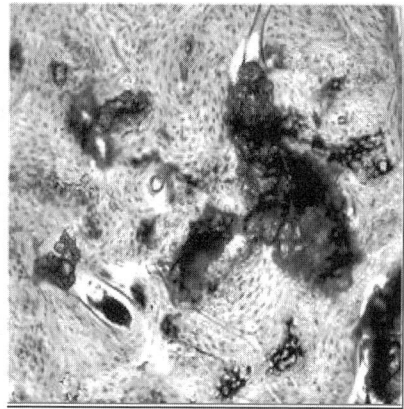

Fig. 13.11. (a) Scanning electron microscopic image (200x); (b) Solo-Chrome R staining images of the experimental samples (20x) showing Bonelike® granules involved in the *de novo* mature bone tissue, (NB - New Bone, MB - Matured Bone, and OC - Osteocytes)

been found. The presence of blood vessels was due to active angiogenesis process that is an extreme important process for bone regeneration [47]. SEM and histology analysis of the ovine model also revealed the high bioactivity of the Bonelike®/chitosane-derived matrix. Through histological analysis, it was possible to see some chitosane-derived matrix after 4 months. Although, it appears in an extensive dissolution features and the granules were surrounding by new bone formation with some degree of maturity.

In conclusion, the Bonelike® graft associated to Floseal®, Normal Gel® or chitosan-derived matrix seemed to serve as an excellent scaffold for bone regeneration [46,47]. In addition, the association of Bonelike® to a resorbable vehicle can act as a controlled release system

to osteoinductive molecules and therefore enhances the osteointegration of Bonelike® graft. It is also much more easy-to-handle and can
be considered as an injectable osteoconductive synthetic bone graft
substitute [46,47].

Push-out Testing

Push-out testing was performed [55] in a rabbit model for characteriz-
ing bone/Bonelike® bonding, and the shear force needed to detach the
implanted material was measured. For this testing, a longitudinal incision was made on the anterior surface, extending from about 10mm
below the knee joint for a distance of 25mm. 4.0mm diameter holes
were drilled through cortex and into medulla and then cylindrical
specimens of 4.0 mm in diameter and 20 mm in length were press-
fitted into the holes. This study was performed using periods of
implantation time up to 16 weeks.

Back-scattering SEM image showed that the place where failure
occurred for implanted samples during push-out testing depended
upon the implantation period. Fig. 13.12 shows the back-scattering
image of the interface between Bonelike® and bone, after 16 weeks
implantation. Breakage occurred essentially through the implant
body after such a long implantation period, which indicates a strong
implant/new bone bonding as well as maturity of newly formed bone
around implant. On the other hand, breakage occurred at the bone/
implant interface at early stages after implantation such as 2 and 4
weeks showing that a weak bone-bonding strength was established,
due to the lack of time for a complete bone formation around
Bonelike® and to its immaturity.

Results of push-out testing showed bonding between Bonelike®
and new bone, ranging from 130-145N after 2 weeks of implantation, similarly to the values achieved for pure phase HA. After the
longest implantation period, 16 weeks, the Bonelike® prepared with
4.0wt% of $CaO-P_2O_5$ based glass showed a higher bonding force of
$606\pm45N$ compared to $459\pm30N$ for sintered pure phase HA.

The enhanced biological response of Bonelike® is a consequence
of the synergistic effect of the presence of secondary phases, β- and

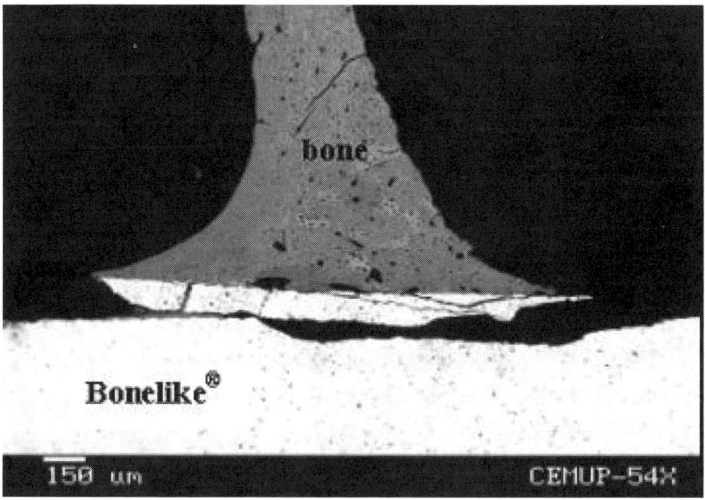

Fig. 13.12. Back-scattering image of interface between Bonelike® material and bone, after 16 weeks implantation

α-TCP phases, and the local release of ions to the surrounded biological medium.

13.2.5 Medical Applications

Bonelike® graft has been successfully applied in several areas of regenerative surgery namely in oral and maxillofacial surgery, implantology and orthopaedics [56-60]. In oral surgery Bonelike® has been used for the regeneration of bone defects after cyst removal and retained tooth extraction in maxillofacial surgery for bone maxilla and mandible reconstruction, in implantology for bone augmentation around implants, ridge augmentation for later implantation and sinus floor elevation, in periodontalogy furcation and intraosseous defects, and in orthopaedic for the regeneration of bone defects caused by trauma, ageing and for the correction of valgus knee using open wedge high tibial osteotomies (HTO).

13.2.5.1 Oral and Maxillofacial Surgery

In oral and maxillofacial surgery [58], Bonelike® has been used to regenerate bone defects after cyst removal in 11 patients, aged between 24 to 53 years with a mean age of 36 years, consisting of 5 men and 6 women. Most cysts of the oral and facial regions under treatment were located within the jaws as an intrabony lesion with a median mandibular cyst, and referred to as a non-odontogenic cyst in the midline aspect of mandible. Sometimes, resection of a large segment of the jaw was necessary to insure complete removal of the lesion. Thereby, the pathological "tissue destruction" process and its suitable surgical removing led to a significant bone loss. The size of the ccavities varied from 3 cm diameter in the minor lesion up to 12 cm of the bigger diameter of the widest cyst. After the cyst were completely removed, the remaining bone cavities were firmly packed with Bonelike® granules mixed with blood and crushed bone remnants compound in an attempt to completely fill bony cavities and "to sculpt" the cortical bone contour.

According to the standard follow-up protocols, radiological examinations were performed and Bonelike®/bone retrieved samples have been analysed histologically using non-decalcified sections obtained perpendicular to bone length axis. Radiographic examination and histological results clearly demonstrated an extensive new bone formation apposed on Bonelike®granules with a significant degree of maturation.

An example of the complete clinical application of the Bonelike® graft is shown in Fig. 13.13. A gingival and periosteal tissue covering the entire maxillary cystic lesion can be observed in Fig. 13.13(a) along with an operative image of maxillary cystic's lesion in Fig. 13.13(b). Post operative radiological analysis shows excellent granules adaptation to the host cavity without material dislocation accompanied by local bone defect regeneration after 3 months (Fig. 13.13c) and after 12 months a complete restoration of the local biofunctionality was achieved (Fig. 13.13d).

These clinical applications in maxillary bone defects indicated perfect bonding between newly-formed bone and Bonelike®granules, along with partially surface biodegradation. This quick and effective osteoconductive response from Bonelike® reduced the time required to reconstruct the bone defected area of patients.

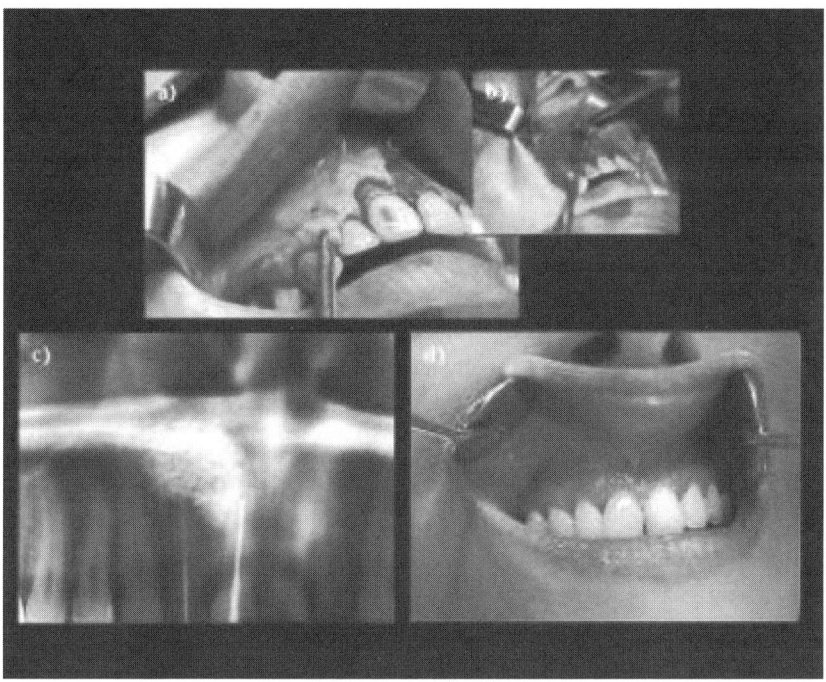

Fig. 13.13. (a) A gingival and periosteal tissue covering the entire maxillary cystic lesion. (b) Operative image of maxillary cystic lesion (c) Post-operative radiograph shows the excellent Bonelike®granules adaptation to the bone cavity without any sings of material dislocation after 3 months, and (d) Complete restoration of the bone defect and biofunctionality may be seen 12 months after implantation

13.2.5.2 Orthopaedics

In orthopaedics applications [60], Bonelike® granules ranging from 500–1000 μm were implanted in the cortical bone of 11 patients undergoing osteotomy in the lateral aspect of the tibia to assess the behaviour. The patients' mean age was 59 years (ranging 48 to 70 years), eight women and three men, all suffering form medical compartment ostoarthritis of the knee. At surgery, a 1×1×1 cm cortical defect was created 3 cm distal to the entry point of the screws, in line with the long axis of tibia. A implanted Bonelike® graft sample was retrieved for histological and SEM analyses during removal of the metallic prosthesis after implantation times of 6, 9 and 12 months.

Based on quantitative histomorphometric studies from approximately 43 assessed slices for the three different implantation periods, the bone implant contact average value varied from 67 ± 10 to $84 \pm 5\%$ through the healing period analysed. The high levels of the percentage bone-to-graft contact demonstrated the osteoconductive capacity of Bonelike® in humans and extensive mature bone formation around the implanted granules as shown in Fig. 13.14(a–c). Bonelike® granules bioresorption was observed which resulted from

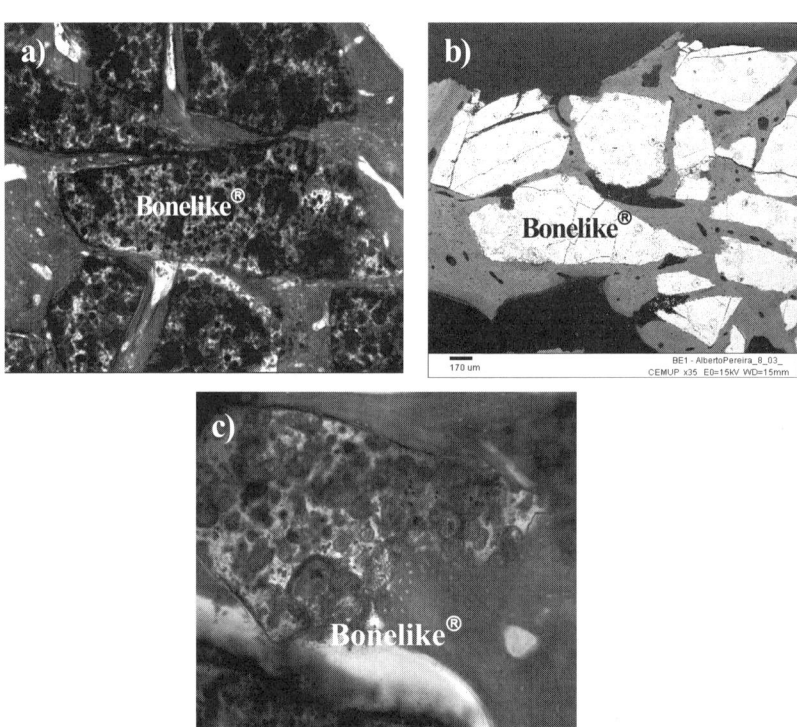

Fig. 13.14. Histological and SEM images showing implantation of Bonelike®granules and new bone formation after 9 months (a, b). Surface resorption of Bonelike®granules observed by histology after 6 months implantation (c), which indicates that this novel bone graft shows controlled biodegradation *in vivo*

the presence of β- and α-TCP phases in its structure, as shown in Fig. 13.14(c). This phenomenon may allow the full regeneration of bone defect area at longer implantations period with complete bioresorption of Bonelike®.

13.3 Summary

Bonelike® represents a new concept for synthetic bone grafts manufacturing and has the ability of mimicking the inorganic chemical composition and structure of natural bone tissues thus enabling enhanced osteointegration. With the increasing demand of synthetic grafts from the market due to their advantages compared to autografts and allografts the application of Bonelike® in regenerative surgery is therefore expected to increase for the benefit of the vast number of patients.

References

1. F.J. Hughes, W. Turner, G. Belibasakis and G. Martuscelli. Effects of growth factors and cytokines on osteoblastic differentiation. Periodontology 2000, 41 (2006) 48.
2. D.W. Sommerfeldt and CT. Rubin. Biology of bone and how it orchestrates the form and function of the skeleton. Eur Spine J. 10 (2001) S86-S95.
3. S. Weiner and W. Traub. Bone structure: from angstroms to microns. FASEB J, 6 (1992) 879.
4. A.M. Parfitt. Pharmacological manipulation of bone remodelling and calcium homeostasis. In. Kanis AJ, ed. Calcium metabolism. Basel: Karger, 1990: p.1-27.
5. J. Hollinger and M.E.K. Wong. The integrated process of hard tissue regeneration with special emphasis on fracture healing. Oral Surg Oral Med Oral Pathol Oral Radiol Endod. 82 (1996) 594.
6. I.H. Kalfas. Principles of bone healing. Neurosurg Focus. 10(2001) 1.
7. I.I. Doron and L.L. Amy. Bone graft substitutes. Operative Tech in Plast and Reconstr. Surg. 9(4) (2003) 151.
8. P.V. Giannoudis, H. Dinopoulos and E. Tsiridis, Bone substitutes: An update, Injury, Int J. Care Injured 365 (2005) 520.
9. E.A.R. Mary and A.Y. Raymond. Bone replacement grafts-The Bone Substitutes. Dent. Clin. North Am. 42(3) (1998) 491.
10. Cato T. Laurencin and Yusuf Khan: Bone grafts and Bone graft substitutes: A brief history (ed. by Cato T. Laurencin) ASTM - International, USA, 2003. p.3.
11. S. Wright. Commentary The Bone-Graft market in Europe, in Emerging Technologies in Orthopedics I: Bone Graft Substitutes. Bone Growth Stimulators and Bone Growth Factors by Datamonitor plc. – Ed. 1999, p. 591.
12. S.D. Boden. Osteoinduction bone graft substitutes: Burden of proof. Bull Amer. Acad. Orthoped Surg. 51(1)(2003) 42.
13. Synthetic bone graft to be tested in revision hip surgery, ApaTech Limited, based in London, UK. News Letter 9th April 2003.

14. Mohamed Attawia, Sudha Kadiyala, Kim Fitzgerald, Karl Kraus and S.P Bruder. Cell-based approaches for one graft substitutes (ed. by Cato T Laurencin) ASTM - International, USA, 2003. p.126.

15. J.D. Santos, G.W. Hastings and J.C. Knowles: Sintered hydroxyapatite compositions and method for the preparation thereof. European Patent WO 0068164, 1999.

16. M.A. Lopes, J.D. Santos, F.J. Monteiro and J.C. Knowles. Glass reinforced hydroxyapatite: a comprehensive study of the effect of glass composition on the crystallography of the composite. J. Biomed. Mater. Res. 39 (1998) 244.

17. M.A. Lopes, F.J. Monteiro and J.D. Santos. Glass-reinforced hydroxyapatite composites: fracture toughness and hardness dependence on microstructural characteristics. Biomaterials 20 (1999) 2085.

18. M.A. Lopes, R.F. Silva, F.J. Monteiro and J.D. Santos. Microstructural dependence of Young's and shear moduli of P_2O_5 glass reinforced hydroxyapatite for biomedical applications. Biomaterials 21(2000) 749.

19. J.D. Santos, R.L. Reis, F.J. Monteiro, J.C. Knowles and G.W. Hastings. Liquid phase sintering of hydroxyapatite by phosphate and silicate glass additions structure and properties of the composites. J. Mater. Sci. Mat. Med. 6 (1995) 348.

20. J.D. Santos, P.L Silva, J.C. Knowles, S. Talal and F.J. Monteiro. Reinforcement of hydroxyapatite by adding P_2O_5- CaO glasses with Na_2O,K_2O and MgO. J. Mater. Sci. Mat. Med. 7(1996) 187.

21. J.E. Davies. The importance and measurement of surface charge species in cell behaviour at the biomaterial interface. In: Ratner BD. Surface characterization of biomaterials. New York: Elsevier; 1988. p. 219–234.

22. B.D. Ratner. Biomaterial surfaces. J. Biomed. Mater. Res. Appl. Biomat. 21 (1987) 59.

23. S.R. Manson, L.A. Harker, B.D. Ratner and A.S. Hoffman. In vivo evaluation of artificial surfaces with a non human primate model of arterial thrombosis. J. Lab. Clin. Med. 95 (1980) 289.

24. F. Grinnell, M. Milamand and P.A. Srere. Studies on cell adhesion. Arch Biochem Biophys 153 (1972) 193.

25. S.K. Chang, O.S. Hum, M.A. Moscarello, A.W. Neumann, W. Zing, M.J. Leutheusser and B. Ruegsegger. Platelet adhesion to solid surfaces: The effect of plasma proteins and substrate wettability. Med. Progr. Technol. 5 (1997) 57.

26. M.A. Lopes, J.C. Knowles, J.D. Santos, Structural insights of glass reinforced hydroxyapatite composites by Rietveld refinement. Biomaterials, 21 (2000)1905.

27. I. Rehman and W. Bonfield, ''Structural characterisation of natural and synthetic bioceramics by photo acoustic-FTIR spectroscopy,'' in Bioceramics, Vol. 8, J. Wilson, L.L. Hench, and D. Greenspan (eds.), Butterworth-Heinmann Ltd., Oxford, 1995, p.163–168.

28. M. Okazaki and M. Sato. Computer graphics of hydroxyapatite and β-tricalcium phosphate. Biomaterials 11 (1990) 573.

29. A. Bigi, G. Falini, E. Foresti, M. Gazzano, A. Ripamonti and N. Roveri. Rietveld structure refinements of calcium hydroxyapatite containing magnesium. Acta Cryst., B52 (1996)87.

30. S. Kotani, Y. Fijita, T. Kitsugi, T. Nakamura, T. Yamamuro, C. Ohtsuki and T. Kokubo. Bone bonding mechanism of β-tricalcium phosphate. J. Biomed. Mat. Res. 25 (1991)1303.

31. M.A. Lopes, F.J. Monteiro, J.D. Santos, A.P. Serro and B. Saramago. Hydrophobicity, surface tension, and zeta potential measurements of glass-reinforced hydroxyapatite composites. Biomed Mater Res, 45(1999) 370.

32. J.D. Santos, J.C Knowles, R.L . Reis, F.J. Monteiro and G.W Hastings. Microstructural Characterrisation of glass reionforced hydroxyapatite composties, Biomaterials 15(1)(1994)5.

33. Y. Yamamuro, L.L. Hench and J. Wilson, CRC Handbook of bioactive ceramics, CRC Press, 1990.

34. M.A. Lopes, F.J. Monteiro and J.D. Santos, Glass reinforced hydroxyapatite composites: Secondary phase proportions and densification effects assessing biocompability, J. Biomed. Mater. Res. (Appl. Biomaterial) 48 (1999) 734.

35. R.W. Rice, Microstructure dependence of mechanical behaviour. In: R.K. MacCrone Editor, *Treatise on materials science and technology* vol. 11 Academic Press, New York (1977), p.200–382.

36. R.A. Hauber and R.M. Anderson. Engineering properties of glass–matrix composites. In: Ceramics and glasses, Engineered Materials Handbook. USA: ASM Publication, p. 858-69.

37. C.J. Kirkpatrick. A critical view of current and proposed methodologies for biocompatibility testing: cytotoxic in vitro. Regulatory Affairs 4 (1992)13.

38. S. Hanson, P.A. Lalor, S.M. Niemi, Ratner BD et al. Testing biomaterials. In : Ratner BD, Hoffman AS editors. Biomaterials Science. An introduction to materials in medicine. Basel: Karger, 1996: p.215.

39. M.A. Lopes, J.C. Knowles, L. Kuru, J.D. Santos, F.J. Monteiro and I. Olsen. Flow cytometry for assessing biocompatibily. J. Biomed. Mat. Res. 41 (1998) 649.

40. M.A. Lopes, J.C. Knowles, J.D. Santos, F.J. Monteiro and I. Olsen. Direct and indirect effects of P_2O_5-glass reinforced hydroxiapatite on the growth and function of osteoblast-like cells. Biomaterials 21 (2000) 1165.

41. M.A. Costa, M. Gutierres, R. Almeida, M.A. Lopes, J.D. Santos and M.H. Fernandes. In vitro mineralisation of human bone marrow cells cultured on Bonelike®. Key. Eng. Mater. 254-256(2004) 821.

42. O. Frank, M. Heim, M. Jakob, A. Barbero, D. Schafer, I.Bendik et al. Real-time quantitative RT-PCR analysis of human bone marrow stromal cells during osteogenic differentiation in vitro. J. Cell. Biochem. 85 (2000) 737.

43. P.J. Marie, M.A. de Vernejoul and A. Lomri. Stimulation of bone formation in osteoporosis patients treated with fluoride associated with increased DNA synthesis by osteoblastic cells in vitro. J. Bone and Mineral Res. 7 (1992) 103.

44. Council of Europe, Convention for the protection of vertebrata animals used for experimental and other scientific purposes (ET 123), Strasbourg, Council of Europe, 1986.

45. European Commission, Directive for the protection of vertebrate animals used for experimental and other scientific purposes (86/609/EEC), Off. J. Eur Comm., L 358, 1, 1986.

46. JV. Lobato, N. Sooraj Hussain, C.M. Botelho, J.M. Rodrigues, A.L. Luis, A.C. Mauricio, M.A. Lopes, J.D. Santos, Assessment of the potential of Bonelike® graft for bone regeneration by using an animal model. Key Eng. Mater. 284 - 286 (2005) 877.

47. JV. Lobato, N. Sooraj Hussain, C.M. Botelho, A.C. Mauricio, A. Afonso, N. Ali and J.D. Santos, Assessment of Bonelike® graft with a resorbable matrix using an animal model. Thin Solid Films 515 (2006) 642.

48. Herbert M. User and Robert B. Nadler, Applications of FloSeal innephron-sparing surgery,Urology 62(2) (2003)342.

49. F.A. Weaver, D.B. Hood, M. Zatina, L. Messina, B. Badduke. Gelatin-thrombin-based hemostatic sealant for intraoperative bleeding in vascular surgery. Ann. Vasc. Surg. 16 (2002) 286.

50. V. Dodane. and V. Vilivalam. Pharmaceutical applications of chitosan. Pharm. Sci. Technol. Today 1 (1998) 246.

51. B. Ettinger, H.K. Genant and C.E. Cann. Long-term estrogen replacement therapy prevents bone loss and fractures. Ann. Intern. Med. 102 (1985) 319.

52. H. Bryant, A.L. Glasebrook, N.N. Yang and M. Sato. An estrogen receptor basis for raloxifene action in bone. J. Steroid Biochem. Mol. Biol. 69 (1999) 37.

53. P.D. Delmas, N.H. Bjarnason, B.H. Mitlak, A.C. Ravoux, A.S. Shah, W.J. Huster, M. Draper and C. Christiansen. Effects of raloxifene on bone mineral density, serum cholesterol concentrations, and uterine endometrium in postmenopausal women. N. Engl. J. Med. 337 (1997) 1641.

54. A.H. Reddi and N.S. Cunningham. Initiation and promotion of bone differen-tiation by bone morphogenic proteins. J. Bone Miner. Res. 8(2) (1993) S499.

55. M.A. Lopes, J.D. Santos, F.J. Monteiro, A. Osaka and C. Ohtsuki. Push-out testing and histological evaluation of glass reinforced hydroxyapatite com-posites implanted in the tibia of rabbits. J. Biomed. Mater. Res. 54 (2001) 463.

56. F. Duarte, J.D Santos and A. Afonso, Medical applications of Bonelike in Maxillofacial Surgery, Mater. Sci. Forum 455-456 (2004) 370.

57. M.A. Costa M. Gutierres, L. Almeida, M.A. Lopes, J.D.Santos, M.H. Fernandes, In vitro mineralisation of human bonemarrow cells cultured on bonelike®. Key Eng. Mater. 254-256(2004) 821.

58. R.C Sousa, J.V. Lobato, N. Sooraj Hussain, M.A Lopes, A.C Mauricio, J.D . Santos , Bone regeneration in maxillofacial surgery using novel Bonelike® synthetic bone graft: radiological and histological analyses, Br. J. Oral. Maxi. Surg. (2006) submitted.

59. M. Gutierres, N. Sooraj Hussain, A. Afonso, L. Almeida, A.T. Cabral, M.A. Lopes and J.D. Santos. Biological behaviour of bonelike® graft Implanted in the tibia of humans. Key Eng. Mater. 284-286(2005) 1041.

60. M. Gutierres, N. Sooraj Hussain, M.A. Lopes, A. Afonso, A. T. Cabral, L. Almeida and J.D. Santos, Histological and scanning electron microscopy analyses of bone/implant interface using the novel Bonelike® synthetic bone graft, J. Ortho. Res. 24 (2006) 953.

14. Machining Cancellous Bone Prior to Prosthetic Implantation

14.1 Introduction

The structure of cancellous bone (i.e. bone with a relative density less than 0.7) is made up of an elaborate sandwich of compact dense bone on the outer shell and a core of porous, cellular material. This configuration minimizes the weight of the bone over a fairly large load-bearing area. As patients become older, weight loss can result in fractures that can be alleviated by using implants at an earlier age. However, the cancellous bone that is replaced is thought to be a primary cause of osteoarthritis in older patients and so implants must match the bone it replaces. When replacing cancellous bone, the area to be removed must be replaced with great care so that the replacement can attach itself to the bone that is still in the body. It is known that nanostructured hard tissue such as bone allows the surfaces of implants to attach themselves to bone a lot quicker than existing surface profiles. This is because living cells have an affinity to nanostructured features. In this chapter, we describe a machining technique that easily removes bone without destroying the natural features of the bone so that implants can attach themselves to the bone on the nanoscale.

One technique that shows much promise in machining bone is ultra high speed milling, this technique has been shown to produce micro and nano scale structures in the same way as a conventional machine tool produces macroscale features. A special requirement of machining at such small scales is the need to increase the rotational speed of the cutting tool. The cutting speed of the cutting tool is given by the following equation:

$$V = r\omega \tag{14.1}$$

Where V is the cutting velocity (m/s), r is the cutting tool radius (m) and ω is the rotational speed in (radians/s). From this relationship it can be seen that as the cutter diameter reduces in size to create micro and nano scale features the rotational speed must dramatically increase to compensate for the loss of cutting speed at the micro and nanoscale. At the present time, the fastest spindle commercially available rotates around 360,000 rpm under load conditions.

Research is currently being undertaken to improve the performance of these spindles where the initial aim is to reach 1,000,000 rpm [1]. Strain rates induced at these high speeds cause chip formation mechanisms to be significantly different than at low speeds. Additionally, it is now possible to experiment at the extreme limits of the fundamental principles of machining at ultra high speed and at the micro and nano scales using the conventional theories of machining. This chapter discusses the use of these theories at the microscale and at high strain rates and discusses the development of a model of initial chip formation during high strain rate deformation at the microscale.

14.2 Structure of Cancellous Bone

At the lower extremities of density, cancellous bone appears to be a complex set of open pores. As density increases the rods of hard tissue spread and flatten, become plate-like and finally fuse to form the dense outlying structure at the surface of the bone. Cancellous bone grows in response to an applied stress. Trabeculae develop along principal stress directions in the loaded bone. These ideas have been measured in vivo of the strains exerted in the cortex of the calcaneus of sheep. The exhibition of directional and anisotropic properties yields special attention from a machining viewpoint. Also, the nature of the microstructure may well exaggerate the wear of very small cutting tools. At the microscale, cancellous bone is a composite of a fibrous, organic matrix of proteins mainly collagen, filled with inorganic calcium compounds such as crystalline hydroxyapatite, $Ca_{10}(PO_4)_6(OH)_2$, and amorphous calcium phosphate, $CaPO_3$. These compounds provide bone with its stiffness. The compositions of compact and cancellous bone are almost the same, i.e. 35% organic matrix, 45% calcium compounds, and the remainder water.

14.3 Theory of Micromachining

Following the development of equations proposed by Shaw [2], these expressions will be applied to a 6 flute end milling cutter with a shank of diameter 1.59 mm, cutting diameter 700 µm, rotated at a speed of 250,000 rpm or 26180 rad/s. The rake angle was $\alpha=7^{\circ}$, clearance angle $\theta=10^{\circ}$ and the shear plane angle $\phi=24^{\circ}$. The material cut was cancellous bovine femur, and the horizontal force F_h was calculated using the equation assuming that the mass of the tool is concentrated at radius, r;

$$F_h = mr\omega^2 \tag{14.2}$$

Where m is the tool mass (kg), and r is the tool radius (m). Coefficients of friction between different materials have been investigated by Bowden and Tabor [3]. They also describe methods used for the determination of the coefficient of friction. Using the method of the inclined plane, the coefficient of friction of cancellous bovine femur on tungsten carbide and steel is in the range $\mu = 0.5\text{–}0.6$ under lubricating conditions, i.e. sliding on a plane coated with a saline solution. Using the following equation,

$$\beta = \tan^{-1}\mu \tag{14.3}$$

The friction angle β can then be determined under these conditions. It was found to be 30.96°. This is in excellent agreement with Merchant and Zlatin's nomograph [2]. The vertical force F_V can be found using the relationship:

$$F_v = \frac{\mu F_h - F_h Tan\,\alpha}{1 + \mu Tan\,\alpha} \tag{14.4}$$

This was found to be 5.25N. Again, referring to the Merchant and Zlatin's [2] nomograph for the coefficient of friction, the value of F_h can be independently predicted to be 5.33N. The force tangential to the tool plane F is found to be:

$$F = F_h Sin\alpha + F_v Cos\alpha \tag{14.5}$$

F was determined as 6.66N. The force normal to the tool plane N is provided using the equation:

$$N = F_h Cos\alpha - F_v Sin\alpha \qquad (14.6)$$

Where N was found to be 11.1N. The force perpendicular to the shear plane F_s can now be determined by,

$$F_s = F_h Cos\phi - F_v Sin\phi \qquad (14.7)$$

And was estimated to be 8.76N. The force normal to the shear plane N_s is given by the equation:

$$N_s = F_v Cos\phi + F_h Sin\phi \qquad (14.8)$$

Where N_s is 9.61N. Now the frictional force F_f is:

$$F_f = F_v Cos\alpha + F_h Sin\alpha \qquad (14.9)$$

F_f is approximately 6.66N. It is possible to check this value with Merchant and Zlatin's [2] nomograph for frictional force. However, the values for F_h and F_v are so small the extreme limits of the nomograph are being tested so it is difficult to give an accurate value for F_f, it is certain this value is below 10N which is in close agreement with the calculated answer. The shear stress τ is found using the following quotient:

$$\tau = \frac{F_s}{A_s} \qquad (14.10)$$

Which has a value of 1.8 GN/m^2. The direct stress ρ is found by applying the relationship,

$$\sigma = \frac{N_s}{A_s} \qquad (14.11)$$

σ is found to be 1.95 GN/m^2. The chip thickness ratio, r, is given by:

$$r = \frac{t}{t_c} \qquad (14.12)$$

Where t is the un-deformed chip thickness (or depth of cut) and t_c is the measured chip thickness. The machining of bone was conducted at such a small scale that it was difficult to measure t. Therefore r was calculated using the equation:

$$r = \frac{\tan\phi}{\cos\alpha + \sin\alpha\tan\phi} \tag{14.13}$$

Which yields r = 0.425 and therefore t = 4.25 μm. This is in excellent agreement with Merchant and Zlatin's [2] nomograph for shear angles and the calculation can be made in confidence. Shear strain γ is found from

$$\gamma = \frac{Cos\alpha}{Sin\phi Cos(\phi-\alpha)} \tag{14.14}$$

γ was found to be 2.55, this can be independently verified from Merchant and Zlatin's nomograph [2] for shear strain which yields a value of 2.51.The cutting velocity V is found using:

$$V = T_{td}\omega \tag{14.15}$$

Where V is 9.1 m/s. The chip velocity is found from applying the following equation:

$$V_c = \frac{Sin\phi}{Cos(\phi-\alpha)} \tag{14.16}$$

Where V_c is equal to 3.9 m/s, this can also be found from

$$V_c = rV \tag{14.17}$$

The two results are in agreement with each other. The shear velocity V_s is given by:

$$V_s = \frac{VCos\alpha}{Cos(\phi-\alpha)} \tag{14.18}$$

Where r V_s is calculated to be 9.5 m/s. V_s can also be found from:

$$V_s = \gamma V Sin\phi \tag{14.19}$$

Again, the two results are in agreement with each other. The strain rate γ is given by:

$$\dot{\gamma} = \frac{VCos\alpha}{\Delta yCos(\phi - \alpha)} \tag{14.20}$$

Where Δy is the shear plane spacing and $\dot{\gamma}$ is found to be 8333 s^{-1}. The feed rate is 1mm/min under experimental conditions and the feed per tooth δ is given by,

$$\delta = \frac{F_r}{N\omega} \tag{14.21}$$

Where N is the number of teeth. Therefore, Δ is 6.66 μm and the scallop height is found by using the following:

$$h = \frac{\delta}{\left(\frac{4T_{td}}{\delta}\right) + \left(\frac{8N}{\pi}\right)} \tag{14.22}$$

Therefore h is calculated to be 1.59×10^{-11}m under the experimental conditions of machining.

14.4 Initial Chip Curl Modeling

Chip curvature is a highly significant parameter in machining operations from which a continuous chip is produced. In this chapter, observations are made on initial chip curl in the simplified case of orthogonal cutting at the micro and nanoscales. The cutting process may be modeled using a simple primary shear plane and frictional sliding of the chip along the rake face. When the region of chip and tool interaction at the rake face is treated as a secondary shear zone and the shear zones are analyzed by means of slip-line field theory, it is predicted that the chip will curl. Thus chip curvature may be interpreted as the consequence of secondary shear. Tight chip curl is usually associated with conditions of good rake face lubrication [4]. At the beginning of the cut, a transient tight curl is often observed, the chip radius increasing as the contact area on the rake face grows to an equilibrium value. Thus it might be suggested that tight curl is an integral part of the primary deformation.

The process of continuous chip formation is not uniquely defined by the boundary conditions in the steady state and that the radius of curl may depend on the build-up of deformation at the beginning of the cut [4]. A treatment of primary chip formation at the micro scale is presented, which considers chip curl as a series of heterogeneous elements in continuous chip formation at the micro scale. The free surface of the chip always displays 'lamellae', which are parallel to the cutting edge. The chip is usually considered to form by a regular series of discrete shear events giving a straight chip made up of small parallel segments. However, no account is taken of the bone material that moves passed the tool between shear events. The following observations follow on from Doyle, Horne, and Tabor's [4] analysis of primary chip formation.

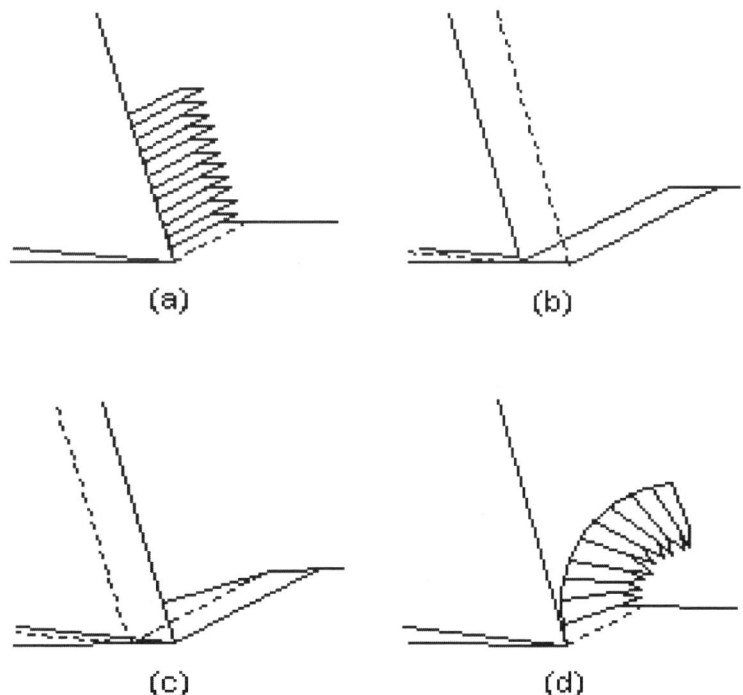

Figure 14.1. Instability during the formation of a chip during micromachining: (a) segmented, continuous chip; (b) chip forming instability due to built-up edge; (c) movement of a built-up edge to form a chip; (d) serrated, continuous chip curl.

Figure 14.1 shows the instabilities during chip formation that gives rise to primary chip curl. The shaded range of Figure 14.1b is the consequence of a built-up edge that very quickly becomes part of the segmented chips shown in Figure 14.1d. This 'material' provides the means to curl the chip and as a consequence of this event, the following model is presented. Previous treatments of chip curl analysis [5] have focused on chip formation with a perfectly stiff cutting tool. However, during the machining of bone it is observed that the cutting tool bends as it cuts [6]. This means that primary chip curl models must account for deflection of the cutting tool by bending during an orthogonal machining operation. Computational approaches to modeling chip formation at the micro and nanoscales have been attempted in recent years by a number of researchers [7, 8], who have used a molecular dynamics simulation approach using stiff cutting tools.

The generation of a transient built-up edge ahead of the cutting tool between shearing events in a bulging-type of motion generates the shape of the segment of the metal chip. This is shown in Figure 14.1c, with the built-up edge forming the 'shaded triangle' above the shear plane. If it is assumed that the built-up edge does not 'escape' under the tool edge, then the areas of the shaded triangles in Figures 14.1b and 14.1c will be equal. The chip moves away from the rake force in a manner shown in Figure 14.1d. The radius of chip curl can be calculated by assuming that the built-up edge in transient and that the element of the 'bulged' material contains a small angle relative to the tool and workpiece. This angle will inevitably change during the bending action of the cutting tool. With reference to Figure 14.2, if we assume that the cutting tool moves from point A to point D then the shear plane AC rotates to position HC as the built-up edge from triangle ABD is pushed into the segment of the chip. At point D, the shear along DF begins and segment DHCF is completed. HC and DF meet at R, the centre of the circle of the chip segment. Since the angle HRD is small, RD may be referred to as the radius of the chip. The clearance angle is θ.

Triangles ABD and HBC are equal in area and the depth of cut FG is equal to d. The spacing between the segments, i.e., the lamellae, is CE, which is equal to BD, which is equal to s. The chip thickness between lamellae, TC, is equal to t, whilst the rake angle SBD is equal to α. The cutting tool bends when machining at the microscale,

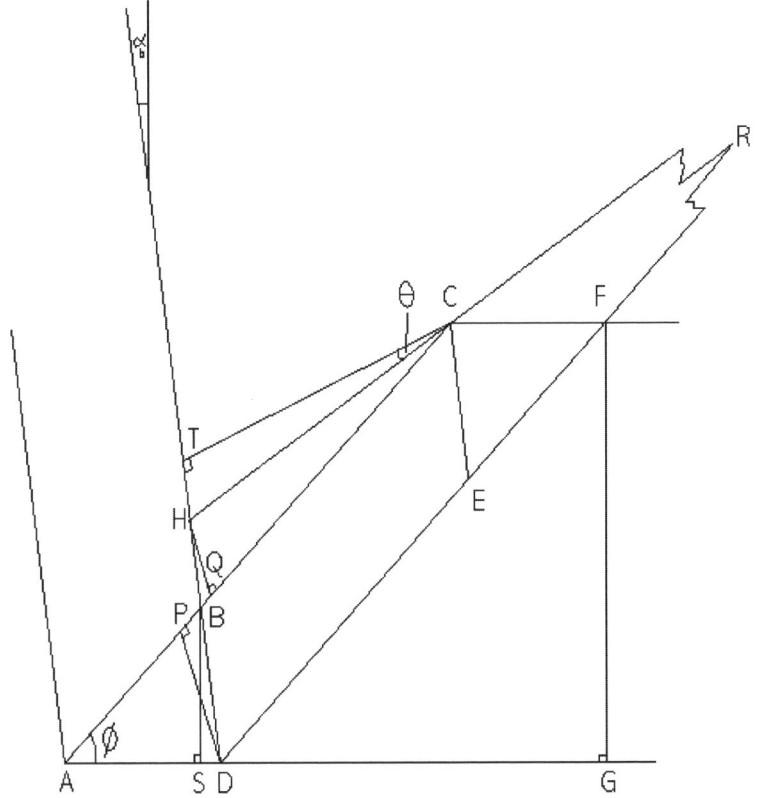

Figure 14.2. Schematic diagram of the geometry of the primary chip that forms a curled chip.

which reduces the effective rake angle to α_b. We know that the chip radius r can be taken as RD, whilst the shear angle subtended is BÂD, or ϕ. The calculation of the chip radius is provided by the following analysis,

$$DP = s.\cos(\phi - \alpha_b) \qquad (14.23)$$

$$AB = \frac{B.S}{\sin\phi} \qquad (14.24)$$

Where,

$$BS = s.\cos\alpha_b \qquad (14.25)$$

Thus,

$$DF = AC = \frac{d}{\sin\phi} \qquad (14.26)$$

And,

$$AB = \frac{s.\cos\alpha_b}{\sin\phi} \qquad (14.27)$$

Now,

$$BC = AC - AB = \frac{d}{\sin\phi} - \frac{s.\cos\alpha_b}{\sin\phi} \qquad (14.28)$$

Therefore,

$$BC = \frac{(d - s.\cos\alpha_b)}{\sin\phi} \qquad (14.29)$$

The areas of $\triangle ABD$ and $\triangle HBC$ are equal, such that,

$$AB.DP = HQ.BC \qquad (14.30)$$

Hence,

$$HQ = \frac{DP.AB}{BC} = \frac{s.\cos\alpha_b.s.\cos(\phi-\alpha_b).\sin\phi}{\sin\phi.(d - s.\cos\alpha_b)} \qquad (14.31)$$

And thus,

$$HQ = \frac{s^2.\cos\alpha_b.\cos(\phi-\alpha_b)}{(d - s.\cos\alpha_b)} \qquad (14.32)$$

Also,

$$BH = \frac{HQ}{\cos(\phi-\alpha_b)} = \frac{s^2.\cos\alpha_b}{(d - s.\cos\alpha_b)} \qquad (14.33)$$

And,

$$DH = BH + BD = s + \frac{s^2.\cos\alpha_b}{(d - s.\cos\alpha_b)} \qquad (14.34)$$

Such that,

$$DH = \frac{s.d}{(d - s.\cos\alpha_b)} \qquad (14.35)$$

Hence,

$$CH = \frac{TC}{\cos\theta} = \frac{t}{\cos\theta} \qquad (14.36)$$

Therefore,

$$\sin H\hat{R}D = \sin H\hat{C}B = \frac{HQ}{HC} = \frac{s^2.\cos\alpha_b.\cos(\phi-\alpha_b).\cos\theta}{t.(d-s.\cos\alpha_b)} \qquad (14.37)$$

And,

$$\sin D\hat{H}R = \sin T\hat{H}C = \cos\theta \qquad (14.38)$$

In triangle HRD,

$$\frac{RD}{\sin D\hat{H}R} = \frac{DH}{\sin H\hat{R}D} \qquad (14.39)$$

Therefore,

$$RD = r = DH \cdot \frac{\sin D\hat{H}R}{\sin H\hat{R}D} = \frac{s.d}{(d-s.\cos\alpha_b)} \cdot \frac{\cos\theta.t.(d-s.\cos\alpha_b)}{s^2.\cos\alpha_b.\cos(\phi-\alpha_b).\cos\theta} \qquad (14.40)$$

Thus,

$$r = \frac{d.t}{s.\cos\alpha_b.\cos(\phi-\alpha_b)} \qquad (14.41)$$

If the width of the lamellae, s, is small compared to the chip thickness, then for continuous machining with a single shear plane,

$$\frac{d}{t} = \frac{\sin\phi}{\cos(\phi-\alpha_b)} \qquad (14.42)$$

Hence,

$$\frac{t}{\cos(\phi-\alpha_b)} = \frac{d}{\sin\phi} \qquad (14.43)$$

And so,

$$r = \frac{d^2}{s.\cos\alpha_b.\sin\phi} \qquad (14.44)$$

Equation 14.44 predicts a positive chip radius at negative rake angles. The approximations considered in this model are appropriate when one considers that the model assumes that a secondary shear plane exists.

14.5 Experimental

14.5.1 Micromachining Apparatus

The machining of bovine femur was performed using a modified machining centre. The bio-machining centre was constructed to incorporate a high-speed air turbine spindle rated to operate at 360,000 rpm under no load conditions. When operating at relatively deep depths of cut, the speed of the spindle decreases to approximately 250,000 rpm. The table of the machine tool was configured to move in x-y-z co-ordinates by attaching a cross-slide powered by a d.c. motor, in all three principal axes. Each motor was controlled by a Motion-master^{TM} controller with a resolution as low as 500 nm. The cutting tools used were coated with diamond. The bio-machining centre is shown in Figure 14.3.

The bovine femur samples were machined at various depths of cut at high speed and were machined in an aqueous saline solution. The cutting tools were inspected at the end of all machining experiments using an Environmental Scanning Electron Microscope. The measured spindle speed was 250,000 rpm during the machining experiments. The depth of cut ranged between 50 μm and 100 μm for all machining experiments. The machining feed rate was conducted at 5 mm/s (0.3 m/min). The microscale cutting tool used was 700 μm in diameter (microscale) and was associated with a cutting speed of 117 m/min and a machining feed rate of 0.3 m/min. The results of the experimental procedures are shown in Table 14.1.

The machined chips were examined in an environmental scanning electron microscope where the lamellar spacing on each chip was determined. Transient chip curl was measured at the first 90° of tight chip curl. The curl radii was then compared with the calculated value derived using the idealized model, taking into account the degree of bending of the cutting tool.

Table 14.1. Experimental data comparing initial chip curl during bio-machining and initial chip curl predicted by the model. The depth of cut was 100 μm.

Rake angle after bending (o)	Shear plane angle (o)	Mean lamellar spacing (μm)	Observed chip curl (mm)	Calculated chip curl (mm)
22	37	0.98	17.55	18.01
15	25	1.55	14.42	14.65
8	18	1.9	16.55	17.1
3	12	2.95	15.82	16.22

14.5.2 Observations of Bone Chips

There are significant differences in the size and shape of chips when machined at medium and high speeds. This is especially so for biological materials such as cancellous bone. Figure 14.4 shows a collection of chips machined from bovine femur. It is seen in Figure 14.4 that many of the particles are in fact chunks of material rather than nicely

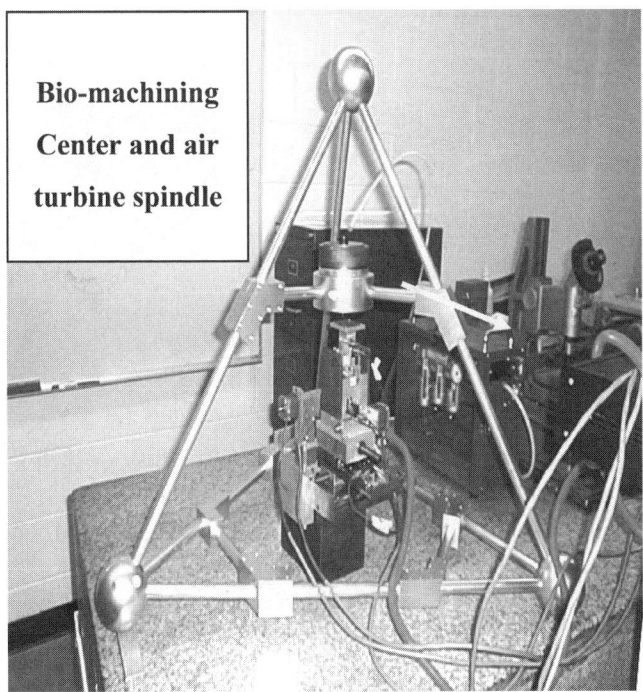

Figure 14.3. Bio-machining centre equipped with a high-speed air turbine spindle for rotating coated micro milling tools.

formed chips. It is possible that the chunks were formerly parts of larger chips that have since broken down and that chip thickness values should be recalculated based on the larger chip size. It can also be seen that the chips in Figure 14.4 are more consistent in terms of length, width, and depth. Their lamellar spacing is also regular in period, which would indicate that cutting conditions at high speed are stable (Fig. 14.5). Single chip formations are shown in Figure 14.6. While the width observed is similar to that for low speed cutting, the chip length of high-speed chips is much shorter than low speed chips. This could be because at low speed the chip has a greater time in contact with bone thereby removing more material, which is reflected in the increased chip length.

One of the major differences observed between low and high speed bio-machining of bone is in the spacing of the lamellae. In low speed cutting, the chip spacing varies by a significant amount.

However, at high cutting speeds the spacing is regular in period. At high speeds this process is accelerated to an extremely high level as the strain rate calculations have shown. In fact experiments show that chip types are similar in other materials such as metals. Figure 14.7 shows a magnified image of a coated cutting tool. The clearance

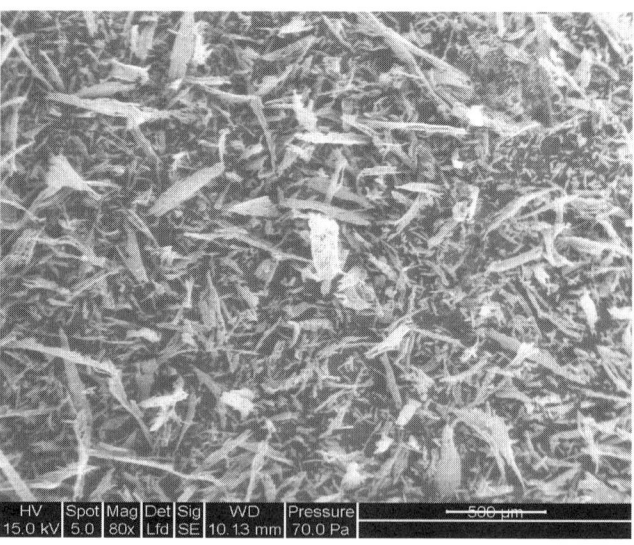

Figure 14.4. Characteristic chip shapes cutting bovine femur at high speeds.

Figure 14.5. Individual chip formation at high speed.

Figure 14.6. Lamellae spacing of bovine femur at high speed.

faces of the flutes of the cutting tool show adherent bone chips with finely striated lamellae, as noted on the left hand side of the tool. Figure 14.8 shows a magnified image of a coated cutting tool detailing the cutting edge and its relationship to the adherent film of bovine femur showing fine striations of lamellae generated at high strain rates.

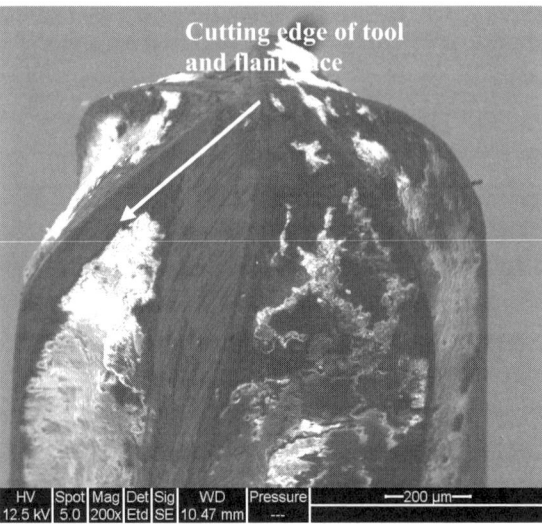

Figure 14.7. Magnified image of the cutting tool showing cutting edges and adhered bone material.

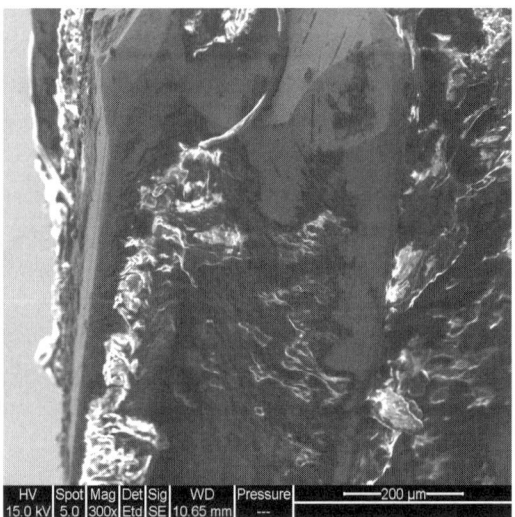

Figure 14.8. View of cutting edges and adherent bovine femur chip showing fine striations of lamellae.

14.5.3 Bio-machining Results

The results of machining bovine femur at the microscale are compared to the model described for primary chip curl during the primary stages of chip formation. It should be noted that all results presented in

Table 14.1 are for bone machined in an aqueous saline environment. Table 14.1 shows the results for biomachining using a variety of rake angles. It should be noted that bending of the cutting tool produces a less acute rake angle when machining takes place. However, the shear plane angle is increased and larger chips are produced.

14.6 Discussion

It can be seen from the above analysis that despite the extremely high strain rates imposed due to high speed cutting, macroscale equations can be applied accurately and produce impressive results. The most significant differences however appear in the following categories: strain rate, scallop height, and chip type. Many of the forces are of a similar order of magnitude offering no significant difference between macro low speed and micro high speed machining. This is important during tool design as small tools must absorb the same impact forces as larger tools do during impact. However, when considering the strain rate it can be seen during micro high speed machining the strain rate is 8333 s^{-1} compared to the macro low speed case of 667 s^{-1}, a 12.5 times increase which relates directly to a 12.5 increase in speed from 20,000 rpm to 250,000 rpm. The increase in strain rate is directly related to the increase in cutting speed, this is expected as the cutter is imparting the strain and therefore a rate of strain to the material. The lamellae spacing Δy in equation 20 has a significant effect on the strain rate, comparing macro and micro scale chips it is found that lamellae are ten times more closely packed in the high speed chips than the low speed chips.

  The purpose of machining is to create surfaces that are useful, hence surface quality should be an important consideration of milling, a measure of this is scallop height. An improvement is seen in the micro high-speed case with a scallop height of $1.58 \times 10^{-11} \text{m}$ compared to $8.9 \times 10^{-9} \text{m}$ for macro slow speeds. Although both values seem insignificant it must be remembered at the micro and nano scales post process finishing is inappropriate, therefore created structures must be produced to specification without further processing. Additionally, owing to the aspect ratio small imperfections become serious defects at small scales. From the calculations it can be seen that there is an improvement in the scallop height, which is not the improvement required when considering the scale order of magnitude has changed by a factor of four. This is because the current spindle speeds reached are not high enough for effective machining, if this speed is increased to 1,000,000 rpm then the orders of magnitude are increased further still.

The experimental results and observations provide an interesting view of machining bone at the microscale. When one considers the approximations made in the derivation of the chip curl model, the experimentally measured results compare well with the calculated chip curl. This indicates that cutting tool bending contributes significantly to initial chip curl prior to any significant frictional interactions on the rake face of the cutting tool. The proposed model describes the initial stages of chip curl quite well. If the description of chip curl is accurate, then continuous chip formation at the microscale needs to be re-investigated. If one considers the movement of the cutting tool (Figure 14.2), from point A towards point D, we expect the shear plane to oscillate between AC and HC depending on the amount of energy required to move the built-up edge into the segment of the subsequent chip. The cycle begins again when accumulated material is deposited on to the edge of the cutting tool then on to the subsequent segment of the chip produced during machining.

14.7 Conclusions

The equations of metal cutting can be applied in the high-speed micro scale environment. The nomographs of Merchant and Zlatin [2] can be applied confirming that future calculations can be compared to these well constructed charts. High strain rates change the mechanism of chip formation thereby altering the shape of the chip. Also, high strain rates appear to provide less dependence on material properties in determining chip formation and shape.

A model of chip curl at the microscale has been developed and agrees well with experimental data. It appears that the bending of the cutting tool contributes significantly to the primary chip prior to significant frictional interactions on the rake face of the cutting tool. It is shown that primary chip curl is initiated by the amount of material deposited onto the cutting tool that manifests itself as a wedge angle that controls the amount of material pushed into the base of the segment of the chip between oscillations of the primary shear plane. Further studies on chip formation at the atomic scale are needed to develop nanomanufacturing processing methods for biological materials such as bone. The future development of this technique lies in the ability to rotate cutting tools at extremely high spindle speeds.

References

1. Kanjarkar K. C., Cui J., Jackson M. J., Hyde L. J., and G. M. Robinson, *Optimum Design And Analysis Of High-Speed Spindles For Nanomachining Applications Using Computational Fluid Dynamics Approach*, Submitted to Applied Mathematical Modeling, May 2004.
2. Milton C. Shaw, *Metal Cutting Principles*, Oxford Science Publications – Series on Advanced Manufacturing, Clarendon Press, University of Oxford, 1996, pp. 18–46.
3. Bowden F. P., and Tabor D., *The Friction & Lubrication Of Solids*, Oxford Science Publications, Clarendon Press, University of Oxford, 2001, pp. 73–75, 83–85.
4. Doyle, E. D., Horne, J. G., and Tabor, D., *Frictional interactions between chip and rake face in continuous chip formation*, Proc. Roy. Soc., of London, A366, 1979, 173–183.
5. Jawahir, I. S., and Zhang, J. P., *An analysis of chip formation, chip curl and development, and chip breaking in orthogonal machining*, Trans. North Amer. Manuf. Res. Inst. – Soc. Manuf. Eng., 23, 1995, 109–114.
6. Kim, C. J., Bono, M., and Ni, J., *Experimental Analysis of chip formation in micro-milling*, Trans. North Amer. Manuf. Res. Inst.-Soc. Manuf. Eng., 30, 2002, 247–254.
7. Komanduri, R., Chandrasekaran, N., and Raff, L. M., *Molecular dynamics simulation of the nanometric cutting of silicon*, Philosophical Magazine, B81, 2001, 1989–2019.
8. Luo, X., Cheng, K., Guo, X., and Holt. R., An investigation into the mechanics of nanometric cutting and the development of its test bed, Int. J. Prod. Res., 41, 2003, 1449–1465.

15. Titanium and Titanium Alloy Applications in Medicine

15.1 Metallurgical Aspects

15.1.1 Introduction

Titanium is a transition metal. It occurs in several minerals including rutile and ilmenite, which are well dispersed over the Earth's crust. Even though titanium is as strong as some steels, its density is only half of that of steel. Titanium is broadly used in a number of fields, including aerospace, power generation, automotive, chemical and petrochemical, sporting goods, dental and medical industries, [1–3]. The large variety of applications is due to its desirable properties, mainly the relative high strength combined with low density and enhanced corrosion resistance, [4]. Among metallic materials, titanium and its alloys are considered the most suitable materials in medical applications because they satisfy the property requirements better than any other competing materials, like stainless steels, Cr-Co alloys, commercially pure (CP) Nb and CP Ta, [5–6]. In terms of biomedical applications, the properties of interest are biocompatibility, corrosion behavior, mechanical behavior, processability and availability, [7–9].

Titanium may be considered as being a relatively new engineering material. It was discovered much later than the other commonly used metals, its commercial application starting in the late 40's, mainly as structural material. Its usage as implant material began in the 60's, [10]. Despite the fact that titanium exhibits superior corrosion resistance and tissue acceptance when compared with stainless steels and Cr-Co-based alloys, its mechanical properties and tribological behavior restrain its use as biomaterial in some cases. This is particularly true when high mechanical strength is necessary, like in hard tissue replacement or under intensive wear use, [11]. To overcome such restrictions, CP titanium was substituted by titanium alloys, particularly, the classic grade 5, i.e. Ti-6Al-4V alloy. The Ti-6Al-4V $\alpha+\beta$ type alloy, the most worldwide utilized titanium alloy, was initially developed for aero-space applications, [12, 13]. Although this type of alloy is considered a good material for surgically implanted parts, recent studies have found that vanadium may react with the tissue of the human body, [2]. In addition, aluminum may be related with neurological disorders and Alzheimer's disease, [2]. To overcome the potential vanadium toxicity, two new vanadium free $\alpha+\beta$ type alloys were developed in the 1980's. Vanadium, a β-stabilizer element, was replaced by niobium and iron, leading to Ti-6Al-7Nb and Ti-5Al-2.5Fe $\alpha+\beta$ type alloys, [4, 6, 14]. While both alloys show mechanical and metallurgical behavior comparable to those of Ti-6Al-4V, a disadvantage is that they all contain aluminum in their compositions.

In recent years, several studies have shown that the elastic behavior of $\alpha+\beta$ type alloys is not fully suitable for orthopedic applications, [15–18]. A number of studies suggest that unsatisfactory load transfer from the implant device to the neighboring bone may result in its degradation, [9]. Also, numerical analyses of hip implants using finite element method indicate that the use of biomaterials with elastic behavior similar to cortical bones improves the distribution of stress around the implanted bone, [19]. While the elastic modulus of a cortical bone is close to 18 GPa, [7], the modulus of Ti–6Al–4V alloy is 110 GPa, [7]. In such a case, the high elastic modulus of the implant material may lead to bone resorption and possible unsuccessful implantation procedure. The elastic behavior mismatch between the implant and the adjacent bone is named 'stress shielding effect', [19].

Since CP titanium and some specific $\alpha+\beta$ type titanium alloys do not completely meet the demands of medical applications, especially concerning mechanical behavior and toxicity to human body, a new class of alloys has been investigated for biomedical applications in the last decade, the β type alloys. After proper heat treatments this type of alloys may exhibit low elastic modulus, very good corrosion resistance, suitable mechanical properties and good biocompatible behavior, as they may be obtained by adding biocompatible alloying elements like Nb, Ta and Zr to titanium, [20–24].

15.1.2 Basic Aspects of Titanium Metallurgy

The microstructure diversity of titanium alloys is a result of an allo-tropic phenomenon. Titanium undergoes an allotropic transformation at 882°C. Below this temperature, it exhibits a hexagonal close-packed (HCP) crystal structure, known as α phase, while at higher temperature it has a body-centered cubic (BCC) structure, β phase. The latter remains stable up to the melting point at 1,670°C, [5]. As titanium is a transition metal, with an incomplete d shell, it may form solid solutions with a number of elements and hence, α and β phase equilibrium temperature may be modified by allowing titanium with interstitial and substitutional elements.

Titanium alloying elements fall into three class: α-stabilizers, β-stabilizers and neutral. While elements defined as α-stabilizers lead to an increase in the allotropic transformation temperature, other elements, described as β-stabilizers provoke a decrease in such a temperature, [25]. When a eutectoid transformation takes place, this β-stabilizer is termed eutectoid β-stabilizer, otherwise, it is called isomorphous β-stabilizer. If no significant change in the allotropic transformation temperature is observed, the alloying element is defined as neutral element. Figure 15.1 shows a schematic representation of types of phase diagram between titanium and its alloys elements, [5, 25].

As a result, titanium alloys with an enormous diversity of compositions are possible. Among α-stabilizer elements are the metals of IIIA and IVA groups (Al and Ga) and the interstitials C, N and O. On the contrary, β-stabilizer elements include the transition elements (V, Ta, Nb, Mo, Mg, Cu, Cr and Fe) and the noble metals.

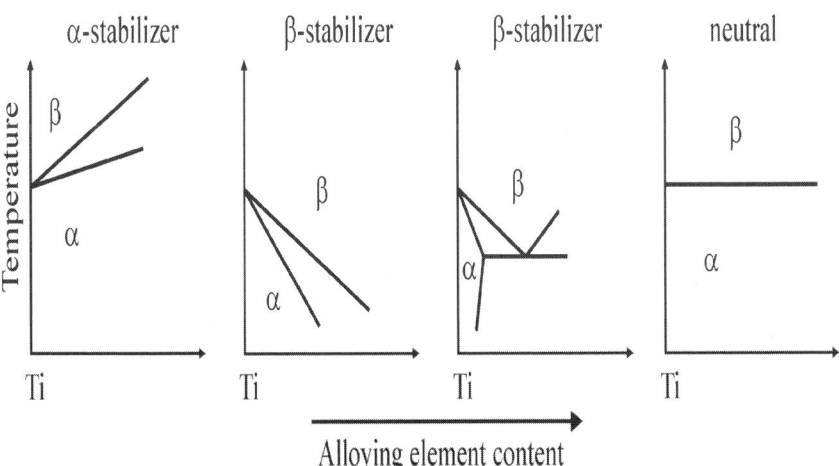

Fig. 15.1. Schematic representation of types of phase diagram between titanium and its alloying elements [5, 25]

Addition of α and β-stabilizer elements to titanium gives rise to a field in the corresponding phase diagram where both α and β phase may coexist. Titanium alloys exhibit a variety of properties, which are connected to chemical composition and metallurgical processing, [26–28]. According to the nature of their microstructure, titanium alloys may be divided as either α alloys, β alloys and α+β alloys, [29]. Beta alloys may be further classified into near β and metastable β alloys.

Alpha titanium alloys are especially formed by CP titanium and alloys with α-stabilizer elements, which present only α phase at room temperature. Such alloys show high creep resistance and are thus suitable for high temperature service. Since no metastable phase remains after cooling from high temperature, no major modification in terms of microstructure and mechanical properties are possible using heat treatments. Finally, as α phase is not subjected to ductile-brittle transition, these alloys are proper for very low temperature applications. Regarding mechanical and metallurgical properties, α alloys present a reasonable level of mechanical strength, high elastic modulus, good fracture toughness and low forgeability, which is due to the HCP crystal structure.

Beta titanium alloys are obtained when a high amount of β-stabilizer elements are added to titanium, which decreases the temperature of the allotropic transformation (α/β transition) of titanium, [30]. If the β-stabilizer content is high enough to reduce the martensitic start temperature (M_S) to temperatures below the room temperature, nucleation and growth of α phase will be very restricted, and hence, metastable β is retained at room temperature under rapid cooling, as depicted in Figure 15.2. This type of titanium alloy may be hardened by using heat treatment procedures, [31]. In some cases, depending upon composition and heat treatment parameters, precipitation of ω phase is possible. However, ω phase may cause embrittlement of a titanium alloy and, in general its precipitation must be avoided, [32]. β titanium alloys are very brittle at cryogenic temperatures and are not meant to be applied at high temperatures, as they show low creep resistance.

Finally, *α+β alloys* include alloys with enough α and β-stabilizers to expand the α+β field to room temperature, [5, 25]. The α and β

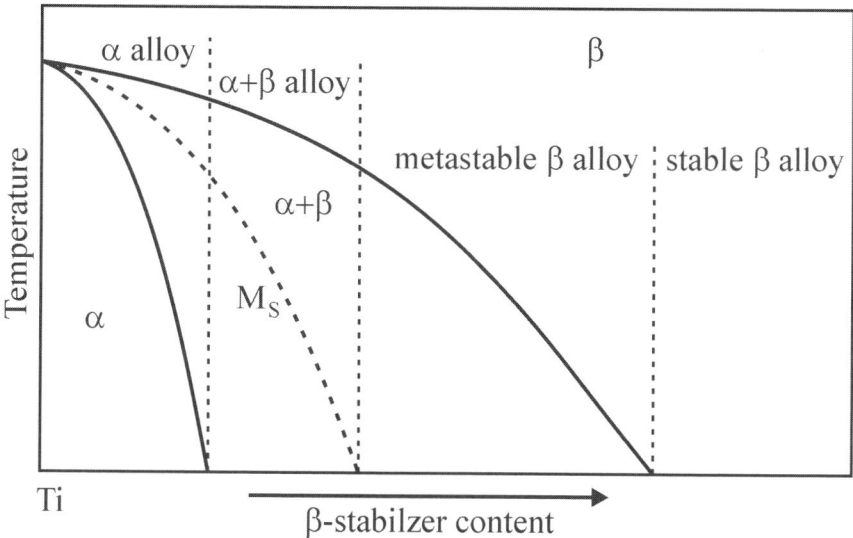

Fig. 15.2. Partial phase diagram of titanium and a β-stabilizer element [5, 6]

phase combination allows one to obtain an optimum balance of properties. The characteristics of both α and β phases may be tailored by applying proper heat treatments and thermo-mechanical processing. A significant assortment of microstructures may be obtained when compared to α type alloys. The Ti-6Al-4V alloy is an example of α+β type alloy. Due to its large availability, very good workability and enhanced mechanical behavior at low temperatures, such an alloy is the most common composition among the titanium alloys and based on these characteristics it is still largely applied as a biomaterial, mainly in orthopedic implant devices. Figure 15.3 depicts the microstructures of β and α+β titanium alloys.

As in the case of iron (steels), allotropic transformation is the main reason for the enormous variety of microstructure in titanium alloys. Titanium alloy microstructures are formed by stable and metastable phases, [33, 34]. In general, for limited β-stabilizer content and depending on cooling conditions, titanium alloys show only α and β phases.

However, if the thermodynamic equilibrium is not reached, metastable phases may be retained at room temperature, mainly, martensitic and ω phases. According to several authors, [35–37], titanium alloys with β-stabilizer elements like Mo, Nb, Ta and V, may form two types of martensitic structures. If the β-stabilizer content is considered low, rapid cooling leads to formation of hexagonal martensite, termed α'.

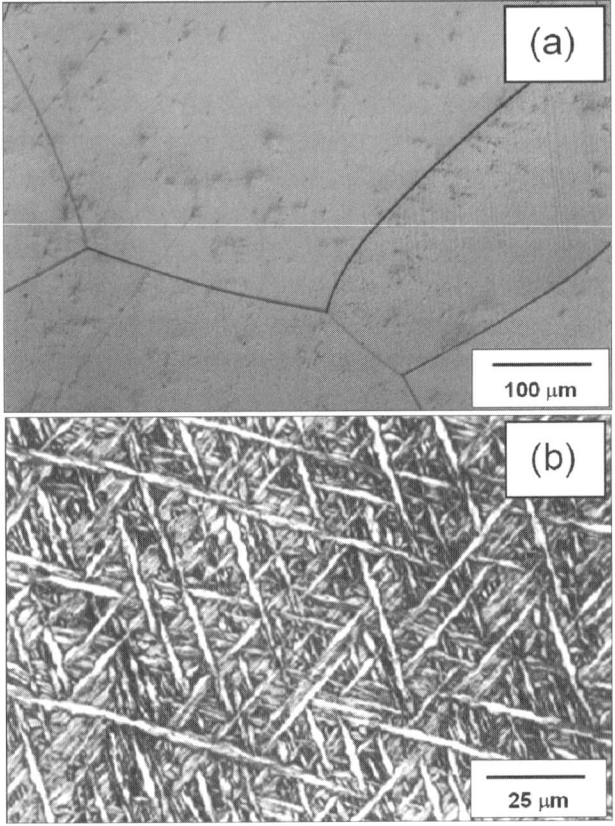

Fig. 15.3. Microstructures of (a) β Ti-35Nb (wt%) and (b) α+β Ti-6Al-7Nb (wt%) alloys cooled in air

When this content is high, α' martensite undergoes a distortion, loses its symmetry and is substituted by orthorhombic martensite, defined as α'', [37].

When titanium alloys with β-stabilizer elements are submitted to rapid cooling from high temperature, the β phase may transform either in martensitic structures or eventually, in metastable ω phase, [35]. Figure 15.4 presents a microstructure of the Ti-25Nb (wt%) after cooling in water and in air showing α'' and ω formation.

Precipitation of ω phase occurs only in a limited range of the alloy element and may arise during the quenching from high temperature (β phase), forming a thermal ω phase. However, ω phase may also form after ageing of a rapid by quenched structure at medium temperatures, resulting in isothermal ω phase, as indicated in Figure 15.5.

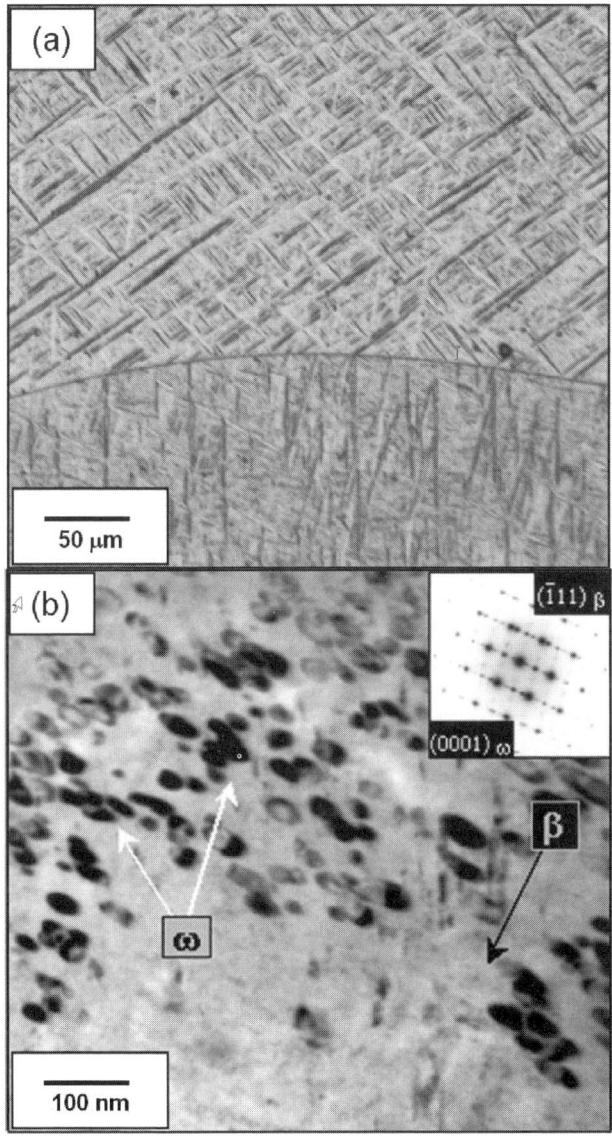

Fig. 15.4. Microstructure of the Ti-25Nb (wt%) alloy: (a) water cooled sample showing martensitic structure (OM analysis) and (b) air cooled sample showing ω phase dispersed in β matrix and respective SADP showing ω and β phases microstructure (TEM analysis)

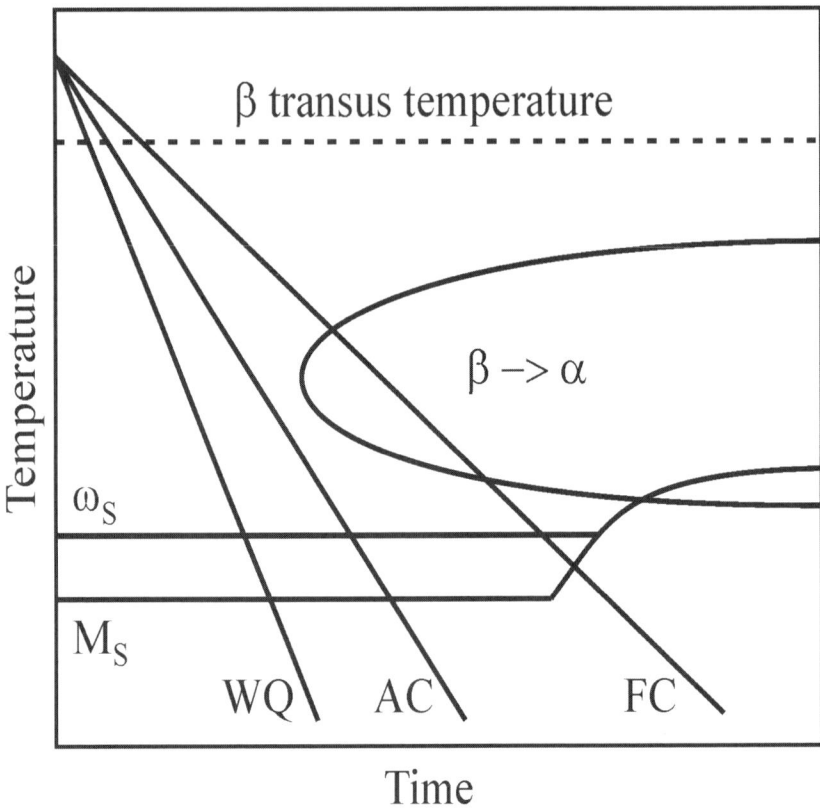

Fig. 15.5. A schematic TTT diagram for β phase transformation in titanium alloys with β- stabilizer elements [35]

15.1.3 Mechanical Behavior

Concerning mechanical behavior, biomedical titanium alloys applied as biomaterial mainly in hard tissue replacement, must exhibit a low elastic modulus combined with enhanced strength, good fatigue resistance and good workability. Mechanical behavior of titanium alloys is directly related to composition and mainly, thermo-mechanical processing. Some mechanical properties of selected titanium-based materials applied as biomaterials are shown in Table 15.1 [38].

Mechanical strength may be increased by adding alloying elements, which may lead to solid-solution strengthening or even, precipitation

Table 15.1. Selected Ti-base materials developed for medical applications [38]

Material	Tensile Strength (MPa)	Yield Strength (MPa)	Elongation (%)	Elastic Modulus (GPa)
α type				
Pure Ti grade 1	240	170	24	102.7
Pure Ti grade 2	345	275	20	102.7
Pure Ti grade 3	450	380	18	103.4
Pure Ti grade 4	550	485	15	104.1
α+β type				
Ti-6Al-4V	895–930	825–869	6–10	110–114
Ti-6Al-4V ELI	860–965	795–875	10–15	101–110
Ti-6Al-7Nb	900–1050	880–950	8.1–15	114
Ti-5Al-2.5Fe	1020	895	15	112
β type				
Ti-13Nb-13Zr	973–1037	836–908	10–16	79–84
Ti-12Mo-6Zr-2Fe	1060–1100	1000–1060	18–22	74–85
Ti-15Mo	874	544	21	78
Ti-15Mo-5Zr-3Al	852–1100	838–1060	18–25	80
Ti-15Mo-2.8Nb-0.2Si	979–999	945–987	16–18	83
Ti-35.3Nb-5.1Ta-7.1Zr	596.7	547.1	19	55
Ti-29Nb-13Ta-4.6Zr	911	864	13.2	80

of second phases. Also, by using ageing processes, metastable structures obtained by rapid quenching from β field may give rise to fine precipitates, which considerably increases mechanical strength.

Titanium alloys present a high strength-to-weight ratio, which is higher than with most of steels. While CP titanium has yield strength between 170 (grade 1) and 485 MPa (grade 4), titanium alloys may present values higher than 1500 MPa [25].

The elastic modulus or Young modulus corresponds to the stiffness of a material and is associated to the way inter-atomic forces vary with distance between atoms in the crystal structure. A comparison between both crystal structures of titanium has led to the conclusion that HCP structure presents higher values of elastic modulus than the BCC structure. Hence, addition of β-stabilizer elements allows β phase stabilization and hence, low elastic modulus alloys. While CP titanium shows elastic modulus values close to 105 GPa, Ti-6Al-4V type α+β alloy presents values between 101 and 110 GPa, type β titanium alloys may present values as low as 55 GPa, [38]. When compared with common alloys used as biomaterials, such 316 L stainless steel (190 GPa) and Co-Cr alloys (210–253), low elastic modulus titanium alloys display a more compatible elastic behavior to that of the human bone [39]. In general, as the elastic modulus decreases, so does the mechanical strength and vice versa.

Analysis of slip systems in different crystal structures reveals that plastic deformation is easier in BCC crystal structure than in HCP structure. It explains the enhanced ductility of β phase when compared to α phase. In a HCP structure the number of slip systems is only three, while this number increases to 12 in the case of BCC structure. In addition, the ease of plastic deformation facility is directly connected to the minimum slip distance, b_{min}, [25], which is given by the interatomic distance divided by the respective lattice parameter. Since, HCP structure exhibits a higher slip distance than BCC structure, it is possible to conclude that the atomic planes slip or the plastic deformation is easier in BCC structure than HCP. Hence, β type alloys present the best formability among the titanium alloys.

15.1.4 Corrosion Behavior

Corrosion resistance is one of the main properties of a metallic material applied in the human body environment and the success of an implant depends on the careful examination of this phenomenon. The performance of an implant is directly related to its ability in functioning in the aggressive body fluids. In general, these fluids consist of a series of acids and certain amount of NaCl. In normal conditions, its pH is 7, however, it may be altered due to immune system response, like in the case of a infection or inflammation. In an event of a corrosion process, the implant component may lose its integrity, leading to a failure. In addition, release of corrosion products may lead to undesirable biological reactions. Certainly, this will depend on the nature of chemical reactions on the implant surface in view of the fact that corrosion is essentially a chemical process.

Titanium shows an excellent corrosion resistance, which is directly related to the formation of a stable and protective oxide layer, essentially TiO_2. The reactivity of titanium can be measured by its standard electrode potential (standard electromotive force (EMF) series), which is -1.63 V [5]. Such a value indicates that titanium has a high chemical reactivity and is easily oxidized, giving rise to a very adherent and thin oxide layer on the titanium surface. This oxide layer passivates the titanium, which results in a protection against further corrosion process as long as this layer is maintained. Actually, formation of passivation films on titanium does not mean cessation of corrosion processes. It means that the corrosion rate will be significantly reduced. Therefore, titanium is corrosion resistant in oxidizing environments but not resistant in reducing medium [5].

In general, anodic polarization testing is an efficient method of analyzing corrosion behavior of metallic material in a corrosive environment. Figure 15.6 depicts the anodic polarization curve for CP titanium and Ti-6Al-4V alloys, showing the applied potential as a function of resulting currenty density (versus saturated calomel electrode (SCE)), obtained with 5g/l NaCl, pH 4 solution as an electrolytic medium at 310 K. The potential was scanned at 0.1 mVs^{-1} [11].

The anodic portion of the polarization curve allows one to evaluate the corrosion behavior of a metallic material in an electrolytic medium. Evaluation is obtained by determining the range of potentials in which passivation films are stable, and also by finding the current density. As usual, polarization tests started at a negative potential of -1.0 V vs. SCE, reaching more positive values.

In such a process, the initial sector of the anodic polarization curve refers to the beginning of a corrosion phenomenon, where the metallic material reacts with the supporting electrolyte, leading to active corrosion.

The following segment is related to the formation of an oxide passivating film, when the electric current stabilization takes place. As the potential increases, the current density also increases and eventually, the rupture of passivation film occurs. At this point, the protective layer loses its efficiency causing pitting corrosion. However,this hypothesis is not confirmed during the reverse scanning of potential.

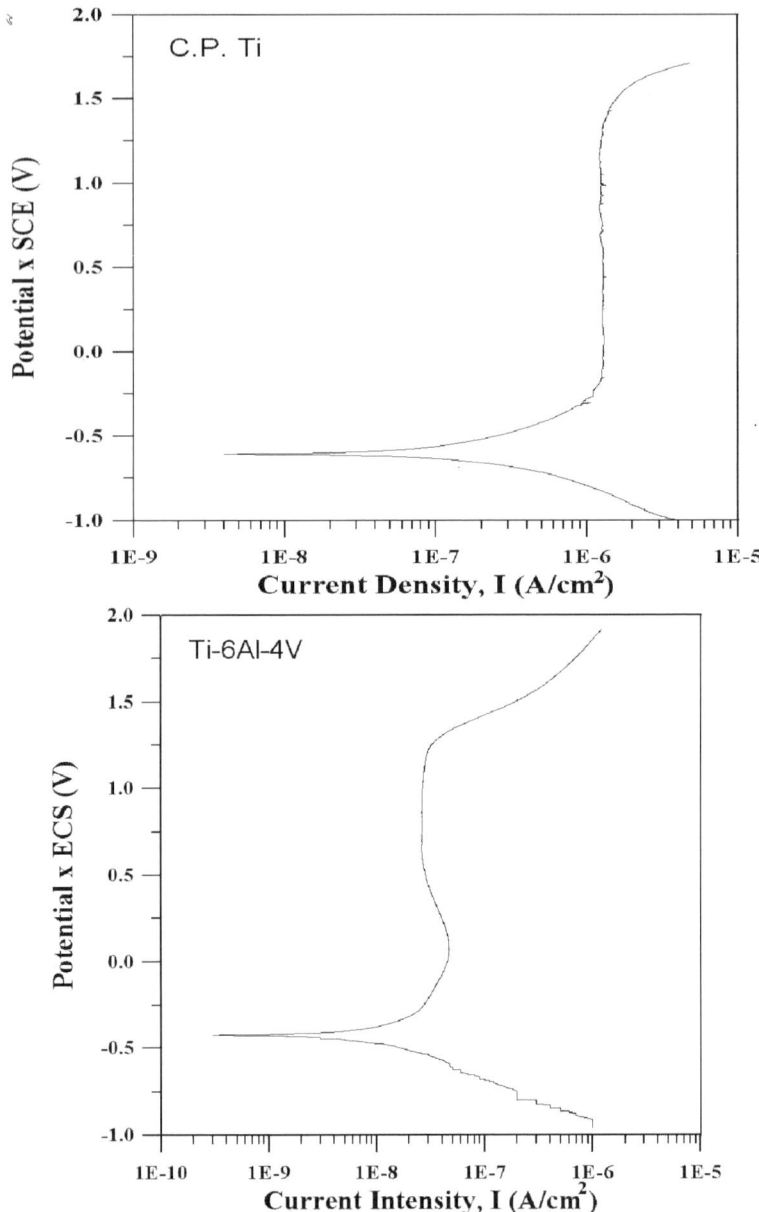

Fig. 15.6. Polarization curves for CP titanium and Ti-6Al-4V alloy (Scan rate of 0.1 mV.s^{-1})

Polarization curves obtained during forward and backward scans of potential are superimposed and no pitting potential is observed, which allows one to conclude that both materials show outstanding resistance to corrosion.

15.2 Principal Requirements of Medical Implants

15.2.1 Introduction

Medical implants are products that have to satisfy functionality demands defined by the working environment-human body. They could be used in almost every organ of the human body. Ideally, they should have biomechanical properties comparable to those of autogenous tissues without any adverse effects. The principal requirements of all medical implants are corrosion resistance, biocompatibility, bioadhesion, biofunctionality, processability and availability. To fulfil these requirments most of the tests are directed into the study extracts from the material, offering screens for genotoxicity, carcinogenicity, reproductive toxicity, cytotoxicity, irritation, sensitivity and sterilization agent residues [40]. The consequences of corrosion are the disintegration of the implant material per se, which weakens the implant, and the harmful effect of corrosion on the surrounding tissues and organs is produced.

Medical implants are regulated and classified in order to ensure safety and effectiveness to the patient. One of the main goals of implant research and development is to predict long-term, in-vivo performance of implants. Lack of useful computer-modeling data about in-vivo performance characteristics makes the evaluation of synergistic contributions of materials, design features and therapeutic drug regimens difficult. The present trends in modern implant surgery are networking various skilled and gifted specialists such as traumatologists, orthopedists, mechanical engineers, pharmacists and others in order to get better results in research, development and implementation into practice.

15.2.2 Metallic Biomaterials

The first metal alloy developed specifically for the human body environment was the 'vanadium steel' that was used to manufacture bone fracture plates (Sherman plates) and screws. Most metals that are used to make alloys for manufacturing implants such as iron (Fe), chromium (Cr), cobalt (Co), nickel (Ni), titanium (Ti), tantalum (Ta), niobium (Nb), molybdenum (Mo), and tungsten (W), can only be tolerated by the body in minute amounts. Sometimes these metallic elements, in naturally occurring forms, are essential in red blood cell functions (Fe) or synthesis of a vitamin B12 (Co), etc. but cannot be tolerated in large amounts in the body [41].

Metallic biomaterials can be divided into four subgroups: stainless steels, the cobalt-based alloys, titanium metals, and miscellaneous others (including tantalum, gold, dental amalgams and other special metals). They are very effective in binding the fractured bone, do not corrode and do not release harmful toxins when exposed to body fluids and therefore can be left inside the body for a long period of time. Their disadvantage is a much larger hardness and stiffness compared to the bone and possibility of interfering with the re-growth of the bone.

15.2.3 The Surface-Tissue Interaction

Good corrosion resistance of titanium depends upon the formation of a solid oxide layer (TiO_2) to a depth of 10 nm. After the implant is inserted, it immediately reacts with body liquids that consist of water molecules, dissolved ions, and proteins as shown in Figure 15.7.

Geometry, roughness and other characteristics of the implant surface also importantly influence the surface-tissue interaction, which is considered to be dynamic. Due to these phenomena, over time new stages of biochemical formations can be developed. In the first few

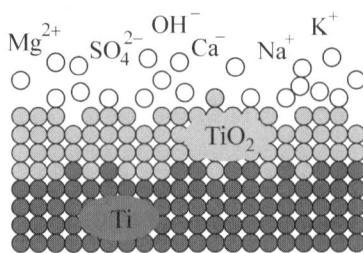

Fig. 15.7. Interaction between titanium and body liquids [42]

seconds after the contact has been made, there is only water, dissolved ions, and free biomolecules in the closest proximity of the surface, but no cells. The composition of the body liquid changes continuously as inflammatory and healing processes proceed, causing changes in the composition of the adsorbed layer of biomolecules on the implant surface until it is balanced. Cells and tissues eventually contact the surface and, depending on the nature of the adsorbed layer, they respond in specific ways that may further modify the adsorbed biomolecules, [43].

Surface roughness also plays an important role in osteointegration. Osteoblast cells are more likely to get attached to rough-sand blasted surfaces, meaning lower cell numbers on the rougher surfaces, decreased rate of cellular proliferation, and increased matrix production compared to smooth surface. Experiments made by Feighan [44] showed that an average roughness increased from 0.5 to 5.9 μm also increase the interfacial shear strength from 0.48 to 3.5 MPa.

15.2.4 Machining of Titanium Alloys

Titanium alloys are among the most widely used and promising materials for medical implants. Selection of titanium alloy for implementation is determined by a combination of the most desirable characteristics including immunity to corrosion, biocompatibility, shear strength, density and osteointegration, [3]. The excellent chemical and corrosion resistance of titanium is to a large extent due to the chemical stability of its solid oxide surface layer to a depth of 10 nm, [45]. Under in vivo conditions the titanium oxide (TiO_2) is the only stable reaction product whose surface acts as catalyst for a number of chemical reactions. However, micromotion at the cement-prosthesis and cement-bone are inevitable and consequently, titanium oxide and titanium alloy particles are released in the cemented joint prosthesis. Sometimes this wear debris accumulates as periprosthetic fluid collections and triggers giant cell response around the implants, [46]. TiO_2 film, such as the ones anodically formed in aqueous electrolytes, consists mainly of anatase and is an n-type semiconductor with a low electronic charge conductivity and a high resistance to anodic current [47].

Processes of machining titanium alloys involve conventional machining operations (turning, face milling, high-speed cutting (HSC), milling, drilling), forming operations (cold and hot forming; hydroforming, forging) and alternative machining operations (laser cutting, waterjet cutting, direct metal laser sintering). Machining operations of titanium alloys are considered to be difficult, due to its relatively high tensile strength, low ductile yield, 50% lower modulus of elasticity (104 GPa) and approximately 80% lower thermal conductivity than that of steel [48]. The lower modulus of elasticity may cause greater 'spring back' and deflection effect of the workpiece. Therefore, more rigid setups and greater clearances for tools are required. In the tool contact zones high pressures and temperatures occur (the tool-to-workpiece interface). The amount of heat removed by laminar chips is only approximately 25%, the rest is removed via the tool. Due to this phenomenon titanium alloys can be machined at comparatively low cutting speeds. At higher temperatures caused by friction the titanium becomes more chemically reactive and there is a tendency for titanium to »weld« to tool bits during machining operations. Overheating of the surface can result in interstitial pickup of oxygen and nitrogen, which will produce a hard and brittle alpha case. Carbides with high WC-Co content (K-grades) and high-speed steels with high cobalt content are suitable for use as cutting materials in titanium machining operations [48]. Turning operations of titanium alloys should have cutting depths as large as possible, cutting speeds V_C from 12 to 80 m/min and approximately 50% lower when high-speed steel (HSS) tools are used. The heat generated should be removed via large volumes of cooling lubricant. Chlorinated cutting fluids are not recommended because titanium can be susceptible to stress corrosion failures in the presence of chlorine. Any type of hot working or forging operation should be carried out below 925°C due to high level of titanium reactivity at high temperature.

Some medical implants are produced modularly, using different materials and processing techniques. For example, the femoral stem as part of the hip endoprosthesis is produced in a combination of casting, forging and milling. The final machining operation is performed on CNC machine using CAD-CAM principle.

Good alternatives to conventional machining techniques are alternative techniques such as waterjet cutting, sintering, or direct metal laser sintering. The latter is a rapid prototyping technique enabling prompt modeling of metal parts with high bulk density on the basis of individual three-dimensional data, including computer tomography models of anatomical structures, [49]. The concept of layer by layer building rather than removing waste material to achieve the desired geometry of a component, opens up endless possibilities of alternative manufacture of medical devices and is more environment friendly.

15.2.5 Surface Treatments and Coatings

Mechanical methods for surface treatment can be divided into methods involving removal of surface material by cutting (machining of the surface), abrasive action (grinding and polishing) and those where the treated material surface is deformed by particle blasting. Chemical methods are based mainly on chemical reactions occurring at the interface between titanium and a solution (solvent cleaning, wet chemical etching, passivation treatments and other chemical surface treatments such as Hydrogen peroxide treatment). Electrochemical surface methods are based on different chemical reactions occurring at an electrically energized surface (electrode) placed in an electrolyte (electropolishing and anodic oxidation or anodizing).

Improving the method for both wear and corrosion resistance of titanium implant surfaces in cases where the protection by natural surface oxide films is insufficient can be done through the deposition of thin films. These coatings should have a sufficiently high adherence to the substrate throughout the range of conditions to which the implant is exposed in service. They must tolerate the stress and strain variations that any particular part of the implant normally impose on the coating. The coating process must not damage the substrate and must not induce failure in the substrate or introduce impurities on the surface, which may change interface properties [47]. Coatings should be wear resistant, barrier layers preventive of substrate metal-ion release, to low-friction haemocompatible, non-thrombogenic surfaces [47]. Such surface modification could be done by various processes such as precipitations from the chemical vapor phase, Sol-Gel coatings, chemical vapor deposition (CVD) or physical vapor deposition (PVD). The properties of PVD coatings are good thickness, roughness, hardness, strength and adhesion as well as structure, morphology, stoichiometry and internal stresses. PVD processes include evaporation, sputtering, ion plating and ion implantation. They are carried out in vacuum, at backpressures of less than 1 Pa, [47]. Speaking of the CVD methods, they involve the reaction of volatile components at the substrate surface to form a solid product. Typical CVD coatings are depositions of TiN, TiC and TiC_xN_{1-x}. The early coatings were deposited onto hard metal tools such as WC-Co. Good coating uniformity is an advantage of the CVD method, lower operating temperatures of PVD method that can be combined in the plasma-assisted CVD process. Biomaterial produced by low temperature CVD and PVD is diamond-like carbon (DLC). DLC coatings can address the main biomechanical problems with the implants currently used, e.g. friction, corrosion and biocompatibility [50].

The concept of bioactive coatings uses a principle of enabling an interfacial chemical bond between the implant and the bone tissue due to a specific biological response, [51]. Surface modifications should provide distinct properties of interaction with cell molecules, which promote the adaptation or in-growth of cells or tissue onto the surface of fixation elements of a medical implant or prevent cell interaction with the implant surface.

Suitable bioactive surface modifications are comparable to those known from the stoichiometrically passivated titanium surface, in terms of high mechanical stability against shearing forces, long-term chemical stability and corrosion resistance in a biocompatible manner. One of the most popular bioactive coatings is hydroxyapatite (HA), which is similar to the mineral phase of natural hard tissue, i.e. about 70% of the mineral fraction of a bone has a HA-like structure. HA can also be regarded as non-resorbable in a physiological environment, as long as it remains crystalline and is of high purity. It is the most stable calcium phosphate phase in aqueous solutions [52]. It has weaker mechanical properties and low resistance to fatigue failure. Surface treatments techniques for HA are plasma spraying (Vacuum Plasma Spraying-VPS) electrophoretic deposition of HA and micro-arc oxidation.

15.2.6 Applications in Practice

Different types of fracture repair mechanisms are known in medical practice. Incomplete fractures such as cracks, which only allow micro-motion between the fracture fragments, heal with a small amount of fracture-line callus, known as primary healing. In contrast, complete fractures that are unstable, and therefore generate macromotion, heal with a voluminous callus stemming from the sides of the bone, known as secondary healing. The treatments can be non-surgical or surgical. Non-surgical treatments are immobilization with plaster or resin casting and bracing with a plastic apparatus. The surgical treatments of bone fractures (osteosynthesis) are divided into external fracture fixation, which does not require opening the fracture site, or internal fracture fixation, which requires opening the fracture. With external fracture fixation, the bone fragments are held in alignment by pins placed through the skin onto the skeleton, structurally supported by external bars. With internal fracture fixation, the bone fragments are held by wires, screws, plates, and/or intramedullary devices [53].

Surgical wires are used to reattach large fragments of bone. They are also used to provide additional stability in long-oblique or spiral fractures of long bones which have already been stabilized by other means. Straight wires are called Steinmann pins. In the case of a pin diameter less than 2.38 mm, they are named Kirschner wires. They are widely used primarily to hold fragments of bones together provisionally or permanently and to guide large screws during insertion. *Screws* are the most widely used devices for fixation of bone fragments, Figure 15.8(a). There are two types of bone screws cortical bone screws, which have small threads, and cancellous screws, which have large threads to get more thread-to-bone contact. They may have either V or buttress threads. According to their ability to penetrate the cortical screws are sub-classified further, into self-tapping and non-self-tapping. The self-tapping screws have cutting flutes that thread the pilot drill-hole during insertion. In contrast, the non-self-tapping screws require a tapped pilot drill-hole for insertion. The bone immediately adjacent to the screw often undergoes necrosis initially, but if the screw is firmly fixed when the bone revascularizes, permanent secure fixation may be achieved, [54]. This is particularly true for titanium alloy screws or screws with a roughened thread surface, with which bone ongrowth results in an increase in removal torque [54]. *Plates* are available in a wide variety of shapes and are intended to facilitate fixation of bone fragments, Figure 15.8(b). They range from the very rigid, intended to produce primary bone healing, to the relatively flexible, intended to facilitate physiological loading of bone. The rigidity and strength of a plate in bending depends on the cross-sectional thickness and material properties of which it is made. Consequently, the weakest region in the plate is the screw hole, especially if the screw hole is left empty, due to a reduction of the cross-sectional area in this region. The effect of the material on the rigidity of the plate is defined by the elastic modulus of the material for bending, and by the shear modulus for twisting, [55]. Thus, given the same dimensions, a titanium alloy plate will be less rigid than a stainless steel plate, since the elastic modulus of each alloy is 110 GPa and 200 GPa, respectively. *Intramedullary devices (IM nails or rods)* are used as internal struts to stabilize long bone fractures, Figure 15.8(c). IM nails are also used for fixation of femoral neck or intertrochanteric bone fractures; however, this application requires the addition of long screws. A whole range of designs is available, going from solid to cylindrical, with shapes such as cloverleaf, diamond, and slotted cylinders. Compared to plates, IM nails are better positioned to resist multidirectional bending than a plate or an external fixator, since they are located in the center of the bone. However, their torsional resistance is less than that of the plate, [55].

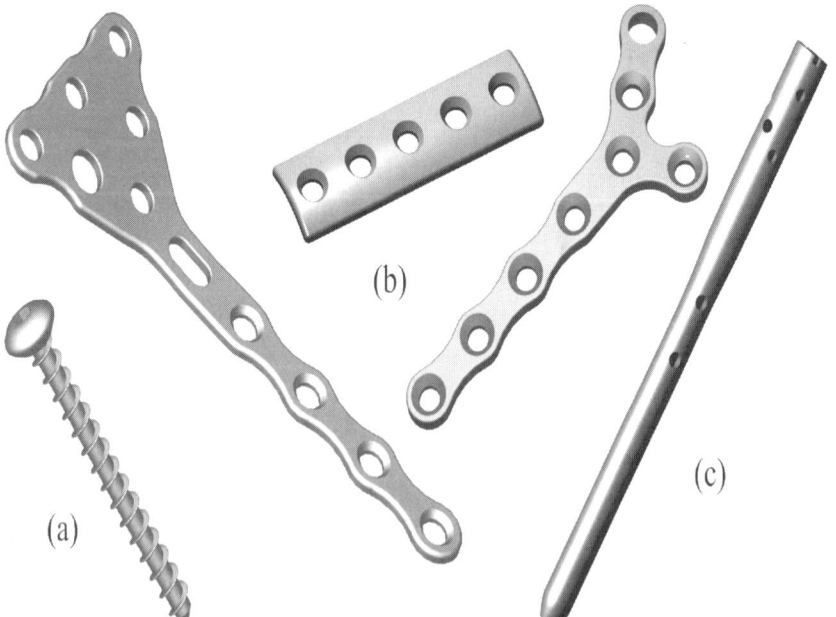

Fig. 15.8. Titanium trauma medical implants

The design of an implant for joint replacement should be based on the kinematics and dynamic load transfer characteristic of the joint. The material properties, shape, and methods used for fixation of the implant to the patient determines the load transfer characteristics. This is one of the most important elements that determines long-term survival of the implant, since bone responds to changes in load transfer with a remodeling process, known as Wolff's law. Overloading the implant-bone interface or shielding it from load transfer may result in bone resorption and subsequent loosening of the implant, [56]. *The endoprosthesis for total hip replacement* consists of a femoral component and an acetabular component, Figure 15.9(b). The femoral stem is divided into head, neck, and shaft. The femoral stem is made of Ti alloy or Co-Cr alloy and is fixed into a reamed medullary canal by cementation or press fitting. The femoral head is made of Co-Cr alloy, aluminum, or zirconium. Although Ti alloy heads function well under clean articulating conditions, they have fallen into disuse because of their low wear resistance to third bodies, e.g., bone or cement particles. The acetabular component is generally made of ultra-high molecular weight polyethylene (UHMWPE). *The prosthesis for total knee joint replacement* consists of femoral, tibial, and/or patellar components, Figure 15.9(a). Compared to the hip joint, the knee joint has a more complicated geometry and movement mechanics, and it is not intrinsically stable. In a normal knee, the center of movement

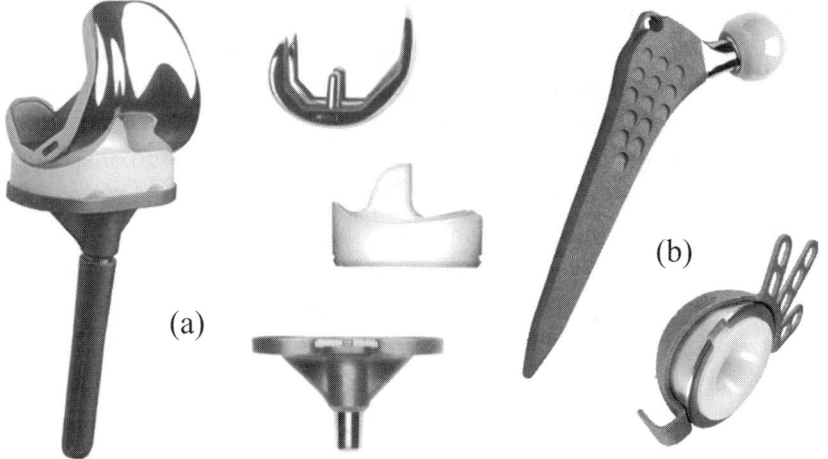

Fig. 15.9. Titanium orthopedics medical devices [58]

is controlled by the geometry of the ligaments. As the knee moves, the ligaments rotate on their bony attachments and the center of movement also moves. The eccentric movement of the knee helps distribute the load throughout the entire joint surface, [57]. Total knee replacements can be implanted with or without cement, the latter relying on porous coating for fixation. The femoral components are typically made of Co-Cr alloy and the monolithic tibial components are made of UHMWPE. In modular components, the tibial polyethylene component assembles onto a titanium alloy tibial tray. The patellar component is made of UHMWPE, and a titanium alloy back is added to components designed for uncemented use.

For maxillofacial osteosynthesis in *the cranio-facial and mandibular areas titanium plate and screw systems* are preferred. In order to make them pliable, many of the plates are made from CP titanium sheet that is in the soft-recrystallized condition. The corresponding screws are either made from CP titanium or alloy and can be as small as 1 mm in diameter, [59].

15.3 Shape Memory Alloys

15.3.1 Introduction

Smart materials have been given a lot of attention mainly for their innovative use in practical applications. One example of such materials is also the family of shape memory alloys (SMA) which are arguably the first well known and used smart material. Shape memory alloys possess a unique property according to which, after being deformed at one temperature, they can recover to their original shape upon being heated to a higher temperature. The effect was first discussed in the 1930s by Ölander [60] and Greninger and Mooradian [61]. The basic phenomenon of the shape memory effect was widely reported a decade later by Russian metallurgist G. Kurdjumov and also by Chang and Read [62]. However, presentation of this property to the wider public came only after the development of the nickel-titanium alloy (nitinol) by Buehler and Wang [63]. Since then, research activity in this field has been intense, and a number of alloys have been investigated, including Ag-Cd, Au-Cd, Cu-Zn, Cu-Zn-Al, Cu-Al-Ni, Cu-Sn,

Cu-Au-Zn, Ni-Al, Ti-Ni, Ti-Ni-Cu, Ni-Ti-Nb, Ti-Pd-Ni, In-Ti, In-Cd and others. Crystallography of shape memory alloys have been studied for the last four decades. Only a fraction of the available literature is listed here [64–73]. Because these materials are relatively new, some of the engineering aspects of the material are still not well understood. Many of the typical engineering descriptors, such as young's modulus and yield strength, do not apply to shape memory alloys since they are very strongly temperature dependent. On the other hand, a new set of descriptors must be introduced, such as stress rate and amnesia. That is why numerous constitutive models have been proposed over the last 20 years to predict thermomechanical behavior [74–87].

15.3.1.1 Thermomechanical Behavior

These materials have been shown to exhibit extremely large, recoverable strains (on the order of 10%), and it is these properties as functions of temperature and stress that allow SMAs to be utilized in many exciting and innovative applications. From a macroscopic point of view, the mechanical behavior of SMAs can be separated into two categories: the *shape memory effect* (SME), where large residual (apparently plastic) strain can be fully recovered upon raising the temperature after loading and unloading cycle; and the *pseudoelasticity* or *superelasticity*, where a very large (apparently plastic) strain is fully recovered after loading and unloading at constant temperature. Both effects are results of a martensite phase transformation. In a stress-free state, an SMA material at high temperatures exists in the parent phase (usually a body-centered cubic crystal structure, also refered as the austenite phase). Upon decreasing the material temperature, the crystal structure undergoes a self-accommodating crystal transformation into martensite phase (usually a face-centered cubic structure). The phase change in the unstressed formation of martensite from austenite is referred to as 'self-accommodating' due to the formation of multiple martensitic variants and twins that prohibits the incurrence of a transformation strain. The martensite variants, evenly distributed throughout material, are all crystallographically equivalent, differing only by habit plane. The process of self-accomodation by twinning allows an SMA material to exhibit large reversible strains with stress.

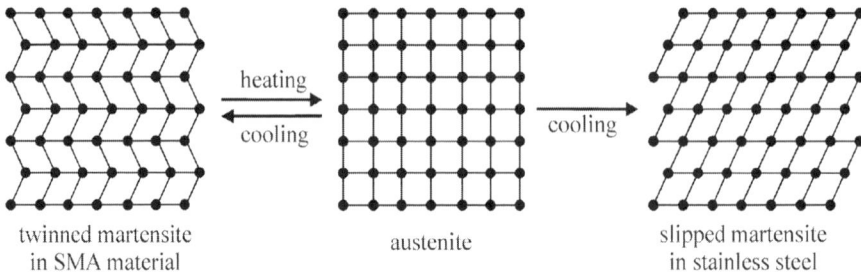

Fig. 15.10. Martensite transformation in shape memory alloys and steels

However, the process of self-accomodation in ordinary materials like stainless steel does not take place by twinning but via a mechanism called slip. Since slip is a permanent or unreversible process, the shape memory effect cannot occur in these materials. The difference between the twinning and slip process is shown in Figure 15.10.

In the stress-free state, an SMA material has four transition temperatures, designated as M_f, M_S, A_S, A_f, i.e. Martensite Finish, Martensite Start, Austenite Start, and Austenite Finish, respectively. In the case of 'Type I' materials, temperatures are arranged in the following manner: $M_f < M_S < A_S < A_f$. A change of temperature within the range $M_S < T < A_S$ induces no phase changes and both phases can coexist within $M_f < T < A_f$. With these four transformation temperatures and the concepts of self-accommodation, the shape memory effect can be adequately explained. As an example let us consider a martensite formed from the parent phase, Figure 15.11(a), cooled under stress-free conditions through M_S and M_f. This material has multiple variants and twins present, Figure 15.11(b), all crystallographically equivalent, but with different orientation (different habit plane indices). When a load applied to this material reaches a certain critical stress, the pairs of martensite twins begin 'de-twinning' to the stress-prefered twins, Figure 15.11(c). It means that the multiple martensite variants begin to convert to a single variant determined by alignment of the habit planes with the direction of loading, Figure 15.11(d). During this process of reorientation, the stress rises very slightly in comparison to the strain. As the single variant of martensite is thermodynamically stable at $T < A_S$, upon unloading there is no conversion to multiple variants and only a small elastic strain is recovered, leaving the material with a large residual strain, Figure 15.11(e). The de-twinned martensite material can recover the entire residual strain by

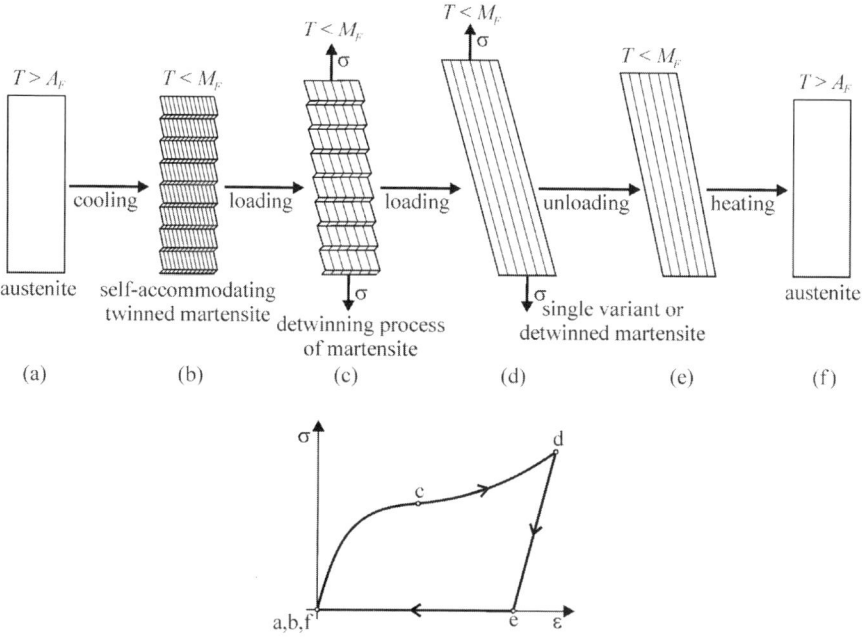

Fig. 15.11. Shape memory effect

simply heating above A_f; the material then transforms to the parent phase, which has no variants, and recovers to its original size and shape, Figure 15.11(f), thus creating the shape memory effect.

The pseudoelastic effect can be explained, if an SMA material is considered to be entirely in the parent phase (with $T > A_f$), Figure 15.12(a). When stress is applied to this material, there is a critical stress at which the crystal phase transformation from austenite to martensite can be induced, Figure 15.12(b). Due to the presence of stress during the transformation, specific martensite variants will be formed preferentially and at the end of transformation, the stress-induced martensite will consist of a single variant of detwinned martensite, Figure 15.12(c). During unloading, a reverse transformation to austenite occurs because of the instability of martensite at $T > A_f$ in the absence of stress, Figure 15.12(e). This recovery of high strain values upon unloading yields a characteristic hysteresis loop, diagram in Figure 15.12, which is known as pseudoelasticity or superelasticity.

Many of the possible medical applications of SMA materials in the 1980's were attempting to use the thermally activated memory effect. However, temperature regions tolerated by the human body are

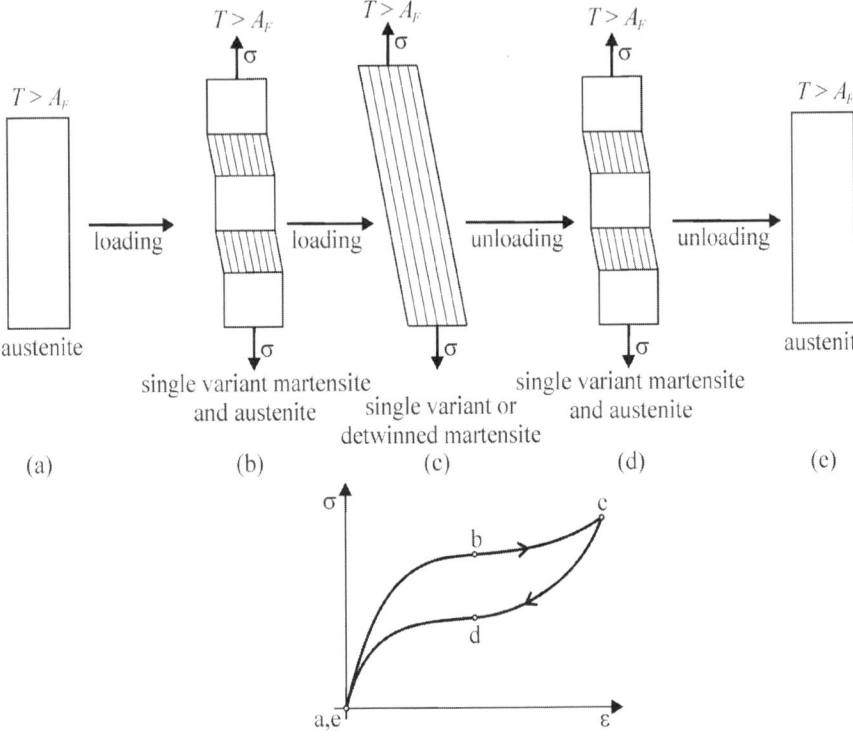

Fig. 15.12. Pseudoelasticity or superelasticity

very limited. Small compositional changes around the 50–50% of Ti-Ni ratio can make dramatic changes in the operating characteristics of the alloy. Therefore very precise control of phase transition temperatures is required. On the other hand, pseudoelasticity is ideally suited to medical applications since the temperature region of optimum effect can easily be located to encompass ambient temperature through body temperature.

15.3.2 Biocompatibility

It is important to understand the direct effects of an individual component of the alloy since it can dissolve in the body due to corrosion and it may cause local and systemic toxicity, carcinogenic effects and immune response. The cytotoxicity of elementary nickel and titanium has been widely researched, especially in the case of nickel, which is a toxic agent and allergen, [88, 89, 90]. Nickel is known to have toxic effects on soft tissue structures at high concentrations and also appears to be harmful to bone structures, but substantially less than cobalt or vanadium, which are also routinely used in implant alloys. Experiments with toxic metal salts in cell cultures have shown decreasing toxicity in the following order: Co > V > Ni > Cr > Ti > Fe, [91]. The dietary exposure to nickel is 160–600 mg/day. Fortunately most of it is eliminated in the feces, urine and sweat. Pure nickel implanted intramuscularly or inside bone has been found to cause severe local tissue irritation and necrosis and high carcinogenic and toxic potencies. Due to corrosion of medical implants, a small amount of these metal ions is also released into distant organs. Toxic poisoning is later caused by the accumulation, processing and subsequent reaction of the host to the corrosion of the Ni-containing implant. Nickel is also one of the structural components of the metalloproteins and can enter the cell via various mechanisms. Most common Ni^{2+} ions can enter the cell utilizing the divalent cation receptor or via the support with Mg^{2+}, which are both present in the plasma membrane. Nickel particles in cells can be phagocytosed, which is enhanced by their crystalline nature, negative surface energy, appropriate particle size (2–4 μm) and low solubility. Other nickel compounds formed in the body are most likely to be $NiCl$ and NiO, and fortunately there is only a small chance that the most toxic and carcinogenic compounds like Ni_3S_2, are to be formed. Nickel in soluble form, such as Ni^{2+} ions, enters through receptors or ion channels and binds to cytoplasmic proteins and does not accumulate in the cell nucleus at concentrations high enough to cause genetic consequences. These soluble Ni^{2+} ions and are rapidly cleaned from the body. However, the insoluble nickel particles containing phagocytotic vesicles fuse with lysosomes, followed by a decrease of phagocytic intravesicular pH, which releases Ni^{2+} ions from nickel containing carrier molecules. The formation of oxygen radicals, DNA damage and thereby inactivation of tumor suppressor genes is contributed by that.

On the other hand, titanium is recognized to be one of the most biocompatible materials due to the ability to form a stable titanium oxide layer on its surface. In an optimal situation, it is capable of excellent osteointegration with the bone and it is able to form a calcium phosphate-rich layer on its surface, Figure 15.13, very similar to

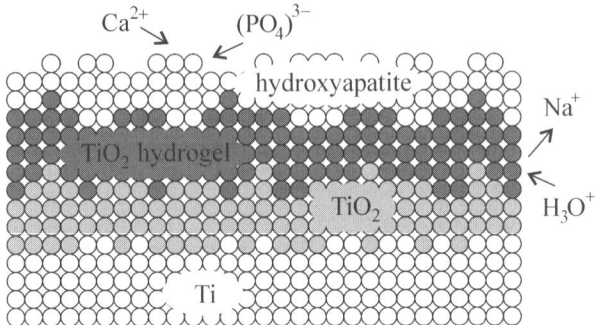

Fig. 15.13. Formation of hydroxyapatite layer on titanium oxide film [92]

hidroxyapatite which also prevents corrosion. Another advantageous property is that in case of damaging the protective layer the titanium oxides and Ca-P layer regenerate.

The properties and biocompatibility of nitinol have their own characteristics which are different from those of nickel or titanium alone. In vitro NiTi biocompatibility studies on the effects of cellular tolerance and its cytotoxicity have been performed on various cell culture models [93, 94]. Human monocytes and microvascular endothelial cells were exposed to pure nickel, pure titanium, stainless steel and nitinol. Nitinol has been shown to release higher concentrations of Ni^{2+} ions in human fibroblast and osteoblast cultures, which did not affect cell growth [95–97]. Metal ion release study also revealed very low concentrations of nickel and titanium that were released from nitinol. Researchers therefore concluded that nitinol is not genotoxic.

For in vivo biocompatibility studies of nitinol effect, different experiments have been done on animals. Several in vivo nitinol biocompatibility studies that were done in the last decade disclosed no allergic reactions, no traces of alloy constituents in the surrounding tissue and no corrosion of implants. Studies of rat tibiae response to NiTi, compared with Ti-6Al-4V and stainless steel, showed that the number and area of bone contacts was low around NiTi implants, but the thickness of contact was equal to that of other implants. Normal new bone formation was seen in rats after 26 weeks after implantation. Good biocompatibility results of NiTi are attributed to the fact that implants are covered by a titanium oxide layer, where only small traces of nickel are being exposed.

15.3.2.1 Corrosion Behavior

The body is a complicated electrochemical system that constitutes an aggressive corrosion environment for implants which are surrounded by bodily fluids of an aerated solution containing 0.9% NaCl, with minor amounts of other salts and organic compounds, serum ions, proteins and cells which all may modify the local corrosion effect. High acidity of certain bodily fluids is especially hostile for metallic implants. Acidity can increase locally in the area adjacent to an implant due to inflammatory response of surrounding tissues mediating hydrogen peroxide and reactive oxygen and nitrogen compounds. The local pH changes for infected tissues or near haematomas are relatively small, however these changes can alter biological processes and thereby the chemistry around the implant. It is known that small point corrosion or pitting prevails on surfaces of metallic implants. Another important feature is roughness of the surface which increases the reacting area of the implant and thereby add to total amount of corrosion. Therefore surface finishing is a major factor in improving corrosion resistance and consequently biocompatibility of medical devices, [98, 99].

Corrosion resistance of SMA has also been studied *in vivo* on animals. Plates and stents have been implanted in dogs and sheeps for several months. Corrosion has been examined under microscope and pitting was established as predominant after the implants were removed. Thus surface treatments and coatings were introduced. The improvement of corrosion resistance was considerable, since pitting decreased in some cases from 100 microns to only 10 microns in diameter.

15.3.3 Surface of Implant

The human response to implanted materials is a property closely related to the implant surface conditions. The major problems associated with the implants currently used are inadequate implant-tissue interface properties. Parameters that characterize surface property are chemical composition, crystallinity and heterogeneity, roughness and wettability or surface free energy that is a parameter important for cell adhesion. Each parameter is of great importance to biological response of the tissue. Another problem is implant sterilization that can remarkably modify desired parameters. Steam and dry sterilization are nowadays replaced by more advanced techniques like hydrogen peroxide plasma, ethylene oxide, and electron and γ-ray irradiation.

The surface of NiTi SMA has revealed a tendency towards preferential oxidation of titanium. This behavior is in agreement with the fact that the free enthalpy of formation of titanium oxides is negative and exceeds in absolute value the enthalpy of formation of nickel oxides by at least two to three times. The result of oxidation is an oxide layer of a thickness between 2–20 nm, which consists mainly of titanium oxides TiO_2, smaller amounts of elemental nickel Ni^{2+}, and low concentrations of nickel oxides NiO. The surface chemistry and the amount of Ni may vary over a wide range, depending on the preparation method. The ratio of Ti/Ni on polished surface is around 5.5, while boiled or autoclaved items in water show decreased concentration of Ni on the surface and the Ti/Ni ratio increases up to 23 to 33, [100]. Different in vitro studies have shown how the physical, chemical and biocompatible properties of the implant surface can be improved, [101–105].

15.3.3.1 Surface Improvements

Some of the most important techniques for improving the properties of Ni-Ti alloy surfaces are: *(1) Surface modification by using energy sources and chemical vapors* like hydroxyapatite, laser and plasma treatment, ion implantation, TiN and TiCN chemical vapor deposits. Hydroxyapatite coatings result in the best known biocompatibility and reveal a tendency to dissolution due to its relative miscibility with body fluids. Ion implantation and laser treatments usually result in surface amorphization that improves corrosion resistance, but the obtained amorphous surface layers are often not uniform. Laser surface melting leads to an increased oxide layer, decrease of Ni dissolution and improvement of the cytocompatibility up to classical Ti level. There is also a possibility that laser melted surfaces may be enriched in nickel, and become harder than bulk and swell. TiN and TiCN coatings are known to improve corrosion resistance but large deformations caused by the shape memory effect may cause cracking of the coating. Therefore, for plates and staples a plasma-polymerized tetrafluoroethylene has been introduced. *(2) Development of bioactive surfaces* is another approach to improve biocompatibility of the SMA. Human plasma fibronectin covalently immobilised to NiTi surface improved the attachment of cells while corrosion rates were reduced drastically. Studies showed NiTi surface improved with this method caused a development of Ca-P layers, which in fact eliminate the need for hydroxyapatite coatings, [102, 106]. *(3) Electrochemical processing for oxidation in air/oxygen* is a most common way of metal surface treatment. The technique combines electrochemical processes and oxidation in various media. Growth of native passive films that are highly adhesive and do not crack or break due to dynamic properties of SMA is promoted with this method. Oxide films obtained in air have different colors, thickness, and adhesive properties, with TiO_2 as a predominant oxide type. *(4) Oxidation of SMA medical devices in water and steam* is also one of the surface improvement techniques. Implants are preliminary chemically etched and boiled in water. The result is a surface with a very low Ni concentration, while etching removes surface material that was exposed to processing procedures and acquired various surface defects and heterogeneity. It also selectively removes nickel and oxidizes titanium. Surfaces obtained after oxidation in steam show better properties than those oxidized in water. *(5) Electrochemical techniques* are commonly used to passivate NiTi surfaces. Surface passivation using electropolishing is often considered as a treatment of first choice just because this technique is used for surface conditioning of stainless steel, Co-Cr alloys, etc. However, the universal techniques developed for surface passivation of various alloys used for medical purposes are not necessary efficient for NiTi.

It should also be noted that the implant surface coatings are not always beneficial. The major problem of titanium based alloys is that the formation of TiO_2, according to the chemical equation $Ti + 2H_2O \rightarrow TiO_2 + 4H^+ + 4e^-$, reduces the pH level at the titanium/coating interface. This means that if the coating is composed of hydroxyapatite, it can dissolve, which gradually leads to detachment of the coating.

15.3.4 Medical Applications

The trends in modern medicine are to use less invasive surgery methods that are performed through small, leak tight portals into the body called trocars. Medical devices made from SMAs use a different physical approach and can pull together, dilate, constrict, push apart and have made difficult or problematic tasks in surgery quite feasible. Therefore unique properties of SMAs are utilized in a wide range of medical applications. Some of the devices used in various medical applications are listed below.

Stents are most rapidly growing cardiovascular SMA cylindrical mesh tubes that are inserted into blood vessels to maintain the inner diameter of a blood vessel. The product has been developed in response to limitations of balloon angioplasty, which resulted in repeated blockages of the vessel in the same area. Ni-Ti alloys have also become the material of choice for superelastic *self-expanding (SE) stents* that are used for a treatment of the superficial femoral artery disease, Figure 15.14(a). The SE nitinol stents are produced in the open state mainly with laser cut tubing and later compressed and inserted into the catheter. They can also be produced from wire and laser welded or coiled striped etched sheet. Before the compression stage, the surface of the stent is electrochemically polished and passivated to prescribed quality. Deployment of the SE stent is made with the catheter. During the operation procedure, when the catheter is in the correct position in the vessel, the SE stent is pushed out and then it expands against the inner diameter of the vessel due to a rise in temperature (thermally triggered device). This opens the iliac artery to aid in the normal flow of blood. The delivery catheter is then removed, leaving the stent within the patient's artery. Recent research has shown that implantation of a self-expanding stent provides better outcomes, for the time being, than balloon angioplasty, [107–109]. *The Simon Inferior Vena Cava (IVC) filter* was the first SMA cardiovascular device. It is used for blood vessel interruption for preventing pulmonary embolism via placement in the vena cava. The Simon filter is filtering clots that travel inside bloodstream, [110]. The device is made of SMA wire curved similarly to an umbrella that traps the clots that are better dissolved in time by the bloodstream. For insertion, the device is exploiting the shape memory effect, i.e. the original form in the martensitic state is deformed and mounted into a catheter. When the device is released, the body's heat causes the filter to return to its predetermined shape. *The Septal Occlusion System* is indicated for use in patients with complex ventricular septal defects (VSD) of significant size to warrant closures that are considered to be at high risk for standard transatrial or transarterial surgical closure based on anatomical conditions and/or based on overall medical condition. The system consists of two primary components; a permanent implant, which is constructed of an SMA wire framework to which polyester fabric is attached, and a coaxial polyurethane catheter designed specifically to facilitate attachment, loading, delivery and deployment to the defect, [111]. The implant is placed by advancing the delivery catheter through blood vessels to the site of the defect inside the heart. The implant remains in the heart and the delivery catheter is removed. Instruments for minimally invasive surgery used in endoscopic surgery could not be feasible without implementation of SMA materials. The most representative instruments such as *guidewires, dilatators* and *retrieval baskets* exploit good kink resistance of SMAs, [112]. *Open-heart stabilizers* are instruments similar to a steerable joint endoscopic camera. In order to perform bypass operations on the open heart stabilizers are used to prevent regional heart movements while performing surgery. Another employment of the unique properties of SMAs such as constant force and superelasticity in heart surgery is a *tissue spreader* used to spread fatty tissue of the heart, Figure 15.14(b).

In general, conventional orthopedic implants by far exceed any other SMA implant in weight or volume. They are used as fracture fixation devices, which may or may not be removed and as joint replacement devices. Bone and nitinol have similar stress-strain characteristics, which makes nitinol a perfect material for production of bone fixation plates, nails and other trauma implants, [113]. In traditional trauma surgery bone plates and nails fixated with screws are used for fixation of broken bones. Shape memory fixators are one step forward applying a necessary constant force to faster fracture healing. *The SMA embracing fixator* consists of a body and sawtooth arms, [114]. It embraces the bone about 2/3 of the circumference, Figure 15.14(c). The free ends of the arms that exceed the semi-circle are bent more medially to match the requirement fixation of a long tubular body whose cross section is not a regular circle. The applied axial compression stress is beneficial for enhancing healing and reducing segmental osteoporosis caused by a stress shielding effect. Its martensitic transformation temperature is 4–7°C and shape recovery temperature is around the body's normal temperature, 37°C. Similar to the embracing fixator is the so called *Swan-Like Memory-Compressive Connector (SMC)* for treatment of fracture and nonunion of upper limb diaphysis. The working principle of the device is similar with one important improvement. The SMC trauma implant is able to put constant axial stress to a fractured bone, [115]. For fixation of tibial and femoral fractures nails fixated with screws are normally used. New *SMA inter-locking intramedullary nails* have many advantages compared to traditional ones. For example, when cooled SMA inter-locking nails are inserted into a cavity, guiding nails are extracted and body heat causes bending of nails into a preset shape applying constant pressure in the axial direction of the fractured bone, [116]. The SMA effect is also used in surgical fixators made from wire. Certain devices that have been developed to fix vertebra in spine fractures are similar to an ordinary staple. *Staple shaped compression medical devices* are also used for internal bone fixation, [117]. The compression staple is one of most simple and broadly used SMA devices in medicine, Figure 15.14(d). Since its introduction in 1981, over a thousand patients have been all successfully treated using this device. *The SMA Patellar Concentrator* was designed to treat patellar fractures, Figure 15.14(e). The device exerts continuous compression for the fixation of patella fracture. The shape of patellar concentrator consists of two basic patellae claws, conjunctive waist and three apex patellae claws. The thickness of the device may vary between 1.8 and 2.2 mm depending on different sizes of the concentrator. In clinical surgery, the claws are unfolded and put over fractured patella. Exposed to body temperature, the device tends to recover to its original state resulting in a recovery compressive force, [118].

Dentists are using devices made from SMA for different purposes. NiTi based SMA material performs exceptionally at high strains in strain-controlled environments, such as exemplified with *dental drills* for root canal procedures. The advantage of these drills is that they can be bent to rather large strains and still accommodate the high cyclic rotations, [111]. Superelastic SMA wires have found wide use as orthodontic wires as well, Figure 15(a). *NiTi orthodontic archwire* was first produced in batches and clinically used in China in the beginning of 1980's, [119]. Due to its unique property-superelasticity, the wire exerts gentle and retentive force to teeth, which is superior to stainless steel wire. Shape memory bracelet do not require as frequent visits to the dentist as the classical ones because of their ability to self adjust. The therapeutic period is therefore cut down by 50%.

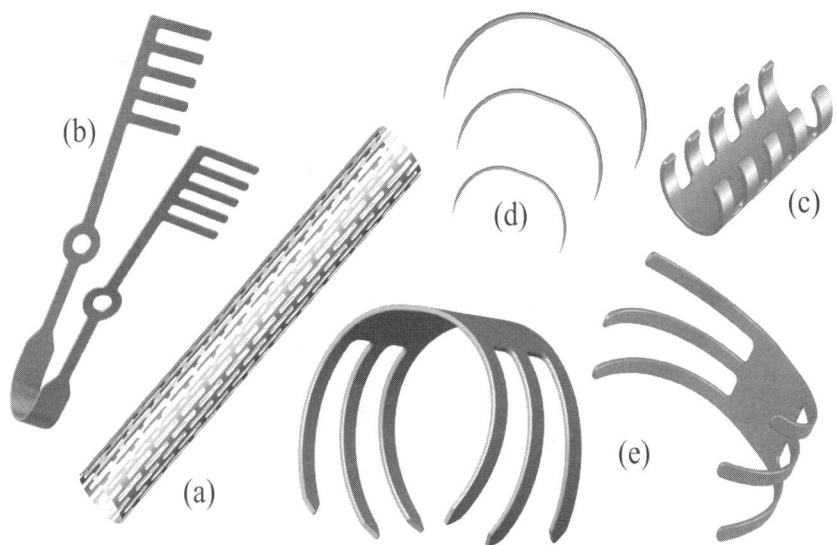

Fig. 15.14. Examples of nitinol medical devices

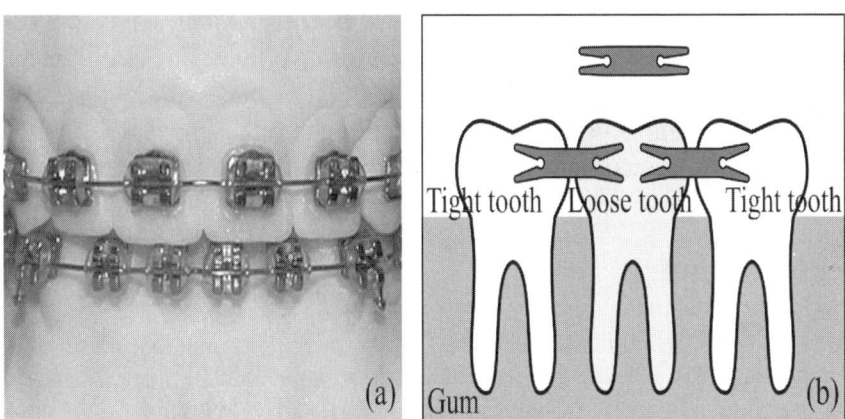

Fig. 15.15. Dental applications of nitinol

Lately a special *fixator for mounting bridgework* has been developed, Figure 15.15(b). A small piece of SMA metal is notch on both sides and placed between teeth and bridgework. As the temperature rises the notched area of metal is expanded on both sides causing a permanent hold of bridgework. The tooth fixator can also be used to prevent a loose tooth from falling out.

15.4 Conclusions

The use of titanium alloys as biomaterials has been growing due to their reduced elastic modulus, superior biocompatibility, high strength to weight ratio and enhanced corrosion resistance when compared to more conventional stainless steel and Co-Cr alloys. Ti-6Al-4V (and Ti-6Al-4V ELI), the most common titanium alloy is still the most extensively used titanium alloy for medical applications. However, V and Al have been found to be toxic to the human body. In this context, β titanium alloys have been studied and developed and due to their high strength, biocompatible behavior, low elastic modulus, superior corrosion resistance and good formability, they are likely to replace the classic α+β type Ti-6Al-4V for medical applications in the near future.

SMA implants and medical devices have been successful because they offer a possibility of performing less invasive surgeries. Nitinol wires in medical instruments are more kink resistant and have smaller diameters compared to stainless steel 316L or polymer devices. Research to develop composite materials, containing SMA that will prove cost efficient and porous SMAs that will enable the transport of body fluids from the outside to inside of the bone is currently underway.

References

1. H. Sibum, Titanium and titanium alloys – from raw material to semi-finished products, Advanced Engineering Materials 5(6) (2003) 393.
2. K. Wang, The use of titanium for medical applications in the USA, Materials Science and Engineering A 213 (1996) 134.
3. H.J. Rack and J.I. Qazi, Titanium alloys for biomedical applications, Materials Science and Engineering C 26 (2006) 1269.

4. M. Niinomi, Recent metallic materials for biomedical applications, Metallurgical and Materials Transactions 33A (2002) 477.
5. G. Lütjering and J.C. Williams, Titanium, Springer-Verlag, Berlin, 2003.
6. M. Long, H.J. Rack, Titanium alloys in total joint replacement – a materials science perspective, Biomaterials, 19 (1998) 1621.
7. K.S. Katti, Biomaterials in total joint replacement, Colloids and Surfaces B: Biointerfaces 39 (2004) 133.
8. J.A. Disegi, Titanium alloys for fracture fixation implants, Injury, International Journal of The Care of the Injured 31 (200) S-D14.
9. G. He and M. Hagiwara, Ti alloy design strategy for biomedical applications, Materials Science and Engineering C 26 (2006) 14.
10. B.P. Bannon and E.E. Mild, Titanium Alloys for Biomaterial Application: An Overview, Titanium Alloys in Surgical Implants, ASTM STP 796, H.A. Luckey and F. Kubli, Jr, Eds., American Society for Testing and materials, 1983, pp. 7–15.
11. V. Oliveira, R.R. Chaves, R. Bertazzoli and R. Caram, Preparation and characterization of Ti-Al-Nb orthopedic implants, Brazilian Journal of Chemical Engineering 17 (1998) 326.
12. R.R. Boyer, Ana overview on the use of titanium in the aerospace industry, Materials Science and Engineering A 213 (1996) 103.
13. J.G. Ferrero, Candidate materials for high-strength fastener applications in both the aerospace and automotive industries, Journal of Materials Engineering and Performance 14 (2005) 691.
14. M. Semlitsch, F. Staub and H. Weber, Titanium-aluminum-niobium alloy, development for biocompatible, high-strength surgical implants, Biomedizinische Technik 30 (1985) 334.
15. T.P. Vail, R.R. Glisson, T.D. Koukoubis and F. Guilak, The effect of hip stem material modulus on surface strain in human femora, Journal of Biomechanics 31 (1998) 619.
16. M. Niinomi, T. Akahori, T. Takeuchi, S. Katsura, H. Fukui and H. Toda, Mechanical properties and cyto-toxicity of new beta type titanium alloy with low melting points for dental applications, Materials Science and Engineering C 25 (2005) 417.
17. M. Kikuchi, M. Takahashi and O. Okuno, Elastic moduli of cast Ti-Au, Ti-Ag, and Ti-Cu alloys, Dental Materials 22 (2006) 641.
18. H.-S. kim, W.-Y. Kim and S.-H. Lim, Microstructure and elastic modulus of Ti-Nb-Si ternary alloys for biomedical applications, Scripta Materialia 54 (2006) 887.
19. S. Gross and E.W. Abel, A finite element analysis of hollow stemmed hip prostheses as a means of reducing stress shielding of the femur, Journal of Biomechanics 34 (2001) 995.
20. Y.L. Hao, M. Niinomi, D. Kuroda, K. Fukunaga, Y.L. Zhou and R. Yang, Aging response of the Young's modulus and mechanical properties of Ti-29Nb-13Ta-4.6Zr, Metallurgical and Materials Transactions 34A (2003) 1007.

21. Y.L. Hao, M. Niinomi, D. Kuroda, K. Fukunaga, Y.L. Zhou, R. Yang and A. Suzuki, Young's modulus and mechanical properties of Ti-29Nb-13Ta-4.6Zr in relation to α'' martensite, Metallurgical and Materials Transactions 33A (2002) 3137.

22. B. Gunawarman, M. Niinomi, T. Akahori, T. Souma, M. Ikeda and H. Toda, Mechanical properties and microstructures of low cost β titanium alloys for healthcare applications, Materials Science and Engineering C 25 (2005) 304.

23. N. Sakaguchi, M. Niinomi, T. Akahori, J. Takeda and H. Toda, Relationship between tensile deformation behavior and microstructure in Ti-Nb-Ta-Zr, Materials Science and Engineering C 25 (2005) 363.

24. D. Kuroda, M. Niinomi, M. Morinaga, Y. Kato and T. Yashiro, Design and mechanical properties of new β type titanium alloys for implant materials, Materials Science and Engineering A 243 (1998) 244.

25. M. Peters, H. Hemptenmacher, J. Kumpfert and C. Leyens, in: C. Leyens and M. Peters (Eds.), Titanium and Titanium Alloys, Wiley-VCH, 2003, p. 1–57.

26. P. Ari-Gur and S.L. Semiatin, Evolution of microstructure, macrotexture and microtexture during hot rolling og Ti-6Al-4V, Materials Science and Engineering A 257 (1998) 118.

27. G. Lütjering, Property optimization through microstructural control in titanium and aluminum alloys, Materials Science and Engineering A 263 (1999) 117.

28. Y.V.R.K., Prasad and T. Seshacharyulu, Processing maps for hot working of titanium alloys, Materials Science and Engineering A 243 (1998) 82.

29. H.L. Freese, M.G. Volas and J.R. Wood, in: D.M. Brunette, P. Tengvall, M. Textor and P. Thomsen (Eds.), Titanium in Medicine, Springer, 2001, p. 25–51.

30. F.H. Froes and H.B. Bomberger, The beta titanium alloys, Journal of Metals 37 (1985) 28.

31. O.P. Karasevskaya, O.M. Ivasishin, S.L. Semiatin and V. Yu. Matviychuk, Deformation behavior of beta-titanium alloys, Materials Science and Engineering A 354 (2003) 121.

32. D.J. Lin, J.H. Chern and C.P. Ju, Effect of omega phase on deformation behavior of Ti-7.5Mo-xFe alloys, Materials Chemistry and Physics 76 (2002) 191.

33. D.L. Moffat and D.C. Larbalestier, The competition between the alpha and omega phases in aged Ti-Nb alloys, Metallurgical Transactions 19A (1988) 1687.

34. H.M. Flower, S.D. Henry and D.R.F. West, The $\beta \leftrightarrows \alpha$ transformation in dilute Ti-Mo alloys, Journal of Materials Science 9 (1974) 57.

35. X. Tang, T. Ahmed and H.J. Rack, Phase transformations in Ti-Nb-Ta and Ti-Nb-Ta-Zr alloys, Journal of Materials Science 35 (2000) 1805.

36. A.V. Dobromyslov and V.A. Elkin, Martensitic transformation and metastable b-phase in binary titanium alloys with d-metals of 4–6 periods, Materials Science and Engineering A 354 (2003) 121.

37. A.V. Dobromyslov and V.A. Elkin, The orthorhombic α"-phase in binary titanium base alloys with d-metals of V-VIII groups, Materials Science and Engineering A (in press).

38. M. Niinomi, Mechanical properties of biomedical titanium alloys, Materials Science and Engineering A 243 (1998) 231.

39. J.B. Brunski, in: B.D. Ratner, A.S. Hoffman, F.J. Schoen and J.E. Lemons, (Eds.), Biomaterials Science – An Introduction to Materials in Medicine, Elsevier Academic Press, San Diego, 2004, p. 137–153.

40. F. Wataria, A. Yokoyamaa, M. Omorib, T. Hiraic, H. Kondoa, M. Uoa, T. Kawasakia. Biocompatibility of materials and development to functionally graded implant for bio-medical application, Composites Science and Technology, Vol. 64 (2004) 893–908.

41. J. Black. Biological Performance of Materials, 2nd ed., New York: M. Dekker, Inc., (1992).

42. T. Kokubo, H.M. Kim, M. Kawashita. Novel bioactive materials with different mechanical properties, Biomaterials 24/13 (2003) 2161–2175.

43. J.B. Park, Y.K. Kim. Metallic Biomaterials, Chapter 37. In The Biomedical Engineering HandBook, 2nd ed., J.D. Bronzino, B. Raton (eds.), CRC Press LLC, (2000).

44. J.E. Feighan, V.M. Goldberg, D. Davy, J.A. Parr, S. Stevenson. The influence of surfaceblasting on the incorporation of titanium-alloy implants in a rabbit intramedullary model, 77A, (1995) 1380–1395.

45. P. Tengvall, I. Lundstrom. Physico-chemical considerations of titanium as a biomaterial, Clin. Materials 9, (1992) 115–134.

46. V.E. Henrich, P.A. Cox. The Surface Science of Metal Oxides. Cambridge University Press, Cambridge (1994).

47. R. Thull, D. Grant. Physical and Chemical Vapor Deposition and Plasma-assisted Techniques for Coating Titanium. In Titanium in Medicine, D.M. Brunette, P. Tengvall, M. Textor, P. Thomsen (eds.), Springer-Verlang Berlin Heidelberg (2001) 284–335.

48. F. Klocke. Manufacturing Technology I, WZL-RWTH, Aachen (2001).

49. D.A. Hollander, M. von Walter, T. Wirtz, R. Sellei, B. S. Rohlfing, O. Paar, H.J. Erli. Structural, mechanical and in vitro characterization of individually structured Ti–6Al–4V produced by direct laser forming, Biomaterials, Vol. 27 (2006) 955–963.

50. M.M. Morshed, B.P. McNamara, D.C. Cameron, M.S.J. Hashmi. Stress and adhesion in DLC coatings on 316L stainless steel deposited by a neutral beam source, J. of Materials Processing Technology, Vol. 143 (2003) 922–926.

51. L.L. Hench, R.J. Splittr, W.C.Allen, T.K. Greenlec. Bonding mechanisms at the interface of ceramic prosthetic materials, J. Biomed. Mater. Res. Symp. 2, (1971) 117–141.

52. K. de Groot, C.P.A.T. Klein, J.G.C. Wolke, J.M.A. de Blieck-Hogervorst. Plasma-sprayed coatings of calcium phosphate, CRC Handbook of Bioactive Ceramics, CRC Press, Boston, 2, (1990) 133–142.

53. A. Hulth. Current concepts of fracture healing, Clin Orthop (1989) 249–265.
54. P. Hutzschenreuter, H. Brümmer. Screw design and stability. In Current concepts of Internal Fixation, H. Uhthoff (ed.), Springer-Verlag, Berlin (1980) 244–250.
55. G.V.B. Cochran. Biomechanics of orthopaedic Structures. In Primer in Orthopaedic Biomechanics, Churchill Livingstone, New York (1982) 143–215.
56. A. Sarmiento, E. Ebramzadeh, W.J. Gogan. Cup containment and orientation in cemented total hip arthroplasties. J. Bone Joint Surg 72B(6), (1990) 996.
57. A.H. Burstein, T.H. Wright. Biomechanics. In J. Insall, R Windsor, W Scott (eds.), Surgery of the knee, 2nd ed., Churchill Livingstone, New York (1993) Vol. 7, 43–62.
58. Lima-Lto SpA, Medical Systems, Via Nazionale 52, 33030 Villanova di San Daniele del Friuli (Udine), Italy. URL: http://www.lima.it/english/medical_syst.html
59. M.S. Perren, O.E.M. Pohler, E. Schneider. Titanium as Implant Material for Osteosynthesis Applications. In Titanium in Medicine, D.M. Brunette, P. Tengvall, M. Textor, P. Thomsen (eds.), Springer-Verlang Berlin Heidelberg (2001) 772–823.
60. A. Olander. An electrochemical investigation of solid cadmium-gold alloys, J. Am. Chem. Soc., 54 (1932) 3819–3833.
61. A.B. Greninger, V.G. Mooradian. Strain transformation in metastable beta copper-zinc and beta copper-tin alloys, AIME 128 (1938) 337–368.
62. L.C. Chang, T.A. Read. Plastic deformation and diffusionless phase changes in metals-the gold-cadmium beta phase, Am. Inst. Min. Metall. Eng., J. Met. 191/1 (1951) 47–52.
63. W.J. Buehler, F.E. Wang. A summary of recent research on the Nitinol alloys and their potential application in ocean engineering, Journal of Ocean Engineering 1 (1967) 105–108.
64. C.M. Wayman. Introduction to the crystallography of martensitic transformations, The Macmillan Company, 1964.
65. K. Otsuka, C.M. Wayman. Shape memory materials, Cambridge University Press, 1998.
66. M.S. Wechsler, D.S. Liberman, T.A. Read. On the theory of the formation of martensite, Trans. AIME 197 (1953) 1503–1515.
67. J.S. Bowles, J.K. Mackenzie. The crystallography of martensite transformations I, Acta Metallurgica 2 (1954) 129–137
68. T. Saburi, C.M. Wayman. Crystallographic similarities in shape memory martensites, Acta Metallurgica 27/6 (1979) 979–995.
69. K. Adachi, J. Perkins, C.M. Wayman. Type II twins in self-accommodating martensite plate variants in a Cu-Zn-Al shape memory alloy, Acta Metallurgica 34/12 (1986) 2471–2485.
70. R.D. James, K.F. Hane. Martensitic transformations and shape-memory materials, Acta Materialia 48/1 (2000) 197–222.

71. Madangopal Krishnan. The self accommodating martensitic microstructure of Ni-Ti shape memory alloys, Acta Materialia 46/4 (1998) 1439–1457.

72. T. Inamura, Y. Kinoshita, J.I. Kim, H.Y. Kim, H. Hosoda, K. Wakashima, S. Miyazaki. Effect of {0 0 1}<1 1 0> texture on superelastic strain of Ti-Nb-Al biomedical shape memory alloys, Materials Science and Engineering A (2006) In Press.

73. K. Bhattacharya. Microstructure of martensite: why it forms and how it gives rise to the shape-memory effect, Oxford Series on Materials Modelling, 1st Ed., Oxford University Press, Oxford 2003.

74. R. Stalmans, L. Delaey, J. Van Humbeeck. Generation of recovery stresses: thermodynamic modelling and experimental verification, J. de Phys. IV 7 (1997) 47–52.

75. G.R. Barsch, J.A. Krumhansl. Twin boundaries in ferroelastic media without interface dislocations, Phys. Rev. Lett. 53/11 (1984) 1069–1072.

76. F. Falk. Model free energy, mechanics, and thermodynamics of shape memory alloys, Acta Metall. 28 (1980) 1773-1780.

77. G.A. Maugin, S. Cadet. Existence of solitary waves in martensitic alloys, Int. J. Eng. Sci. 29/2 (1991) 243–258.

78. L.C. Brinson, R. Lammering. Finite element analysis of the behavior of shape memory alloys and their applications, Int. J. Solids and Struct., 30/23 (1993) 3261–3280.

79. Y. Ivshin, T.J. Pence. A thermomechanical model for a one variant shape memory material, J. Intell. Mat. Syst. and Struct., 5/7 (1993) 455–473.

80. C. Liang, C.A. Rogers. One-dimensional thermomechanical constitutive relations for shape memory materials, J. Intell. Mater. Syst. and Struct. 1/2 (1990) 207–234.

81. J.G. Boyd, D.C. Lagoudas. Thermomechanical response of shape memory composites, J. Intell. Mater. Syst. and Struct. 5 (1994) 333-346.

82. K. Tanaka. A thermomechanical sketch of shape memory effect: One-dimensional tensile behavior, Res Mech. 18 (1986) 251-263.

83. L.C. Brinson. One-dimensional constitutive behavior of shape memory alloys: thermomechanical derivation with non-constant material functions and redefined martensite internal variable, J. Intell. Mater. Syst. and Struct. 4 (1993) 229–242.

84. J. Lubliner, F. Auricchio. Generalized plasticity and shape-memory alloys, Int. J. Solids and Structures, 33/7 (1996) 991–1003.

85. V.P. Panoskaltsis, S. Bahuguna, D. Soldatos. On the thermomechanical modeling of shape memory alloys, Int. J. Non-Linear Mech. 39/5 (2004) 709–722

86. Q.P. Sun, K.C. Hwang. Micromechanics constitutive description of thermoelastic martensitic transformations, Advances in Applied Mechanics 31 (1994) 249–298.

87. F. Kosel, T. Videnic. Generalized plasticity and uniaxial constrained recovery in shape memory alloys, Mech. Adv. Mater. Struc., 14/1 (2007) 3–12.

88. E. Denkhaus, K. Salnikow; Nickel essentiality, toxicity, and carcinogenicity, Critical Reviews in Oncology/Hematology 42 (2002) 35–56.
89. E. Nieboer RT. Tom, WE. Sanford. Nickel metabolism in man and animals. In: Nickel and Its Role in Biology: Metal Ions in Biological Systems, Vol 23 (Sigel H, ed). New York: Marcel Dekker, 1988, 91–121.
90. G. Glenn E. Fletcher, Franco D. Rossetto, John Turnbull, and Evert Nieboer; Toxicity, Uptake, and Mutagenicity of Particulate and Soluble Nickel Compounds, 1994, Environ Health Perspect 102(Suppl 3), 69–79.
91. A. Yamamoto, R. Honma, M. Sumita. Cytotoxicity evaluation of 43 metal salts using murine fibroblasts and osteoblastic cells, Journal of Biomedical Materials Research, 39 (1998) 331–340.
92. C. Combes, C. Rey, M. Freche. XPS and IR study of dicalcium phosphate dihydrate nucleation on titanium surfaces, Colloids and Surfaces B: Biointerfaces, 11/1–2 (1998) 15–27.
93. C. Shih, S. Lin, K. Chung, Y. Chen, Y. Su, S. Lai, G.Wu, C. Kwok and K. Chung. The cytotoxicity of corrosion products of Nitinol stent wires on cultured smooth muscle cells. J. Biomed. Mater. Res. 52 (2000) 395–403
94. D.J. Wever, A.G. Veldhuizen, M.M. Sanders, J.M Schakenraad, J.R. Horn. Cytotoxic, allergic and genotoxic activity of a nickel-titanium alloy, Biomaterials, 18 (1997) 1115–1120.
95. I.C. Wataha, P.E. Lockwood, M. Marek, M. Ghazi. Ability of Ni-containing biomedical alloys to activate monocytes and endothelial cells in vitro, Journal of Biomedical Materials Research, 45 (1999) 251–257.
96. J. Ryhänen, E. Niemi, W. Serlo, E. Niemelä, P. Sandvik, H. Pernu, T. Salo. Biocompatibility of nickel-titanium shape memory metal and its corrosion behavior in human cell cultures. J Biomed Mat Res 35 (1997) 451–457.
97. C. Wirth, V. Comte, C. Lagneau, P. Exbrayat, M. Lissac, N. Jaffrezic-Renault, L. Ponsonnet. Nitinol surface roughness modulates in vitro cell response: a comparison between fibroblasts and osteoblasts, Materials Science and Engineering: C, 25 (2005) 51–60.
98. C. Trepanier, T. Leung, M. Tabrizian, L'H. Yahia, J. Bienvenu, J. Tanguay, D. Piron and L. Bilodeau. Preliminary investigation of the effect of surface treatment on biological response to shape memory NiTi stents. J. Biomed. Mater. Res. 48 (1999) 165–171.
99. S.A. Shabalovskaya. Surface, corrosion and biocompatibility aspects of Nitinol as an implant material. Bio-Medical Materials and Engineering 12 (2002) 69–109.
100. S.A. Shabalovskaya, On the nature of the biocompatibility and on medical applications of NiTi shape memory and superelastic alloys. Biomed.Mater.Eng. 6 (1996) 267–289.
101. V.M. Frauchiger, F. Schlottig, B. Gasser, M. Textor. Anodic plasma-chemical treatment of CP titanium surfaces for biomedical applications, Biomaterials, 25 (2004) 593–606.

102. X. Lu, Z. Zhao, Y. Leng. Biomimetic calcium phosphate coatings on nitric-acid-treated titanium surfaces, Materials Science and Engineering: C, In Press (2006).

103. J. Park, D.J. Kim, Y.K. Kim, K.H. Lee, K.H. Lee, H. Lee, S. Ahn. Improvement of the biocompatibility and mechanical properties of surgical tools with TiN coating by PACVD, Thin Solid Films, Volume 435 Issues 1–2 (2003) 102–107.

104. N. Shevchenko, M.T. Pham, M.F. Maitz. Studies of surface modified NiTi alloy, Applied Surface Science, 235 (2004) 126–131.

105. K. Endo. Chemical modification of metallic implant surfaces with biofunctional proteins (Part 1). Molecular structure and biological activity of a modified NiTi alloy surface. Dent. Mater. J. 14 (1995) 185–198.

106. F. Liu, F. Wang, T. Shimizu, K. Igarashi, L. Zhao. Hydroxyapatite formation on oxide films containing Ca and P by hydrothermal treatment, Ceramics International, Volume 32/5 (2006) 527–531.

107. M. Schillinger, S. Sabeti, C. Loewe. Balloon angioplasty versus implantation of nitinol stents in the superficial femoral artery, Journal of Vascular Surgery, Volume 44 Issue 3 (2006) 684.

108. B. Rapp. Nitinol for stents, Materials Today, Vol. 7 Issue 5 (2004) 13.

109. S. Tyagi, S. Singh, S. Mukhopadhyay, U. A. Kaul. Self- and balloon-expandable stent implantation for severe native coarctation of aorta in adults, American Heart Journal, Volume 146 Issue 5 (2003) 920–928.

110. M. Simon, R. Kaplow, E. Salzman, D. Freiman. A vena cava filter using thermal shape memory alloy Experimental aspects, Radiology, Volume 125 (1977) 87–94.

111. T. Duerig, A. Pelton, D. Stöckel. An overview of nitinol medical applications, Materials Science and Engineering, A273–275 (1999) 149–160.

112. H. Fischer, B. Vogel, A. Grünhagen, K.P. Brhel, M. Kaiser. Applications of Shape-Memory Alloys in Medical Instruments, Materials Science Forum, V 394-395 (2002) 9–16.

113. A.R. Pelton, D. Stöckel, T.W. Duerig. Medical Uses of Nitinol, Materials Science Forum. V. 327-328 (2000) 63–70.

114. K. Dai, X. Wu, X. Zu. An Investigation of the Selective Stress-Shielding Effect of Shape-Memory Sawtooth-Arm Embracing Fixator, Materials Science Forum, V. 394-395 (2002) 17–24.

115. C. Zhang, S. Xu, J. Wang, B. Yu, Q. Zhang. Design and Clinical Applications of Swan-Like Memory-Compressive Connector for Upper-Limb Diaphysis, Materials Science Forum, V. 394-395 (2002) 33–36.

116. G. Da, T. Wang, Y. Liu, C. Wang. Surgical Treatment of Tibial and Femoral Factures with TiNi Shape-Memory Alloy Interlocking Intramedullary Nails, Materials Science Forum, V. 394-395 (2002) 37–40.

117. C. Song, T.G. Frank, P.A. Campbell, A. Cuschieri. Thermal Modelling of Shape –Memory Alloy Fixator for Minimal-Access Surgery, Materials Science Forum, V. 394-395 (2002) 53–56.

118. S. Xu, C. Zhang, S. Li, J. Su, J. Wang. Three-Dimensional Finite Element Analysis of Nitinol Patellar Concentrator, Materials Science Forum, V. 394–395 (2002) 45–48.
119. Y.Chu, K. Dai, M. Zhu, X. Mi. Medical Application of NiTi Shape Memory Alloy in China, Materials Science Forum, V. 327-328 (2000) 55–62.

INDEX

A

AFM, see Atomic force microscopy (AFM)
Alumina, 223–224, 461–462
Aluminium oxide, 52, 219–220
  nanometric machining, 2–3
Armchair nanotubes, 444
Atomic force microscopy (AFM), 387, 464–467

B

BEG, see Bias-enhanced growth (BEG)
BEN, see Bias-enhanced nucleation (BEN)
Beryllium mask materials, 291Bias-enhanced growth (BEG), 247–248
Bias-enhanced nucleation (BEN), 156, 165–166, 188, 247
  CVD diamond technology, 149–150
  Deposition routes, 244–248
BioMEMS devices, 245
Bonding bridges, 21–22, 41–43
Boundary conditions, mechanical micromachining, 519

C

Cancellous bone, 513–531
Carbon nanotubes, see Carbon nanomaterials
Chemical vapor deposition (CVD)
  diamond technology
  advantages/disadvantages, 154
  bias-enhanced nucleation, 165

DC plasma-enhanced, 151–152
  filament assembly modification, 159–160
  hetero-epitaxial growth, 164
  historical developments, 196
  homo-epitaxial growth, 164
  hot filament, 152
  materials, substrates, 154–156
  metallic (Mo) wires, 172
  metastable diamond growth, 147–149
  microwave plasma-enhanced, 152
  modified hot filament CVD, 159–160
  Mo/Si substrate, 156–157
  nanocrystalline diamond, 243–244
  nucleation and growth, diamond, 152–153, 162–163
  performance studies, 181–188
  plasma-enhanced CVD, 151–152
  pretreatment, substrates, 156
  process condition, 160–162
  process types, 141–143
  properties, diamond, 144
  RF plasma-enhanced CVD, 151
  Si/Mo substrate, 156–157
  substrates, 154–159
  synthesis, diamond, 143–149
  technology development, 149–154
  temperature influence, 166–171
  three-diamond substrates, 171–172
  time-modulated CVD, 188–196
  WC-Co, 157–159, 172–175
Chip formation
  nanometric machining, 9–11
Chiral nanotubes, 444
Control, nanometric machining, 11
Conventional machining comparison, 17

Cubic, boron nitride, 209, 217
Cutting edge radius, 4, 8, 10, 12–17
Cutting tools, 2, 141–144, 171, 180, 187, 224–229, 514, 520, 524, 531

D

DC plasma-enhanced CVD, 151–152,
Decomposition, adsorbed species, 142
Deterministic mechanical nanometric machining, 2
Diamond technology, CVD
  advantages/disadvantages, 154
  basics, 148–149
  bias-enhanced nucleation, 165–166
  DC plasma-enhanced CVD, 151–152
  filament assembly modification, 159–160
  heteroepitaxial growth, 164
  historical developments, 145–146
  homoepitaxial growth, 164
  hot filament CVD, 152–154
  materials, substrates, 154–156
  metastable diamond growth, 146–149, 162
  microdrills, 261
  microwave plasma-enhanced CVD, 152
  modified hot filament CVD, 261
  Mo/Si substrate,156–157
  nucleation and growth, diamond, 152, 162–171
  performance studies, 181–188
  pretreatment, substrates, 156
  process conditions, 160–162
  process types, 141–143
  properties, diamond, 144
  RF plasma-enhanced CVD, 151
  Si/Mo substrate, 156–157
  substrates, 154–159
  synthesis, diamond, 143–149
  technology development, 149–154
  temperature influence, 166–171
  three-dimensional substrates, 171–172

WC-Co, 339, 157–159, 172–180, 228
Dip pens, (dip coating is present in text), 50, 51
Direct-current (DC) plasma-enhanced CVD, 151
Displacement, nanometric machining, 2(not found with Displacement)
Dissolution models, 30–31(not exact term found)
Ductile regime, 15–16(not exact term found)

E

EDM, see Electric discharge machining (EDM)
Electric discharge machining (EDM), 2, 152(not exact term found)
Electroplating, 141
Electrostatic forces, 54, (exact term not found)
ELID, see Electrolytic in-process dressing (ELID)
EMC, see Electromagnetic compatibility (EMC) standards
EMI, see Electromagnetic interfacing (EMI) standards
Etching, 1-46, 99-118
Excimer lasers, 2
Experimental approaches
  mechanical micromachining, 524–525
Exposure, x-ray lithographic microfabrication, 2

F

FEA, see Finite element analysis (FEA) model
Field effect transistors (FETs) 148
Filament assembly modification, 159–160
Film formation, 25, 32, 142
Finite element analysis (FEA) model, 534

Five-axis CNC machining centers, 548 (exact term not found)
Fluid flow analysis, 480 (only Fluid Flow occur)
FRF, see Frequency response function (FRF)
FTS, see Fast Tool Servo (FTS) system
Future directions, 44–45
    diamond nanogrinding, 143(exact term not found)
    machining, 138, 182, 253, 518, 632, 683 not in context with main entry
    mechanical micromachining, 2
    nanomanufacturing, 3
    x-ray lithographic microfabrication, 2

G

Growth, diamond, 145–48, 151–53, 162–64

H

HEMA, see Hydroxyethylmethacrylate (HEMA)
Heteroepitaxial growth, 164
HFCVD, see Hot filament CVD (HFCVD)
High-speed air turbines, 524–25
Historical developments, CVD diamond technology, 149–50
Hot filament CVD (HFCVD) 152–54, 159–61
Hydroxyethylmethacrylate (HEMA), 385

I

IADF, see Ion angular distribution function (IADF)
IBARE, see Ion bean-assisted radical etching (IBARE)

IBE, see Ion beam etching (IBE)
IEDF, see Ion energy distribution function (IEDF)
Implementation, nanometric machining, 2–5
Infrastructure, commercialization issues, 2
Initial chip curl modeling, 518
Injection compression molding, 216
Injection molding, 216

L

LEED, see Low-energy electron diffraction (LEED) study
LIGA process, 2
Lithographic method, 2
Lithographic processes, see also X-ray lithographic microfabrication
LODTM, see Large Optics Diamond Turning Machine (LODTM)

M

Machining, see also High-aspect ratio microstructures
Manipulative techniques, nanofabrication, 2
Masks, materials, 2
MD, see Molecular dynamics (MD) simulation
MEMO, see Methacryloxypropyl trimethoxy silane (MEMO)
Metal cutting chip formation, 9, 12
Metallic (Mo) wires, 172
Metastable diamond growth, 146–47
Micro- and nanogrinding, see also Diamond nanogrinding
Microcrystalline diamond (MCD) films, 189, 243–46, 250, 264
Microfabrication, see also Laser-based micro- and nanofabrication
Microfabrication, x-ray lithography, 2
Microfluidic devices, 263
Micromilling, 5

Microwave plasma CVD (MPCVD), 190–91, 194–95, 254–56
MIMIC, see Micromolding in capillaries,
Minimum undeformed chip thickness, 11–13
Modeling, see also Simulation
Mode shapes, tetrahedral structures, 164
Modified hot filament CVD, 152, 190, 196, 245, 254, 261
Mounted wheels, 204
MPCVD, see Microwave plasma CVD (MPCVD)
Multi-wall carbon nanotubes (MWCNTs), 443–44, 449, 451, 454, 456–58, 466

N

Nano- and microgrinding, see also Diamond nanogrinding
Nanocrystalline diamond, 243–65
Nanocrystalline diamond (NCD) films, 243–47, 263
Nanofabrication, see also Laser-based micro- and nanofabrication
Nanogrinding, see also Diamond nanogrinding; Micro- and nanogrinding
Nanomanufacturing, 3, 531
Nanometric machining 2–5, 9–11, 14–18
Non-mechanical nanometric machining, 2
Nucleation, 162–67

O

Optics, laser-based micro- and nanofabrication, 315–17

P

Partial ductile mode grinding,16, 18
PAS, see Polyalkensulfone (PAS)

PDMS, see Polydimethylsiloxane (PDMS)
PMI, see Polymethacrylimide (PMI)
PMMA, see Polymethylmethacrylate (PMMA)
Polymethylmethacrylate (PMMA), 384, 395, 400
Polytetrafluoroethylene (PTFE), 85, 92
POM, see Polyoxymethylene (POM)
Porous nanogrinding tools, 554
Precision micro- and nanogrinding, see Micro- and nanogrinding
Properties, diamond, 144
PSU, see Polysulfone (PSU)
PTFE, see Polytetrafluoroethylene (PTFE)
Pyrex glass, 66–67

Q

Quartz, 209, 214, 217, 221, 247

R

Reactive ion etching (RIE), see also Deep reactive ion etching (DRIE)
Resists, 110
RF plasma-enhanced CVD, 151
RIE, see Reactive ion etching (RIE)

S

Shielding gas, 303, 313, 460
Silicate, 461
Silicon carbide, 16, 52, 156, 203, 209, 220
Standardization, commercialization issues, 280, 313
Synthesis, diamond, 143–149

T

Testing, commercialization issues, 468–469
Tetrahedral structures, 164, 245
Three-dimensional substrates, 171–180

Tilting, 101–106 Time-modulated
    CVD (TMCVD), 249–256
Tip angles, 168, 173–174, 180–181,
    466–467

U

Ultraprecision machine tools, 11–12

V

Volume transport, 455

W

WC-Co, 157–159, 172–180, 186, 261
Workpiece material properties, 14–18

X

X-ray diffraction, 464, 487
X-ray lithographic microfabrication,
    2, 453
X-ray lithography, 2
X-ray masks, 2

Y

Young's modulus, 242, 491, 493, 541,
    555,

Dr. Mark J. Jackson

Professor of Mechanical Engineering
College of Technology
Purdue University
United States of America

C. Eng., Engineering Council of London, U.K., 1998
M. A. Status, Natural Sciences, University of Cambridge, U.K., 1998
Ph. D., Mechanical Engineering, Liverpool, U.K., 1995
M. Eng., Mechanical & Manufacturing Engineering, Liverpool, U.K., 1991
O.N.D., Mechanical Engineering, Halton College, U.K., 1986
O.N.C. Part I, Mechanical Engineering, Halton College, U.K., 1984

Doctor Jackson began his engineering career in 1983 when he studied for his O.N.C. part I examinations and his first-year apprenticeship-training course in mechanical engineering. After gaining his Ordinary National Diploma in Engineering with distinctions and I.C.I. prize for achievement, he read for a degree in mechanical and manufacturing engineering at Liverpool Polytechnic and spent periods in industry working for I.C.I. Pharmaceuticals, Unilever Industries, and Anglo Blackwells. After graduating with a Master of Engineering (M. Eng.) degree with Distinction under the supervision of Professor Jack Schofield, M.B.E., Doctor Jackson subsequently read for a Doctor of Philosophy (Ph. D.) degree at Liverpool in the field of materials engineering focusing primarily on microstructure-property relationships in vitreous-bonded abrasive materials under the supervision of Professor Benjamin Mills. He was subsequently employed by Unicorn Abrasives' Central Research & Development Laboratory (Saint-Gobain Abrasives' Group) as materials technologist, then technical manager, responsible for product and new business development in Europe, and university liaison projects concerned with abrasive process development. Doctor Jackson then became a research fellow at the Cavendish Laboratory, University of Cambridge, working with Professor John Field, O.B.E., F.R.S., on impact fracture and friction of diamond before becoming a lecturer in engineering at the University of Liverpool in 1998. At

Liverpool, Dr. Jackson established research in the field of micromachining using mechanical tools, laser beams, and abrasive particles. At Liverpool, he attracted a number of research grants concerned with developing innovative manufacturing processes for which he was jointly awarded an Innovative Manufacturing Technology Center from the Engineering and Physical Sciences Research Council in November 2001. In 2002, he became associate professor of mechanical engineering and faculty associate in the Center for Manufacturing Research, and Center for Electric Power at Tennessee Technological University (an associated university of Oak Ridge National Laboratory), and a faculty associate at Oak Ridge National Laboratory. Dr. Jackson was the academic adviser to the Formula SAE Team at Tennessee Technological University. In 2004 he moved to Purdue University as Associate Professor of Mechanical Engineering in the College of Technology.

Doctor Jackson is active in research work concerned with understanding the properties of materials in the field of microscale metal cutting, micro- and nanoabrasive machining, and laser micro machining. He is also involved in developing next generation manufacturing processes and biomedical engineering. Doctor Jackson has directed, co-directed, and managed research grants funded by the Medical Research Council, Engineering and Physical Sciences Research Council, The Royal Society of London, The Royal Academy of Engineering (London), European Union, Ministry of Defense (London), Atomic Weapons Research Establishment, National Science Foundation, N.A.S.A., U. S. Department of Energy (through Oak Ridge National Laboratory), Y12 National Security Complex at Oak Ridge, Tennessee, and Industrial Companies, which has generated research income in excess of $15 million. Dr. Jackson has organized many conferences and currently served as General Chair of the International Surface Engineering Congress and is Deputy President of the World Academy of Materials and Manufacturing Engineering. He has authored and co-authored over 250 publications in archived journals and refereed conference proceedings, has written a book on "micro and nanomanufacturing", is guest editor to a number of refereed journals, and is currently editing a book on "commercializing micro- and nanotechnology products". He is the editor of the "*International Journal of Nanomanufacturing*", associate editor of the "*International Journal of Molecular* Engineering", and is on the editorial board of the "*International Journal of Machining and Machinability of Materials*" and the "*International Journal of Manufacturing Research*".

Dr. Waqar Ahmed

Professor of Nanotechnology
NIBEC
University of Ulster
United Kingdom

C. Chem., Royal Society of Chemistry of London, U.K., 2002
Fellow, Royal Society of Chemistry of London, U.K., 2002
C. Eng., Engineering Council of London, U.K., 1998
Fellow, Institution of Materials, Minerals, and Mining, London, U.K., 1998
Ph. D., Chemistry, University of Salford, U.K., 1985
B. Sc. (Hons.), Chemistry and Biochemistry, University of Salford, U.K., 1982

Waqar Ahmed is Professor of Nanotechnology and Advanced Materials in the School of Electrical and Mechanical Engineering at the University of Ulster, Northern Ireland in the U.K. He obtained his B. Sc. in Science and Ph. D. in Physical Chemistry from Salford University. His research involved "Low Pressure Chemical Vapor Deposition of In-situ Doped Polycrystalline Silicon for Microelectronic Devices" sponsored by G.E.C. Research. Along with his supervisor, Professor Hitchman, he developed kinetic models for obtaining a better fundamental understanding of in-situ doped LPCVD process employed in semiconductor industry. The modeling work was coupled experimental work involved in-situ analysis of the gas-phase species involved in the CVD process. He spent his early career in the semiconductor industry working for GEC Hirst Research Center, Wembley, U.K., Ferranti Electronics, and INMOS, working on practical applications of CVD, ion-implantation, oxidation, and diffusion processes for device manufacturers in high-pressure production environments. At G.E.C. he developed Ultra High Vacuum CVD Processes for deposition of large grained polysilicon and doped polysilicon for liquid crystal displays working primarily on developing an understanding of the kinetics of these processes and developing industrial solutions to within-wafer uniformity problems associated with in-situ doping processes. This was followed by a spell at VSW Scientific Instruments as Product Manager of the Molecular Beam Division developing novel molecular beam scattering systems for leading academics in the field such as Professor Sir David King (Cambridge), Professor Neville Richardson (Liverpool IRC in Surface Science), Professor H. Reider (Berlin) and Professor John Foord (Oxford). He then set up the Surface Engineering Division with Professor David Armour and Professor Derek Arnell at Salford University as Operations Manager in order commercialize the surface engineering activity within the University and co-ordinated the North West Surface Engineering Initiative involving leading research centers around the North West Region including IRC in Surface Science, Liverpool; Daresbury Research Laboratories, Advanced Manufacturing Research Institute, Manchester Corrosion Protection Center, and PREST funded by the DTI. He has acted as a consultant to Daresbury Laboratory for commercialization of some of its research activity.

He has published over 300 papers, and has authored or co-authored 6 book chapters. He, and his co-workers, has also contributed to the wider industrial adoption of surface coating solutions, through research into modeling of gas phase processes in CVD and studies of tribological behavior. His move to Ulster University at the end of 2005 involved a widening of his group's research, to include the development of nanoparticles, nanocomposite coatings, laser assisted processes, and functional thin films and devices. He has been active in the international vacuum coating scene for many years, including involvement in organising EU – Two Stroke Engine Workshops, the International Surface Engineering Congress 2004, 2005 and 2006, NanoMATS 2005 in Aveiro, Portugal, International Surface Engineering Conference in Newcastle 1995, and Thin Films 2006 in Singapore. He has acted as a Proceedings Co-Editor and Program Chair for ISEC 2005 and 2006. He has served on the editorial board of Vacuum and is currently editor of the International Journal of Nanomanufacturing and editorial board member for International Journal of Surface Science and Engineering. He is a Fellow of the Institute of Materials, Minerals and Mining (FIMMM) and Fellow of the Royal Society of Chemistry.

Professor Sir Harold Kroto, FRS, was born in 1939 in Wisbech, Cambridgeshire, and brought up in Bolton, Lancashire. He graduated in Chemistry at the University of Sheffield in 1961 and in 1964 received his PhD there for research with R. N. Dixon on high-resolution electronic spectra of free radicals produced by flash photolysis. After two years post-doctoral research in electronic and microwave spectroscopy at the National Research Council in Ottawa, Canada, he spent one year at Bell Laboratories, New Jersey, studying liquid phase interactions by Raman spectroscopy and he also carried out studies in Quantum Chemistry. He started his academic career at the University of Sussex (Brighton) in 1967, where he became a professor in 1985 and in 1991 he was made a Royal Society Research Professor. The research program at Sussex has covered several interdisciplinary areas. One area focused on the creation and spectroscopic characterization of new molecules, in particular, unstable species and reaction intermediates which contained labile multiple bonds. During a project that explored the possible source of these carbon chains in space, laboratory experiments that simulated the chemical reactions in the shells of red giant carbon stars were carried out which serendipitously uncovered the existence of C_{60} Buckminsterfullerene. In follow-up investigations of this original discovery the molecule was isolated independently at Sussex and structurally characterised. The presently active research program derives directly from the earlier work on C_{60} and focuses on the implications of the discovery for several areas of fundamental chemistry as well as the way in which it has revolutionized our perspective on carbon based materials. The research encompasses the basic chemistry of the fullerenes, fundamental studies of carbon and metal clusters as well as carbon microparticles and nanotubes. Work on various aspects of interstellar and circumstellar molecules and dust is also in progress. Some parts of the research have been successful due to their interdisciplinary nature and this has been the result of synergistic collaborations involving primarily: colleagues J. F. Nixon, R. Taylor and D. R. M. Walton at Sussex, T. Oka at NRC (Canada), and R. F. Curl and R. E. Smalley at Rice University (Texas). Since 1990 he has been chairman of the editorial board of the Chemical Society Reviews. Sir Harold has received the following awards in the course of his career: 1981-82, Tildon Lecturer (Royal Society of Chemistry); 1990, Elected Fellow of the Royal Society; 1991, Royal Society Research Fellowship; 1992, International Prize for New Materials (American Physical Society), Italgas Prize for Innovation in Chemistry, Université Libre de Bruxelles (Doctor Honoris Causa), University of Stockholm (Doctor Honoris Causa), Longstaff Medal of the Royal Society of Chemistry, Academia Europea (Member); 1993, University of Limburg (Doctor Honoris Causa); 1994, Hewlett Packard Europhysics Prize, Moet Hennessy and Louis Vuitton Science Pour l'Art Prize; 1995, University of Sheffield (Honorary Degree), University of Kingston (Honorary Degree); and 1996, Knighthood for services to chemistry and the Nobel Prize for Chemistry. Professor Sir Harold W. Kroto, FRS, is currently at Florida State University, Tallahassee, Florida, USA.